C0-ATQ-518

Exploration of the Polar Upper Atmosphere

NATO ADVANCED STUDY INSTITUTES SERIES

Proceedings of the Advanced Study Institute Programme, which aims at the dissemination of advanced knowledge and the formation of contacts among scientists from different countries

The series is published by an international board of publishers in conjunction with NATO Scientific Affairs Division

A	Life Sciences	Plenum Publishing Corporation
B	Physics	London and New York
C	Mathematical and Physical Sciences	D. Reidel Publishing Company Dordrecht, Boston and London
D	Behavioural and Social Sciences	Sijthoff & Noordhoff International Publishers
E	Applied Sciences	Alphen aan den Rijn and Germantown U.S.A.

Series C – Mathematical and Physical Sciences

Volume 64 – Exploration of the Polar Upper Atmosphere

Exploration of the Polar Upper Atmosphere

Proceedings of the NATO Advanced Study Institute held at Lillehammer, Norway, May 5-16, 1980

edited by

C. S. DEEHR
Geophysical Institute, University of Alaska, Fairbanks, Alaska, U.S.A.

and

J. A. HOLTET
Institute of Physics, University of Oslo, Oslo, Norway

D. Reidel Publishing Company

Dordrecht : Holland / Boston : U.S.A. / London : England

Published in cooperation with NATO Scientific Affairs Division

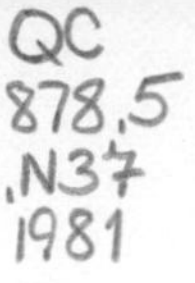

Library of Congress Cataloging in Publication Data

Nato Advanced Study Institute, Lillehammer, Norway, 1980
Exploration of the polar upper atmosphere.

(NATO advanced study institutes series : Series C, Mathematical and physical sciences ; v. 64)
Includes index.
1. Atmosphere, Upper–Polar regions–Congresses. I. Deehr, C. S. II. Holtet, Jan A. III. Title. IV. Series.
QC878.5.N37 1980 551.5′14′0911 80-29055
ISBN 90-277-1225-5

Published by D. Reidel Publishing Company
P.O. Box 17, 3300 AA Dordrecht, Holland

Sold and distributed in the U.S.A. and Canada
by Kluwer Boston Inc.,
190 Old Derby Street, Hingham, MA 02043, U.S.A.

In all countries, sold and distributed
by Kluwer Academic Publishers Group,
P.O. Box 322, 3300 AH Dordrecht, Holland

D. Reidel Publishing Company is a member of the Kluwer Group

Printed in The Netherlands

CONTENTS

PREFACE

This book is an ordered collection of tutorial lectures on the physical processes in the polar upper atmosphere given at the NATO Advanced Study Institute (ASI) on "The Exploration of the Polar Upper Atmosphere" held at Lillehammer, Norway, May 5-16, 1980.

The polar cap is an important part of the high latitude atmosphere not only because of circulation and horizontal transport in the neutral atmosphere and convection in the ionosphere, but also because of its unique energy sources and sinks. In addition, solar wind plasma is led into the upper atmosphere by the geomagnetic field at the poles, and the polar cap is, as stated by Tutorial Leader Roederer in this volume, "the place where outer space meets earth".

The atmosphere at lower latitudes is well-known to the ground-based observer, and the advent of satellite observations was simply the beginning of a new perspective. The exploration of the atmosphere at polar latitudes, however, proceeded in quite the opposite manner, and satellite maps of the polar caps may be compared with a relatively meagre set of ground-based data. Recent efforts to extend the polar observations from the ground have resulted in the need for a review of the physical principles and processes occurring in the polar upper atmosphere. The interdisciplinary nature of these efforts led to the emphasis here on a tutorial program.

The ASI was structured according to the physical regions of the atmosphere and the means of exploring them. Nine sessions were each organized by a tutorial leader to describe the region of interest both to the advanced student and workers in the field of magnetospheric and upper atmospheric physics.

Consistent with the spirit of polar exploration was the setting of the ASI in the native countryside of many accomplished polar explorers. Consistent also with the breadth of interest

C. S. Deehr and J. A. Holtet (eds.), Exploration of the Polar Upper Atmosphere, ix–x.

and talent of men such as Nansen, were the extracurricular presentations such as a tour of the Folk Museum of Gudbrands Valley, "Maihaugen", a lecture on neurological functions affecting auroral observations by J. Roederer, an organ concert in Lillehammer Church by V. Vasyliunas, and the first edition of a musical work for piano and string quartet set to the words of a poem by the Norwegian poet Caspari called "Nordlyset" (The Northern Lights) by W. Stoffregen. Activities in a lighter vein included the installation of W.R. Piggot as Grand Master of the International Union for Merging of the Poles (IUMP). He then appointed as Knights Errant Egeland, Roederer, Schove, Stoffregen and Vasyliunas, for their scientific and humanistic contributions to The International Magnetic Field.

The success of this ASI was due in large part to the unheralded efforts of the Secretariat: Anne-Sophie Andresen and Elisabeth Iversen.

Charles Deehr
Fairbanks

Jan Holtet
Oslo

August 1980

Nordlys
Willy Stoffregen
adagio
Piano
dolce
decresc - -
rit

PARTICIPANTS

ANDERSON, D.N.
NOAA/ERL/SEL, Department of Commerce
Boulder, Colorado 80302
USA

ANDRE, D.
Max-Planck-Institut für Aeronomie
Postfach 20, 3411 Katlenburg-Lindau
FRG

BANKS, P.
Physics Department
Utah State University
Logan. Utah 84322
USA

BERKEY, F.T.
Center for Atmospheric and Space Sciences
UMC-41
Utah State University
Logan, Utah 84322
USA

BERTHELIER, A.
LGE - CNRS
4 Avenue de Neptune, 94100 St Maur
France

BREKKE, A.
The Auroral Observatory
P.O.Box 953, N-9001 Tromsø
Norway

BÖSINGER, T.
University of Oulu, Department of Physics,
P.O.Box 131, SF-90570 Oulu 57
Finland

CHRISTIANSEN, P.J.
Department of Physics, MAPS
University of Sussex
Falmer, Brighton BN1 9QM
United Kingdom

COFFEY, M.T.
NCAR
P.O.Box 3000, Boulder, Colorado 80307
USA

COGGER, L.L.
Department of Physics
University of Calgary
Calgary, Alberta
Canada T2N 1N4

COLLIS, P.
Environmental Sciences Department
University of Lancaster
Bailrigg, Lancaster
United Kingdom

CREUTZBERG, F.
Herzberg Institute of Astrophysics
National Research Council
Ottawa, Ontario
Canada K1A OR6

DEEHR, C.S.
Geophysical Institute
University of Alaska
Fairbanks, Alaska 99701
USA

DØHL, J.
Institute of Physics
University of Oslo
P.O.Box 1038 Blindern, Oslo 3
Norway

EGELAND, A.
Institute of Physics
University of Oslo
P.O.Box 1038, Blindern, Oslo 3
Norway

EMERY, B.A.
National Center for Atmospheric Research
P.O.Box 300, Boulder, Colorado 80307
USA

EVANS, R.E.
Physics Department
University of Leicester
Leicester
United Kingdom

FEYNMAN. J.
Department of Physics
Boston College
Chestnut Hill, Massachusetts 02167
USA

FRIIS-CHRISTENSEN, E.
Geophysical Department
Meteorological Institute
Lyngbyvej 100, DK-2100 København
Denmark

GELLER, M.A.
NASA/Goddard Space Flight Center
Laboratory for Planetary Atmospheres.
Atmospheric Chemistry Branch,
Greenbelt. Maryland 20771
USA

GENDRIN, R.
OPN/RPE - CNET
92131 Issy-les-Moulineaux
France

GERARD, J.-C.
Institut d'Astrophysique
Université de Liege
Avenue de Cointe, B-4200 Liege
Belgium

GRANDAL, B.
NDRE
P.O.Box 25, N-2007 Kjeller
Norway

HAERENDEL, G.
Max-Planck-Institut für Physik und Astrophys.
Institut für Extraterrestrische Physik
8046 Garching b. München
FRG

HALDOUPIS, C.I.
Department of Physics
University of Crete
Iraklion, Crete
Greece

HARANG, O.
The Auroral Observatory
P.O.Box 953, N-9001 Tromsø
Norway

NIELSEN, E.
Max-Planck-Institut für Aeronomie
Postfach 20. 3411 Katlenburg-Lindau
FRG

OLESEN, J.K.
Ionosphere Laboratory of the Danish Met. Inst.
Technical University of Denmark
DK-2800 Lyngby
Denmark

PETTERSEN, Ø.
The Auroral Observatory
P.O.Box 953, N.9001 Tromsø
Norway

PIGGOTT, W.R.
21 Hillingdon Rd.
Uxbridge, Middelsex UB10 OAD
United Kingdom

POTEMRA, T.A.
The Johns Hopkins University
Applied Physics Laboratory
Laurel, Maryland 20810
USA

RANTA, H.
Geophysical Observatory
SF-99600 Sodankylä
Finland

REAGAN, J.B.
Lockheed Palo Alto Research Lab.
Dept. 52-12, B/205
3251 Hanover street, Palo Alto, California 94304
USA

REES, M.H.
Geophysical Institute
University of Alaska
Fairbanks, Alaska 99701
USA

RODGER, A.
British Antarctic Survey
Madlingley Road, Cambridge CB3 OET
United Kingdom

ROEDERER, J.G.
Geophysical Institute
University of Alaska
Fairbanks, Alaska 99701
USA

ROSENBERG, T.J.
Institute for Physical Science and Technology
University of Maryland
College Park, Maryland 20742
USA

SANDHOLT, P.E.
The Auroral Observatory
P.O.Box 953, N-9001 Tromsø
Norway

SCHMERLING, E.R.
Space Plasma Physics
Solar Terrestrial Division
NASA, Washington D.C. 20546
USA

SCHOVE, D.J.
St. David's College
Beckenham, Kent BR3 3BQ
United Kingdom

SCHUCHARD, K.G.H.
Institut für Astrophysik und Extraterrestrische Forschung
Universität Bonn
Auf dem Hügel 71, 53 Bonn 1
FRG

SHEPHERD, G.G.
Centre for Research in Experimental Space Sci.
York University
4700 Keele St., Downsview, Ontario
Canada M3J 1P3

SINGH, V.
Physics Institute
Odense University
DK-5230 Odense M.
Denmark

SIVJEE, G.G.
Geophysical Institute
University of Alaska
Fairbanks, Alaska 99701
USA

SMITH, A.J.
Britisch Antarctic Survey
Madingley Road, Cambridge CB3 OET
United Kingdom

SMITH, R.W.
School of Physical Science
Ulster Polytechnic
Jordanstown, Co. Antrim
United Kingdom

STAMNES, K.
Geophysical Institute
University of Alaska
Fairbanks, Alaska 99701
USA

STEEN, Å.
Kiruna Geophysical Institute
P.O.Box 704, S-98127 Kiruna
Sweden

STOFFREGEN, W.
Alvägen 29A
S-75245 Uppsala
Sweden

STUBBE, P.
Max-Planck-Institut für Aeronomie
3411 Katlenburg-Lindau 3
FRG

SWEENEY, P.
16 Jordanstown Road
Newtown Abbey, Co. Antrim
United Kingdom

THOMAS, L.
Rutherford and Appleton Laboratory
Ditton Park, Slough SL3 9JX
United Kingdom

THOMAS, R.W.
Department of Physics
The University
Southamton S09 5NH
United Kingdom

THORNE, R.M.
Department of Atmospheric Sciences
University of California
Los Angeles, California 90024
USA

HENRIKSEN, K.
The Auroral Observatory
P.O.Box 953, N-9001 Tromsø
Norway

HOLT, O.
The Auroral Observatory
P.O.Box 953, N-9001 Tromsø
Norway

HOLTET, J.A.
Institute of Physics
University of Oslo
P.O.Box 1038 Blindern, Oslo 3
Norway

HUGHES, T.J.
Herzberg Institute of Astrophysics
National Research Council
RM 2029, Sussex Drive 100
Ottawa, Ontario
Canada K2A 2V1

HULTQVIST, B.
Kiruna Geophysical Institute
P.O.Box 704, S-98127 Kiruna
Sweden

HUNSUCKER, R.
Geophysical Institute
University of Alaska
Fairbanks, Alaska 99701
USA

IMHOF, W.L.
Lockheed Palo Alto Research Lab.
Space Science Laboratory
3251 Hanover Street, Palo Alto, Calif. 94304
USA

IVERSEN, I.B.
Danish Space Research Institute
Lundtoftevej 7, DK-2800 Lyngby
Denmark

JACOBSEN, T.A.
NDRE
P.O.Box 25, N-2007 Kjeller
Norway

JENSEN, V.N.
Ionosphere Laboratory
Danish Meteorological Institute
Technical University of Denmark
DK-2800 Lyngby
Denmark

JONES, T.B.
Physics Department
The University
Leicester LE1 7RH
United Kingdom

JØRGENSEN, T.S.
Meteorological Institute
Lyngbyvej 100, DK-2100 Copenhagen
Denmark

KEYS, J.G.
Physics & Engineering Laboratory
D.S.I.R.
Omakau, Central Otago
New Zealand

KLUMPAR, D.M.
Center of Space Sciences, F02.2
The University of Texas at Dallas
P.O.Box 688, Richardson, Texas 75080
USA

KOEHLER, R.A.
Center for Research in Exp. Space Science
York University
4700 Keele St., Downsview, Ontario
Canada M3J 1P3

KOFMAN, W.
Centre National de la Recherche Scientifique
CEPHAG
P.O.Box 46, 38402 Saint Martin d'Heres
France

KOHL, H.
Max-Planck-Institut für Aeronomie
3411 Katlenburg-Lindau 3,
FRG

LEJEUNE, G.
CEPHAG - ENSIEG,
P.O.Box 46, 38402 Saint Martin d'Heres
France

LLEWELLYN, E.J.
Institute of Space and Atmospheric Studies
University of Saskatchewan
Saskatoon, Saskatchewan
Canada

MANTAS, G.P.
Department of Physics
University of Crete,
Iraklion, Crete
Greece

MAZAUDIER, C.
HMA/RPE
38-40 Avenue du General Leclec
92131 Issy-les-Moulineaux
France

MAYR, H.G.
Goddard Space Flight Center, Code 621
Greenbelt, Maryland 20771
USA

MENG, C.-I.
Applied Physics Laboratory/JHU
Laurel, Maryland 20810
USA

MIZERA, P.F.
The Aerospace Corporation A6/2447
P.O.Box 92957, Los Angeles, California 90009
USA

MYRABØ, H.Kr.
NDRE
P.O.Box 25, N-2007 Kjeller
Norway

MÅSEIDE, K.
Institute of Physics
University of Oslo
P.O.Box 1038 Blindern, Oslo 3
Norway

NEUBERT, T.
Danish Space Research Institute
Lundtoftevej 7, DK-2800 Lyngby
Denmark

THRANE, E.V.
NDRE
P.O.Box 25, N-2007 Kjeller
Norway

TULUNAY, Y.
O.D.T.U. Fizik Bölümü
Ankara
Turkey

UNGSTRUP, E.
Danish Space Research Institute
Lundtoftevej 7, DK-2800 Lyngby
Denmark

VASYLIUNAS, V.M.
Max-Planck-Institut für Aeronomie
3411 Katlenburg-Lindau 3
FRG

VICKREY, J.F.
Radio Physics Laboratory
SRI International
Menlo Park, California 94025
USA

VILLAIN, J.-P.
Max-Planck-Institut für Aeronomie
3411 Katlenburg-Lindau 3
FRG

VOLLAND, H.
Radioastronomisches Inst. der Universität Bonn
5300 Bonn - Endenich, Auf dem Hügel 71
FRG

VONDRAK, R.R.
Radio Physics Laboratory
SRI International
Menlo Park, California 94025
USA

WANNBERG, G.
Kiruna Geophysical Institute
Box 704, S-981 27 Kiruna
Sweden

WOOLISCROFT, L.J.C.
Department of Physics
The University of Sheffield
Sheffield S3 7RH
United Kingdom

WILSON, C.R.
Geophysical Institute
University of Alaska
Fairbanks, Alaska 99701
USA

ZI, MIN YUN
Max-Planck-Institut für Aeronomie
3411 Katlenburg-Lindau 3
FRG

ZUBER, A.
Meteorological Institute
Arrheniuslaboratory
106 91 Stockholm
Sweden

MIDDLE ATMOSPHERE DYNAMICS AND COMPOSITION

Marvin A. Geller

NASA/Goddard Space Flight Center, Laboratory for Planetary Atmospheres, Atmospheric Chemistry Branch, Greenbelt, MD 20771

INTRODUCTION

In this paper we have adopted the term "middle atmosphere" to describe that region of the earth's atmosphere that lies above the tropopause but below an altitude of 100 km. Looking at Fig. 1, we see that the altitude extent of the middle atmosphere varies with latitude and season such that, for example, the lower boundary of the middle atmosphere, the tropopause, is at about 17 km (about 80 mb) at the equator and at about 8 or 9 km at the pole (about 300 mb). Also, we see from this latitude-height temperature figure that other factors must be considered to explain the zonally-averaged temperature structure of the middle atmosphere beside radiative energy inputs. For instance, we see that in both the winter and summer, the equatorial lower stratosphere temperatures are cooler than at middle latitudes and that mesopause temperatures are warmer during the darkness of winter than during the continual sunlight of summer. We will see that the additional physics needed to explain these features involve the dynamics of the middle atmosphere.

In the following then, we will briefly touch on a few aspects of middle atmosphere dynamics and composition concentrating on only those topics that are important at middle and high latitudes.

MIDDLE ATMOSPHERE DYNAMICS

The steady-state mean zonal wind (the east-west flow averaged around latitude circles) is in approximate geostrophic equilibrium away from the equator. Thus, the mean zonal wind field shown in Fig. 2 is related by the thermal wind equation to the zonally averaged temperature field shown in Fig. 1. We see from

C. S. Deehr and J. A. Holtet (eds.), Exploration of the Polar Upper Atmosphere, 1–16.

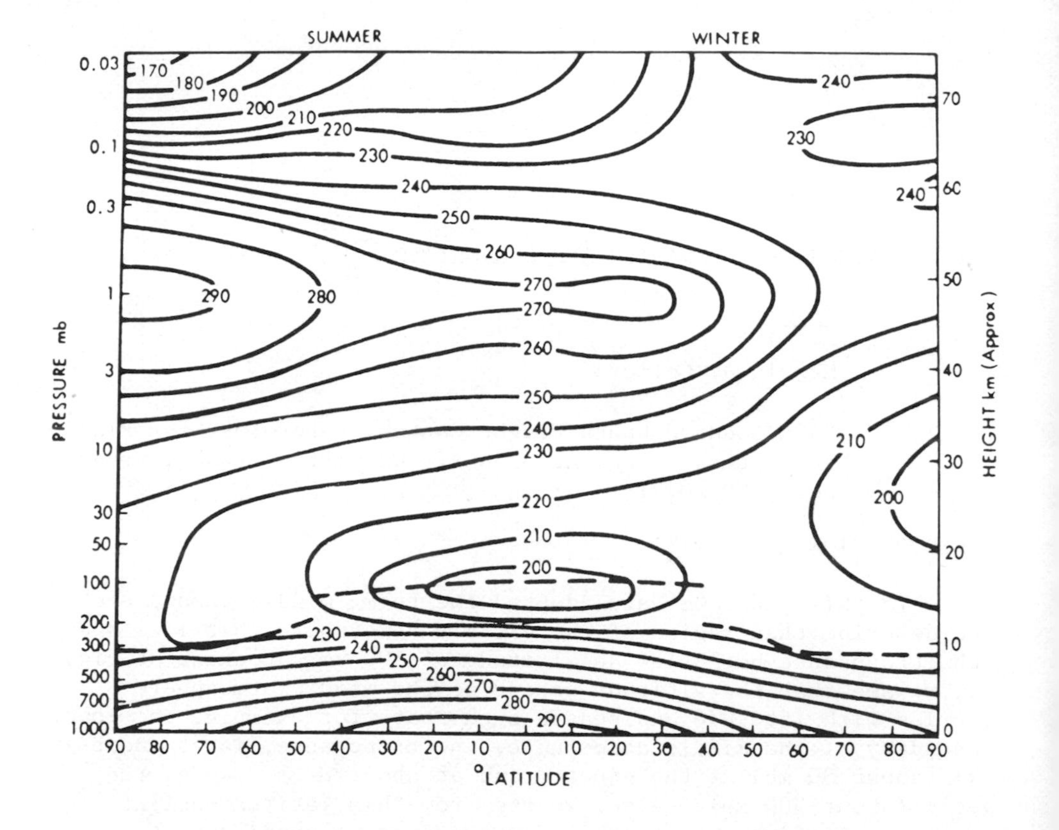

Fig. 1. Schematic latitude-height section of zonal mean temperature (K) at the solstices. Dashed line indicates tropopause level. (from (1))

Fig. 2 the westerly (west to east) middle atmospheric jet in winter and the slightly weaker easterly (east to west) jet in summer. In the following, we will briefly discuss the physics of this summer to winter reversal of the middle atmosphere jet as well as the effects of planetary waves on the mean zonal flow that sometimes produce sudden stratospheric warming events.

The radiative heating of the middle atmosphere is given by the net effect of the absorption of solar ultraviolet energy and the subsequent infrared emission to space from ozone, carbon dioxide and water vapor. This leads to a strong north-south differential in the heating of the middle atmosphere with net heating rates of about $10^{o}K\ day^{-1}$ at the summer polar stratopause where there is continual sunlight and a cooling of the same amount at the winter polar stratopause where there is continual darkness.

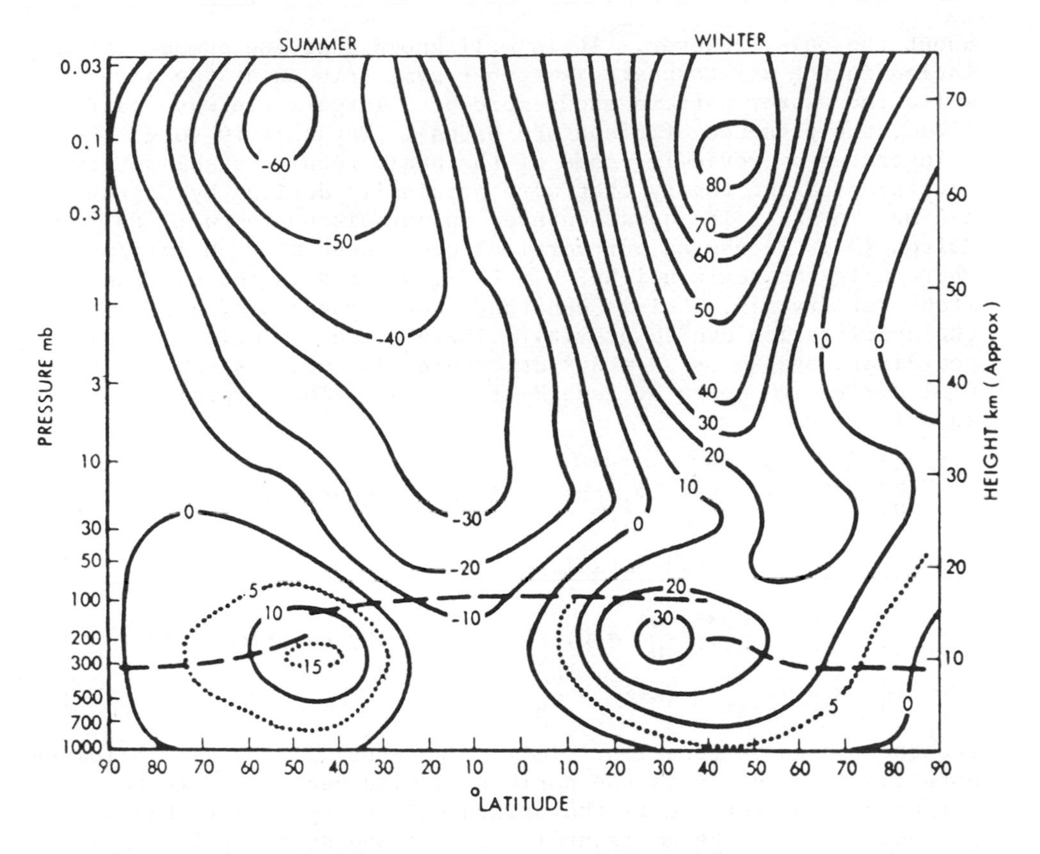

Fig. 2. Schematic latitude-height section of the mean zonal wind ($m\ s^{-1}$) at the solstices. Positive (negative) values indicate westerlies (easterlies). (from (1))

This heating distribution produces rising motion over the summer pole and sinking motion over the winter pole with a summer to winter flow at levels above the stratopause level and a winter to summer flow below as is required by continuity. Given the fact that the Coriolis force will deflect this flow to the right in the Northern Hemisphere and to the left in the Southern Hemisphere, a westerly middle atmosphere jet is produced in the winter hemisphere with an easterly jet in the summer hemisphere. This rising motion in the vicinity of the summer mesopause gives rise to the observed low temperatures there through expansion cooling while the sinking motions in the winter mesopause region give rise to the observed high temperatures there (see Fig. 1). This simple picture gives a qualitative understanding of the zonally averaged middle atmosphere circulation.

The observed middle atmosphere circulation is not symmetric

about the pole, however. It is well known that the summer circulation in the stratosphere is rather symmetric about the pole while the winter pattern shows more of a large-scale wavy pattern (i.e., the presence of planetary waves). Why this is so can be illustrated by reviewing some of the basic results of planetary wave propagation theory that were originally derived by Charney and Drazin (2). The notation used in our discussion will follow Holton (3). We use as our vertical coordinate $z^* = -H \ln(p/p_o)$ where H is the scale height, $H = RT_o/g$ where R is the gas constant for dry air, T_o is a constant mean temperature, and g is the acceleration due to gravity. The vertical velocity in this coordinate system is $w^* = dz^*/dt$, where d/dt denotes the substantial derivative operator (see Holton, (4)). The governing equations are

$$\frac{d\vec{v}}{dt} + f\,\hat{k} \times \vec{v} = -\vec{\nabla}\Phi, \text{ the horizontal equation of motion;} \quad (1)$$

$$\frac{\partial \Phi}{\partial z^*} = \frac{RT}{H}, \text{ the hydrostatic equation;} \quad (2)$$

$$\frac{\partial u}{\partial x} + \frac{\partial v}{\partial y} + \frac{\partial w^*}{\partial z^*} - \frac{w^*}{H} = 0, \text{ the continuity equation; and} \quad (3)$$

$$\left(\frac{\partial}{\partial t} + \vec{v}\cdot\vec{\nabla}\right)\frac{\partial \Phi}{\partial z^*} + w^* N^2 = \frac{\kappa \dot{Q}}{H}, \quad (4)$$

the thermodynamic equation. In writing eqns. (1-4), the following notation is used. $\vec{v}$ is the horizontal wind vector. f is the Coriolis parameter. $\hat{k}$ is the upward unit vector. Φ is the geopotential. $\vec{\nabla}$ is the horizontal gradient operator. T is temperature. u is the westerly wind component, and v is the southerly wind component. N^2 is the Brunt-Vaisala frequency squared, $N^2 = R/H\,(\partial T/\partial z^* + \kappa T/H)$, where $\kappa = R/c_p$, c_p being the specific heat of dry air at constant pressure. $\dot{Q}$ is the diabatic heating rate per unit mass. For large-scale adiabatic motions on a mid-latitude β-plane (a Cartesian geometry in which the curvature effects of the earth's spherical geometry are ignored, but the Coriolis parameter depends on latitude in linear fashion), eqns. (1-4), can be combined to give the quasi-geostrophic potential vorticity equation,

$$\left(\frac{\partial}{\partial t} + \vec{v}_g\cdot\vec{\nabla}\right)q = 0, \quad (5)$$

where $\vec{v}_g$ is the geostrophic wind and

$$q = \nabla^2\psi + f + e^{z^*/H}\frac{f_o^2}{N^2}\frac{\partial}{\partial z^*}\left(e^{-z^*/H}\frac{\partial \psi}{\partial z^*}\right). \quad (6)$$

In eqn. (6), $f = f_o + \beta y$ where f_o is the value of the Coriolis parameter at the middle latitude of the β-plane, β is the rate at which the Coriolis parameter varies in the northward direction.

y is the distance measured northward from the central latitude of the β-plane, and ψ is the streamfunction as defined by $\psi = \Phi/f_o$.

If we assume a mean zonal flow $\bar{u}(z^*)$ and linearize eqn. (5) letting $\psi = -\bar{u}y + \psi'$, one gets

$$\left(\frac{\partial}{\partial t} + \bar{u}\frac{\partial}{\partial x}\right)q' + \beta\frac{\partial \psi'}{\partial x} = 0 \qquad (7)$$

$$\text{where } q' = \nabla^2\psi' + e^{z^*/H}\frac{f_o^2}{N^2}\frac{\partial}{\partial z^*}\left(e^{-z^*/H}\frac{\partial \psi'}{\partial z^*}\right). \qquad (8)$$

We then investigate the vertical propagation of wave disturbances by putting the following expression for ψ' into (7) and (8).

$$\psi'(x,y,z^*,t) = \Psi(z^*)e^{i(kx + \ell y - kct) + \frac{z^*}{2H}} \qquad (9)$$

$$\text{which gives } \frac{d^2\Psi}{dz^{*2}} + m^2\Psi = 0 \qquad (10)$$

$$\text{where } m^2 = \frac{N^2}{f_o^2}\left\{\frac{\beta}{\bar{u}-c} - (k^2 + \ell^2)\right\} - \frac{1}{4H^2} \qquad (11)$$

From eqns. (10) and (11) it can be seen that for vertically propagating solutions to exist, m^2 must be positive which for stationary waves (c = 0) requires that

$$0 < \bar{u} < \frac{\beta}{k^2 + \ell^2 + \frac{f_o^2}{4N^2H^2}} \equiv U_c \qquad (12)$$

where U_c can be seen to be the Rossby wave phase speed. Charney and Drazin (2) then argued that the continents and oceans are distributed around the globe in a way that the principal forcing of stationary planetary waves occurs with k = ℓ = 2 which implies that for a β-plane centered at 45°, $U_c \approx 40\ ms^{-1}$. Note also that since U_c decreases with increasing k and ℓ the larger-scale waves propagate upward more readily (i.e. they can propagate their energy upwards through greater westerly mean zonal winds).

From this it is easy to see the reason why the summer mean circulation in the middle atmosphere is approximately symmetric since stationary planetary waves cannot propagate through the summer easterlies. Eqn. (12) implies that they also cannot propagate through the strong winter westerlies, but later work by Dickinson (5) and others has indicated that these waves can propagate upward in the winter hemisphere given the wind state as pictured in Fig. 2 that varies with latitude and altitude.

The most spectacular manisfestation of planetary waves in

the middle atmosphere takes place during a sudden stratospheric warming. This is illustrated in Fig. 3. Fig. 3 shows the pre-warming situation at 50 mb on January 25, 1957. Note that there exists a cold vortex over the pole with troughs extending across the North American and Eurasian continents. There are warm temperatures associated with the Aleutian anticyclone. On February 4 we see an intensification of this flow pattern with a simultaneous growth of the warm pockets. On February 9 the flow has broken up into pairs of cyclonic and anticyclonic vortices, and the zonal current has weakened. The pool of cold air that previously lay over the pole has split and moved to middle latitudes with the two warm pockets that originally existed having migrated northward to form a single warm region over the pole. The effect of this stratospheric warming on the 50 mb mean zonal wind and the zonally-averaged temperature is to decelerate the mean zonal westerlies, eventually giving easterlies there and to give rise to much warmer polar temperatures. In the Southern Hemisphere, warming events are also observed in which there are stratospheric temperature increases that are equivalent to those observed in the Northern Hemisphere, but reversals of the winter polar westerlies are not found to occur there (see (7)).

Labitske (8) has used rocket measurements of temperature and high altitude satellite radiance data to trace the effects of stratospheric warmings into the upper mesosphere as well as to demonstrate how these effects depend on latitude. Her study indicated that there is a compensation effect during a stratospheric warming in both altitude and latitude so that during the time when the high latitude stratosphere is warming, the high latitude mesosphere and the low latitude stratosphere are cooling, and the low latitude mesosphere is warming. Labitske (8) suggests that the term "stratospheric warming" is a misleading one given that opposite effects occur simultaneously in different altitude and latitude regions. Having looked into the morphology of stratospheric warmings as well as some of the properties of extratropical planetary waves, let us now look into how these fit together into our present concept of the physics of stratospheric warming events.

We start with proving a result that was originally shown by Eliassen and Palm (9) that upward propagating stationary planetary waves transport heat northward. We start with the geostrophic relation,

$$v = \frac{1}{f_o} \frac{\partial \Phi}{\partial x}; \text{ the hydrostatic} \tag{13}$$

relation, eqn. (2); and the steady state thermodynamic equation,

$$\bar{u} \frac{\partial}{\partial x} \frac{\partial \Phi}{\partial z*} + w*N^2 = 0 \tag{14}$$

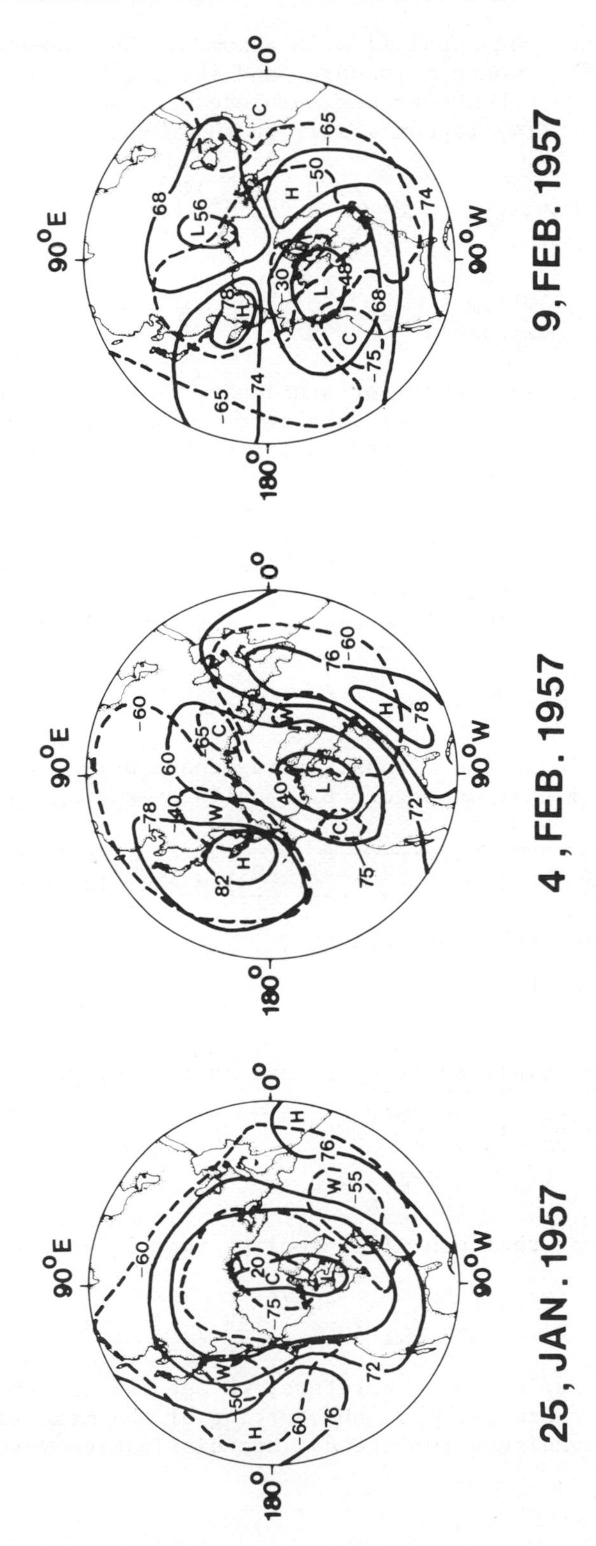

Fig. 3. 50 mb charts showing evolution of the sudden stratospheric warming of early 1957. Contours (full line) in 100's of feet; isotherms (dashed lines) in deg C (adapted from (6)).

where a basic state mean zonal flow is assumed. The upward energy flux is $\langle -\omega\Phi \rangle$, where ω is dp/dt and the bracket notation indicates an average taken over one complete wave period and wavelength. Using eqn. (14) it can easily be shown that

$$- \langle \omega\Phi \rangle = \frac{p}{H} \langle w^*\Phi \rangle = \frac{p}{H} \left\langle - \frac{\bar{u} \frac{\partial}{\partial x}(\frac{\partial \Phi}{\partial z^*})\Phi}{N^2} \right\rangle$$
$$= \frac{\bar{u}}{N^2} \frac{p}{H} \left\langle \frac{\partial \Phi}{\partial x} \frac{\partial \Phi}{\partial z^*} \right\rangle = \frac{u f_o pR}{N^2 H^2} \langle vT \rangle . \qquad (15)$$

We have already seen that for stationary waves to propagate energy upward, $\bar{u}$ must be greater than zero. Thus, stationary waves that propagate energy upward must also transport heat northward, but how does this heat transport as well as the wave's momentum transport affect the mean zonal flow? To look at this we will prove a version of the noninteraction theorem that was originally shown by Charney and Drazin (1) and was generalized by Boyd (10) and Andrews and McIntyre (11). We start with eqn. (5), the conservation of potential vorticity. The linearized version of this is

$$\frac{\partial q'}{\partial t} + \bar{u} \frac{\partial q'}{\partial x} + \frac{\partial \psi'}{\partial x} \frac{\partial \bar{q}}{\partial y} = 0 \qquad (16)$$

where the overbar indicates a zonal average, and primed variables are deviations from that average. By zonally averaging eqn. (5), one gets

$$\frac{\partial \bar{q}}{\partial t} = -\overline{(\vec{v}_g \cdot \vec{\nabla})q} = - \overline{(- \frac{\partial \psi'}{\partial y} \frac{\partial q'}{\partial x} + \frac{\partial \psi'}{\partial x} \frac{\partial q'}{\partial y})} = - \frac{\partial}{\partial y} \overline{(q' \frac{\partial \psi'}{\partial x})} \qquad (17)$$

Now, assuming a wave behavior such that

$$\psi', q' \propto e^{ik(x-ct)} \qquad (18)$$

in eqn. (16) and multiplying it by ψ' and averaging, we get

$$\overline{\psi' \cdot \{(\bar{u} - c) \frac{\partial q'}{\partial x} + \frac{\partial \bar{q}}{\partial y} \frac{\partial \psi'}{\partial x}\}} = 0 \qquad (19)$$

Thus, if $\bar{u} - c \neq 0$, $\overline{\psi' q'_x} = \overline{q' \psi'_x} = 0$, (20)

in which case we see that eqn. (17) implies that $\partial \bar{q}/\partial t = 0$, but since

$$\frac{\partial \bar{q}}{\partial y} = \beta - \frac{\partial^2 \bar{u}}{\partial y^2} + e^{z^*/H} \frac{f_o^2}{N^2} \frac{\partial}{\partial z^*} (e^{-z^*/H} \frac{\partial \bar{u}}{\partial z}), \qquad (21)$$

this implies that $\partial \bar{u}/\partial t = 0$. Therefore, we see that in the absence of critical lines where $\bar{u}-c = 0$, the forcing of the mean flow by the perturbations vanishes for steady, non-dissipative waves.

Utilizing what we have derived about planetary wave propagation and its effect on the mean zonal flow, we can now discuss the basis of the currently accepted theory of stratospheric warmings that was first presented by Matsuno (12). The course of events begins with an increase in the amplitude of the tropospheric planetary waves for some reason, perhaps due to the resonance mechanism of Tung and Lindzen (13). This increase in the planetary wave amplitudes at low levels gives an increased northward heat transport at low levels which gives rise to a rising motion at high latitudes and a sinking motion at low latitudes. The Coriolis torque deflects the equatorward moving air to the right (for the Northern Hemisphere) which gives rise to decreased westerlies at high levels. If this is as far as the situation evolves, only a minor warming takes place. If, however, the Coriolis torque is sufficient at high altitudes to produce easterlies, a critical level is established for stationary planetary waves, i.e., $\bar{u} = c = 0$ at some altitude. Then, we have the situation pictured in Fig. 4. The planetary wave energy cannot penetrate this critical level. Thus, there is northward heat transport only below the critical level. This imbalance in the heat transport with height gives rise to an even more marked secondary circulation, as pictured, whose upward motion gives rise to expansion cooling that balances some but not all of the northward heat transport below the critical level, and causes a cooling at high latitudes above the critical level. The absorption of planetary wave momentum at the critical level gives rise to a downward propagation of the situation. Note how this mechanism depends on the existence of conditions that violate the noninteraction theorem, i.e., wave transience, dissipation and the presence of critical levels. Note also how this mechanism gives rise to warming and cooling patterns that are opposite at high and low latitudes as well as at lower and higher altitudes in agreement with observations.

MIDDLE ATMOSPHERE COMPOSITION

The upper boundary of the middle atmosphere approximately coincides with the altitude of the turbopause (about 105 km). Thus, the major gaseous constituents are well mixed through the middle atmosphere. There are many minor constituents of the middle atmosphere that are not well-mixed, however. In the following, we will discuss the chemistry of two of these trace constituents together with the importance of transport in determining their distribution. These are ozone which plays a dominant role in the radiation balance of the middle atmosphere and nitric oxide which plays a dominant role in determining the structure of the lower ionosphere.

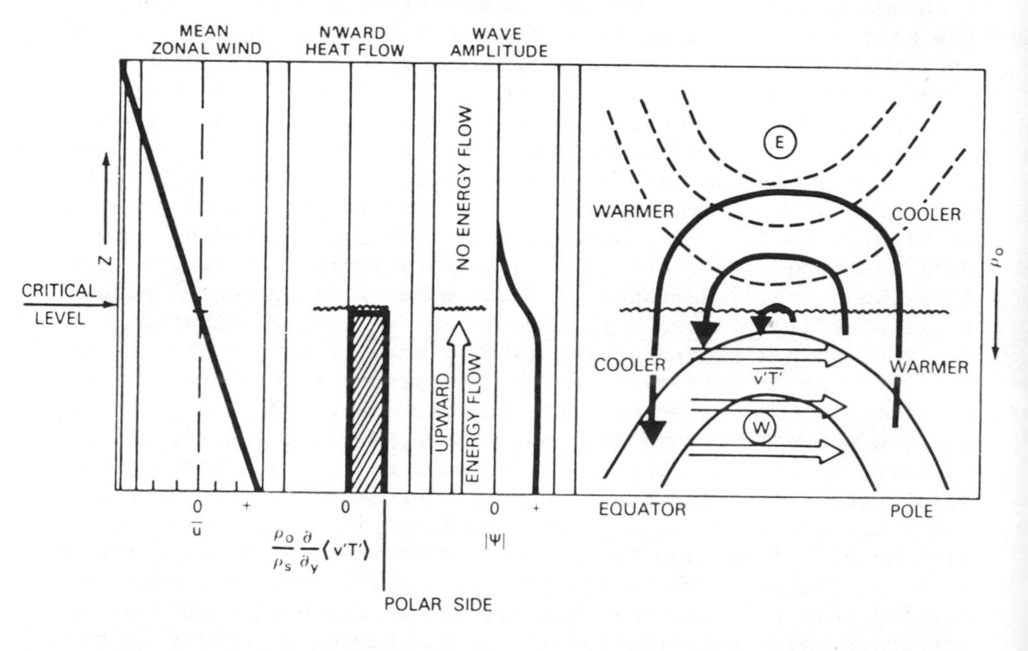

Fig. 4. Idealized critical layer properties for a stratospheric warming. From left to right the figure shows the mean zonal wind at midchannel, the meridional derivative of the northward heat flow as a function of height along the polar side of the channel, and the wave amplitude and energy flow at midchannel. The diagram on the far right shows the two-dimensional idealized structure of a critical layer. Open arrows indicate the direction of eddy heat transport. Solid arrows show cross-channel and vertical flow patterns, the induced secondary circulation. Contours indicate zonal winds; dashed lines are easterly, and solid lines are westerly (from (14)).

Ozone

Ozone is formed in the upper stratosphere and lower mesosphere by the following reactions:

$$O_2 + h\nu(\lambda \leq 242 \text{ nm}) \rightarrow O + O \tag{22}$$

and

$$O_2 + O + M \rightarrow O_3 + M \tag{23}$$

where M is any major atmospheric constituent, i.e., O_2 or N_2. Ozone and atomic oxygen can be destroyed in a pure oxygen atmosphere by the reactions:

$$O_3 + h\nu(\lambda \leq 1,180 \text{ nm}) \rightarrow O_2 + O \quad (24)$$

$$O + O_3 \rightarrow 2\ O_2, \text{ and} \quad (25)$$

$$O + O + M \rightarrow O_2 + M. \quad (26)$$

Reactions (22-26) constitute the so-called Chapman (15) set of reactions. In recent years, it has become obvious that reactions 24-26 are not sufficient ozone destruction mechanisms to account for observed ozone distributions. This has led to considerations of some catalytic cycles for ozone and atomic oxygen loss involving oxides of nitrogen, oxides of hydrogen, and halocarbons which appear to not only give sufficient ozone destruction, but also point out the danger of man's activities leading to enhanced concentrations of middle atmosphere minor constituents that catalytically destroy ozone.

Since ozone formation depends on the amount of atomic oxygen production by solar ultraviolet dissociation of molecular oxygen, one would expect maximum ozone amounts to be found in regions and at times when solar radiation is maximum, i.e., at the subsolar point in the tropics. Observations indicate that this is not so, however. Fig. 5 shows the zonally-averaged distribution of total ozone as a function of latitude and time of year. What one sees is that maximum ozone amounts occur in spring and at high latitudes with minimum ozone amounts present in regions that are thought to be where production is maximum.

Brewer (17) and Dobson (18) noting this discrepancy, suggested that there exists a mean meridional circulation of the type that is pictured in Fig. 6. Note that at high latitudes this circulation transports ozone down from its source region near the stratopause leading to enhanced ozone amounts while at low latitudes the rising motion inhibits the transport of ozone downward away from its source leading to smaller ozone amounts. The Brewer-Dobson circulation also helps to explain why there is so little water vapor in the stratosphere since the upward velocities carry the water vapor upward through the cold tropical tropopause where much of it is condensed out.

The Brewer-Dobson circulation does not appear to agree with the mean meridional circulation in the lower stratosphere inferred from observations by Vincent (20) where a two-celled circulation is derived with rising motion in the tropics, downward motion in middle latitudes and rising motion in the polar regions. The resolution of this apparent paradox is found in an extension of the previously shown noninteraction theorem. We note that the proof of this theorem depends on the potential vorticity being a conserved quantity, but since ozone is also a conserved quantity in the lower stratosphere it follows that in the absence of critical levels and dissipation that the ozone transports by steady vertically propagating planetary waves will not give rise to changes in the zonally averaged ozone distribution. This is in fact what is shown by observational studies (21) and model results (22) where the ozone transports by the eddies are nearly balanced by the transports by the mean meridional circulation. Thus, while the eddies help to produce a complicated appearing mean meridional circulation, the net effect of the eddies plus the differential heating in the troposphere is a Brewer-Dobson type circulation with air parcels having a net upward motion at low latitudes and having a net downward motion at high latitudes in the lower stratosphere.

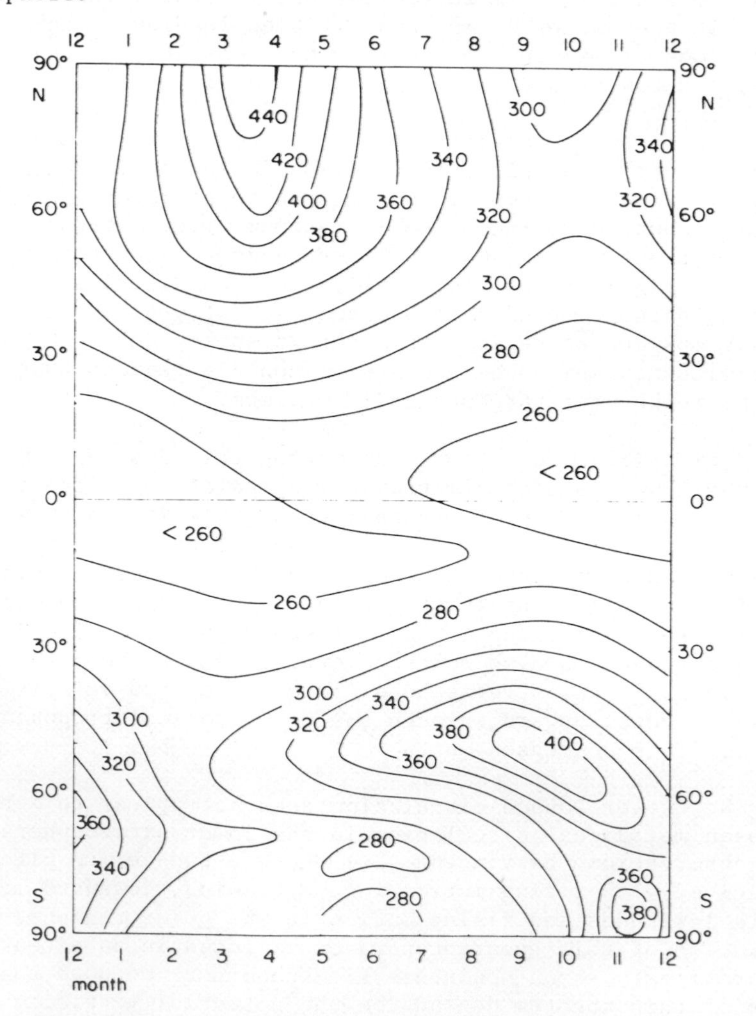

Fig. 5. The observed behavior of columnar ozone (Dobson units) as a function of latitude and season (from (16)).

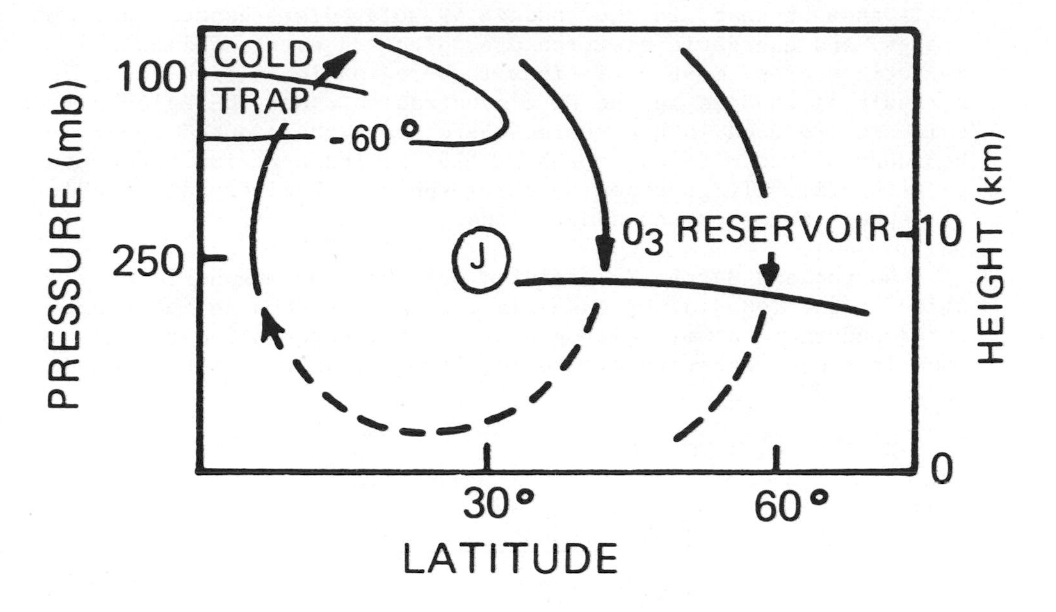

Fig. 6. The Brewer-Dobson model of the mean meridional circulation. Heavy lines denote the tropopause (from (19)).

Since stratospheric warmings disrupt the normal circulation, we expect there to be a marked change from the normal winter ozone distribution during a warming episode. Ghazi (23) has used satellite observations of total ozone together with stratospheric temperature to relate the response of total ozone to the stratospheric warming of January 1971. What he finds is that there is an increase in total ozone at high latitudes accompanying this warming event. This agrees with our picture of stratospheric warmings being related to a strong northward eddy heat transport which is only partially balanced by the expansion cooling induced by the planetary wave activity below the critical level. Apparently, the increased ozone amounts at high latitudes are the result of the accompanying northward eddy ozone transport[2] which is only partially balanced by the induced rising motion.

Nitric Oxide

Nitric oxide is a very important minor constituent in the upper part of the middle atmosphere since it is the principal constituent in the D region that can be ionized by solar radiation. Thus, ionization in the quiet D region begins with the action of solar Lyman-α radiation ionizing nitric oxide.

$$NO + h\nu(121.6\text{ nm}) \rightarrow NO^+ + e^- . \qquad (27)$$

It is thought that, in the absence of solar disturbances such as flares, and energetic electron precipitation events following magnetic storms, most variations in D-region ionization occur as a result of changes in the NO concentration. Now, D-region nitric oxide is produced in the thermosphere above the adopted upper boundary of the middle atmosphere (25) so its D-region concentration is primarily governed by thermospheric chemistry and middle atmosphere transport considerations.

Norton and Barth (26) pointed out that thermospheric nitric oxide production did not take place primarily through the reaction of ground state atomic nitrogen $N(^4S)$, but through the reaction of the first excited state of atomic nitrogen, $N(^2D)$, with molecular oxygen, i.e.,

$$N(^2D) + O_2 \rightarrow NO + O \qquad (28)$$

Thus, the production of NO in the thermosphere is a function of the production of $N(^2D)$. These are as follows (27):

$$N_2 + e^-(\text{fast}) \rightarrow N(^2D) + N(^4S) + e^- \qquad (29)$$

$$N_2 + h\nu(80\text{-}100\ \text{nm}) \rightarrow N(^2D) + N(^4S) \qquad (30)$$

$$N_2^+ + O \rightarrow NO^+ + N(^2D) \qquad (31)$$

and $$NO^+ + e^- \rightarrow N(^2D) + O \qquad (32)$$

At least half of the atomic nitrogen product of the dissociative recombination of NO^+ is thought to be in the $N(^2D)$ state rather than $N(^4S)$. One sees then that the amount of $N(^2D)$ and hence the production of NO, depends on the energetic electron flux (Reaction 29), the level of ionization (Reactions 31 and 32) and the EUV flux. This is reflected in the observations of nitric oxide at an altitude of 105 km (28) that show more NO at high latitudes where energetic electron fluxes are maximum than at middle and low latitudes. Moreover, it is found that the amount of high latitude nitric oxide increases after days of high magnetic activity (28), (29).

There is clear evidence that D-region ionization responds to both energetic particle fluxes and to meteorological disturbances. For instance, Bourne and Hewitt (30) have done a superposed epoch analysis of ionospheric absorption (a measure of the D-region electron density) at Lindau keyed to the geomagnetic disturbance index Kp which illustrates that the ionospheric absorption (and by inference, the electron density) are maximum a few days after Kp is maximum and minimum a few days after Kp is minimum. The work of Spjeldvik and Thorne (31) and Larsen _et al._ (32) has made a convincing case that the enhancement in D-region electron density following magnetic storms is due to increased precipitation of energetic electrons; however, it is also possible that some of the increase in electron density is due to increased NO amounts. Also, the decreased electron density amounts following very quiet days are very likely due to decreased NO concentrations. Shapley and Beynon (33) have used superposed epoch analysis of ionospheric absorption at Lindau keyed to stratospheric temperature at Berlin, to show that the ionospheric absorption at Lindau is a maximum when the stratospheric temperature is a maximum at Berlin; i.e., during a stratospheric warming. What we are probably seeing in these two results are the net effects of time varying ionization by precipitating energetic electrons, time varying thermospheric NO production that is a function of energetic particle fluxes, and the time varying advection of NO to mid-latitudes from the NO-rich auroral zone. Note that we have seen previously that the circulation that gives rise to a stratospheric warming, gives rise to a mesosphere-lower thermosphere circulation that is equatorward (see Fig. 4) which will transport the NO-rich air to mid-latitudes. Correlations between equatorward flow in the lower thermosphere and NO concentrations (as evidenced by D-region electron densities) during winter have been found previously (34), (35). One thing which is not clear, however, is the respective role of the various mechanisms that contribute to the downward transport of NO from the region of its production in the lower thermosphere to the D-region.

CONCLUDING DISCUSSION

The middle atmosphere is a complex region where both dynamics and composition are affected by tropospheric forcing from below and by energetic inputs from above. Phenomena that occur in one region of the middle atmosphere can be coupled to other altitude regions above or below. We have illustrated that stratospheric warmings are connected to tropospheric planetary wave fluxes and that the resulting middle atmosphere circulation can alter the ozone amounts in the stratosphere and the nitric oxide amounts in the mesosphere and lower thermosphere.

The topics treated in this paper were chosen to be illustrative of middle atmosphere processes rather than being chosen for a comprehensive review of high latitude processes in the middle atmosphere. Interesting phenomena such as the marked enhancements of high latitude mesospheric heating (36) and the high latitude depletions of ozone (37) that occur during polar cap absorption events were not discussed at all, for example.

FOOTNOTES

(1) The same general process is also thought to produce the reversed temperature gradient in the lower stratosphere; that is to say, the large release of latent heat in the tropical troposphere gives rise to upward motion that extends into the stratosphere giving rise to expansion cooling there with downward motion and compressional heating at higher latitudes.

(2) This can also be interpreted by noting that the Lagrangian circulation during a stratospheric warming (24) acts to intensify the Brewer-Dobson circulation in the lower stratosphere.

ACKNOWLEDGEMENTS

The preparation of this paper was primarily done while the author was on the faculty of the University of Miami, supported by grant NSF ATM-7908352 of the Climate Dynamics Research Section.

REFERENCES

(1) Murgatroyd, R.J., 1969: (G.A. Corby, Ed.), London, Roy. Meteor. Soc., 159.
(2) Charney, J.G. and P.G. Drazin, 1961: J. Geophys. Res., 66, 83.
(3) Holton, J.R., 1975: pp 218 in The Dynamic Meteorology of the Stratosphere and Mesosphere, American Meteorological Society.
(4) Holton, J.R., 1972: pp 319 in An introduction to Dynamic Meteorology, Academic Press.
(5) Dickinson, R.E., 1968: J. Atmos. Sci., 25, 984.
(6) Reed, R.J., J. Wolfe, and H. Nishimoto, 1963: J. Atmos. Sci., 20, 256.
(7) Barnett, J.J., 1975: Nature, 255, 387.
(8) Labitske, K., 1972: J. Atmos. Sci., 29,1395.
(9) Eliassen, A. and E. Palm, 1960: Geophys. Publ. 22, No. 3.
(10) Boyd, J., 1976: J. Atmos. Sci., 33, 2285.
(11) Andrews, D.G. and M.E. NcIntyre, 1976; J. Atmos. Sci., 33, 2031.
(12) Matsuno, T., 1971: J. Atmos. Sci., 28, 1479.
(13) Tung, K.K. and R.S. Lindzen, 1979. Mon. Wea. Rev., 107, 735.
(14) Schoeberl, M.R., 1978: Rev. Geophys. Space Phys., 16. 735.
(15) Chapman, S., 1930: Mem. Roy. Meteorol. Soc., 3. 103.
(16) Dutsch, H.V., 1971: Advances in Geophys., 15, 219.
(17) Brewer, A.W., 1949: Quart. J. Roy. Meteor. Soc., 75. 351.
(18) Dobson, G.M.B., 1956: Proc. Roy. Soc. Lond., A236, 187.
(19) Wallace, J.M., 1978: A Lagrangian view, p. 25 in "Advanced Study Program Summer Lecture Notes", NCAR, Boulder, Colorado.
(20) Vincent, D.G., 1968: Quart. J. Roy. Meteor. Soc., 94. 333.
(21) Miller, A.J., R.M. Nagatani. K.B. Labitske, E. Klinker, K. Rose, and D.F. Heath. 1977: Proc. Joint. Symp. Atmos. Ozone, 135.
(22) Cunnold, D.M., F.N. Alyea, and R.G. Prinn, 1979: PAGEOPH (in press).
(23) Ghazi, A., 1974: J. Atmos. Sci., 31, 2197.
(24) Matsuno, T. and K, Nakamura, 1979; J. Atmos. Sci., 35, 640.
(25) Strobel, D.F., 1971: J. Geophys. Res., 76, 8384.
(26) Norton, R.B. and C.A. Barth. 1970: J. Geophys. Res., 75, 3903.
(27) Oran, E.S., P.S. Julienne, and D.F. Strobel, 1975: J. Geophys. Res., 80, 3068.
(28) Cravens, T.E. and A.I. Stewart, 1978: J. Geophys. Res., 83. 2446.
(29) Gerard, J. and C.A. Barth, 1977: J. Geophys. Res., 82, 2446.
(30) Bourne, I.A. and L.W. Hewitt, 1968: J. Atmos. Terr. Phys., 30, 1381.
(31) Spjeldvik, W.N. and R.M. Thorne, 1975: J. Atmos. Terr. Phys., 37, 777.
(32) Larsen, T.R., T.A. Potemra, W.I. Imhoff, and J.B. Reagan. 1977: J. Geophys. Res., 82, 1519,
(33) Shapley, A.H. and W.J.G. Beynon, 1965: Nature, 206, 1242.
(34) Geller, M.A., G.C. Hess, and D. Wratt, 1976: J. Atmos. Terr. Phys., 38, 287.
(35) Hess, G.C. and M.A. Geller, 1978: J. Atmos. Terr. Phys., 40, 895.
(36) Banks, P.M., 1979: J. Geophys. Res., 84, 6709.
(37) Heath, D., A.J. Krueger, and P.J. Crutzen, 1977: Science, 197, 886.

DYNAMICS OF THE THERMOSPHERE DURING QUIET AND DISTURBED CONDITIONS

H. Volland

Radioastronomical Institute
University of Bonn
53 Bonn, W.- Germany

ABSTRACT

The positive temperature gradient and molecular dissipation mechanisms at thermospheric heights result in the nearly linear response of that region to external forcing. A simplified theory of the propagation and generation of acoustic-gravity waves and of planetary waves is outlined using the linearization of the hydrodynamic equations and the superposition of harmonic waves via a Fourier series of a Fourier integral. The main energy sources at thermospheric heights are solar XUV, wave dissipation, Joule heating, and Ampere forcing. Basic wind configurations are then calculated which are generated by the predominant, harmonically-varying heat and momentum sources of magnetospheric origin.

1. INTRODUCTION

The thermosphere is that region above approximately 85 km altitude which is characterized by monotonically increasing kinetic temperature. It reacts to external and internal heat and momentum sources in a manner which in many respects is quite different from that of the lower atmosphere. In addition, solar radiation reaching the lower atmosphere and the ground is rather constant, but solar XUV radiation absorbed within thermosphere is highly variable depending on solar activity. Internal heat redistribution due to latent heat and turbulence which is so dominant within the lower atmosphere is nearly absent within the thermosphere. On the other hand, the electric field of magnetosperic origin is a heat and momentum source for the thermospheric gas due to interaction with the ionospheric plasma. The Coriolis force which is an important virtual force for lower atmospheric

C. S. Deehr and J. A. Holtet (eds.), Exploration of the Polar Upper Atmosphere, 17–30.

dynamics is of minor importance within the thermosphere where ion drag and molecular viscosity dominate. The lower atmosphere is in a state of instability which gives rise to complicated non-linear wave interactions. The thermosphere, however, is rather stable, behaving like a low pass filter by suppressing waves of small periods and small horizontal scales, thereby acting like a damped oscillator system. It is that property which makes large scale dynamics of planetary and tidal waves the prominent feature of the thermosphere. In this paper, I want to discuss some basic ideas about wave dynamics at thermospheric heights with special emphasis to their generation at high latitudes.

2. THEORY OF THERMOSPHERIC WAVES

The stable configuration of the thermosphere with its positive temperature gradient together with efficient molecular dissipation mechanisms like viscosity and heat conduction are responsible for the nearly linear response of the thermosphere to external forcing. Since dissipation prevents an unlimited increase of wave amplitudes, linearization of the hydrodynamic equations is a reasonable approximation to describe wave propagation. Thus, any disturbance of arbitrary time dependence (including pulse type disturbances) can be simulated by the superposition of harmonic waves via a Fourier series or a Fourier integral. It is possible, therfore, to describe the reaction of the thermosphere to harmonically varying energy and momentum sources.

The time dependence of harmonic waves is

$$e^{-i\omega t} \tag{1}$$

with $\omega = 2\pi f$ the angular frequency and f the frequency of the wave.

2.1 Acoustic-gravity waves

Acoustic-gravity waves have periods of the order of milliseconds to hours. They are of small or medium scales with horizontal wave lengths of up to several hundreds of km. Because of these properties, the Coriolis force and the spherical form of the earth can be neglected. The wave parameters of relative pressure, p/p_o and vertical wind, W of plane waves propagating within a cartesian plane in a nondissipative isothermal atmosphere with an angle of incidence θ with respect to the vertical are proportional to

$$\left.\begin{matrix} p/p_0 \\ w \end{matrix}\right\} \propto \exp\{i(k_x x + k_y y \pm k_z z - \ \ t) + z/(2H)\} \qquad (2)$$

$k_h = \sqrt{k_x^2 + k_y^2} = k_o \sin\theta$ horizontal wave number

$k_o = \omega/c$ (c speed of sound)

H the pressure scale height

The vertical wave number is related to the horizontal wave number according to Hines (1960)

$$k_z = \frac{1}{c}\sqrt{\omega^2 - \omega_a^2 - (\omega^2 - \omega_p^2)k_h{}^2/k_o{}^2} \qquad (3)$$

with

$\omega_a = \frac{\gamma g}{2c}$ the Brunt-Väisälä frequency

$\omega_b = \frac{2\sqrt{(\gamma-1)}}{\gamma}\omega_a$ the acoustic cutoff frequency

γ the ratio of the specific heats

g the gravitational acceleration

Acoustic waves are defined as waves with frequencies $|\omega| > \omega_a$. For $|\omega| >> \omega_a$, (3) may be approximated by

$$k_z \simeq \sqrt{k_o^2 - k_h^2} = k_o \cos\theta \ . \qquad (4)$$

These waves are nondispersive and isotropic having a phase velocity

$$v = \omega|\sqrt{k_h^2 + k_z^2} \simeq c. \qquad (5)$$

Gravity waves are defined as waves with frequencies $|\omega| < \omega_b$. For $|\omega| << \omega_b$, (3) becomes

$$k_z \simeq \frac{1}{c}\sqrt{-\omega_a^2 + \omega_b^2 k_h^2/k_o^2}. \qquad (6)$$

These waves are anisotrophic because the restoring force, which in the case of acoustic waves is the compressibility of air, is gravitational force in the vertical direction in the case of gravity waves. The phase velocity of gravity waves depends, therefore, on the direction of propagation. For horizontal wave numbers

$$k_h < \frac{\omega_a k_o}{\omega_b} \tag{7}$$

they become evanescent waves. That means they lose their wave characteristics in the vertical direction, and their amplitudes decay exponentially. These waves cannot transport wave energy in the vertical direction.

For

$$k_h \geqq \frac{\omega_a k_o}{\omega_b} , \tag{8}$$

the waves become propagating waves. Their vertical phase velocity, is opposite to their vertical group velocity v_{gr}:

$$v = \omega/k_z; \quad v_{gr} = \frac{\partial\omega}{\partial k_z} \quad -k_z{}^3/(k_h^2\omega_b^2) \tag{9}$$

This leads to the curious result that the vertical phase progression of a gravity wave is opposite to the direction of its wave energy transport (Hines, 1960).

2.2 Planetary and tidal waves

Planetary waves have horizontal scales of the order of the dimensions of the earth, and their periods are of the order of one solar day or larger. Therefore, tidal waves are a type of planetary wave. The Coriolis force as well as the sphericity of the earth have to be taken into account for these waves. From the theory of planetary waves within a nondissipative isothermal atmosphere, it follows (e.g., Longuet-Higgins, 1968; Volland and Mayr, 1977) that the x-dependence in (2) has to be replaced by a longitudinal (λ) dependence:

$$e^{ik_x x} \rightarrow e^{im\lambda} \quad (m = 0, 1, 2 \ldots \text{ zonal wave number}) \tag{10}$$

while the y- dependence has to be replaced by Hough functions which are functions of latitude ϕ

$$e^{ik_y y} \rightarrow {}^m_n(\phi) \quad (n \text{ meridional wave number} \tag{11}$$

Only individual modes of wave numbers (m,n) can be generated.

The vertical wave number k_z corresponds to k_z in (6):

$$(k_z)^m_n = \frac{1}{2H}\sqrt{\frac{4(\gamma-1)\ H}{\gamma\ \ h^m_n} - 1} \tag{12}$$

where h_n^m is a separation constant called the equivalent depth. Comparison with (6) yields

$$k_h = \frac{\omega}{\sqrt{gh_n^m}} \tag{13}$$

From the theory of oceanic waves in a flat ocean of depth h, it is well known that their phase velocity is $v = (gh)^{1/2}$. This is the origin of the name "equivalent depth". However, while the ocean depth h is always a real number, the equivalent depth has no direct physical meaning, and it can be positive or negative. If it is negative, the wave behaves like an evanescent wave.

2.3 Wave dissipation

Molecular heat conduction and viscosity play an increasingly dominant role for wave dissipation at thermospheric heights. That effect can be very clearly seen in the case of heat conduction. The characteristic time of vertical heat transport is

$$\tau \simeq \frac{H^2 \rho c_p}{K} \simeq \begin{cases} 30 \text{ h} & 120 \text{ km} \\ 1 \text{ h} \quad \text{at} & 200 \text{ km} \\ 8 \text{ min} & 300 \text{ km} \end{cases} \tag{14}$$

(K is the coefficient of heat conduction; c_p is the specific heat at constant pressure). If the wave period exceeds the characteristic time, then heat conduction tends to destroy the vertical wave structure, and ordered wave energy is transferred into unordered heat energy. The result is an increase of the vertical wavelength and a decrease of the wave amplitude with height.

The effect of molecular viscosity is very similar, the principal physical difference being that mechanical momentum is exchanged vertically while in the case of heat conduction, heat is exchanged vertically.

Collisions between ions and neutrals are responsible for a horizontal exchange of momentum between ionospheric plasma and the neutral wind. Since ions are bound to the geomagnetic field lines above about 150 km altitude ion drag is strongest for zonal and is weakest for meridional winds at the equator. However, for planetary waves of low wave numbers (or large scales), the horizontal distribution of ion drag is smeared out so that the assumption of a constant effective number of collisions frequency is already a fairly good approximation.

The dissipation effects of heat conduction, viscosity and ion drag can be parameterized to a first order approximation by a complex vertical wave number:

$$k_z \simeq \frac{1}{2H} (\alpha + i\beta) \qquad (15)$$

which yields for upgoing waves a vertical dependence in (2) of

$$\exp \{i\frac{\alpha z}{2H} - \frac{(\beta-1)}{2H} z\} \qquad (16)$$

At altitudes above about 150 km, all planetary waves have numbers $\beta > 1$ increasing with altitude, and α decreasing with altitude so that the amplitudes of free waves decrease and the vertical wavelengths, $\lambda_z = 4\pi H/\alpha$, increase with altitude.

Ion drag tends to turn the wind toward the direction of the negative pressure gradient (see Fig. 1). Since ion drag exceeds the Coriolis force, at least at F layer heights, the Coriolis force plays only a minor role there. It can be shown that the Hough functions in (11), which are the eigenfunctions of the lower atmosphere, approach the spherical functions P_n^m at thermospheric heights (e.g., Volland and Mayr, 1977).

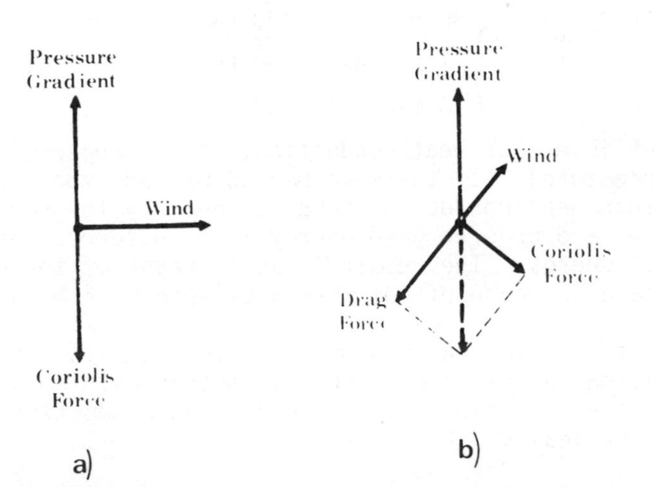

Figure 1. Geostrophic wind (a) and ageostrophic wind due to ion drag (b).

Within the dynamo region below about 150 km, the coupling between plasma and neutral gas becomes more complicated. Here, the feedback from the electric current system to the neutral gas is not negligible (e.g., Volland and Grellmann, 1978).

2.4 Wave generation

There exist four sources to drive thermospheric winds:

a) the direct solar XUV heat input
b) heating by wave dissipation due to waves of lower atmospheric origin
c) Joule heating due to magnetospheric electric fields
d) Ampere forcing due to magnetospheric electric fields.

In the framework of our approximation, the wave parameters of the individual modes are

$$\left.\begin{matrix} p \\ w \end{matrix}\right\} \simeq \left\{\begin{matrix} p_n^m(z) \\ w_n^m(z) \end{matrix}\right\} P_n^m(\phi)\ e^{im\lambda\ -i\omega t} \tag{17}$$

where the $p_n^m(z)$ and $w_n^m(z)$ are the height-dependent structure functions of the pressure and the vertical wind.

The horizontal wind $\underline{U}$ can be separated into curl-free and source-free components:

$$\underline{U} = a\{-\phi_n^m(z)\ \nabla + \psi_n^m(z)\ \nabla x\underline{r}\} P_n^m(\phi)\ e^{im\lambda\ -\ i\omega t} \tag{18}$$

where the $\phi_n^m(z)$ and $\psi_n^m(z)$ are the height structure functions of the wind, and $\underline{r}$ is the unit vector in the r-direction. and $\underline{r}$ is the unit vector in the r-direction. The external Ampere force can be separated in the same manner:

$$\underline{m} = a(-\ \nabla D + \nabla x C\underline{r}) \tag{19}$$

The functions ϕ, ψ, D, and C as well as the heat source Q can be represented by a series of spherical functions. It can be shown that, assymptotically at high altitudes, the pressure and the horizontal wind functions of the individual modes are related to the corresponding terms of the external forces as

$$\frac{p_n^m}{p_o} = -\frac{ia^2\omega_k Q_n^m}{n(n+1)g^2H^2} - \frac{aD_n^m}{gH}$$

$$\phi_n^m = \frac{a\ Q_n^m}{n(n+1)gH} \tag{20}$$

$$\psi_n^m = -\frac{C_n^m}{i\omega_k}$$

where a is the earth's radius, and $\omega_k = \omega + i\nu$ is a complex frequency representing wave dissipation by an effective Rayleigh friction term (Volland, 1979). Eq. (20) shows that a heat source can only generate a curl-free wind, while a mechanical Ampere force can only generate a source-free wind. Exact numerical calculations give basically the same results (Mayr and Harris, 1978). Pressure-and curl-free wind decrease as n(n+1) which indicates the suppression of small scale structures at thermospheric heights.

3. HEAT INPUT

3.1 Solar radiation

The horizontal distribution of solar XUV heat input is approximately proportional to the solar zenith angle. This may be approximated by the lowest order terms in a series of spherical functions:

$$Q_{EUV} \simeq \bar{Q}(z) \quad P_0^0 - 0.8 \cos\omega_A t \, P_1^0 - (0.63 - 0.15 \cos\omega 2 \, _A t) \, P^0$$

$$- (2 \, P_1^1 - 0.5 \cos\omega_A t \, p_2^1) \cos\tau \quad - 0.31 \, P_2^2 \tau \cos 2 \; . \quad (21)$$

where $\omega_A = 2\pi/1$ year is the angular frequency of one year, τ is the local time, and P_n^m are spherical functions in Neumann's normalization. The term $P_0^0 = 1$ describes the global average. The term P_1^0 is responsible for an annual variation, the term P_2^0 describes the zonally averaged heating difference between lower and higher latitudes including a semi-annual term, the terms with cos τ are diurnal variations, those with cos 2τ are semi-diurnal variations. The factor $\bar{Q}$ only slightly depends on height above about 100 km and has the value $\bar{Q} \simeq 3$ Watt/kg.

3.2 Heating due to wave dissipation

Waves generated within the lower and middle atmosphere and propagating upward are dissipated within the upper atmosphere. It is not well known which type of waves contributes mainly to that kind of heat input. Gravity waves (Hines, 1965), infrasound waves (Rind, 1977), and semi-diurnal tides (Lindzen and Blake, 1970) are likely candidates. The horizontal and vertical distribution of this heat input is also not known. An estimate of the global mean, based on mean exospheric temperatures extrapolated to zero solar activity, gives the number

$$\bar{Q}_{diss} \simeq 0.6 \; \bar{Q} \; P_0^0 \qquad (22)$$

This heat distribution transformed into geographic coordinates and developed into a series of spherical functions can be written as (Volland, 1979).

where 0.6 $\bar{Q} \simeq 2$ Watt/kg for $\bar{Q}$ from (21) (Volland and Mayr, 1977). Thus, the heating due to wave dissipation could increase the mean exospheric temperature by several hundred degrees Kelvin even during extremely quiet solar activity.

3.3 Joule heating

The system consisting of the solar wind magnetosphere, and the ionosphere behaves like a hydromagnetic generator (Hartmann flow) in which kinetic energy of the solar wind is transferred into electric currents at dynamo layer heights. These are Pedersen currents which heat the neutral gas via Joule heating. This Joule heating is averaged over local time and plotted in Fig. 2 versus invariant colatitude for quiet and for moderately disturbed conditions.

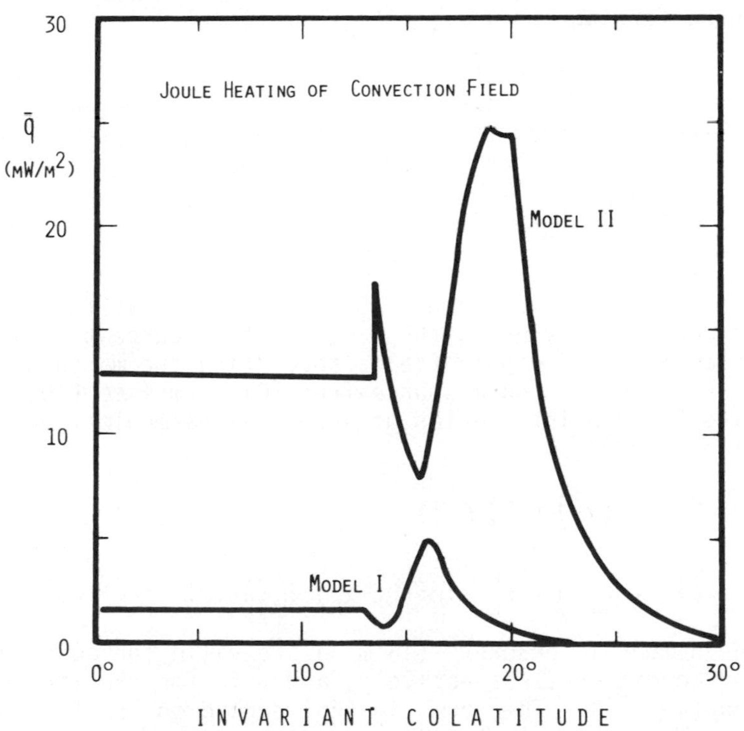

Figure 2. Joule heating due to the convection electric field. The local time-averaged term is shown vs. invariant colatitude for quiet conditions (model I) and for moderately disturbed conditions (model II).

This heat distribution transformed into geographic coordinates and developed into a series of spherical function can be written as (Volland 1979):

$$Q_{Joule} \simeq \bar{Q}\{ \begin{pmatrix} 0.12 \\ 1.2 \end{pmatrix} P_0^0 + \begin{pmatrix} 0.5 \\ 4.9 \end{pmatrix} P_2^0 + \begin{pmatrix} 0.1 \\ 1.0 \end{pmatrix} P_2^1 \cos(\lambda-\lambda_B).\} \quad (23)$$

where $\bar{Q}$ is from (21), and λ_B = 270°E is the geographic longitude of the location of the geomagnetic pole on the northern hemisphere. The last term in (23) is due to the transformation from geomagnetic into geographic coordinates. The upper row of numbers is valid during quiet conditions, the lower row is valid during moderately disturbed conditions. Clearly, mean Joule heating during moderately disturbed conditions exceeds the XUV heat input. Although the higher order terms in (23) have rather large coefficients, their effect on thermospheric dynamics is minor due to the low pass filter effect of the thermosphere.

3.4 Ampere forcing

The electric field of magnetospheric origin drives the ions within the ionosphere. These ions collide with the neutral gas and act like a momentum (or Ampere) force to generate neutral winds. That momentum force is described by

$$\underline{m} = \frac{\underline{j} \times \underline{B}_o}{\rho} \ (m/s^2) \quad (24)$$

where $\underline{i}$ is the electric current, $\underline{B}_o$ the geomagnetic field, and ρ the mean density. By using the same electric current as in (23), applying an inclined dipole field, separating the momentum force into a curl-free (D) and a source-free (C) term (see (19)), and developing C and D into series of spherical harmonics, one arrives at (Volland, 1979)

$$\begin{aligned} D &\simeq - \begin{pmatrix} 3 \\ 13 \end{pmatrix} \nu P_1^1 \cos\tau \quad . \ . \ . \ . \ . \\ C &\simeq - \begin{pmatrix} 5 \\ 13 \end{pmatrix} \nu P_2^0 \sin(\Omega t+\lambda_B) + \begin{pmatrix} 9 \\ 44 \end{pmatrix} \nu P_2^1 \sin\tau \ ... \end{aligned} \quad (25)$$

where the number in brackets (in m/s) are valid for quiet and for disturbed conditions, respectively, and ν is for the ion-neutral collision (in s^{-1}). The terms in (25) depend on local time τ and on universal time t. That last dependence arises from the transformation from geomagnetic into geographic coordinates. The higher order terms, in particular all terms depending on the Hall conductivity, have been ignored. These terms should be taken into consideration in order to describe the influence of Ampere forcing on thermosphere dynamics more exactly.

4. BASIC WIND CONFIGURATIONS

4.1 Curl-free winds

Figure 3 shows schematically the horizontal wind pattern and the meridional cross-section of the curl-free wind cells generated by the corresponding Joule heating terms, (2,0) and (2,1) in (23). The (0,0) term in (23) describes the global upwelling of air without horizontal movement. The (2,0) term describes the transport of excessive heat deposited at high latitudes to lower latitudes. The return flow is below about 130 km altitude which is near the lower limit of Joule heating input.

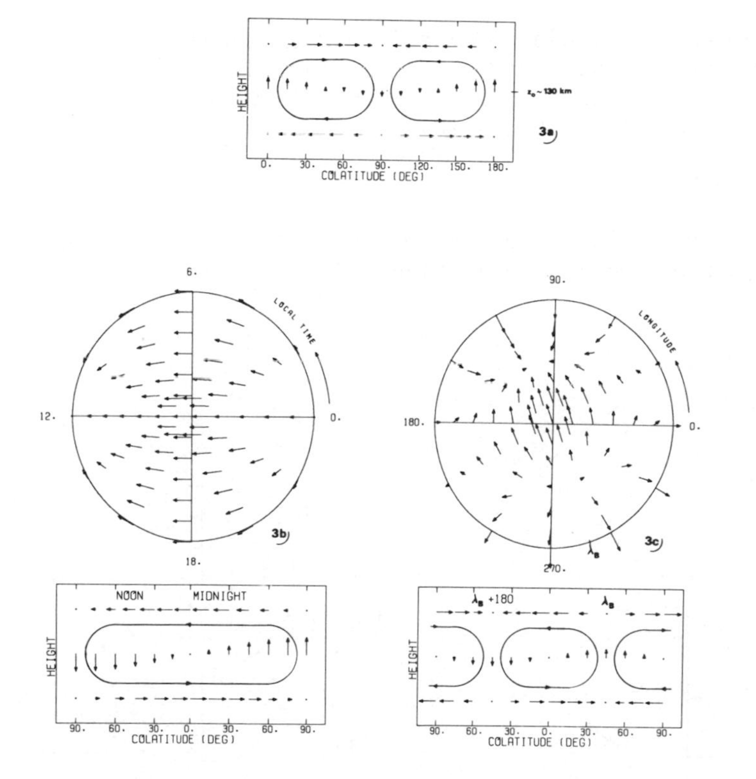

Figure 3. Curl-free winds due to Joule heating.
(a) Meridional cross section (not to scale) of the wind cell driven by the (2,0) term in (23).
(b) Meridional cross section and horizontal structure on the northern hemisphere vs. local time of the diurnal component (1,1) not included in (23)).
(c) Same as Fig. 3a, except for the (2,1) component in (23) vs. longitude.

The (2,1) term depends on longitude, and shows the longitudinal modulation of the zonal wind system due to the offset of the auroral ovals with respect to the geographic axis.

The curl-free wind systems of Figure 3 are superimposed on winds generated by XUV radiation (eq. 21). While the (0,0) term in (23) during quiet conditions adds to the (0,0) terms in (21) and (22) to generate the global mean temperature and pressure, the (0,0) term in (23) during disturbed conditions temporarily increases the global mean temperature and pressure. The (2,0) term in (23) is opposite to the corresponding (2,0) term in (21) during equinox conditions. These terms nearly cancel each other during quiet conditions. During disturbed conditions, the Joule heating term dominates.

More detailed calculations dealing with zonally-averaged wind systems during quiet and during disturbed conditions are in basic agreement with this picture (Dickinson et al., 1975; Straus et al., 1975; Straus, 1978).

4.2 Source-free winds

The two horizontal wind systems generated by Ampere forcing (25) are shown schematically in Figure 4.

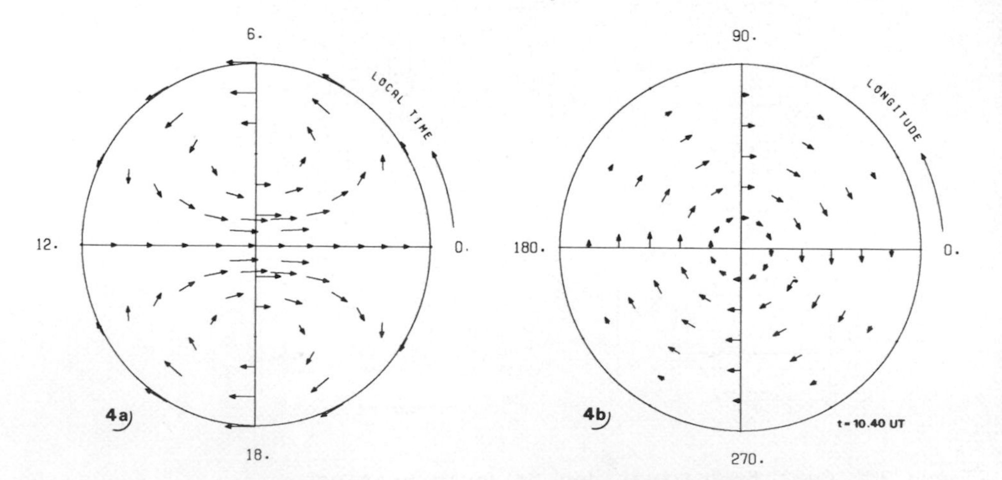

Figure 4. Source-free winds due to Ampere forcing
(a) Horizontal wind structure in the northern hemisphere vs. local time of the (2,1) component in (25).
(b) Same as in Fig. 4a, except for the (2,0) component in (25) vs. longitude at 10h 24 m UT.

Here, the (2,1) term describes only roughly the two cell configuration on one hemisphere. These cells follows rather closely the plasma drift due to the electric convection field with its reversal near - 75° geomagnetic latitude (Fedder and Banks 1972; Maeda, 1976, 1977; Mayr and Harris, 1978). The (2,0) term in (25) which depends on universal time has not yet been detected.

Although the source-free winds may reach rather large magnitudes locally, they are of no great significance for the composition effects at thermospheric heights, because they are not accompanied by large-scale vertically-oriented wind cells (Mayr, et al., 1978).

The sum of all the wind systems described above is the observed wind. For numerical calculations dealing with that complicated interplay of various heat and momentum sources cf. Straus and Schulz (1976) and Mayr et al. (1978).

4.3 Local disturbances

The time-dependent heat and momentum input within the auroral zones during geomagnetically disturbed conditions generates a broad spectrum of acoustic-gravity waves and planetary waves. Because of the low pass filter effect of the thermosphere, planetary waves with the lowest wave numbers are mainly responsible for the global heat redistribution at thermospheric heights (Richmond, 1978, 1979). Impulse type disturbances on a planetary scale can be simulated most simply by the superposition of linear waves as described in section 2.2 in terms of Fourier integrals (Volland and Mayr, 1971). Smaller scale disturbances which generate mainly short periodic acoustic-gravity waves are more conveniently described as plane waves in a cartesian coordinate system neglecting the wave guide properties of the atmosphere (e.g., Testud, 1970; Klostermeyer, 1972; Richmond and Matsushita, 1975; Yeh and Liu, 1974). The thermosphere acts like a height-dependent band pass for short period gravity waves by selecting waves with periods of the order of about 20 to 60 minutes and horizontal wavelengths of the order of about 200 to 600 km (Liu and Klostermeyer, 1975). This filter effect depends also on the direction of propagation, on the prevailing wind, and on the location of the source.

Besides the auroral zones as sources of acoustic (Wilson, 1972) and gravity (Testud, 1970) waves, solar eclipses may generate gravity waves (Schödel et al., 1973a). There is also evidence for a tropospheric origin of acoustic (e.g., Gossard and Hooke, 1975) and gravity waves (e.g., Hines, 1967; Schödel et al., 1973b). Travelling ionospheric disturbances (TID) at ionospheric F layer heights are caused by gravity waves (e.g., Gossard and Hooke, 1975).

REFERENCES

Dickinson, R. E., Ridley, E. C., and Roble, R. G., 1975. J. Atm. Sci., 32, 1737.
Fedder, J. A., and Banks, P. M., 1972. J. Geophys. Res., 77, 2328.
Gossard, E. E., and Hooke, W. H., 1975. Waves in the Atmosphere, Elsevier, Amsterdam.
Hines, C. O., 1960. Can. J. Phys., 38, 1441.
Hines, C. O., 1965. J. Geophys. Res., 70, 177.
Hines, C. O., 1967. J. Geophys. Res., 72, 1877.
Klostermeyer, J., 1972. J. Atm. Terr. Phys., 34, 765.
Lindzen, R. S., and Blake, D., 1970. J. Geophys. Res., 75, 6868.
Liu, C. H., and Klostermeyer, J., 1975. J. Atm. Terr. Phys., 37, 1099.
Longuet-Higgins, M. S., 1968. Phil. Trans. Roy. Soc. London, 262, 511.
Maeda, H., 1976. J. Atm. Terr. Phys., 38, 197.
Maeda, H., 1977. J. Atm. Terr. Phys., 39, 849.
Mayr, H. G., and Harris, I., 1978. J. Geophys. Res., 83, 3327.
Mayr, H. G., Harris, I., and Spencer, N. W., 1978. Rev. Geophys Space Phys., 16, 535.
Richmond, A. D., 1978. J. Geophys. Res., 83, 4131.
Richmond, A. D., 1979. J. Geophys. Res., 84, 5259.
Richmond, A. D., and Matushita, S., 1975. J. Geophys. Res., 80, 2839.
Rind, D., 1977. J. Atm. Terr. Phys., 39, 445.
Schödel, J. P., Klostermeyer, J., and Röttger, J., 1973a. Nature, 245, 87.
Schödel, J. P., Klostermeyer, J., Rottger, J., and Stilke, G. 1973b. J. Geophys., 39, 1063.
Straus, J. M., 1978. Rev. Geophys. Space Phys., 15, 156.
Straus, J. M., and Schulz, M., 1976. J. Geophys Res., 81, 5822.
Straus, J. M., Creekmore, R. M., Harris, R. M., Ching, B. K., and Chiu, Y. T., 1975. J. Atm. Terr. Phys., 37, 1545.
Volland, H., 1979. J. Atm. Terr. Phys., 41, 853.
Volland, H., and Grellmann, L., 1978. J. Geophys. Res., 83, 3699.
Volland, H., and Mayr, H. G., 1971. J. Geophys. Res., 76, 3764.
Volland, H., and Mayr, H. G., 1977. Rev. Geophys. Space Phys., 15, 203.
Testud, J., 1970. J. Atm. Terr. Phys., 32, 1793.
Wilson, C. R., 1972. J. Geophys. Res., 77, 1820.
Yeh, K. C., and Liu, C. H., 1974. Rev. Geophys. Space Phys., 12, 193.

WIND INDUCED COMPOSITION EFFECTS AT HIGH LATITUDES

H. G. Mayr and I. Harris

NASA Goddard Space Flight Center
Greenbelt, Maryland USA

Abstract

With the diffusive separation of the upper atmosphere horizontal motions lead to horizontal differentiation in composition. Thermally driven winds effectively transport the lighter and minor species from the warmer toward the cooler regions on the globe. Due to the abundance of atomic oxygen a back pressure develops which significantly dampens the wind velocities and increases the temperature contrast. During summer and under magnetically disturbed conditions, the observed large temperature enhancements and depletions in O and He are prominent manifestations of these processes at polar latitudes. As a feedback mechanism, advective mass transport completely changes the "primary" composition effects induced by eddy diffusion. Owing to the rotational nature of EXB momentum coupling its signature in composition is comparatively weak, and clearly distinguishable from the effects of Joule dissipation.

INTRODUCTION

H. Volland discussed the dynamics of the upper atmosphere and theoretical concepts such as perturbation theory and wave analysis. In this paper we shall outline the temperature and composition structures of that region and provide some interpretation in terms of transport processes.

C. S. Deehr and J. A. Holtet (eds.), Exploration of the Polar Upper Atmosphere, 31–54.

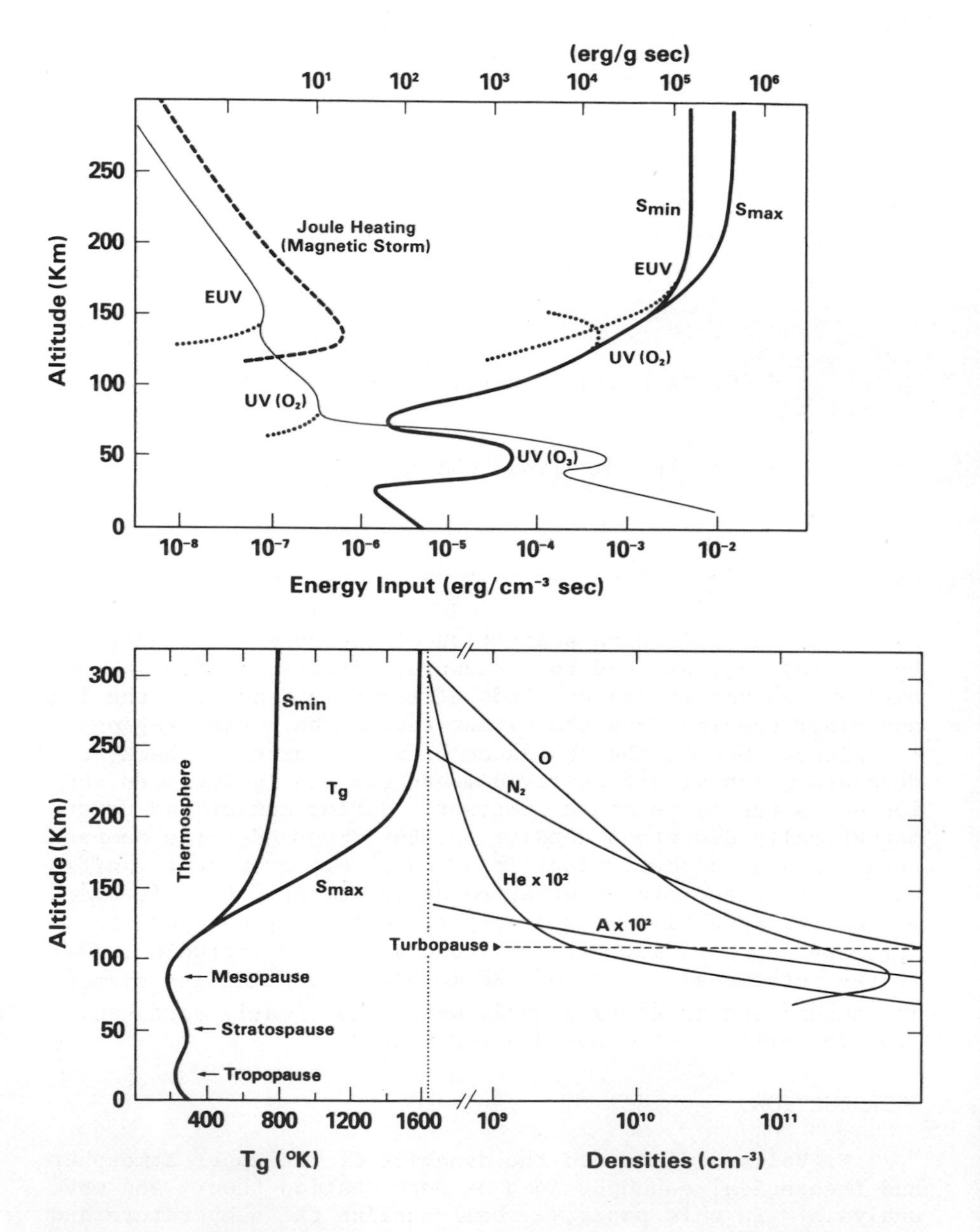

Figure 1. Typical height distributions for the heating rates due to solar radiation and Joule dissipation (during disturbed conditions). The solar radiative input per mass of the absorbing atmosphere is shown in heavy solid lines for minimum (S_{min}) and maximum (S_{max}) solar activity conditions. Average temperature and density distributions are also presented.

For orientation, the basic conditions in the terrestrial atmosphere are summarized in Figure 1. The upper atmosphere or thermosphere receives most of its energy through EUV radiation (24, 44) which is small compared with the UV sources from dissociation of O_2 and O_3 at lower altitudes. In general, energy transfer downwards is insignificant for the energetics of the lower atmosphere, while even a small leakage due to upward propagating waves may have profound effects on the thermosphere. In terms of specific heating rates relative to the low ambient density, the EUV input is comparatively large and accounts for the large temperatures in that region. Solar wind energy is transmitted through the magnetosphere in the form of precipitating particles and electric fields (1, 6, 7, 11, 17, 35, 36, 42). In magnitude this source is comparable to the energy from EUV at high latitudes and it is significant for the global energy budget during magnetic storms. Most of the thermal energy deposited in the upper atmosphere is ultimately carried through vertical heat conduction to lower altitudes where radiative cooling becomes important; the large temperature gradient above the mesopause is a prominent signature of this process. Composition measurements in the lower thermosphere suggest that the turbopause is located near 110 km. Above that height, molecular diffusion (in contrast to mixing in the lower atmosphere) dominates such that major atmospheric species tend to follow height distributions determined by mass and temperature. This diffusive separation in composition has profound effects on the formation of the ionosphere and considerably complicates the redistribution of energy and mass in the thermosphere.

Based on theoretical results we shall discuss a few examples where wind induced diffusion is a major factor in our understanding of the upper atmosphere: (1) Seasonal anomalies in temperature and composition. (2) Variations in eddy diffusion and the resulting feed back by horizontal transport. (3) Magnetic storm effects. (4) Atmospheric signatures due to Joule heating and electric field momentum coupling.

SEASONAL VARIATIONS

We shall discuss only the gross features in the seasonal variations which are probably understood. In detail, though, many of the phenomena still challenge our ability to quantitatively describe the thermosphere. Because of the long time scale involved eddy transport is important but not fully accounted for. Dynamic coupling from the lower atmosphere can be effective but its nature is not well known. The significance of magnetospheric energy sources has not been fully determined.

Observed temperature variations (3, 9) in the thermosphere are shown in Figure 2. At low latitudes the diurnal tide dominates, while in the polar regions one observes a large temperature difference between summer and winter which reflects the differing amounts of solar radiation that are absorbed in both hemispheres. In contrast to this behavior, the measured He concentration in Figure 3 shows a large increase toward the winter hemisphere. Moreover, such a trend is observed in the F region ionization and gas temperature near the mesopause which, during the course of the year, reach maximum values in winter.

The basic ideas that led to an understanding of these phenomena go back to Kellogg (30) who postulated, from an analogy

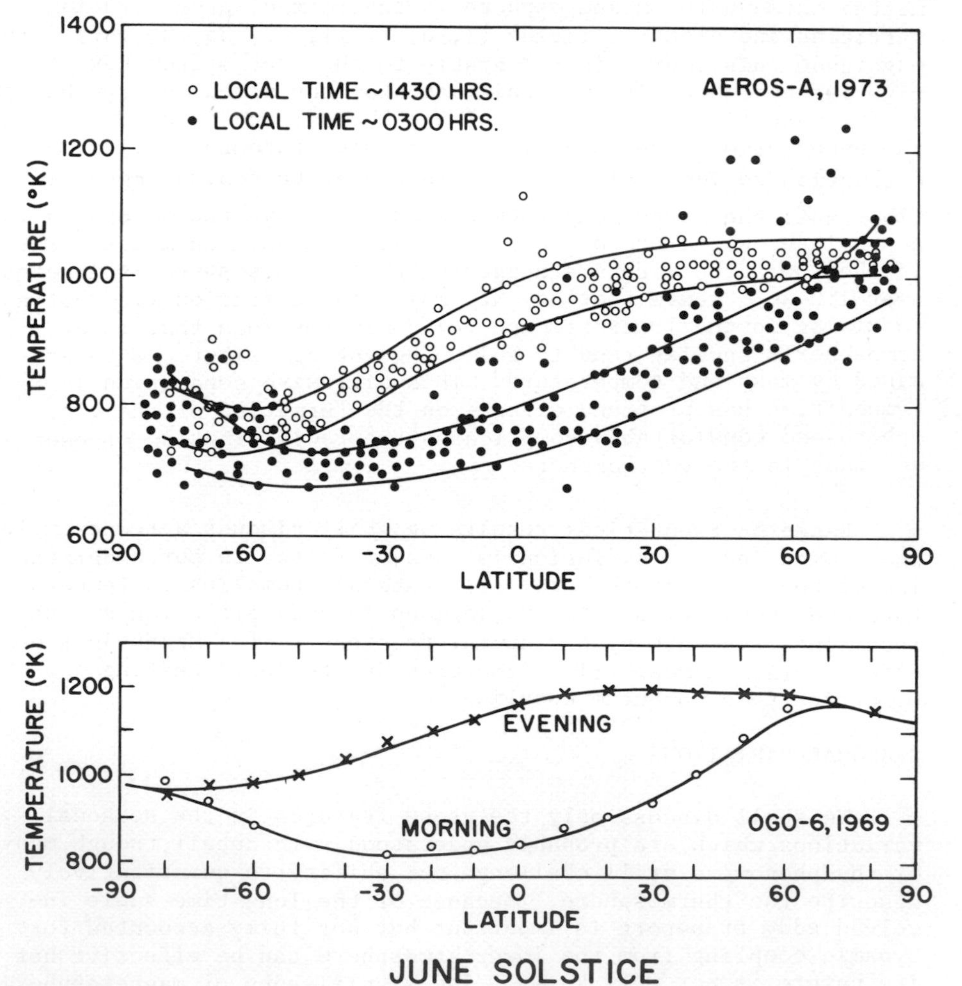

Figure 2. Temperature distributions during solstice observed from N_2 and airglow measurements on AEROS-A (Chandra and Spencer, 1975).

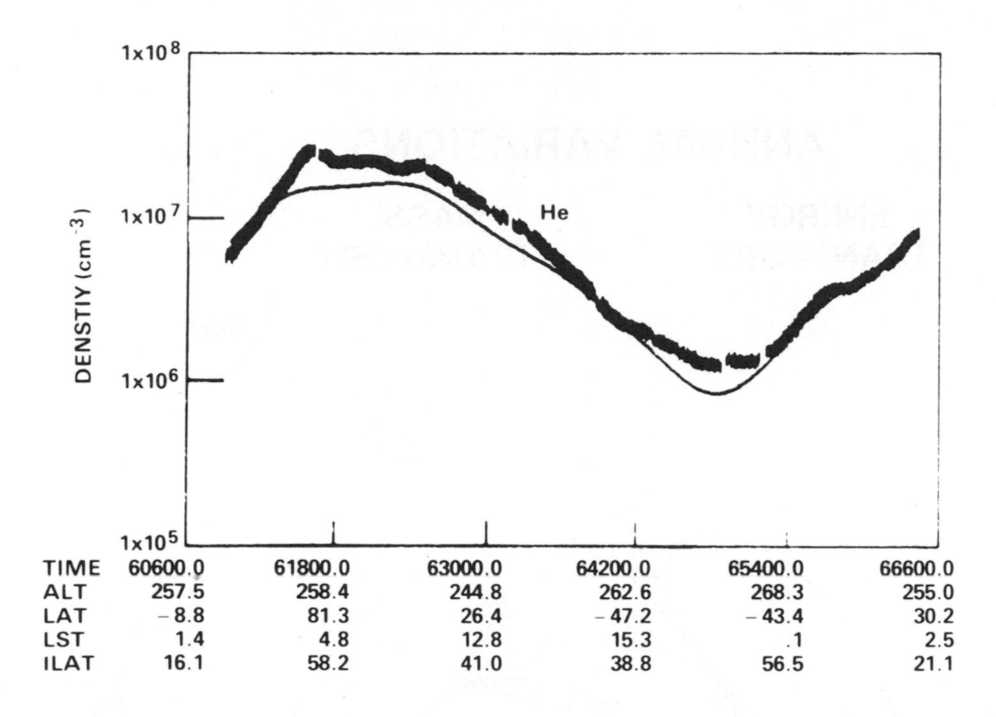

Figure 3. He distributions during solstice measured from AE-C in comparison with the empirical OGO-6 model (continuous line). Courtesy of A. E. Hedin.

with the global redistribution of ozone in the stratosphere, that the meridional circulation should transport significant amounts of oxygen from the summer toward the winter hemisphere. Kellogg furthermore proposed that accumulation of oxygen in winter and subsequent three body recombination ($O + O + N_2 \rightarrow O_2 + N_2$) could release chemical energy contributing to the temperature anomaly near the mesopause. Interpreting the F-region winter anomaly, King (31) presented quantitative evidence which strongly suggested that the relative abundance of O was indeed enhanced in winter and he invoked the circulation concept to explain that. Johnson and Gottlieb (27) estimated that meridional winds can account for the observed winter He bulge (18, 20, 29, 41).

Figure 4 illustrates in simplified geometry the zonally symmetric circulation during solstice conditions. Due to the inclination of the Sun, the summer hemisphere is heated preferentially thus setting up temperature and pressure gradients across the equator that drive winds toward the winter hemisphere. These winds transport significant amounts of energy and mass. Energy transport is primarily effective in the form of adiabatic heating

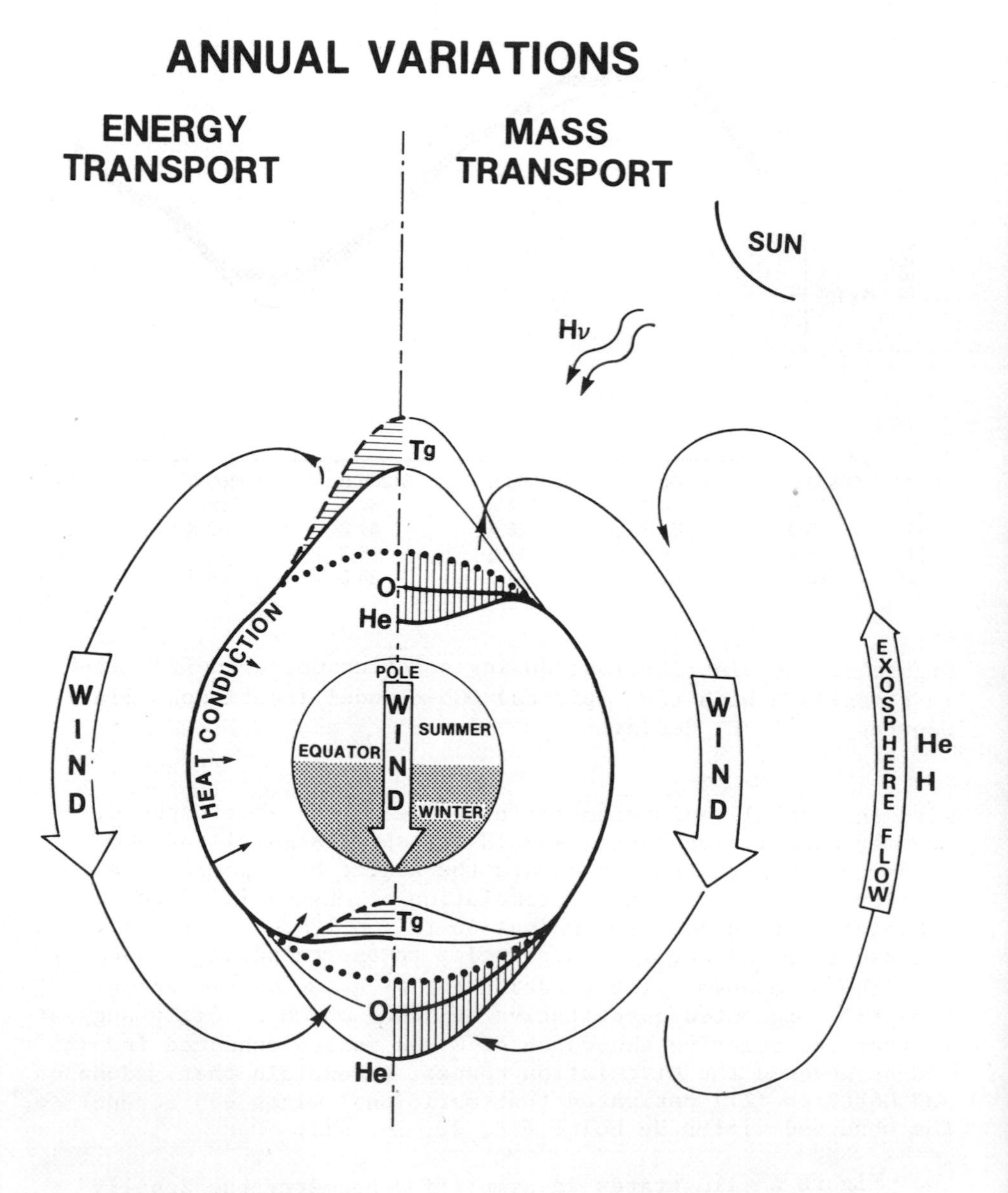

Figure 4. Schematic illustration of the seasonal variations. Dotted lines represent the global average. Solid lines illustrate temperature and density variations under consideration of horizontal transport. Shaded areas represent the energy (T_g) and mass (O, He) redistribution by winds and exospheric flow.

(winter) and cooling (summer) which lowers the temperature contrast between both seasons. Mass transport is complicated by the interaction of three processes: (a) Thermal expansion and contraction leads to a density increase or decrease in the warmer or cooler regions of the atmosphere respectively; this process is most effective for the heavier constituents. (b) Dragged along by the motions of the major gas, the height integrated horizontal transport of a minor species, He for example, is proportional to its large scale height. This mass advection induces vertical diffusion which changes the density distribution so that the He concentration decreases in summer and increases in winter, opposite to thermal expansion and contraction. Wind induced diffusion is important for the lighter species including atomic oxygen. (c) Mass transport by exospheric flow can be compared with fast horizontal diffusion. In that process, particles move in ballistic trajectories from the high temperature-or high density-regions toward the cooler and more tenuous regions on the globe. The resulting flow velocity is highly temperature dependent ($T^{5/2}$) and becomes increasingly more important for smaller masses so that the process significantly affects He and essentially governs the H distribution.

The annual variations in the thermosphere are illustrated with results (Figure 5) from a three dimensional multiconstituent model (38) which are in qualitative agreement with observations (19, 20). Summer to winter variations at the poles are characterized by a large temperature contrast, a winter oxygen bulge below 350 km which is the principle cause of the seasonal anomaly in the F-region and a large (factor of 40) increase in the He concentration from summer to winter. In the lower thermosphere near the mesopause the winter anomaly in temperature is reproduced. Meridional wind velocities of about 40 m/sec (at the equator) are shown for the upper thermosphere. As signatures of the temperature anomaly, the meridional winds reverse direction near 130 km and the zonal winds reverse near 110 km. Below 80 km, the zonal winds change direction again, consistent with geostrophic conditions for a temperature maximum in summer.

Figure 6 illustrates in more detail some of the processes that contribute to the seasonal temperature variations in general and the formation of the mesopause anomaly in particular. Computed relative amplitudes at the poles are labelled with plus or minus signs to indicate a maximum in summer or winter respectively. The energy from EUV (thin line) is the principle source for the temperature variations in the upper thermosphere. Atomic oxygen transported toward the winter hemisphere recombines and releases energy that significantly contributes to the winter anomaly near 90 km (minus sign). This process is in part compensated by the O_2 dissociation and heating from the Schumann Runge

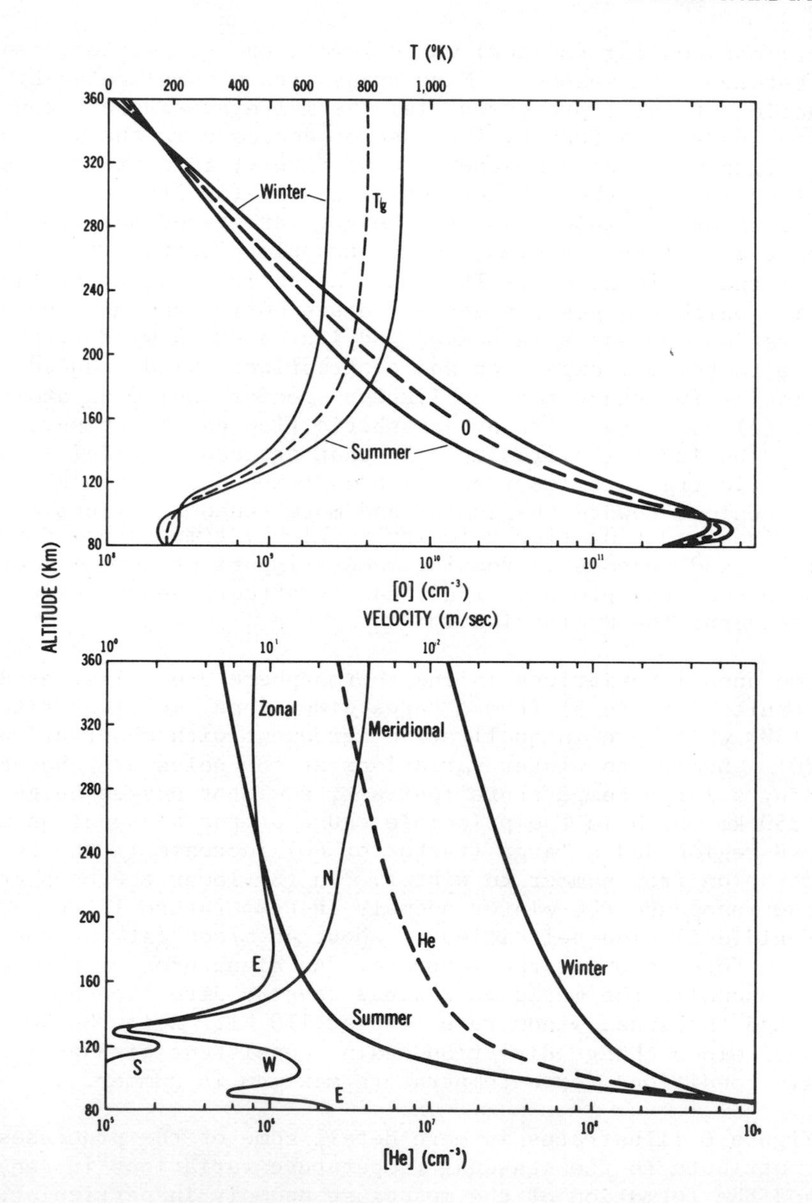

Figure 5. Annual tide computed from a three dimensional theoretical model. Summer and winter profiles refer to the poles. The meridional wind velocity is taken from the equator during June solstice, N and S indicating directions from the north and south respectively. During that season the zonal wind velocities are taken from mid-latitudes, E and W indicating directions from the east and west respectively. The results refer to the fundamental asymmetric component P_1^o.

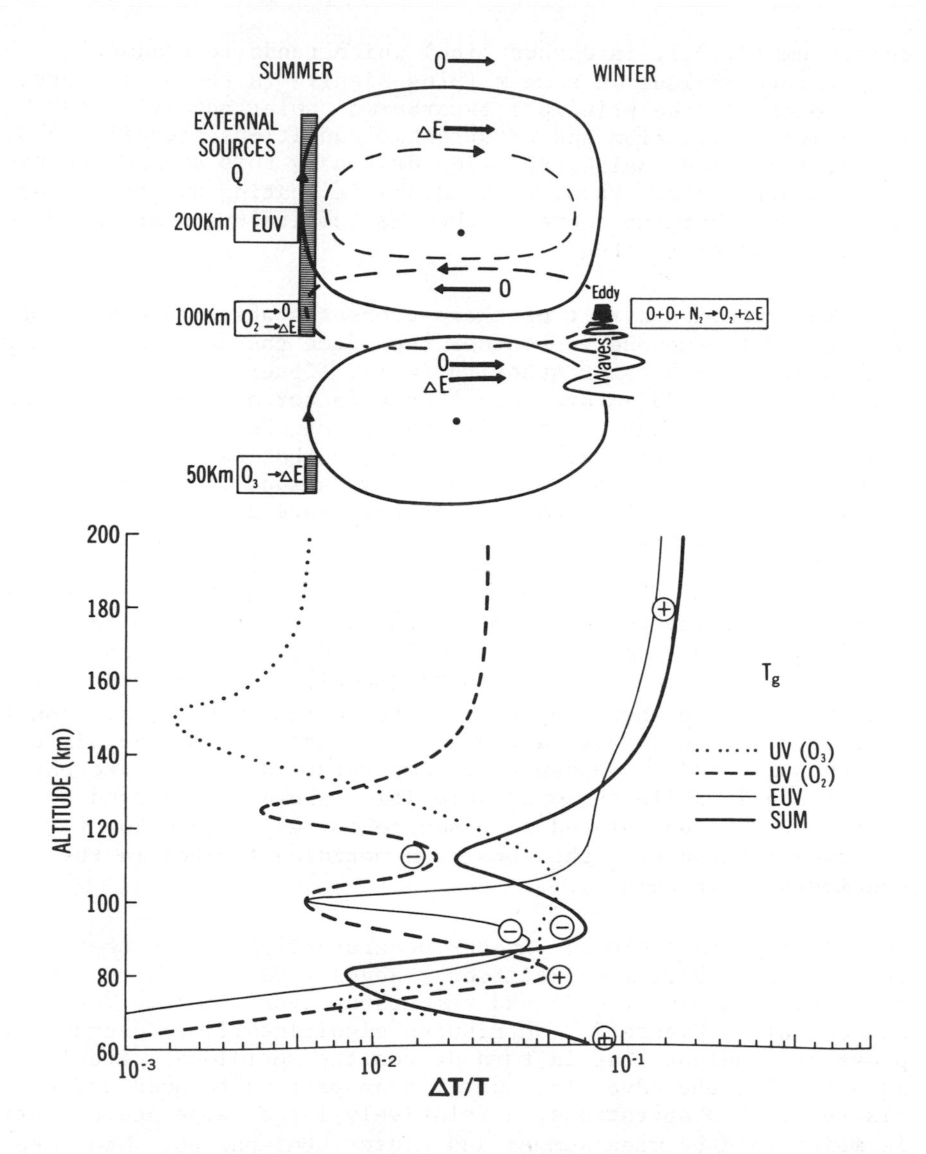

Figure 6. Schematic illustration of the sources that contribute to the annual tide. Advection of energy (ΔE) and atomic oxygen (O) contribute significantly to the energy budget with release of chemical energy in three body recombination being important in the winter hemisphere. To reproduce the magnitude of the mesospheric temperature anomaly an additional "wave source" near 90 km is postulated which is probably associated with planetary waves. The lower half of the figure shows relative temperature amplitudes due to EUV, UV (O_2) and UV (O_3) excitation. Plus or minus signs indicate a temperature maximum in the summer-or winter-hemisphere respectively.

continium (UV(O_2), in dashed line) which tends to produce a temperature maximum in summer (plus sign). In the mesosphere, where ozone is the principle absorber of solar radiation (UV(O_3)), radiative equilibrium and geostrophic conditions prevail. Above 80 km, the meridional circulation driven by this source, however, becomes effective. Thus, both adiabatic heating and recombination energy contribute to increase the gas temperature near the mesopause in winter (dotted line).

The combined effect of these processes which involve mesopheric and thermospheric sources reproduce the temperature anomaly qualitatively - but not quantitatively. Observed temperature variations near 90 km are more than a factor of two larger than predicted. Although the precise nature of the missing process is unknown there is circumstantial evidence that wave energy is important. It was shown originally by Matsuno (33) that planetary waves from the troposphere can propagate during winter into the upper mesophere where wind shear activity (indicative of turbulence) is also observed to be enhanced (52). By adopting an adhoc "wave source" near the mesopause which peaks in the winter hemisphere we are able to produce the desired increase in the magnitude of the temperature anomaly. Moreover, it turns out that the associated changes (in the model) are also important for the thermosphere where in every instance the theoretical results improve relative to observations. The magnitudes of the winter oxygen bulge and the exospheric temperature amplitude decrease significantly while the wind velocities above 200 km tend to increase. The postulated wave source is responsible for the observed reversals in the zonal and meridional winds in the lower thermosphere (Figure 5).

An important element in our understanding of the thermosphere is the feedback from composition changes (due to wind induced diffusion) on wind fields and temperature variations. The winter oxygen bulge (Figure 5) contributes significantly to decrease the pressure gradient that in turn drives the meridional circulation. As a result, the advective energy transport is reduced and, consistent with observations, a relatively large temperature contrast is maintained between summer and winter hemispheres. Model calculations show specifically that in the upper thermosphere the relative amplitude of the annual temperature variation is <u>reduced</u> by more than a factor of two and the meridional wind speeds are <u>increased</u> by about the same factor if one neglects to account for wind-induced diffusion. This feed back is also important for magnetic storms where it tends to confine the temperature enhancements to high latitudes.

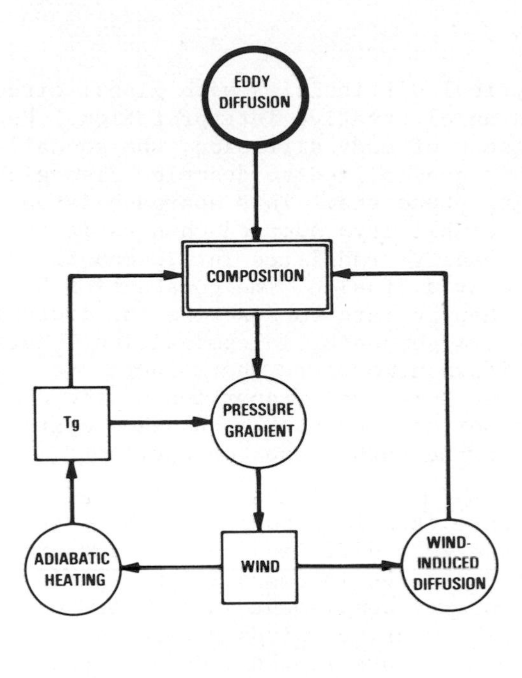

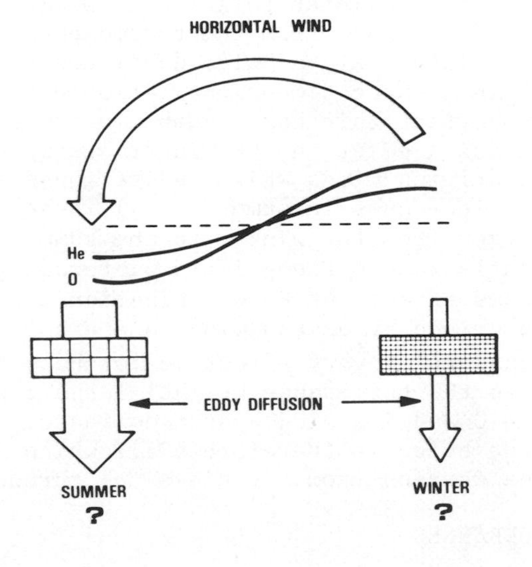

Figure 7. Schematic diagram for the composition effects due to variations in the eddy diffusion coefficient which is (arbitrarily) assumed to be higher in summer than in winter. The feedback associated with horizontal transport is illustrated.

EDDY DIFFUSION

The analytical difficulties with global circulation models have attracted an alternative interpretation. Recognizing the overall importance of eddy diffusion, the so-called turbopause concept was thus generalized to describe also globally nonuniform and time varying phenomena. This approach is very attractive because of its simplicity: Adopted changes in the turbopause height can be readily translated into composition changes. However, with an eddy diffusion time constant of 5 - 10 days static models are not appropriate to describe the diurnal variations and the magnetic storm component in composition. Moreover, in a globally nonuniform atmosphere where there are long term differences between the turbulent properties of the summer and winter hemispheres, important feedback mechanisms operate which are not accounted for in the turbopause concept.

To illustrate this last point we assume that global scale variations in the eddy diffusion coefficient are solely responsible for the changes in composition. It is thereby understood that this process may complement in one form or another the effects of thermally driven winds earlier discussed. For the condition shown in Figure 7, the eddy diffusion coefficient is assumed to decrease in winter relative to summer. As a result fewer oxygen atoms diffuse from the thermosphere down into the mesophere and a winter oxygen bulge developes that pervades the thermosphere. In a three dimensional atmosphere where horizontal transport is important this has a number of consequences (37): (a) The oxygen enhancement in the winter hemisphere represents a pressure bulge driving winds which redistribute energy so that the temperature increases in summer. (c) These winds also transport atomic oxygen from the winter to the summer hemisphere and thus significantly reduce the primary winter oxygen bulge. (c) Similar processes affect the He distribution except that the return flow is driven by oxygen winds and exospheric flow. Such horizontal transport is very effective for He which maintains (with respect to the turbopause height) large concentrations at altitudes where O and its wind system dominate. Relative to O the resulting He bulge is therefore small, which is opposite to the picture one obtains from the classical turbopause concept.

MAGNETIC DISTURBANCES

Energy deposition from the magnetosphere due to particle precipitation and Joule heating is a permanent feature of the upper atmosphere at high latitudes. During substorms and magnetic

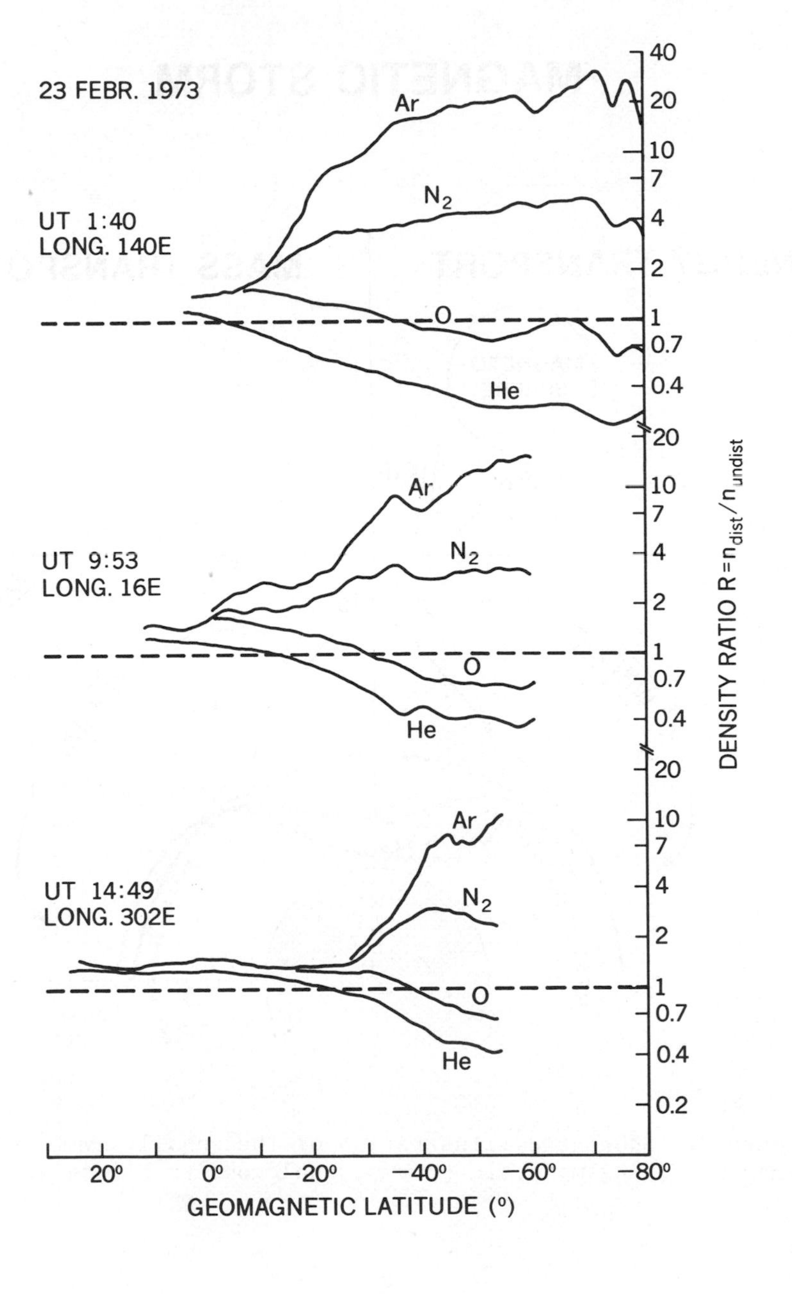

Figure 8. Composition measurements from the gas analyzer on ESRO-4 (Trinks et al., 1975) during the magnetic storm of February 23, 1973.

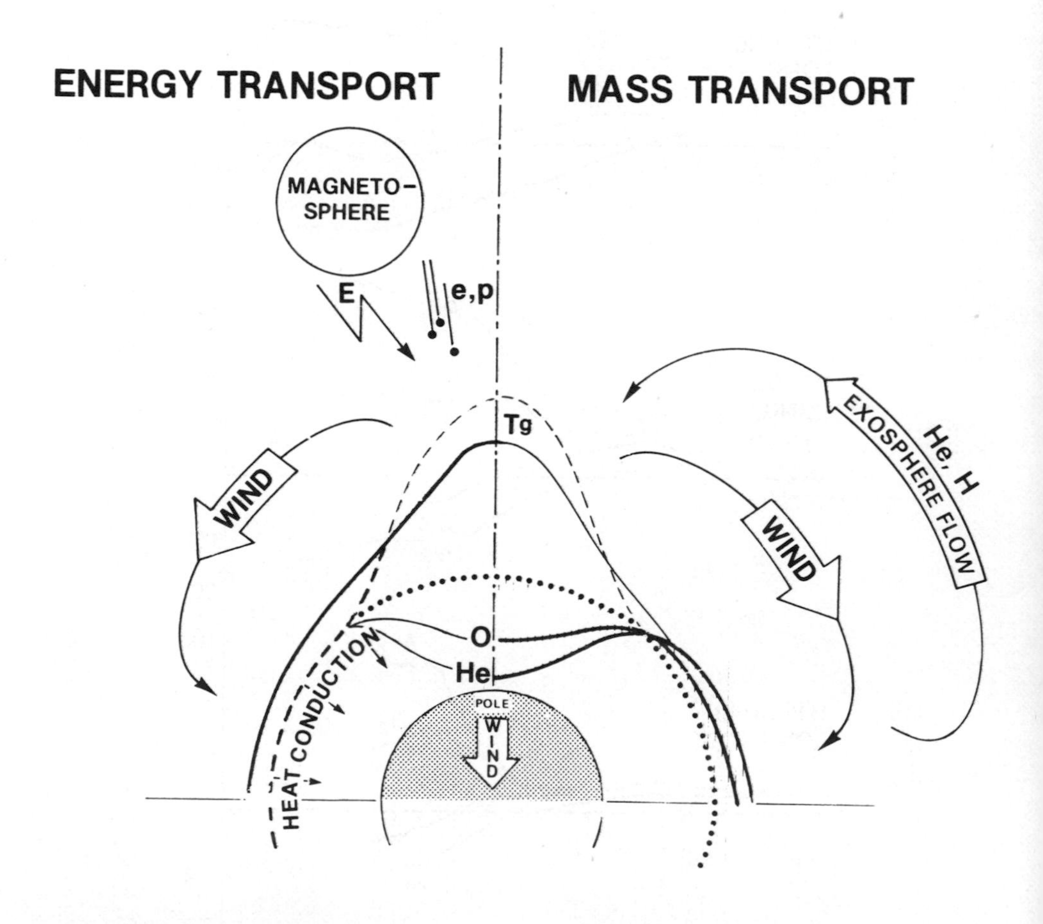

Figure 9. Schematic illustration of the zonally symmetric dynamics during magnetic storms, analogous to Figure 4.

storms, the auroral oval expands and dramatic effects are observed. From satellite drag measurements it is known that the mass density of the thermosphere increases almost uniformally on a global scale, the equatorial region being only slightly less effected than the polar region (26,49). Composition measurements (25, 40, 46) such as shown in Figure 8 (47) reveal that under disturbed conditions large N_2 and Ar enhancements are associated with depletions of O and He at high latitudes. Observations of meridional wind velocities at mid-latitudes reveal that during magnetic storms either the poleward EUV driven winds are weakened on the dayside or the corresponding equatorward winds are enhanced at night (4, 16, 21).

The results of simplified circulation models suggest that we have at least a qualitative understanding of the important processes. For the diffusive thermosphere energy and mass transport are illustrated in Figure 9. High latitude heating by particles and Joule dissipation increases the atmospheric temperature (and pressure) at high latitudes. As a result, winds are set up that redistribute energy and through adiabatic compression raise the equatorial temperature. In parallel, wind induced diffusion redistributes mass such that O and He are depleted in the polar region and accumulate at low latitudes. There is a geometry factor involved in that the area of upwelling near the poles (where mass flow diverges) is much smaller than the horizontal extent near the equator over which subsidence (convergence) occurs. Considering flow continuity, the transport effects on the minor species are hence less pronounced and of opposite direction at low latitudes than at high latitudes.

In Figure 10 the transport effects on the vertical distributions of O and He are shown in relation to thermal expansion. A posteriori, "equivalent temperature" amplitudes are defined which describe the density scale heights under the (incorrect) assumption of diffusive equilibrium. Deviations from the actual kinetic temperature (solid lines) are thus a direct measure for the importance of wind induced diffusion. Two theoretical results are presented illustrating magnetic storm and seasonal variations at high latitudes. Throughout most of the thermosphere, the transport effects are very large except at high altitudes where the collision numbers become small. Due to a skin effect, among other reasons, wind induced diffusion is relatively more important at lower altitudes for the seasonal variations than during magnetic storms.

Considering the different sources that come into play during magnetically disturbed conditions (6, 8, 12, 13, 28, 32, 39, 40, 43) (Figure 11) one expects a complex behavior in ionospheric storms. This is even true for the combined effects of thermospheric winds taken alone which affect the plasma through momentum transfer and indirectly through changes in neutral composition, both interactions having vastly different time constants. Results from a self-consistent hybrid model illustrate the process for 60° and 20° (Figure 12). For a storm related heat source Q which energizes the thermosphere over a time period of about 4 days, we show (1) the computed relative variations of O and N_2 normalized to quiet conditions, (2) the component of the meridional wind velocity parallel to the magnetic field B, (3) the ion density (O^+) at various altitudes and (4) the height, h_{F2}, of the F_2 peak. The response time of the wind field (relative to Q) is very short compared with that of the neutral composition; thus the changes in O/N_2 (which chemically affect the O^+ concentration) are still large when the wind velocity returns to zero. This time delay, which is even more important for shorter storms, accounts for the characteristic sequence between the positive and negative phases in the ionospheric response. At lower latitudes the composition changes are smaller, thermospheric winds are relatively more effective and the positive phase prevails at the height of the F_2 peak and above. Such properties are observed in ionospheric storms (34) but are further complicated by electric fields, electron impact ionization and plasma heating.

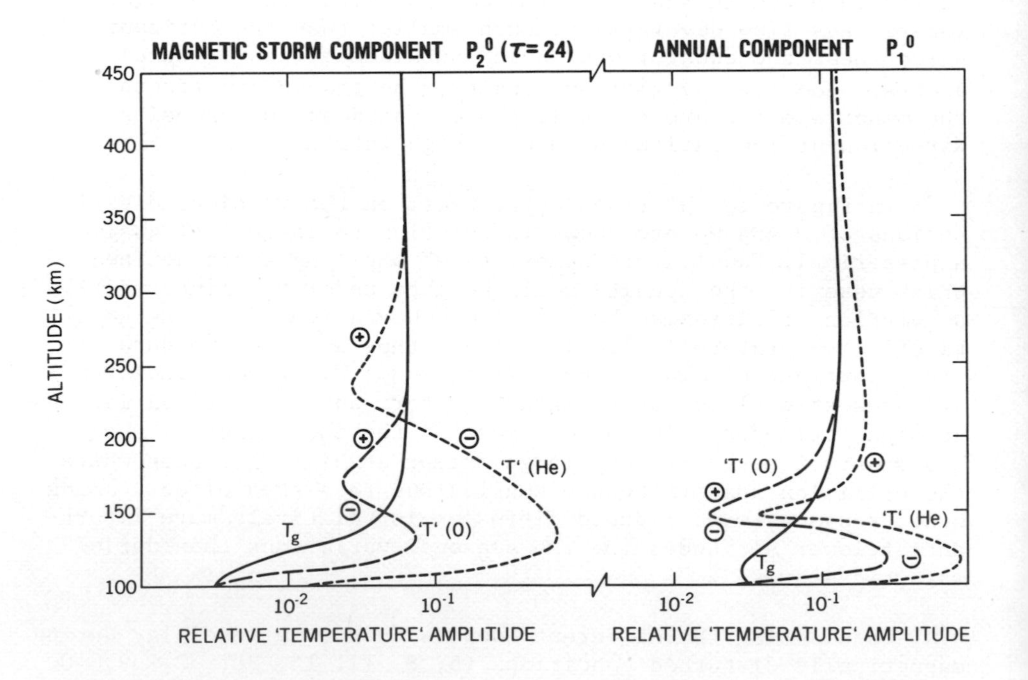

Figure 10. Effective temperature amplitudes for He and O in comparison with the kinetic temperature, T_g, computed for a magnetic storm and for the annual variations in the thermosphere. The differences are a measure for the importance of wind induced diffusion. Plus or minus signs indicate variations in phase or out of phase with the heat input.

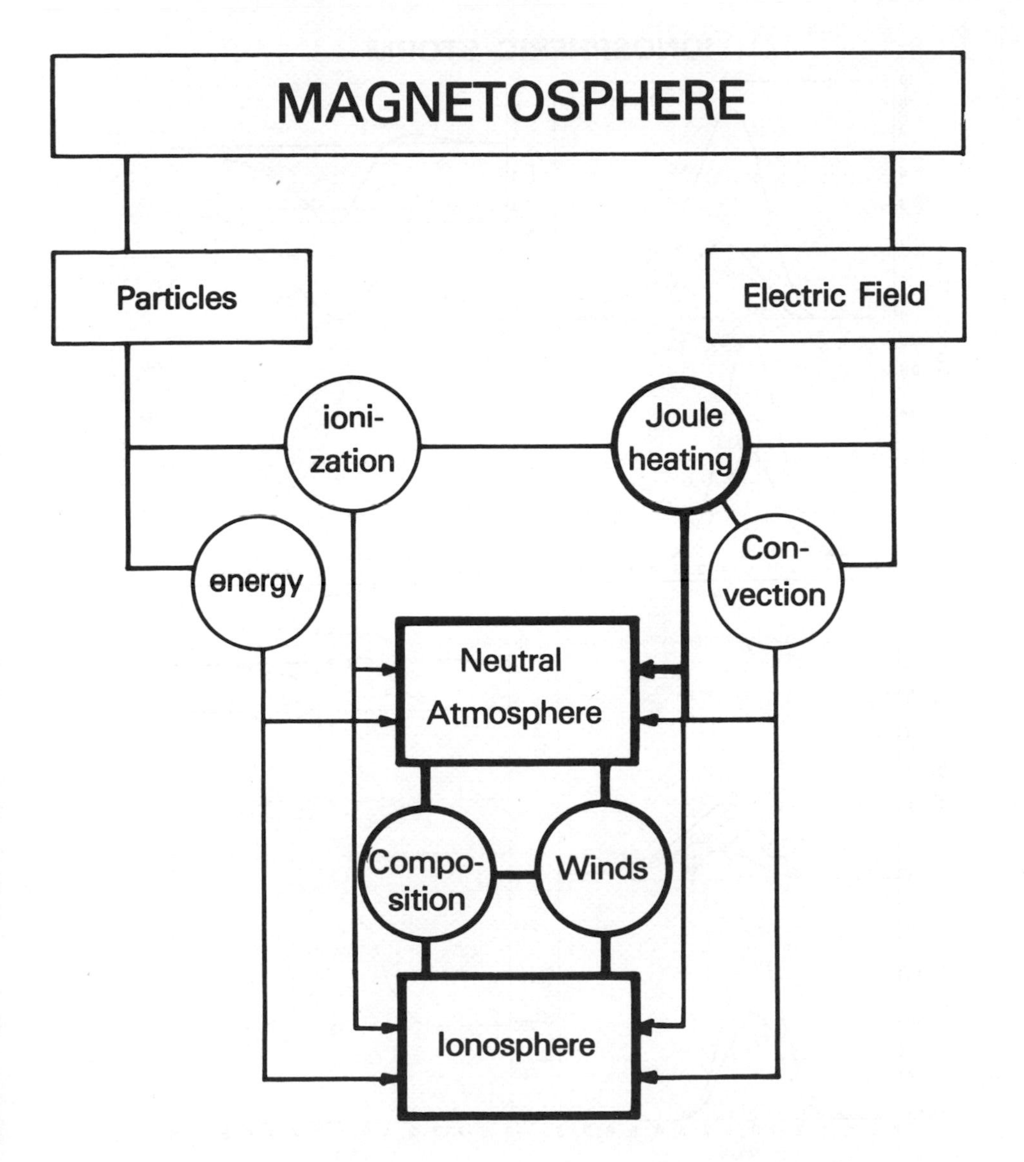

Figure 11. Block diagram illustrating the important physical variables and processes that affect the ionosphere during magnetic storms.

ELECTRIC FIELD MOMENTUM COUPLING

Neutral winds speeds up to 1 km/sec are observed at high latitudes and have been attributed to momentum coupling from ExB drifts that in turn are induced by electrostatic fields of

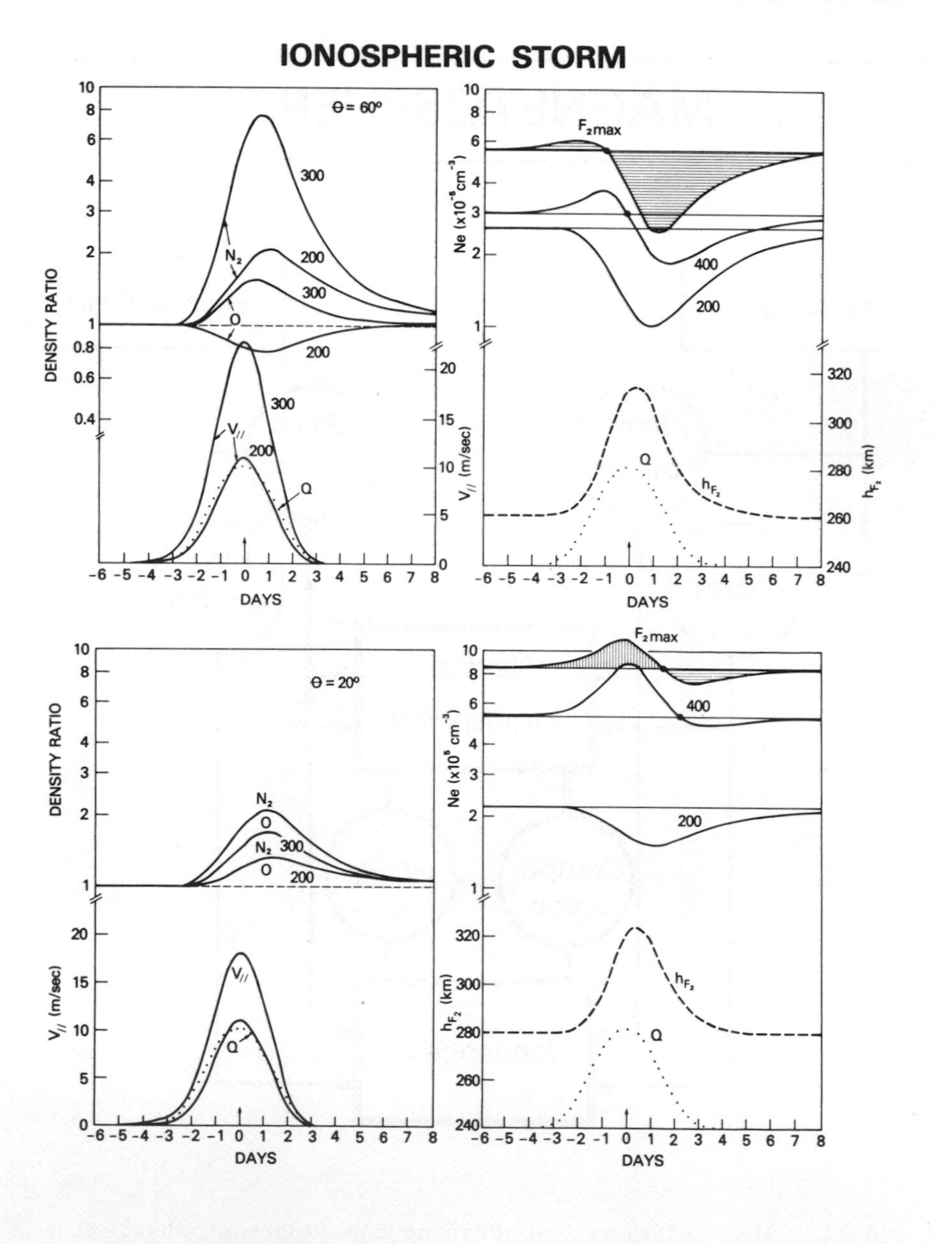

Figure 12. Computer simulation of interaction processes through winds and composition changes which, as a result of Joule heating, affect the ionosphere during magnetic storms (emphasized in Figure 11). The adopted heat source is shown in dotten lines. Note the time delay between the density ratios O/N_2 (relative to quiet conditions) and winds, which accounts for the sequence of positive and negative phases in the ionospheric storm.

magnetospheric origin (1, 2, 4, 5, 14, 15, 22, 23, 45, 51). In thermosphere dynamics one is accustomed to associate winds with energy and mass transport. Energization of the atmosphere by absorption of solar radiation and Joule heating, for example, induces winds that transport amounts of energy comparable to those initiating the motions. But important as these thermally driven winds are for the redistribution of energy and mass their magnitudes are only on the order of 100 m/sec which is usually small compared with the above quoted wind velocities induced by electric field momentum coupling.

The solution of this apparent conflict focuses on an important distinction between forcing functions due to heating, by Joule dissipation for example, and ExB momentum coupling, which produce completely different classes of wind fields one with a relatively large divergence the other one representing primarily a curl field. Although in reality both processes are closely related, for illustrative purposes they are described here separately. Results from a three dimensional model are shown (Figure 13) in the form of amplitudes and phases for the diurnal component (period of 24 hours) with a wave number n=9 (corresponding to a characteristic horizontal dimension of 2000 km) which is representative of the gross electric field pattern at high latitudes. To elucidate the distinguishing characteristics, both forcing functions (for heating and momentum source) are chosen to produce the same wind velocities at the poles (250 m/sec). From this comparison it is apparent that the momentum source is more than 20 times less "effective" for the density variations (and a factor of 100 times less for the temperature) than Joule heating. Moreover one finds that under Joule heating the temperature and heavier atmospheric species N_2 and Ar vary <u>out of phase</u> with O and He, while for the electric field momentum source the lighter species O and He vary <u>in phase</u> with N_2 and Ar.

Figure 14a aids in interpreting these results. Except for the factor B^{-2} the ion drift velocity $V_i=(E \times B)\ B^{-2}$ represents a curl field. By collisional momentum transfer, a corresponding curl component is imposed on the neutral wind velocity, and the Coriolis force in turn couples momentum into a divergence field. That divergence component would be relatively large if it were not for the thermodynamic feedback due to energy transport. But the resulting back pressure counteracts and significantly dampens the divergence field to the degree that its contribution to the total wind velocity becomes very small, implying small temperature and density variations. This situation is in contrast to thermally driven winds where under the influence of ion drag and viscosity the flow proceeds along pressure gradients; the generated divergence field is then relatively large, hence the total wind field is much more efficient in changing temperature and density. This situation may be likened to the mixing of paint,

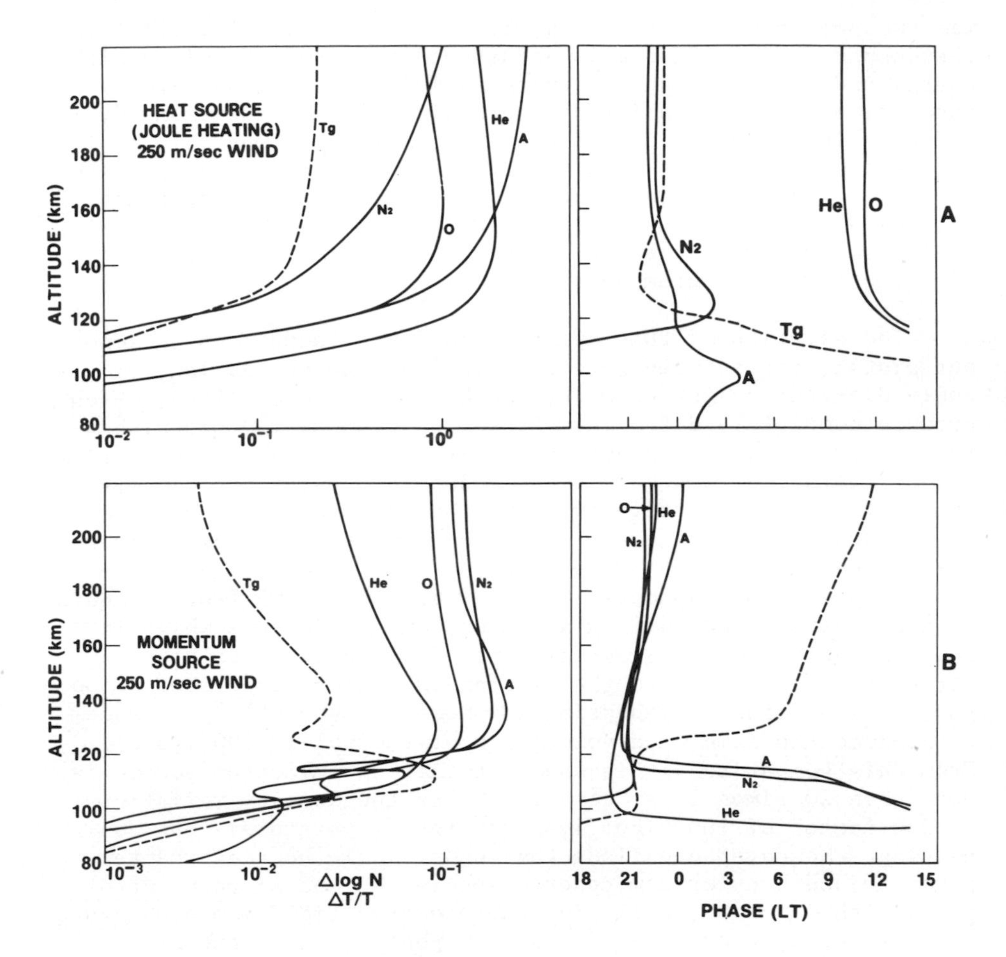

Figure 13. Theoretical results illustrating the differences between temperature and composition signatures due to Joule heating (A) and momentum source (B). In both calculations the polar wind velocities were the same.

where the steady rotation (curl field) is relatively ineffective compared to shaking which produces a divergence field.

The large differences between heat and momentum source signatures in the phase relations of temperature and composition are heuristically discussed in Figure 14b. Given an external heat source Q (e.g. Joule heating), the atmospheric pressure increases and that in turn drives a divergence field which removes energy ($-\Delta Q$) and mass ($-\Delta M$). The net energy change ($Q-\Delta Q$) is usually dominated by the external source. As a result, there is an increase (positive) in temperature ($\Delta T>0$) and density of the major gas ($\Delta N_2>0$). For the minor and lighter gasses (such as

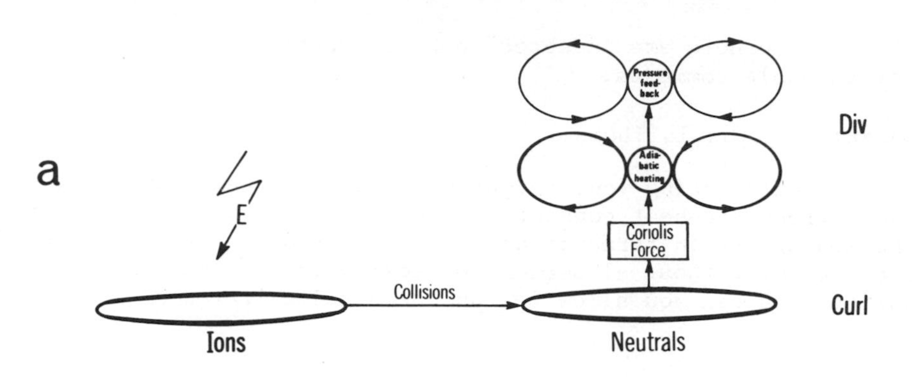

b

JOULE HEATING

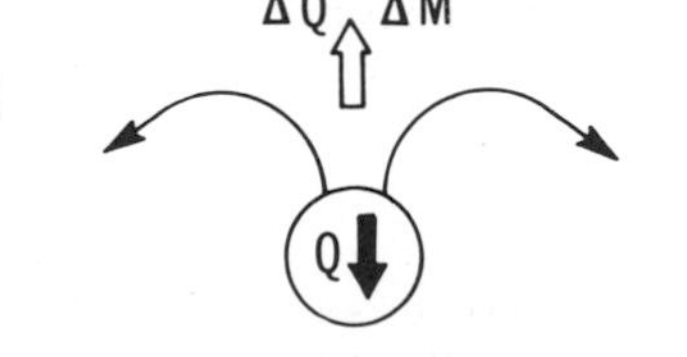

MOMENTUM SOURCE

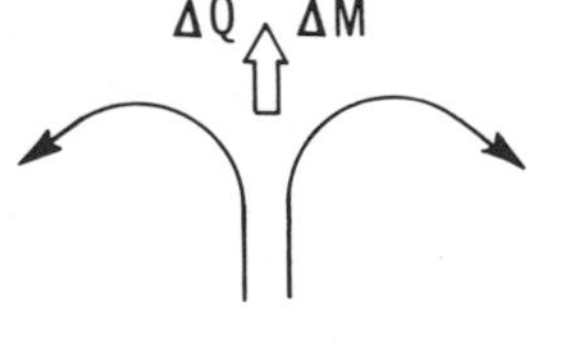

$Q - \Delta Q > 0 \longrightarrow \Delta T > 0 \longrightarrow \Delta N_2 > 0$ $\quad$ $-\Delta Q < 0 \longrightarrow \Delta T < 0 \longrightarrow \Delta N_2 < 0$

$-\Delta M < 0 \longrightarrow \Delta He < 0;\ \Delta O < 0$ $\quad$ $-\Delta M < 0 \longrightarrow \Delta He < 0;\ \Delta O < 0$

Figure 14. Heuristic interpretation of the results shown in Figure 13. (a) ExB drift induces via collisions a curl component in the neutral atmosphere. Due to the Coriolis force a divergence component is generated which causes temperature and density variations. (b) The differences in phase relationships between Joule heating and momentum source signatures.

He) wind induced diffusion effectively removes mass ($-\Delta M$), analogous to the energy loss ($-\Delta Q$). In the absence of a particle source the density therefore decreases ($\Delta He<0$) bringing forth the inverse relationship between heavier and lighter gases. With a momentum source that somehow (and inefficiently) drives a divergence field, a major distinction is that the external heat source Q is zero. All other conditions being the same, the wind induced energy loss ($-\Delta Q$) then causes the temperature ($\Delta T<0$) and major gas ($\Delta N_2<0$) to decrease, in phase with the corresponding decrease of the lighter gas ($\Delta He<0$). As was seen earlier, the Joule heating signature is observed in the composition at high latitudes during magnetic storms.

Acknowledgements:

The authors are indebted to R. E. Hartle and N. W. Spencer for valuable comments.

Question by Dr. L. Thomas:

In calculating your height variations of temperature, did you express the heat conduction from the lower thermosphere into the mesosphere in terms of an eddy (diffusion) heat conduction coefficient without allowance for energy input from eddy dissipation, and did you allow for any seasonal difference?

Answer:

We have not accounted explicitly for eddy dissipation. Some of the "wave energy" that was postulated in describing the mesospheric winter anomaly is probably due to this process. Considering seasonal variations in the eddy diffusion coefficient with inclusion of horizontal transport, our analysis--seen in the light of composition measurements--favors the turbulence to be enhanced in winter over that in summer.

Question by Dr. R. W. Smith

Many of the matters considered in your discussion dealt with global scale energy sources and response. Could you comment on whether the deviations from diffusive equilibrium are likely to be more marked when the same principles are applied to a localized source?

Answer:

That depends on the time scale. But under otherwise similar conditions transport processes are indeed more significant for localized disturbances. Energy transport preferentially dampens small scale perturbations so that the large scale density and temperature variations tend to prevail. On the other hand, wind induced diffusion enhances the horizontal differentiation in composition which has the tendency to conserve small scale structures.

References

(1) Banks, P. M.: 1977, J. Atmos. Terr. Phys., 39, pp. 179.
(2) Banks, P. M., and Doupnik, J. R.: 1975, J. Atmos. Terr. Phys., 37, pp. 951.
(3) Blamont, J. E., Luton, J. M., and Nisbet, J. S.: 1974, Radio Sci., 9, pp. 247.
(4) Brekke, A., Doupnik, J. R., and Banks, P. M.: 1974, J. Geophys. Res. Lett., 2, pp. 3773.
(5) Burch, J. L., Fields, S. A., Hanson, W. B., Heelis, R. A., Hoffman, R. A., and Janetzke, R. W.: 1976a, J. Geophys. Res. 81, pp. 2223.
(6) Burch, J. L., Lennertsson, W., Hanson, W. B., Heelis, R. A., Hoffman, J. H., and Hoffman, R. A.: 1976b, J. Geophys. Res., 81, pp.3886.
(7) Cauffman, D. P., and Gurnett, D. A.: 1972, Space Sci. Rev., 13, pp. 369.
(8) Chandra, S., and Herman, J. R.: 1969, Planet. Space Sci., 17, pp. 841.
(9) Chandra, S., and Spencer, N. W.,: 1975, J. Geophys. Res., 80, pp. 3615.
(10) Duncan, R. A.: 1969, J. Atmos. Terr. Phys., 31, pp. 59.
(11) Evans, D. S., Maynard, N. C., Troim, J., Jacobsen, T., and Egeland, A.: 1977, J. Geophys. Res., 82, pp. 2235.
(12) Evans, J. V.: 1973, J. Atmos. Terr. Phys., 35, pp. 593.
(13) Evans, J. V.: 1975, Rev. Geophys. Space Phys., 13, pp. 887.
(14) Fedder, J. A., and Banks, P. M.: 1972, J. Geophys. Res., 77, pp. 2328.
(15) Hanson, W. B. and Heelis, R. A.: 1975, Space Sci. Instrum., 1, pp. 493.
(16) Hays, P. B., Nagy, A. F., and Robel, R. G.: 1969, J. Geophys. Res., 74, pp. 4162.
(17) Hays, P. B., Jones, R. A., and Rees, M. H.: 1973, Planet. Space Sci., 21, pp. 559.
(18) Hedin, A. E., Mayr, H. G., Reber, C. A., Spencer, N. W., and Carignan, G. P.: 1974, J. Geophys. Res., 79, pp. 215.
(19) Hedin, A. E., Salah, J. E., Evans, J. V., Reber, C. A., Newton, G. P., Spencer, N. W., Kayser, D. C., Alcayde, D., Bauer, P., Cogger L., and McClure, J. P.: 1977, J. Geophys. Res., 82, pp. 2139.
(20) Hedin, A. E., Bauer, P., Mayr, H. G., Carignan, G. R., Brace, L. H., Brinton, H. C., Parks, A. D., and Pelz, D. T.: 1977, J. Geophys. Res., 82, pp. 3183.
(21) Hernandez, G., and Roble, R. G.: 1976b, J. Geophys. Res., 81, pp. 5173.
(22) Heelis, R. A., Hanson, W. B., and Burch, J. L.: 1976, J. Geophys. Res., 81, pp. 3803.
(23) Heppner, J. P.: 1977, J. Geophys. Res., 82, pp. 1115.
(24) Hinteregger, H. E.: 1976, J. Atmos. Terr. Phys., 38, pp. 791.

(25) Jacchia, L. G., Slowey, J. W., and von Zahn, U.: 1976, J. Geophys. Res., 81, pp. 36.
(26) Jacchia, L. G., Slowey J. W., and Cerniani, F.; 1967, J. Geophys. Res., 72, pp. 1423.
(27) Johnson, F. S., and Gottlieb, B.; 1970, Planet. Space Sci., 18, pp. 1707.
(28) Jones, K. L., and Rishbeth, H,: 1971, J. Atmos. Terr. Phys., 33, pp. 391.
(29) Keating, G. M., and Prior, E. J.: 1968, Space Res., 8, pp. 982.
(30) Kellogg, W. W.: 1961, J. Meteorol., 18, pp. 373.
(31) King, G.A.M.:1964, J. Atmos. Sci., 21, pp. 231.
(32) Kohl, H., and King, J. W.: 1967, J. Atmos. Terr. Phys. 29, pp. 1045.
(33) Matsuno, T.: 1975, paper presented at IAGA/IAMAP Conference, Int. Assoc. of Geomagn. and Aeron., Int. Assoc. of Meteorol. and Atmos. Physl, Grenoble, France.
(34) Matsushita, S. A.: 1959, J. Geophys. Res., 64, pp. 305.
(35) Maynard, N. C.: 1974, J. Geophys. Res., 79, pp. 4620.
(36) Maynard, N. C., Evans, D. S., Maehlum B., and Egeland, A.: 1977, J. Geophys. Res., 82, pp. 2227.
(37) Mayr, H. G. and Harris, H.: 1977, Geophys. Res. Lett., 4, pp. 25.
(38) Mayr, H. G., Harris, I., and Spencer, N. W.: 1978, Rev. Geophys. Space Phys., 16, pp. 539.
(39) Miller, N., Grebowsky, J. M., Mayr, H. G., Harris, I., and Tulunay, Y.: 1979, J. Geophys. Res., 84, pp. 6493.
(40) Prölss, G. W., and von Zahn, U.: 1974, J. Geophys. Res., 79, pp. 2535.
(41) Reber, C. A., Cooley, J. E., and Harpold, D. H.: 1968, Space Res., 8, pp. 993.
(42) Roble, R. G. and Matsushita, S.: 1975, Radio Sci., 10, pp. 389.
(43) Seaton, M. J.: 1956, J. Atmos. Terr. Phys., 8, pp. 122.
(44) Stolarski, R. S., Hays, P. B., and Roble, R. G.: 1975, Geophys. Res., 80, pp. 2266.
(45) Straus, J. M.: 1978, Rev. Geophys. Space Phys., 16, pp. 183.
(46) Taeusch, D. R., Carignan G. R., and Reber, C. A.: 1971, J. Geophys. Res., 76, pp. 8318.
(47) Trinks, M., Fricke, K. H., Lairx, U., Prolss, G. W., and von Bahn, U.: 1975, J. Geophys. Res., 80, pp. 4571.
(48) Volland, H.: 1976, J. Geophys. Res., 81, pp. 1621.
(49) Volland, H., and Mayr, H. G.: 1971, J. Geophys. Res., 76, pp. 3764.
(50) Wedde, T., Doupnik, J. R., and Banks, P. M.: 1977, J. Geophys. Res., 82, pp. 2743.
(51) Wu, S. T., Matsushita, S., and DeVries, L. L.: 1974, Planet. Space Sci., 22, pp. 1036.
(52) Zimmerman, A. P., Faire, A. C., and Murphy, A. E.: 1972, Space Res., 12, pp. 615.

THE POLAR F-REGION - THEORY

H. Kohl

Max-Planck-Institut für Aeronomie,
D-3411 Katlenburg-Lindau 3, F.R.G.

Abstract: The polar ionosphere is basically different from the mid-latitude ionosphere, because it is tied up with the magnetosphere. Particle fluxes and electric fields originating in the magnetosphere penetrate into the ionosphere and affect production, loss and transport of ionization there. Some of the polar F-region phenomena, as troughs, polar cap ionization etc. are discussed in the light of these processes.

1. INTRODUCTION

The term "theory" in the title simply means that a number of typical polar F-layer phenomena will not be only described, but it will be discussed what might cause them. There is a fundamental difference between the polar and the mid-latitude ionosphere, one obvious difference is that the polar ionosphere is much more variable in space and time. The reason is that there are a number of effects which are only active at polar regions. We may discuss them in comparison with mid-latitudes looking at the usual F-layer continuity equation

$$\frac{\partial N}{\partial t} = q - \beta N + \mathrm{div}(N\underline{v}) \ ,$$

where N is ion/electron density, q is production rate, β is loss rate, $\underline{v}$ is plasma velocity. Firstly the production rate q in the mid-latitude ionosphere is almost completely dependent on solar XUV radiation, while at polar latitudes particles from the magnetosphere can be a major source of electron/ion production; during the polar night they are even the only one.

C. S. Deehr and J. A. Holtet (eds.), Exploration of the Polar Upper Atmosphere, 55–65.

The transport term $div(N\underline{v})$ is of particular importance, because at high latitudes the plasma velocity $\underline{v}$ can become very large· 1 km/s and more compared to about 100 m/s in mid-latitudes. In general, $\underline{v}$ is the sum of three velocities: diffusion, motion induced by neutral air wind, and $(\underline{E} \times \underline{B})/B^2$ drift caused by electric fields. The latter one is dominating in polar regions, because electric fields extending down from the magnetosphere and secondary fields set up in the ionosphere have the order of 50 mV/m, so that $(\underline{E} \times \underline{B})/B^2 \approx 1$ km/s. In one other respect the transport term has to be treated different. For mid-latitudes one usually considers the horizontal derivatives to be negligible and leaves only $\partial(Nv_z)/\partial z$. At high latitudes, however, $\underline{v}$ is so high and horizontal gradients can be so large that horizontal transport has to be included.

The loss rate β is sometimes highly increased. The reason is that the cross-section for ion/neutral reactions depends on the relative speed of the colliding particles, and it does not matter whether this speed is brought about by high temperature or by a fast flow of the ions through the neutral gas. The above mentioned velocity of ≈ 1 km/s would as reaction rates are concerned correspond to a temperature increase of about 1000 K, which in general makes reactions faster.

In the next chapter these points will be described in more detail, and then the behaviour of the polar F-region will be discussed.

2. PROCESSES TYPICAL FOR THE POLAR IONOSPHERE

Fig. 1 (Banks et al. 1974a) shows calculated ionization rates of electron fluxes for a number of energies. Particles with higher energy penetrate to lower altitudes. This is not surprising, because for every ion/electron pair on the average 35 eV are taken away from the energetic particle, so that low energy electrons rather quickly lose all their energy. It can also be seen from Fig. 1 that particles with higher energy produce less ionization at higher altitudes than low energy particles. This happens, because the ionization cross-section decreases for higher particle energies, as for fast particles the Coulomb force has less time to drag away an electron from a neutral molecule. Fig. 1 also shows that for the F-layer, which only is of interest here, particles with energies of 1 keV and below are most effective in producing ionization. It may be noted that this model calculation by Banks et al. (1974) was done in a very sophisticated way, as it includes various effects, e.g. ionization by secondary electrons, change of pitch angle due to collisions and also energy loss of the incident particles by exciting neutrals.

It may be noted that this model calculation by Banks et al.

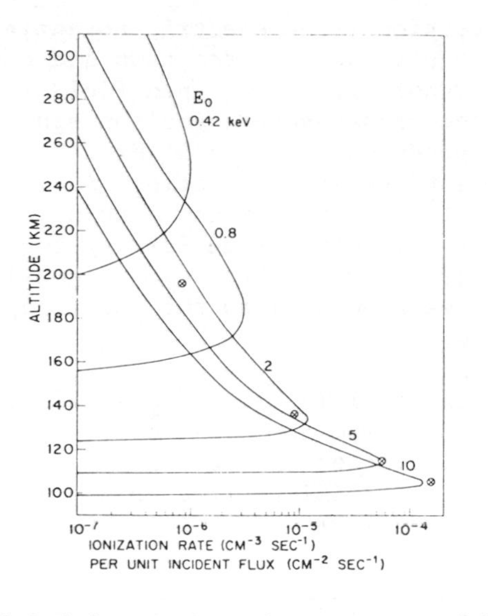

Fig. 1
Ionization rates of electrons of different energies as a function of height after Banks et al. (1974a).

(1974a) was done in a very sophisticated way, as it includes various effects, e.g. ionization by secondary electrons, change of pitch angle due to collisions and also energy loss of the incident particles by exciting neutrals.

Fig. 2 (Spiro et al. 1978) shows a pattern of what is called magnetospheric convection. The F-region takes part in this convection as long as magnetic field lines can be considered as equipotentials. The left hand side of the diagram shows the projection of the flow lines on the northern hemisphere as it is seen by an observer rotating with the earth, and the right hand side the same pattern seen from a non-rotating position. This simply means that on the right hand side the constant angular velocity of the earth is added.

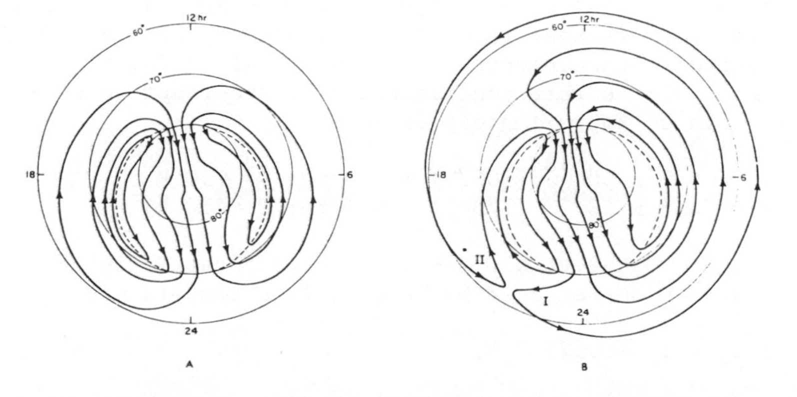

Fig. 2 Magnetospheric convection pattern excluding (left) and including (right) corotation potential after Spiro et al. (1978).

A primary cause for the convection is an electric potential difference of several tens of kilovolts between the dawn and dusk side of the polar cap, which is transferred along open field lines from the solar wind plasma into the magnetosphere and ionosphere. The corresponding electric field produces an $\underline{E} \times \underline{B}/B^2$ drift of the order of .5 to 1 km/s in the anti-sunward direction. At the border of the polar cap the convection swings around flowing back to the dayside along morning- and eveningside forming two closed loops. This region of sunward flow can be roughly identified with the auroral oval, while we have defined the polar cap as the region of anti-sunward flow.

The sunward flow in the auroral oval is also an $\underline{E} \times \underline{B}/B^2$ drift. The corresponding electric field is determined by several factors. Firstly, within a conducting sheet as the ionosphere an electric field can not be confined to a limited area as the polar cap but will continue into the auroral oval. This spreading out is influenced by a change in electrical conductivity, the auroral zone being highly conducting as compared to the polar cap. Furthermore, the electric field is modified by electric currents flowing from the magnetosphere through the ionosphere and back again.

It should be noted that a pattern like in Fig. 2 represents a schematic model. Various similar models are presented in the literature showing different features, but they all consist of two closed loops with clockwise (evening) and counterclockwise (morning) circulation.

As was mentioned in the introduction one of the consequences of this fast convection is a possible increase of ion loss rate (Banks et al. 1974b). This happens because the reaction cross-section generally depends on the relative speed between the colliding ion and neutral. In the F-layer the most important reactions for the loss of O^+ are $O^+ + O_2 \rightarrow O_2^+ + O$ and $O^+ + N_2 \rightarrow NO^+ + N$, where the molecular ions further recombine according to $M^+ + e \rightarrow M$. According to laboratory measurements (McFarland et al. 1973) the temperature dependence of the rate coefficient k of the reaction $O^+ + N_2 \rightarrow NO^+ + N$ can be approximately expressed by

$$k = 1.2 \times 10^{-18} \; (300/T) \; m^3 s^{-1} \text{ for } T < 750 \text{ k}$$
$$k = 8 \times 10^{-20} \; (T/300)^2 \; m^3 s^{-1} \text{ for } T > 750 \text{ k}.$$

According to Banks et al. (1974b) the temperature in the above equations can be replaced by an effective temperature T_{eff}

$$T_{eff} = T_o + .329 \, E_\perp^2,$$

which takes into account the effect of the increasing relative ion/neutral velocity due to $\underline{E} \times \underline{B}/B^2$ drifts. Assuming T_o= 1000 K k increases by a factor of 3.3 for E = 50 mV/m. For the reaction $O^+ + O_2 \rightarrow O_2{}^+ + O$ the corresponding effect is relatively small, because the rate coefficient does not very much depend on temperature (relative speed) in the relevant range.

The consequence is not only that the F-layer ionization is reduced, but also that the ion composition changes and NO^+ ions become more and more important, whereas at midlatitudes O^+ is the dominant ion in the F-layer. Fig. 3 taken from Schunk et al. (1975) shows the result of a model calculation. It is clearly seen that below 250 km NO^+ has become the dominant ion. The electron density profile shows almost constant density from about 150 km to 350 km altitude. Such a profile looks peculiar from the mid-latitude point of view, however, the Chatanika radar has, indeed, observed this kind of N(z) variation. The reduction in F peak density for the case of Fig. 3 compared to the E = 0 case is about a factor of three.

Another reason for an increased loss rate of O^+ may be due to the excitation of vibrational states of N_2 which is caused by precipitating electrons or by secondary electrons released by them. The rate coefficient k for vibrationally excited N_2 can easily increase by a factor of four compared to the normal case.

After having described a number of effects that influence production, loss and tranport of ionization in the high latitude F-region the next section will discuss observations.

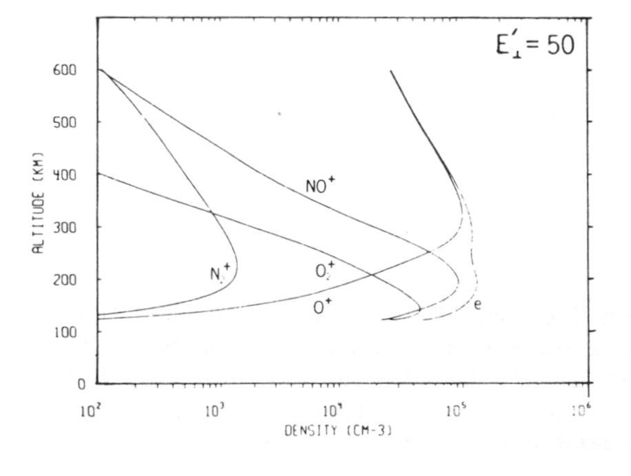

Fig. 3
Ion and electron density profiles for excessive loss rates due to fast convection after Schunk et al. (1975).

3. DISCUSSION OF SOME POLAR F-REGION PHENOMENA

Fig. 4 shows observations of electric field and low and high energy particle fluxes made by INJUN 5 on a pass from the morning- to the eveningside through the auroral oval and across the polar cap (Frank and Gurnett 1971). The dashed line in the uppermost part of the figure is a reference line for the electric field, which is given by the induced field in the frame of the moving satellite. The sunward motion in the oval and the anti-sunward motion in the polar cap are clearly seen. If the point of reversal is used as a definition of the polar cap/auroral oval border (vertical dashed line), then the energetic electrons obviously appear equatorward of this boundary, while the softer particles which are responsible for F-layer ionization, are penetrating on either side of it. The straight line of dots between the two strong spikes represents the lower limit of the instruments sensitivity.

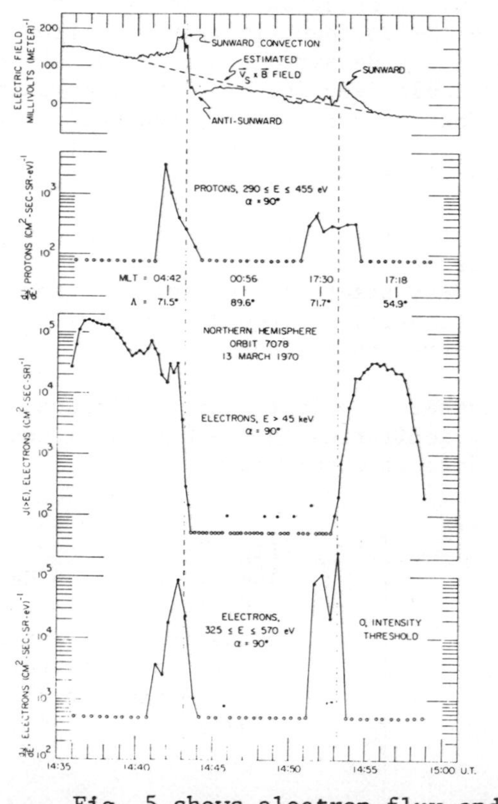

Fig. 4
Particle fluxes and electric field for an INJUN 5 pass from the morning to the evening side across the polar cap after Frank and Gurnett (1971).

Fig. 5 shows electron flux and F-layer electron density measurements made by ISIS 2 on a pass form the noonside to the midnightside (Whitteker et al. 1978). Taking the trapping boundary of the 22 keV electrons as the border of the polar cap we again see that soft particle fluxes appear also polewards of that point.

The rather strong .15 keV electron flux between $\Lambda \approx 76°$ and $\Lambda \approx 80°$ on the noonside is of particular interest. This area is called the cleft and is a region where the magnetic field lines are linked to what is now called the boundary layer, which is a layer just inside the sunward side of the magnetospheric boundary. It is believed that here solar wind plasma, or better magnetosheath plasma, is entering directly the upper atmosphere.

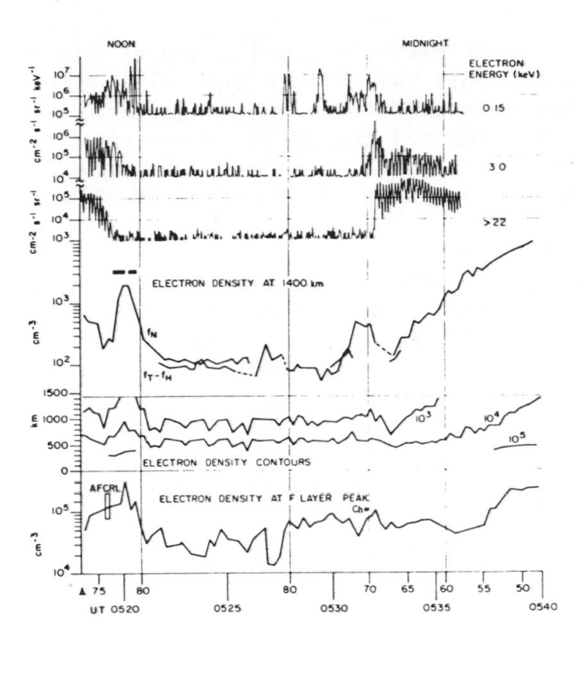

Fig. 5
F-layer electron densities and energetic particle fluxes for an ISIS 2 pass through the winter polar ionosphere after Whitteker et al. (1978).

This is the plasma we observe in the .15 keV channel in Fig. 5, and it produces a peak of ionization in the F-layer, as can be seen in the lower part of the figure.

However, there are problems with this explanation. Firstly, why is this peak becoming more pronouced with height? One possible reason is that the F-layer scale height becomes very large because of high electron temperatures (Whitteker et al. 1978). The other problem is related to the convection velocity. If it is about 1 km/s as we have anticipated before, the electrons produced in the some hundreds of kilometers broad cleft region would be swept away into the polar cap in some hundreds of seconds. This statement can be generalized. If you anywhere in the convection system observe an electron density, it was at a place 100 km away 100 s before and 1000 km away 1000 s before (1 km/s drift speed assumed and the time short compared to the life time of electron/ion pairs).

This problem is related to the question: what causes the ionization above the polar cap? Looking at Fig. 5 we see that the soft electron precipitation between $\Lambda = 80°$ (noon) and $\Lambda = 80°$ (midnight) is about two orders of magnitude less than that at the cleft. The electron concentration, however, varies only by about one order of magnitude. A possible explanation is that most of the ionization above the polar cap is flowing in from the cleft. But then one would not expect such a steep gradient of electron density at the poleward side of the cleft.

A general answer that may be given, and which may or may not be right, is that things are highly variable in the polar ionosphere. Thus, it may be that sometimes this density gradient at the cleft does not exist, or sometimes the drift component across the cleft is much less than 1 km/s what may happen, when the drift is more tangential to parts of the polar cap boundary (Heelis et al. 1976).

Leaving this question open we proceed to the next problem, which is the so-called trough. Fig. 6 shows F-layer peak electron densities measured by ISIS 2 on the midnight meridian (Chacko 1978). While the height of the layer remains almost constant the peak electron density decreases reaching a minimum at about 60° geomagn. lat. and increases again abruptly towards higher latitudes. The poleward edge of this trough roughly coincides with the equatorward edge of the auroral oval. In fact, such a trough is observed all around the auroral oval with a frequency of occurrence of almost 100% at midnight and about 50% at noon.

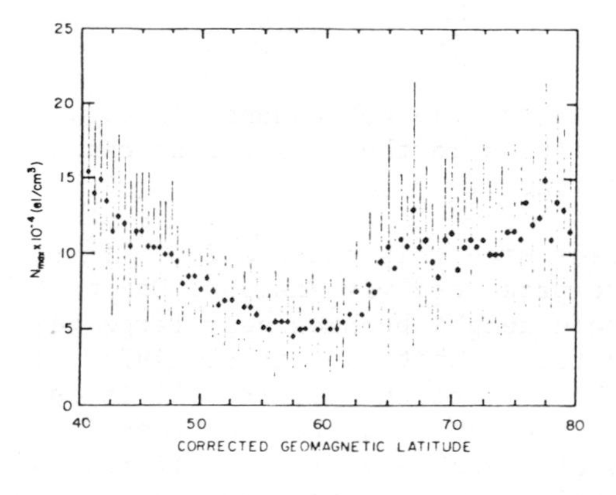

Fig. 6 Electron density and height of the F-layer peak from thirty passes of ISIS 2 on the nightside after Chako (1978).

A number of suggestions have been made in order to explain the existence of troughs. The proposed mechanisms include excessive loss rates as a consequence of vibrational excitation of nitrogen, neutral wind effects etc. One process shall be discussed here in more detail without putting any preference on it. The idea is that the trough is a consequence of convection. Let us look again on Fig. 2. The eveningside of the right hand diagram shows a peculiarity. Following the outermost flow line towards the night side there is a stagnation point at about 22 L.T., where the line swings back to the noonside. The reason is that away from the pole the electric field induced velocity becomes smaller, while the corotation velocity becomes larger. Both velocities are in opposite directions on the eveningside, so somewhere they cancel each other. This area is shown in the left hand side of Fig. 7 in a schematic way. If is clear that between the flow towards midnight and the backward flow there is a stagnation line where the velocity is

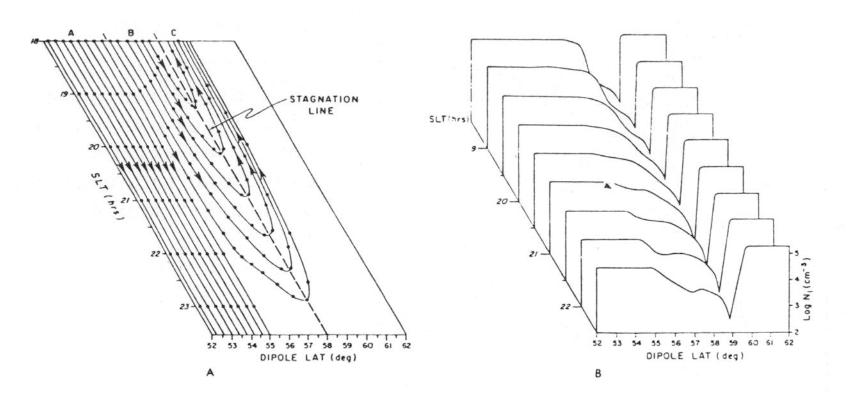

Fig. 7 Left: Schematic diagram of convection at the eveningside. The dots denote one hour intervals in real time. Right: Result of a model calculation showing the development of the trough (Spiro et al. 1980).

zero, and that away from this line the velocity is increasing but in opposite directions. In Fig. 7 the dots on the flow lines denote time intervals of one hour. On the flow lines at low latitudes the dots coincide with local time. That means that the plasma is just corotating with the earth and there is no additional convection. On flow lines further polwards the drift speed is much smaller; on the first flow line that swings back it takes 11 hours for the plasma to move from the 18 h-meridian to the stagnation point, and this offers a possible explanation of the trough formation. F-layer plasma that is observed in the trough region has been created in the sunlit dayside ionosphere and its density has very much decayed during the long travel time to the nightside.

The right hand side of Fig. 7 shows the result of a model calculation by Spiro et al. (1978). The development of the trough in local time is clearly seen. However, the increase in electron density at the poleward edge of the trough was just arbitrarily assumed; the calculations could only be done down to the minimum of the trough, because of the limitations of the method.

A given electron concentration had been assumed at the 18 L.T. meridian and then tubes of ionization have been followed on their way calculating the decay of ionization but assuming no production source. But, of course, the increase of electron concentration at the poleward edge of the trough can only be simulated by taking into account the particle ionization in the oval.

Such more general calculations have been performed, for instance by Knudsen et al. (1977) and by Watkins (1978). These authors included photoionization on the sunlit dayside as well as

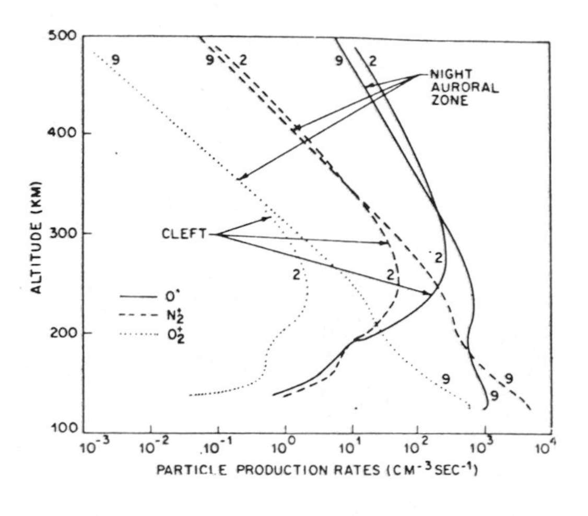

Fig. 8 Ionization production rates due to precipitating electrons in the cleft and midnight auroral oval after Knudsen et al. (1977).

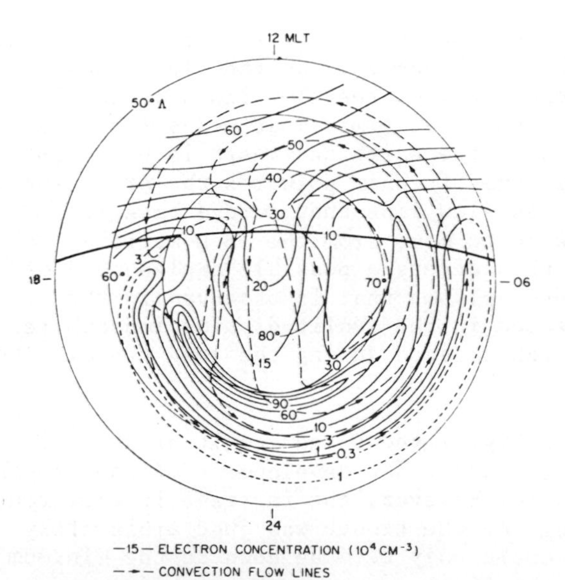

Fig. 9 Results of a model calculation by Knudsen et al. (1977).

ionization by energetic particles in the auroral oval and above the polar cap and, of course, also a convection model, which was slightly different from that used by Spiro et al. (1978). Furthermore, excessive loss rates according to Banks et al. (1974) were taken into account.

Fig. 8 shows the production rates of precipitating electrons used by Knudsen et al. (1977), which were calculated from typical particle measurements using the method of Banks et al. (1974a).

At 300 km altitude and above the production rates are very similar in the cleft and in the auroral oval, while at lower altitudes ionization is much stronger in the oval because of the higher energy of the particles impinging there.

Fig. 9 shows the result of the model calculation by Knudsen et al. (1977), which refers to winter time. From the solid line towards the dayside there is still photoionization leading for fairly high electron concentrations. Above polar cap we find electron concentrations of more than $10^{11} m^{-3}$ resulting from plasma transport from the dayside as well as from particle precipitation. In the auroral we have high plasma densities as a consequence of precipitation, and equatorwards of the oval we have a trough extending from the evening- to morningside. This trough is also a result of very slow convection at the eveningside, but different from Spiro et al. (1978) there are slow flow lines from the evening trough to the morningside.

There are a number of phenomena in the polar F-region that could not be dealt with in this short overlook. Among there are seasonal variations, the universal Time effect, the influence of neutral air winds and others. I would, therefore, like to direct the attention of the reader to some exemplary work by Watkins (1978), Roble et al. (1977) and Hays et al. (1979).

REFERENCES

Banks, P.M., Chappell, C.R., and Nagy, A.F. 1974a, J. Geophys. Res., 79, 1459-1470.
Banks, P.M., Schunk, R.W., and Raitt, W.J. 1974b, Geophys. Res. Let., 1, 239.
Chacko, C.C. 1978, J. Geophys. Res., 83, 5733.
Frank, L.A., and Gurnett, D.A. 1971, J.Geophys. Res., 76, 6829.
Hays, P.B., Meriwether, J.W., and Roble, R.G. 1979, J. Geophys. Res., 84, 1905.
Heelis, R.A., Hanson, W.B., and Burch, J.L. 1976, J. Geophys. Res. 81, 3803
Knudsen, W.C., Banks, P.M., Winningham, J.D., and Klumpar, D.M. 1977, J. Geophys. Res., 82, 4784.
McFarland, M., Albritton, D.L., Fehsenfeld, F.C., Ferguson, E.E., and Schmeltekopf, A.L. 1973, J. Chem. Phys., 59, 6620.
Roble, R.G., and Rees, M.H. 1977, Planet. Space Sci., 25, 991.
Schunk, R.W., Raitt, W.J., and Banks, P.M. 1975, J. Geophys. Res., 80, 3121.
Spiro, R.W., Heelis, R.A., and Hanson, W.B. 1978, J. Geophys. Res., 83, 4255.
Watkins, B.J. 1978, Planet. Space Sci., 26, 559.
Whitteker, J.H., Shepherd, G.G., Anger, C.D., Burrows, J.R.,Wallis, D.D., Klumpar, D.M., and Walker, J.K. 1978, J. Geophys. Res., 83, 1503.

RADIO OBSERVATIONS OF THE AURORAL F-REGION

T.B. Jones

Physics Department, Leicester University, U.K.

Radio wave measurements of the parameters of the polar F-region are reviewed. A long series of observations of the bottom and top sides of the layer have been made by means of ground based and satellite borne ionosondes respectively. From these data, the basic features and behaviour of the region have been determined. In recent years new techniques, such as incoherent scatter radar, have added to our knowledge, particularly regarding transient behaviour, temperature, plasma flow and the electric fields in the layer. Observations by other radio methods, such as satellite beacon measurements of total electron content and scintillation regions, artificial modification of the F-layer by high power radio waves and studies of movements and travelling disturbances are also discussed. Satellite observations of the neutral thermosphere have added greatly to our understanding of the F-region production and loss processes but a discussion of these observations lies outside the scope of the present review.

1. INTRODUCTION

Early measurements of the polar ionosphere were somewhat intermittent; however, since the IGY extensive routine ionosonde observations of the high latitude ionosphere have been undertaken at stations in both the southern and northern polar regions. From this large collection of data the general behaviour of the high latitude F-region has been established and a number of anomalous features recognised. In the mid 1960's sounding of the topside of the layer began using ionosondes carried on board satellites, such as those of the Alouette series. These measurements confirmed many of the features already deduced from the

C. S. Deehr and J. A. Holtet (eds.), Exploration of the Polar Upper Atmosphere, 67–82.

ground based data and were particularly valuable for studies of the high latitude trough and the polar cap. It was from these satellite observations, combined with measurements of charged particles fluxes, that the characteristic behaviour of the high latitude ionosphere was eventually explained.

Satellites carrying radio beacons have played an important role in determining the total content of the F-layer and have been particularly valuable in studying the small scale structures that give rise to scintillation effects in the radio signal. Riometers have also been employed for total content studies although they have proved to be most useful for studies of the lower ionosphere.

In the early 1970's an incoherent scatter radar was installed at Chatanika, Alaska, and immediately produced a wealth of information regarding the high latitude ionosphere. New types of observations were possible, including measurements of electron and ion densities and temperatures, plasma drift velocity and electric fields. The new EISCAT installation in Northern Scandinavia will contribute further to studies of this kind. At the present time a number of different measuring facilities are being established in the vicinity of the EISCAT system which should produce a coordinated programme of the high latitude ionosphere research.

In recent years the advent of new experimental techniques and the special attention given to coordinated research programmes have lead to a considerable improvement in our knowledge of the auroral F-region. This paper attempts to review the results of these observations and to highlight some of the problems still outstanding.

2. VARIATIONS OF F-LAYER IONIZATION PROFILE AND MAXIMUM ELECTRON DENSITY

2.1 High Latitude Ionosonde Observations

In most of the early observations the ground based ionosonde was the principal experimental technique employed. However, from the mid 1960's satellite borne ionosondes were available and data from these instruments greatly enhanced our understanding of the features of the polar F-layer.

At high latitudes the solar zenith angle variations are quite different from those at mid and temperate latitudes. Consequently the high latitude ionosphere was expected to exhibit significant differences from its low latitude counterpart. Some of the objectives of the early work were:

(a) to establish whether or not the ionosphere existed during the long polar night,
(b) to investigate the diurnal variation for the polar day during which the zenith angle changes are very small, and
(c) to compare the arctic and antarctic ionospheres.

It soon became apparent that the ionosphere was present throughout the year and, moreover, the ionograms contained complicated features not observed at mid and low latitudes. Even the determination of foF_2 at an auroral station during the dark (winter) period is complicated by several factors. One of the most serious difficulties arises from the presence of F-region echoes that are similar in appearance to the normal F-region but which behave quite differently. These "auroral or oblique" F-echoes are often "spread" in height and can have a greater critical frequency than the normal F-echo. Moreover, these oblique echoes are often observed to move towards the observing station. The occurrence of these features is related to auroral oval phenomena and tφ the high latitude trough in F-region ionization. Several features of the high latitude ionosphere derived from both bottom and topside ionograms, are now discussed in detail.

2.2 Influence of the Auroral Oval

In the auroral oval ionization is produced at all levels of the ionosphere by particle precipitation and at F-region heights low energy (< 1 keV) particles play a dominant role. The influence of the ionization changes is seen on ionograms in the form of a relatively low F-region trace, often spread in height range, which extends to much higher frequencies than the normal F-layer. The existence of this irregularity zone was confirmed using an airborne ionosonde during flights across the auroral oval (1,2). The authors referred to the region of irregularity as the "F-layer irregularity zone" (FLIZ). Particle ionization gives rise to strong field aligned irregularities which are readily detected at oblique incidence by an ionosonde. Thus, as the auroral oval is approached from the equatorial side the normal ionograms begin to show additional F-echoes corresponding to the oblique reflections. The slant range decreases as the feature moves towards the observing station and produces a dense, abnormally low F-layer when overhead. The field aligned structures are less easily detected when the station is poleward of the FLIZ so that the reappearance of the normal ionogram is quite rapid. Since some low energy particle drizzle is present even on the poleward side of the auroral zone, the ionograms here often show more "spread" features than those on the equatorial side of the FLIZ. These changes can indicate when the FLIZ has passed over a station. A typical FLIZ sequence is reproduced in Fig.1. Note the difference between ionograms A,B,C and F and the

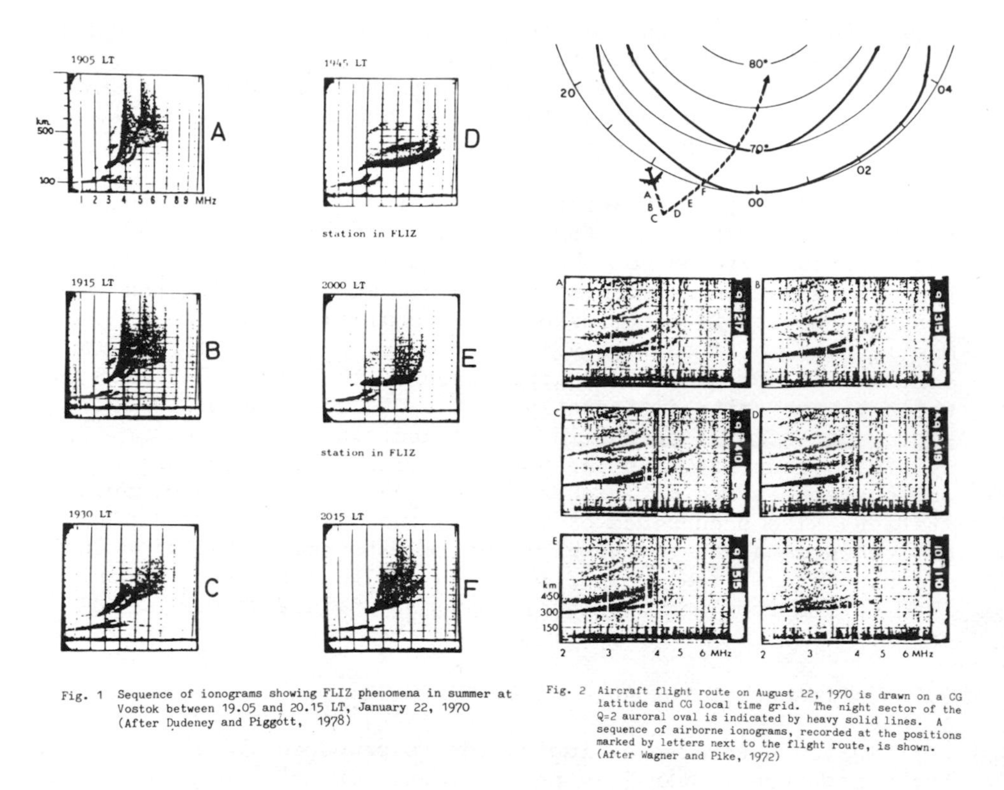

Fig. 1 Sequence of ionograms showing FLIZ phenomena in summer at Vostok between 19.05 and 20.15 LT, January 22, 1970 (After Dudeney and Piggott, 1978)

Fig. 2 Aircraft flight route on August 22, 1970 is drawn on a CG latitude and CG local time grid. The night sector of the Q=2 auroral oval is indicated by heavy solid lines. A sequence of airborne ionograms, recorded at the positions marked by letters next to the flight route, is shown. (After Wagner and Pike, 1972)

two which are in the precipitation zone, D and E.

A particularly impressive series of observations of the FLIZ region have been presented (2) from experiments using a flying ionosonde. Fig.2 shows the aircraft's path and the sequence of ionograms obtained. As the aircraft moves away from the FLIZ the slant range of the oblique echo increases. However, when the aircraft turns and flies towards the FLIZ the range decreases until finally the oblique echo coincides with the original vertical F-region trace. The resulting 'thick' F-layer trace is typical of the data obtained when the irregularity zone is overhead. The irregularity zone is associated with the poleward edge of the high latitude trough.

Thomas et al (3), using topside ionograms taken from the Alouette satellite, showed that the enhancements of the topside ionosphere formed a 'plasma ring' encircling the geomagnetic pole which was similar to the spread F oval. Using the airborne ionosonde Pike (1) obtained a series of ionograms at a range of locations through the day sector of the auroral oval. Comparison of these data with the Alouette electron density observations showed that the topside polar F-layer plasma ring existed in the bottom side as the F-layer irregularity zone.

2.3 F-region Trough Phenomena

One of the major features of the high latitude F-region is the decrease in electron density known as the F-layer trough. This feature was first recognised from topside ionograms made by the Alouette I satellite (4). Ionograms from the same satellite (5) showed that spread F features maximized in an oval shaped region which encircled the geomagnetic pole. Akasofu (6) and others later identified this region as being coincident with the auroral oval, hence the region is commonly designated "spread F oval". Typical examples of Alouette data, plotted in the form of isoionic contour heights against latitude, are reproduced in Fig.3. The high latitude trough between approximately Λ 60 and 75° is clearly visible as is the electron density enhancement on the high latitude side of the trough (7).

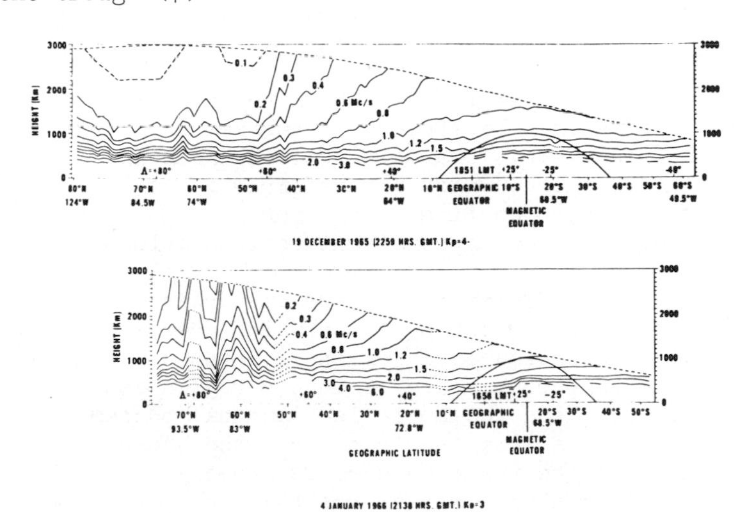

Fig. 3 Contours of constant plasma frequency as a function of height and geographic latitude determined from Alouette II sounder data. (After Nelms and Lockwood, 1967)

The trough feature has been observed on ground based ionograms by many workers (8,9). The feature is characterised by the formation of an oblique echo from the poleward wall of the trough, frequentlyreferred to as the "replacement" layer. A detailed study of the antarctic trough has been made (10,11) using an ionosonde located on an ice shelf which formed a natural interferometer produced by reflections from the bottom of the ice shelf. A typical example of Bowman's results is shown in Fig.4 in which the surfaces of constant ionization are drawn with equal horizontal and vertical scales and the contours are of equivalent plasma frequencies. The feature represents conditions for a North/South cross section of the trough in the early evening. The F-layer

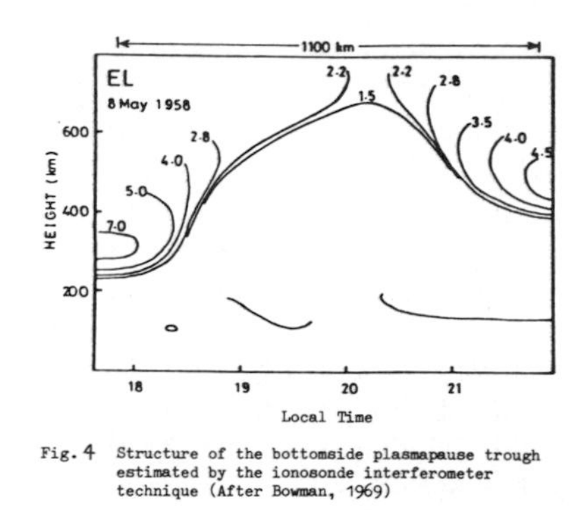

Fig. 4 Structure of the bottomside plasmapause trough estimated by the ionosonde interferometer technique (After Bowman, 1969)

structure on the poleward side of the trough is generated by ionizing particles and its critical frequency varies very slowly with time, as seen in any one sector, or with season. It does, however, show some variation with the solar cycle. In contrast, the F-layer on the equatorward side of the trough varies rapidly with time of day and season. This behaviour reflects the importance of the particle ionization within the polar oval.

2.4 Slant E_s and Lacuna

Plasma instabilities have been reported in the E and lower F layers at auroral latitudes. Ionograms from the ground based stations (12) exhibit characteristic changes in the lower F-region and in the E-region, which are referred to as "slant E_s". These features have been confirmed by aircraft studies (2). Similar features are also observed at antarctic stations (13).

The first indication of these disturbances is seen on the ionogram as a weak slant E_s trace which grows out of the normal E-layer. This indicates a region of intense turbulence in the E-region which produces the oblique E-region echoes. When the oval moves overhead the so-called 'lacuna' phenomenon suddenly appears. At this time the trace from the upper part of the E-layer and from the lower F-layer is greatly attenuated and may disappear altogether. Any traces that are visible show considerable spreading in height. The attenuation is most commonly observed on waves reflected from the peak of the F_1 down to the upper part of the E-layer, although occasionally even the F_2 layer is also affected. The presence of the lower part of the E-layer in its normal form indicates that the loss of signal (lacuna) is not due to conventional absorption in the D-layer. At the end of the event the E and F_1 echoes suddenly reappear in their undisturbed form.

It is generally accepted that the lacuna arises from plasma instabilities (12). In the auroral zone the electric fields are high and may be sufficiently large to drive the plasma near to its limit of stability. Any small local increase in field will then induce a plasma instability which can scatter energy from the radio waves passing through the unstable region. Thus, a wave reflected above the disturbed region will be attenuated by this scattering mechanism. A similar phenomenon, called wide band attenuation has been noted during artificial modification of the ionosphere at mid latitudes with high power radio waves (14). Similar attenuation effects have recently been induced in the high latitude ionosphere by means of the new MPI Lindau heating facility at Ramfjord. A discussion of these new results which employed the Leicester University HF Doppler experiment as the diagnostic are discussed by Dr. Stubbe elsewhere in this volume.

2.5 Magnetic Noon Behaviour

The F-layer, in both Northern and Southern polar regions, exhibits anomalous behaviour in the magnetic noon sector. This was reported by Oguti et al (15) who noted that the F_2 layer critical frequency can jump suddenly by several MHz, remaining large for one to five hours before reverting to its normal value. The anomaly seldom occurs outside $\pm$ 2 hours from local magnetic noon and was a prominent feature of the diurnal variations at the polar stations during the IGY. For example, at South Pole station the highest median value of f_oF_2 for the day occurred at magnetic noon, see Fig. 5. The anomaly is most evident between invariant latitudes of 74 to 78° but examples have been noted at stations as low as 60° invariant latitude during periods of intense magnetic activity. The effect is most pronounced in the equinox months.

Deshpande (16) has suggested that the enhancements are related to the position of the cusps in the magnetosphere. Whittaker (17) has reviewed the ionospheric effects associated with the magnetospheric cleft using both ground based and satellite data. This confirms the previous work of Titheridge (18) who shows that the plasma temperature above the cleft ionization is abnormally large as indicated by the increase in scale height seen in topside ionograms. He also stresses that the effects above hm F_2 are much larger than those seen in the variation of the critical frequency. The phenomenon can thus be present without any obvious change in f_oF_2.

Heikkila et al (19) have shown that electrons and protons with energies less than 1 keV can enter the earth's atmosphere through the high latitude cusp in the magnetosphere. These particles reach the ionosphere at invariant latitudes between 70 and 80° and can contribute significantly to the F-layer ionization content. Isis II topside ionograms show that cusp related electron density enhancements maximize at about 77.5°N geomagnetic latitude near local noon during quiet magnetosphere conditions. The cusp enhancement is approximately a factor of 2 and 14 at the heights of h_{MAX} and 1400 km respectively. This requires in situ production at the layer maximum and above, coupled to an upward expansion due to the high plasma temperatures found beneath the cusp (20). The enhanced ionization can produce complicated features on ground based ionograms such as those discussed by Stiles et al (21).

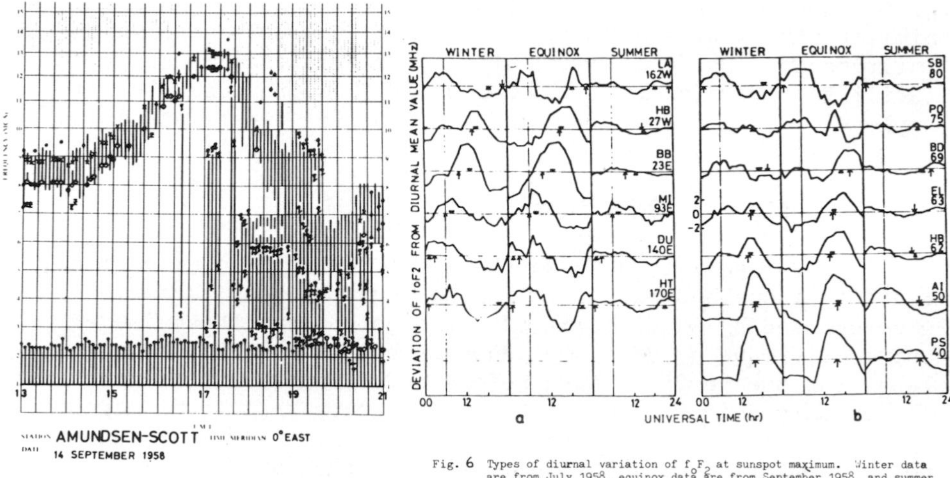

Fig. 5 Expanded time scale f plot showing a typical magnetic noon enhancement at South Pole station recorded on September 14, 1958 (After Dudeney and Piggott, 1978)

Fig. 6 Types of diurnal variation of f_oF_2 at sunspot maximum. Winter data are from July 1958, equinox data are from September 1958, and summer data are from December 1958. The arrows indicate the UT of local solar noon, and the triangles that of local magnetic noon. (After Dudeney and Piggott, 1978)

2.6 General Morphology of F-layer

The general features of the F-layer at arctic and antarctic stations are found to be similar in both hemispheres (22) but the antarctic F-region exhibits some interesting anomalies (23). In winter the f_oF_2 reaches its maximum value at local noon with low values at night. In summer, however, low values are recorded during the day with high values at night.

Another remarkable feature of the diurnal variation of f_oF_2 over the polar cap is the apparent dependence on universal time rather than on local solar time (24,25). The height of the F-layer also exhibits a similar UT dependence (26,27).

The occurrence of the layer maximum at various times of day can be explained by reference to the physical mechanisms which produce the ionization. At local noon, production by solar ultra-violet radiation is a maximum and consequently a maximum in electron density is observed at this time at latitudes up to about 75°. If particles entering the ionosphere through the magnetospheric cleft contribute significantly to the ionisation density, then their contribution would be most significant near local magnetic noon. This would account for high values of f_oF_2 near magnetic noon. The UT dependence of f_oF_2 is related to ionization changes produced by neutral winds. The direction of the winds varies according to UT and so a UT dependence is imposed on the ionization density.

The three maxima in f_oF_2 are clearly evident in the data reproduced in Fig 6a and b. In Fig.6(a) variations are shown for selected stations in the latitude range 65 to 78° ranked in order of geographic longitude and plotted in UT. In Fig.6(b) the stations are chosen so that 06.00 UT is well separated from local noon and the stations are ranked in invariant latitude. Clearly defined maxima in f_oF_2 at local magnetic noon are present in the data from Little America, South Pole, Byrd and Hallett. The 06.00 UT maximum is a prominent feature in the Scott Base, South Pole and Little America data. The relative magnitudes of the maxima show a seasonal and latitudinal dependence. Similar behaviour also occurs in hm F_2. For a detailed discussion of these Antarctic anomalies the reader is referred to the review paper by Dudeney et al (28).

3. F-REGION IRREGULARITIES

The high latitude F-region contains irregularities in electron density varying from small scale structures (29) to large features such as the polar trough (4). When a radio wave from an extra-terrestrial source passes through a region of the ionosphere containing irregularities in electron density the propagation characteristics of the wave are modified. The layer acts as a diffracting screen (30) and the emergent wave's phase and amplitude vary with time according to the properties of the irregularities forming the diffracting region. By measuring these 'scintillation' effects on waves originating from extra terrestrial sources or satellite borne transmitters, a detailed examination of irregularities in the polar F-layer has been possible (31,32).

Intense scintillations are observed over the entire polar caps. In the northern hemisphere the effects appear to be most intense within the auroral oval and at latitudes greater than 80°N (33). The results obtained by Aarons (31) are shown in Fig.7.

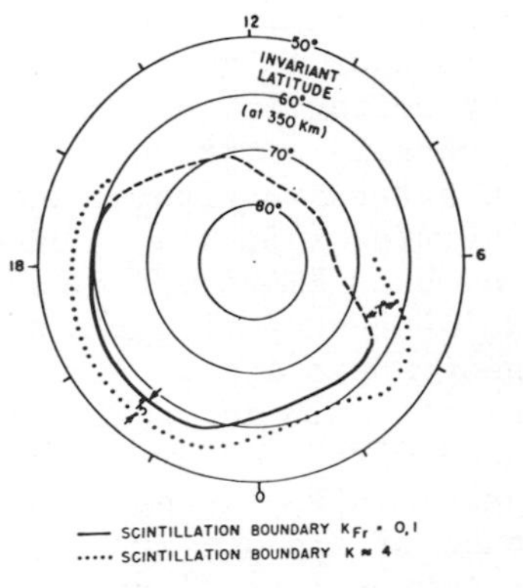

Fig. 7 Boundaries of the F layer irregularity region (After Aarons, 197?)

In the southern polar regions a more complicated geographical distribution is observed and no clear enhancement of scintillation appears in any particular sector of the auroral zone (34).

Contours of constant scintillation have been found to be orientated parallel to lines of constant L values (35). The region of intense scintillation is sharply limited on its equatorward side, this limit being known as the scintillation boundary (36). This boundary does not always correspond to the auroral oval boundary or to the boundary of the F-layer irregularity zone (FLIZ) (29). However, similarities between the boundary of the scintillation region and the lowest latitudes where HF radio backscatter signals are received from field aligned irregularities have been reported (37). The scintillation boundary is frequently co-located with the polar trough with the southern boundary at about 60° invariant latitude in the evening and at about 70° in the morning. Martin et al (38) find that the irregularities producing scintillation lie both poleward and equatorward of the visible aurora. The scintillations peak when the radio wave path passes through the visible aurora and are produced by elongated elliptical irregularities in the electron density distribution.

There appears to be some diurnal variation in the scintillation intensity with more intense effects observed during nighttime. The scintillation also exhibits a clear dependence on magnetic activity with scintillations increasing with Kp especially during daytime. The height of the irregularities has been determined from satellite beacon observations and in the polar regions these lie in the range 200 to 500 km. Some variation in the height of the irregularities is observed depending on time of day, season and Kp.

The electron density irregularities can also be investigated by radar methods. Frequencies from HF to UHF will be scattered from an irregularity provided the electron density gradients associated with the irregularity are large enough and that the radio wave direction is orthogonal to the geomagnetic field line at the irregularity. The radar reflections are called "radar aurora" echoes since it was initially thought that the scattering regions were co-located with the visible aurora. It is now evident that the radar aurora(E-region) is associated with the presence of large electric fields which can produce plasma instabilities (39,40). Most investigations of the radar aurorahave been made at VHF and UHF frequencies (41,42) since it is easy to generate a narrow radar beam which is not affected by refraction effects in the ionosphere. These frequencies are generally scattered by irregularities at E-region heights and an extensive literature exists on the behaviour of these irregularities (43,44). A new generation of twin common volume radars such as STARE and SABRE (45) are able to measure plasma flows and hence the electric field intensities simultaneously over a wide geographic area. Less attention has been given to radar studies of radio aurora in the F-region due to the difficulties associated with the lower radio frequencies required. However, some measurements have been reported (46,47) which confirm the presence of field aligned structures in the F-region which produce 'auroral type' echoes. These features are thought to be associated with spread F features noted on ionograms. The generation of the F-region field aligned irregularities is not well understood although plasma instabilities associated with strong electric fields are thought to be an important factor in their generation. The relationship between the auroral radar features in the E and F region is not well established, however a comparison of F-region plasma drifts and E-region irregularities in the auroral zone has been presented (48).

In addition to the field aligned irregularities described above, travelling disturbances(TIDs) are also observed in the polar F-layer. These wave-like features have periods ranging from 20 min to one or two hours and velocities in the range from a few metres per sec to about 1 km per sec. Observation of TIDs at high latitudes are rather sparse but a typical record obtained from the Leicester-Tromso University HF Doppler experiment at Tromso is reproduced in Fig.8(a). The TIDs are produced by the propagation of internal gravity waves in the upper atmosphere (49) which distort the isoionic contours. It is these distortions that are observed by radio techniques such as the HF Doppler experiment (50). The auroral zone is thought to be a major source of gravity waves since energy from Joule heating by the electro jets and from particle precipitation events, is dissipated in the form of internal gravity waves. High latitude observations enable the features of the waves to be determined before they have been modified by dissipation and dispersion effects which can greatly influence the wave as it propagates to great distances. Recent observations of TIDs at Tromso (51) have confirmed the presence of both medium and large scale disturbances and the vector velocities of the TIDs have also been measured. The direction of propagation exhibits a strong dependence on the direction of the neutral wind, the TID direction being in anti-phase to

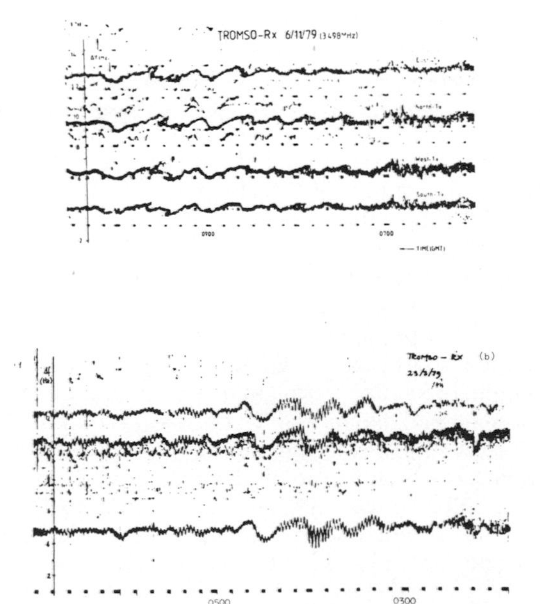

Fig. 8 (a) Medium scale travelling irregularities observed at Tromso
(b) Short period oscillations superimposed on medium scale disturbances, Tromso

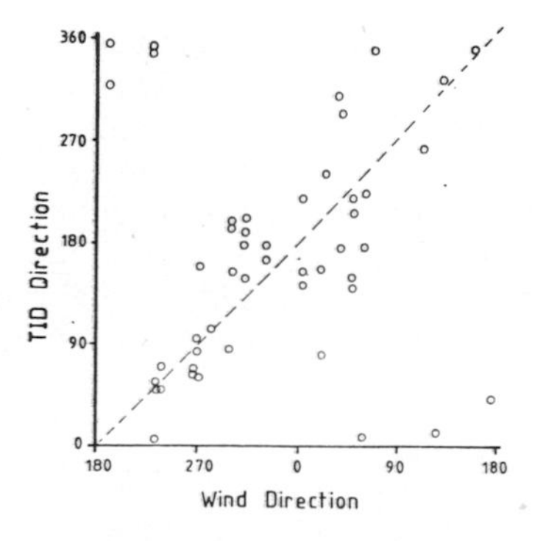

Fig. 9 Plot of TID and corresponding neutral wind directions for Tromso 1979 data, showing the 180 degree phase difference in the two parameters.

that of the wind as indicated in Fig.9. The filtering effect of the neutral wind is clearly evident and it appears that sources outside the auroral zone have to exist in order to account for the northward travelling irregularities observed.

A common feature of the Doppler observations at high latitudes is the presence of short period oscillations ($\tau \sim 1 - 2$ min), see Fig 8(b). These events occur simultaneously over large areas of the ionosphere and are sometimes exactly correlated with similar oscillations observed in the magnetometer records. Although the periods correspond to those of acoustic waves there is no evidence that these are propagating waves. It appears that the features are produced by bulk movements of the F-layer ionization, possibly due to changes in the magnetic field intensity.

In addition to the naturally occurring disturbances, it is possible to induce disturbances artificially in the polar F-region by techniques such as high power radio wave heating and the release of ion clouds from rockets. Some of these man-made disturbances are discussed in another chapter of this volume.

4. INCOHERENT SCATTER OBSERVATIONS

The F-layer observations so far discussed have been undertaken by coherent reflection or scatter of radio waves. Thus the information obtained relates only to the electron density distribution in the region which affects the radio wave characteristics. The free electrons in a plasma are capable of scattering electromagnetic radiation incident on them. The first suggestion that it would be possible to detect such signals from the electrons in the ionosphere was made by Gordon in the early 1950's (52) and the first experimental results from such an experiment were reported by Bowles (53). The early results did not agree with free electron scattering theory due to the influence of the positive ions in the ionosphere. A revised scattering theory was presented (54,55) which satisfactorily accounted for the observed results. The intensity and distribution of the spectrum of the scattered signal depends on the density and temperature of both electrons and ions and the incoherent scatter technique is thus capable of measuring a much greater number of ionospheric parameters than the conventional coherent type of radio experiment. The principal parameters measured by the incoherent scatter radars both directly and as derived quantities are shown in Table 1. It is not possible in this presentation to discuss the technique of incoherent scatter in detail but reference can be made to several review papers (56,57,58

The first incoherent scatter observations of the high latitude F-region were made using the facility at Chatanika (59,60). Electron density height profiles obtained with the radar during

Table 1 List of directly measured and derived parameters from the Incoherent Scatter technique. Most of the parameters can be measured at a range of heights.

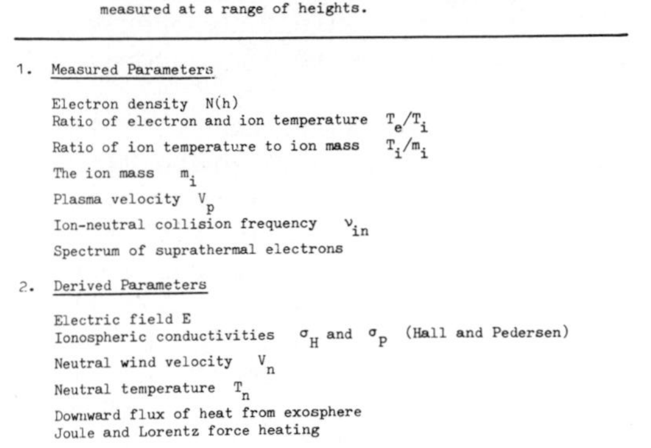

1. Measured Parameters

Electron density N(h)
Ratio of electron and ion temperature T_e/T_i
Ratio of ion temperature to ion mass T_i/m_i
The ion mass m_i
Plasma velocity V_p
Ion-neutral collision frequency ν_{in}
Spectrum of suprathermal electrons

2. Derived Parameters

Electric field E
Ionospheric conductivities σ_H and σ_P (Hall and Pedersen)
Neutral wind velocity V_n
Neutral temperature T_n
Downward flux of heat from exosphere
Joule and Lorentz force heating

disturbed conditions are reproduced in Fig.13(a,b). High energy particle precipitation was observed during the large magnetic storm of August 1972 and the unusual feature in the first data set is the well defined D-layer. In the later measurements this peak has disappeared and a considerable enhancement occurs above the F-layer peak. More typical conditions are shown in Fig.13(b) where no ledges are present below the E-layer maximum. The first profile corresponds to diffuse auroral conditions existing before a large discrete electron precipitation event. The broken curves represent the measured profile during the event and illustrate the large, rapid changes that can occur in electron density during this type of disturbance (61). Further studies of electron density together with ion velocity measurements during an isolated sub-storm have been reported (62). During auroral disturbances considerable amounts of energy can be deposited into the high latitude ionosphere. The Chatanika facility has been extensively

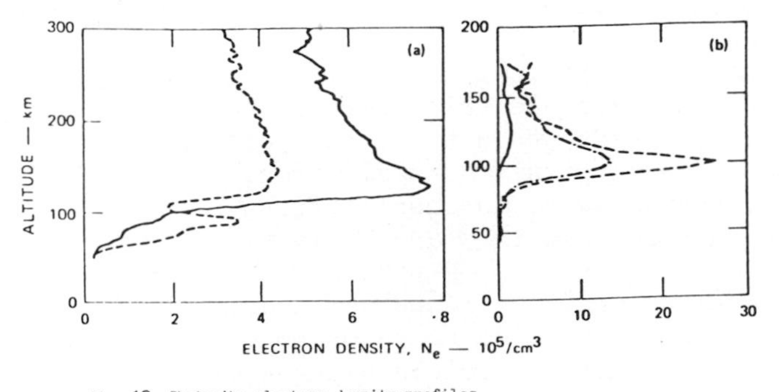

Fig. 10 Chatanika electron density profiles
(a) August 4, 1972, during a large PCA event, dashed line 22.10 - 22.20 UT, solid line 22.40 - 22.49 UT.
(b) March 21, 1973 before and during large discrete electron precipitation event, solid line 08.04.37 - 08.07.12 UT during diffuse aurora before event, dot dashed line 08.08.51 - 08.08.57 UT and dashed line 08.08.57 - 08.09.03 UT indicate the rapid time variations during the event. (After Wickwar, 1975)

used for studies of energy input due to particles and Joule heating (62). Brekke (63) has used the Chatanika facility to determine the relative importance of Joule and Lorentz force heating in the generation of gravity waves in the auroral electrojet regions.

A comparison of electron (Te) and ion (Ti) temperatures measured using the Chatanika facility, with probe measurements on the OGO 6 satellite has been presented elsewhere (65). The results indicate that the radar observations are in very good agreement with the satellite values even in the auroral zone. The influence of anomalous transport coefficients on the thermal structure of the storm time ionosphere has been investigated (66) and can strongly influence the thermal structure of the disturbed polar ionosphere. The anomalies are a consequence of plasma turbulences which are a well known feature of the polar F-layer.

The Chatanika facility has also been used for measurements of auroral zone electric fields, conductivities and plasma flows. The techniques employed for measuring these derived parameters are discussed elsewhere (67). The distribution of these parameters has been successfully measured in the invariant latitude range 63 to 68°N for a wide range of geophysical activity. Kamide et al (68) report downward field aligned currents with densities of 10^{-6} to 10^{-5} A m^{-2} during the evening sector with a reversal of the field direction in the morning sector. During periods of increased geomagnetic activity the observed current densities are enhanced and at times the current flows are reversed.

A new incoherent scatter radar, EISCAT is now being constructed in Northern Scandinavia. This system consists of a monostatic VHF radar and a tristatic UHF radar. It represents a powerful and flexible tool for measurements within the approximate latitude range 60 to 80°N (L = 6 to 10). EISCAT and its scientific objectives have been discussed elsewhere (69). Among the topics to be investigated are:

(a) Ionospheric storm effects.
(b) Magnetospheric electric field intensities.
(c) Field aligned currents and the polar wind.
(d) Measurements within auroral forms.
(e) Atmospheric waves and oscillations.
(f) Seasonal changes in D and F layer.

The research areas outlined above are not exhaustive but give some indication of the wide range of ionospheric and magnetospheric investigations that are possible using EISCAT.

REFERENCES

1. Pike, C.P.: 1971, J. Geophys. Res. 76, pp 7745.

2. Wagner, R.A. and Pike, C.P.: 1972, AGARD Conf. Proc. CP97, ch.4.
3. Thomas, J.O. and Andrews, M.K.: 1969, Planet.Space Sci. 17,pp 433.
4. Muldrew, D.B.: 1965, J. Geophys. Res. 70, pp 2635.
5. Petrie, L.E.: 1966, in Spread F and its effects on radio communications (Technivision Press).
6. Akasofu, S.I.: 1977, Physics of Magnetospheric Substorms (Reidel Pub. Co.).
7. Nelms, G.L. and Lockwood, G.K.E.: 1967, in Space Research VII (Pub. North Holland).
8. Bellchambers, W.H. and Piggott, W.R.: 1960, Proc. Roy. Soc. A256, pp 200.
9. Bellchambers, W.H., Barclay, L.W. and Piggott, W.R.: in The Roy. Soc. IGY Expedition, Halley Bay 1955-9, 2, pp 179.
10. Bowman, G.G.: 1968, J. Atmos. Terr. Phys. 30, pp 1115.
11. Bowman, G.G.: 1969, Planet. Space Sci. 17, pp 777.
12. Olesen, J.K.: 1975, Interpretation of high latitude ionograms, Danish Meteorological Institute, Lyngby.
13. King, G.A.M. and Savage, P.J.: 1973, J. Atmos. Terr. Phys. 35, pp 363.
14. Utlaut, W.F. and Violette, E.J.: 1974, Radio Sci. 9, pp 895.
15. Oguti, T. and Marubashi, K.: 1966, Rep. Ions. Space Res. Jap. 20, pp 96.
16. Deshpande, M.R.: 1973, Nature, 244, pp 109.
17. Whittaker, J.H.: 1976, J. Geophys. Res., 81, pp 1279.
18. Titheridge, J.E.: 1976, J. Geophys. Res., 81, pp 3221.
19. Heikkila, W.J. and Winningham, J.D.: 1971, J. Geophys. Res. 76, pp 883.
20. Chacko. C.C. and Mendillo, M.: 1977, J. Geophys. Res. 82, pp 4757.
21. Stiles, G.S., Hones, E.W. and Winningham, J.D.: 1977, J. Geophys. 82, pp 67.
22. Shapley, A.H.: 1956, IGY Geophys. Monogr. Ser. (AGU) 1, pp 86.
23. Knecht, R.W., McDuffie, R.E. and Aono, Y.: 1961, Ann. Int. Geophys. Year XI, pp 220.
24. Piggott, W.R. and Shapley, A.H.: 1962, Antarctic Res. Geophys. Monogr. 7, pp 111.
25. Duncan, R.A.: 1962, J. Geophys. Res. 67, pp 1823.
26. Patton, D.E., Peterson, V.L., Stonehocker, G.H. and Wright, J.W.: 1965, Geomag. & Aeronomy, Antarctic Res. Series, 4, pp 47.
27. Millman, G.H.: 1972, AGARD Conf. Proc. CP97, ch. 31.
28. Dudeney, J.R. and Piggott, W.R.: 1978, in Upper Atmospheric Research in Antarctica (Ed. Lanzerotti & Park) Pub. Am. Geophys. Union, 29, pp 200.
29. Dyson, P.L.: 1969, J. Geophys. Res. 74, pp 6291.
30. Ratcliffe, J.E.: 1956, Rep. Prog. Phys. 19, pp 188.
31. Aarons, J.: 1972, AGARD Conf. Proc. No 97, ch. 17.
32. Liszka, L.: 1972, AGARD Conf. Proc. CP 97, ch. 16.
33. Frihagen, J.: 1971, J. Atmos. Terr. Phys. 33, pp 21.
34. Titheridge, J.E. and Stuart, G.F.: 1968, J. Atmos. Terr. Phys. 30, pp 85.
35. Liszka, L.: 1964, Arkiv. f. Geofysik, 4, pp 453.

36. Aarons, J. and Allen, R.S.: 1971, J. Geophys. Res. 76, pp 170.
37. Aarons, J.: 1972, AGARD Conf. Proc. No 97, ch. 29.
38. Martin, E. and Aarons, J.: 1977, J. Geophys. Res. 82, pp 2717.
39. Farley, D.T.: 1963, J. Geophys. Res. 68, pp 6083.
40. Rogister, A. and D'Angelo, N.: 1970, J. Geophys. Res. 75, pp 3879.
41. Abel, W.J. and Newell, R.E.: 1969, J. Geophys. Res. 74, pp 231.
42. Chesnut, W.G., Hodges, J.C. and Leaderbrand, R.L.: 1968, SRI Project Report 5535, Stanford Research Inst., California.
43. Greenwald, R.A., Ecklund, W.L. and Balsley, B.B.: 1973, J. Geophys. Res. 78, pp 8193.
44. Moorcroft, D.R. and Tsunoda, R.T.: 1978, J. Geophys. Res. 83, pp 1482.
45. Greenwald, R.A., Weiss, W., Nielsen, E. and Thomson, N.R.: 1978, Radio Sci. 13, pp 1021.
46. Katz, A.H.: 1972, AGARD Conf. Proc. CP 97, ch. 30.
47. Millman, G.H.: 1972, AGARD Conf. Proc. CP 97, ch. 31.
48. Ecklund, W.L., Balsley, B.B. and Carter, D.A.: 1977, J. Geophys. Res. 82, pp 195.
49. Georges, T.M.: 1967, ESSA Tech. Rep. EIR 57-ITSA 54, US Dept. Commerce.
50. Davies, K.: 1962, J. Geophys. Res. 67, pp 4909.
51. Jones, T.B. and Holt, O.: 1980 (unpublished data).
52. Gordon, W.E.: 1958, Proc. IRE 46, pp 1824.
53. Bowles, K.L.: 1958, Phys. Res. Lett. 1, pp 454.
54. Dougherty, J.P. and Farley, D.T.: 1960, Proc. Roy. Soc.A259,pp 79
55. Hagfors, T.: 1961, J. Geophys. Res. 66, pp 1699.
56. Evans, J.V.: 1969, Proc. IEEE, 57, pp 496.
57. Beynon, W.J.G. and Williams, P.J.S.: 1978, Rep on Prog. in Phys. 41, pp 909.
58. Evans, J.V., Oliver, W.L. and Salah, J.E.: 1979, J. Atmos. Terr. Phys. 41, pp 259.
59. Leaderbrand, R.L., Baron, M.J., Petriceks, J. and Bates, H.F.: 1972, Radio Sci. 7, pp 747.
60. Rino, C.L.: 1972, Radio Sci. 7, pp 1049.
61. Wickwar, V.H.: 1974 in Atmos. of Earth & Planets (Ed.McCormack) Reidel Pub. Co. pp 110.
62. Watkins,B.J. and Belon,A.E.: 1978, J.Atmos.Terr.Phys. 40, pp 559
63. Hunsucker, R.D.: 1977, J. Geophys. Res. 82, pp 4826.
64. Brekke, A.: 1979, J. Atmos. Terr. Phys. 41, pp 475.
65. McClure,J.,Hanson,W.,Nagy,A., Cicerone,R., Brace,L, Baron,M, Bauer,P., Carlson,H., Evans,J., Taylor,G., and Woodman,R.: 1973, J. Geophys. Res. 78, pp 197.
66. Fontheim,E., Ong.R., Roble,R., Mayr,H., Baron,M., Heogy, W., Wickwar,V., Vondrak,R., and Ionson, J.: 1978, J. Geophys. Res. 83, pp 4831.
67. Horowitz, J.L.,Doupnick,J.R. and Banks,P.M.: 1978, J. Geophys. Res. 83, pp 1463.
68. Kamide, Y and Horowitz, J.L.: 1978, J.Geophys.Res. 83, pp 1063.
69. Rishbeth, H.: 1978, Proc.Esrange Symp.,Ajaccio (ESA SP-135) 85.

MODIFICATION OF THE F REGION BY POWERFUL RADIO WAVES

P. Stubbe and H. Kopka

Max-Planck-Institut für Aeronomie
3411 Katlenburg-Lindau 3, F. R. Germany

ABSTRACT: Following a brief description of the Tromsø heating facility, the expected F region modification effects are discussed. A short review is given of the theories explaining these effects, with emphasis on those dealing with instabilities excited by powerful electromagnetic waves and phenomena caused by these instabilities.

1. INTRODUCTION

The Max-Planck-Institut für Aeronomie, in collaboration with the University of Tromsø, is presently building an ionospheric modification installation in the auroral zone, near Tromsø/Norway. This Heating facility generates a CW power of 1.5 MW in the frequency range 2.5 - 8.0 MHz. The antenna gain is 24 dBi, corresponding to a beam width of 14.5 deg. The effective radiated power (ERP), i.e. the product of transmitted power and antenna gain, amounts to about 360 MW. For further details see Stubbe and Kopka (1979). The facility will be completed in summer 1980. Preliminary experiments with highly reduced power have been performed since autumn 1979, mainly with respect to VLF generation by modulation of the polar electrojet (see below).

Modification of the ionosphere by powerful radio waves serves different scientific purposes,

(a) to learn about the natural ionosphere: The electron temperature is changed in a quantifyable and reproducible fashion, and it is observed how the ionosphere reacts upon this imposed change. Thereby, several rate coefficients of aeronomic interest can be determined as a function of electron tem-

C. S. Deehr and J. A. Holtet (eds.), Exploration of the Polar Upper Atmosphere, 83–98.

perature, e.g. chemical reaction rates and energy transfer rates.

(b) to use the ionosphere as a natural plasma laboratory and to perform controlled plasma experiments. For most of the plasma phenomena in question, the characteristic scale length is much less than the ionospheric scale length. Thus, the ionospheric plasma is usually a very homogeneous system, much more so than typical laboratory plasmas.

(c) to study nonlinear electromagnetic wave phenomena: If two or more strong radio waves are transmitted, or if a modulated wave is transmitted, waves at combination frequencies are generated within the ionosphere due to nonlinear interaction. This process is particularly strong in the presence of a polar electrojet current. It may be utilized to generate ULF-waves, i.e. micropulsations, and ELF/VLF-waves in the whistler mode. These waves, as they propagate along the magnetic field into the magnetosphere, may resonate with energetic particles, thereby causing particle precipitation.

In this paper, we shall restrict ourselves to F region modification effects.

2. PARAMETRIC DECAY INSTABILITY

The key to an understanding of many modification effects expected in our experiments and obtained in previous experiments at Boulder (Utlaut 1973) and Arecibo (Gordon and Carlson 1973) lies in the excitation of the parametric decay instability by the heating wave. A plasma instability is termed parametric if its excitation is due to an oscillating electric field. In the decay instability, the electromagnetic heating wave decays into a Langmuir wave and an ion acoustic wave. Like any other parametric three-wave instability, it satisfies the frequency and wave number selection rules

$$\omega_o = \omega_1 + \omega_2, \quad \underline{k}_o = \underline{k}_1 + \underline{k}_2 \tag{1}$$

where the suffices 0,1,2 relate to the electromagnetic mother wave, the Langmuir wave, and the ion acoustic wave, respectively.

2.1 Threshold and growth rate

A natural way to obtain the threshold and the initial growth rate is to derive the electrostatic dispersion relation for a plasma under the action of a monochromatic pump field (Aliev et al. 1966, Kaw et al. 1969, Porkolab 1972) and to find solutions with Im $\omega > 0$ (Fejer et al. 1972a). Another method, physically equivalent, is to employ energy balance considerations (Fejer et al. 1972b). The latter method has the advantage of affording a better physical insight.

We consider two plane waves with fields $\underline{E}_o$, $\underline{E}_1$, wave vectors $\underline{k}_o$, $\underline{k}_1$, and frequencies ω_o, ω_1. The Lorentz force acting on a particle with charge q contains a nonlinear contribution, $\underline{F}_{NL}$ (the so-called ponderomotive force), because $\underline{E}_o$, $\underline{E}_1$ depend on the particle position $\underline{r}$, and $\underline{r}$, in turn, depends on $\underline{E}_o$, $\underline{E}_1$. For sufficiently small displacements,

$$\underline{F}_{NL} \approx q[(\underline{r}_o+\underline{r}_1)\cdot\nabla](\underline{E}_o+\underline{E}_1) + q(\dot{\underline{r}}_o+\dot{\underline{r}}_1) \times (\underline{B}_o+\underline{B}_1) \quad (2)$$

with $\underline{r}_o$, $\underline{r}_1$ the displacements caused by waves 1, 2, respectively. Fejer (1979, and references cited therein) shows that, for the case $\omega_o \approx \omega_1$, $\underline{F}_{NL}$ has the following spectral component at the difference frequency $\omega_o - \omega_1$,

$$\underline{F}_{NLD} \approx [\dot{\underline{r}}_1\cdot(m\,\ddot{\underline{r}}_o - q\,\dot{\underline{r}}_o \times \underline{B})] \cdot (\underline{k}_o - \underline{k}_1)/\omega_o \quad (3)$$

with m the particle mass and $\underline{B}$ the external magnetic field. Since r_o and r_1 are inversely proportional to m, the action of the ponderomotive force on the ions can usually be disregarded. Equ. (3) is valid if the thermal electron velocity is much smaller than the phase velocities of waves 0 and 1. The ponderomotive force $\underline{F}_{NL}$ has the effect to drive electrons from higher to lower time averaged electric fields. This can be clearly seen if (2) is applied to the the case of vanishing external magnetic field, B=0, yielding

$$\underline{F}_{NL} = -\frac{e^2}{2m\omega_o^2}\nabla\langle(\underline{E}_o+\underline{E}_1)^2\rangle \quad (2a)$$

We now interpret wave 0 as the electromagnetic mother wave, and 1 as a Langmuir wave which is initially present as a noise component. Furthermore, we introduce the notations $\omega_2 = \omega_o - \omega_1$, $\underline{k}_2 = \underline{k}_o - \underline{k}_1$. The force $\underline{F}_{NLD}$ generates plasma density perturbations at ω_2 and $\underline{k}_2$, $N_2(\omega_2, \underline{k}_2)$. By interaction of the pump with these density perturbations, a current, $j_1(\omega_1, \underline{k}_1)$, is produced at ω_1 and $\underline{k}_1$. The time average of $-\underline{j}_1\cdot\underline{E}_1$ represents the energy per unit volume and time transferred from the pump to the Langmuir wave. According to Fejer (1979),

$$\langle -\underline{j}_1\cdot\underline{E}_1\rangle = \frac{\varepsilon_o^2\omega_p^4}{8NKT_i\omega_o^3}(\underline{E}_o\cdot\underline{E}_1)^2\, B\,(\alpha_i, T_e/T_i) \quad (4)$$

with ω_p the plasma frequency, N the unperturbed electron density, and $\alpha_i = (\omega_2/k_2)\,(m_i/2KT_i)^{1/2}$. The function B is positive for $\alpha_i>0$ $(B(-\alpha_i) = -B(\alpha_i))$, i.e. for $\omega_o > \omega_1$. In the parameter range $1 \leq T_e/T_i \lesssim 3$, B has a maximum value of approximately 0.6 (Fejer 1979, Fig. 3). Equ. (4) is valid if the external magnetic field vanishes, or if $\underline{E}_o \parallel \underline{B}$, which is the case for an ordinary electromagnetic wave in the vicinity of its reflection point, and $\underline{k}_1 \parallel \underline{B}$. Energy loss of the Langmuir wave is given by

$$L = \frac{\varepsilon_o}{2} E_1^2 \nu \qquad (5)$$

with ν the energy damping decrement. By equating (4) and (5), taking $\underline{E}_o \parallel \underline{E}_1$, $\omega_o = \omega_p$, and B = 0.6, the threshold follows as

$$E_{oth}^2 \approx 6.7 \frac{NKT_i}{\varepsilon_o} \frac{\nu}{\omega_p} \qquad (6)$$

For T_i = 1000 K and N ranging from 10^5 to 10^6 cm^{-3}, E_{oth} is approximately 0.1 to 0.5 V/m. The Tromsø heating facility will exceed these values by about a factor of ten ($E_o \lesssim 3V/m$).

The energy growth rate following from (4) and (5) with (6) is

$$\gamma = \nu \left[\left(\frac{E_o}{E_{oth}}\right)^2 - 1\right] \qquad (7)$$

A typical value for γ is of the order 10^3 s^{-1}, depending, of course, on N and E_o/E_{oth}.

In the derivation outlined above, it was tacitly assumed that energy transfer occurs only between the pump and the Langmuir wave with frequency $\omega_1 = \omega_o - \omega_2$. This is justified as long as $\omega_o + \omega_2$, which is damped rather than excited, lies outside the resonance width, i.e. as long as $2\omega_2$ exceeds the bandwidth of the Langmuir wave excited at ω_1, which is of the order 1 kHz.

2.2 Height range of occurrence

In order to estimate in which height range the parametric instability can be excited, we assume a linear electron density profile

$$N_e \sim \omega_p^2 = \omega_o^2 \, [1+(h-h_o)/H] \qquad (8)$$

which is normalized in such a way that an o-mode pump wave with frequency ω_o is reflected at altitude h_o for vertical incidence. Furthermore, we use the dispersion relation for Langmuir waves in the simplified form

$$\omega_1^2 \approx \omega_p^2 \, (1+3k_1^2\lambda_D^2) \qquad (9)$$

Combining (8) and (9) and using the frequency selection rule (1) in the form $\omega_o = \omega_1$, i.e. neglecting ω_2, we obtain

$$3\, k_1^2\lambda_D^2 = (h_o - h)/H \qquad (10)$$

For Landau damping not to be excessive, $k_1\lambda_D$ should fall into the interval $0 \leqslant k_1\lambda_D < 0.2$. Thus, from (10),

$$h_o - 0.12\,H < h < h_o \qquad (11)$$

This is an altitude interval approximately 10 km thick, situated immediately below the reflection altitude of an o-mode pump wave at vertical incidence. An x-mode wave is usually reflected at lower altitudes and can, therefore, not excite the decay instability. It may, however, excite Bernstein modes (Fejer et al. 1972a).

The altitude range described by (11) contains all wavelengths down to 10-20 cm. For some applications it is useful to know the interval in which Langmuir waves with a given k_1 are excited. Its height is specified by (10), its width is given by $\Delta h = (\partial\omega_1/\partial h)^{-1}\Delta\omega_1$. From (8) and (9), $\partial\omega_1/\partial h \approx \omega_o/2H$. Taking $\Delta\omega_1$ to be of the order of the ion acoustic frequency for k_1, we find

$$\Delta h \approx \frac{2H}{\omega_o} k_1 \left[\frac{K(T_e+3T_i)}{m_i} \right]^{12} \qquad (12)$$

To give an example, $H = 100$ km, $\omega_o = 2\pi\cdot 5$MHz, $\lambda_1 = 2\pi/k_1 = 50$ cm, $T_e = 2000$ K, $T_i = 1000$ K, $m_i = m(O^+)$, we obtain $\Delta h \approx 130$ m.

2.3 Saturation

Several mechanisms have been discussed in literature to evaluate the level up to which the parametrically excited plasma waves grow.

a) Convection (Arnush et al. 1974): In an inhomogenous ionosphere, saturation can occur by the linear process of propagation of Langmuir waves out of the excitation region. However, the saturation levels predicted by this theory are so high that nonlinear processes have to play a major role. Nonetheless, the convection mechanism should form a part of any complete saturation theory.

b) Enhanced collisionless damping (Kruer et al. 1972): The parametrically excited plasma waves modify the velocity distribution by trapping particles such that collionless damping is enhanced. This conclusion was drawn from computer simulation experiments performed for a homogenous plasma. Since plasma inhomogeneities change the phase velocity and, thereby, affect the wave-particle interaction mechanism, it is not obvious that these results can be directly applied to an ionospheric situation. In any case, this mechanism is important for high pump field strengths only, as shown by the simulation experiments ($E_o^2 \gtrsim NKT_e/\varepsilon_o$).

c) Orbit perturbation (Weinstock et al. 1973): The electric fields associated with the parametrically excited plasma waves give rise to particle scattering with the effect that energy is removed from the waves and fed into the particles. The theory predicts that this could be the dominating mechanism if the ratio of collisional to Landau damping is of order unity or less, which is the case at the lower end of the height interval given by (11). Furthermore, if this mechanism dominates, the saturation spectrum should become broad in angle, and even isotropic for large pump powers (see their equ. (50)).

d) Cascading decay instability (Fejer et al. 1973; Perkins et al. 1974; Chen et al. 1975): The Langmuir wave excited by the decay instability can become a mother wave in a secondary decay instability, giving rise to a Langmuir wave with slightly lower frequency and an ion acoustic wave. This process can go on until the last excited Langmuir wave does not exceed the instability threshold. This saturation mechanism is the only one to predict the existence of satellite Langmuir waves, thereby providing a discriminating feature which can be experimentally verified. The satellite wave of n-th order is shifted in frequency by approximately $(1-2n)\,\omega_2$ from the pump frequency ω_o. Another discriminating feature is that, according to this theory, the angular spectrum is narrow, with a width not exceeding 25° with respect to the external magnetic field for typical pump powers. The width is expected to increase with increasing pump power.

2.4 Purely growing mode

The decay instability also exists in the so-called purely growing mode, in which the second decay product is not a propagating low frequency wave, but a non-propagating spatially periodic structure, i.e. Re $\omega_2 = o$. Since the real parts of $\omega_o - \omega_2$ and $\omega_o + \omega_2$ become identical in this case, there are two Langmuir waves involved into which pump energy is fed. Starting with a standing Langmuir wave, $E_1 \sim \sin(\omega_1 t - \alpha) \cdot \sin \underline{k}_1 x$, and a homogeneous pump, $E_o \sim \sin \omega_o t$, Fejer (1979) arrives at the following expression for the energy transfer form the pump to the standing Langmuir wave,

$$\langle -\underline{j}_1 \cdot \underline{E}_1 \rangle = \frac{\varepsilon_o^2 \omega_p^4}{8NK(T_e+T_i)\omega_o^3} \sin\alpha \cos\alpha \qquad (13)$$

In contrast to (4), the power transfer now depends on the phase relation between the pump and the standing Langmuir wave. The factor $\sin\alpha \cos\alpha$ has a maximum value of 1/2 at $\alpha = 45°$ or $225°$. With the energy loss rate $L = \varepsilon_o E_1^2 \nu/4$, the resulting threshold is given by

$$E_{oth}^2 = 4\,\frac{T_e+T_i}{T_i}\,\frac{NKT_i}{\varepsilon_o}\,\frac{\nu}{\omega_p} \qquad (14)$$

Since the factor $4(T_e+T_i)/T_i$, which is at least 8, exceeds the factor 6.7 occurring in equ. (6), the decay instability is excited before the purely growing instability. The energy transfer from the pump to the Langmuir waves excited by the decay instability weakens the pump. Thus, although the thresholds given by (6) and (14) are not too different, the excitation levels may be significantly different.

2.5 Anomalous absorption

The process just mentioned, i.e. energy transfer from the electromagnetic heating wave to the parametrically excited electrostatic waves, gives rise to anomalous absorption of the heating wave which may significantly exceed collisional absorption. Anomalous absorption should be taken into account in a complete saturation theory by self-consistently calculating the electric field strength of the pump wave (DuBois 1973). Useful approximate expressions for the attenuation of the pump wave due to anomalous absorption are given by Perkins et al. (1974).

2.6 Generation of energetic electrons

Electrons can gain (or lose) a significant amount of energy be interacting with Langmuir waves if their velocity v lies in the vicinity of the phase velocity v_p. If the energy change of an electron occurs smoothly, the process can be described as diffusion in velocity space. This was done by Weinstock (1975) and Nicholson (1977). Graham et al. (1976a) clearly worked out the limits of applicability of diffusion theory and demonstrated that for typical modification conditions energy exchange occurs abruptly rather than smoothly. Thus, a velocity diffusion approach is usually of limited value.

Generation of energetic electrons gives rise to the excitation of airglow lines, preferentially the red 6300 Å oxygen line (1.96 eV), to a much lesser extent the green 5577 Å oxygen line (4.17 eV), as reported by Haslett et al. (1974). The unique possibility of switching airglow on and off whenever wanted allows to directly measure the airglow quenching rate. Furthermore, by tracing an airglow patch after switch-off, the neutral wind velocity can be directly measured. This method should be sufficiently accurate for large wind velocities and should, therefore, be appropriate for the Tromsø heating experiments.

Haslett et al. (1974) estimated that the downward flux of electrons with energies exceeding 1.96 eV was of the order 10^8 cm^{-2} sec^{-1} in the Boulder experiments. Such a large flux would very rapidly empty the interaction region, unless compensated by a counter flow of thermal electrons. The corresponding upward drift velocity should be very significant, up to 100 m/sec. This process is relevant if the interaction region and the airglow excitation region do not coincide. Attempts should be made to measure these velocities above and below the interaction region.

It should be mentioned that the theories mentioned above are able to explain sufficiently large electron fluxes to account for the observed airglow intensities. In addition, Langmuir waves generated by another process (see Section 4) could significantly contribute to the generation of energetic electrons, particularly if the heating frequency ω_o equals $2\omega_L$ (ω_L = gyrofrequency).

2.7 Modification of incoherent backscatter spectrum

A radar (ω_R, $\underline{k}_R$) detects electrostatic waves (ω, $\underline{k}$) if the Bragg condition $\underline{k}_R' - \underline{k}_R = \pm \underline{k}$ is satisfied, with $\underline{k}_R'$ the wave vector of the scattered radar wave and + or - for a downcoming or upgoing plasma wave. The corresponding radar Doppler shift is $\Delta\omega_R = \pm\, \omega(\underline{k})$. The backscatter spectrum a radar should measure in the presence of parametrically excited Langmuir and ion acoustic waves is sketched in Fig. 1.

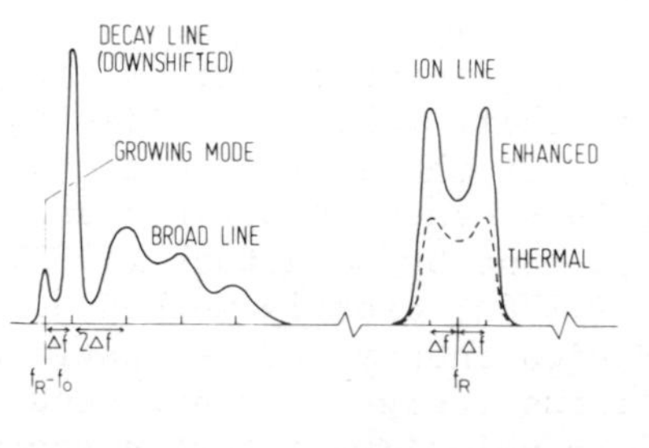

Fig. 1 Schematic illustration of modified incoherent backscatter spectrum. Typical values of Δf for the EISCAT UHF (933 MHz) and VHF (224 MHz) radar are 8 and 2 MHz, respectively

Spectra of this kind have indeed been observed in the Arecibo experiments (e.g. Kantor 1974), including the broad line. Since this line can only be explained by saturation process d) (see Section 2.3), its presence is strongly in support of this process in comparison to the others, at least for low pump powers.

It is still a matter of debate what the Arecibo radar at Θ = 45° (Θ = angle between $\underline{k}_R$ and $\underline{B}$, monostatic operation) really sees, either nonlinearly stabilized Langmuir waves at Θ = 45°, or Langmuir waves enhanced by a sub-threshold pump in a stable plasma at Θ = 45°, or unstable Langmuir waves originally generated at lower Θ with subsequent propagation and refraction to reach Θ = 45°, thereby becoming detectable by the radar. Observation under Θ = 45° is clearly not an optimum situation. Theory (e.g. Perkins et al. 1974) predicts effects several orders of magnitude higher for small Θ. By combining the Tromsø heating facility and the two EISCAT radars (monostatic at 224 MHz, tristatic at 933 MHz, all 4 antennas widely steerable), the accessible angular range is 0° < Θ < 40°. Furthermore, simultaneous observations under different Θ's and k's are possible. Thus, exciting experiments are to be expected. Measurements, which a evidently required, are
- full angular spectrum to compare with theory;
- spectra for different k's to obtain returns from different altitudes within the excitation region;

- spectra for different pump powers to compare the importance of different saturation processes (e.g. the broad line should become smaller in relation to the decay line, and the angular spectrum should become much wider, if saturation processes b) and c) gain importance with increasing pump power).

One should also exploit the diagnostic information contained in the new backscatter lines. After heater switch-off, the damping of the decay line is a measure of the electron collision frequency and, thereby, of the electron temperature. At day, when photoelectrons are present, damping is collisionless, and its rate allows to draw conclusions on the photoelectron spectrum (Yngvesson et al. 1968). The frequency difference between the decay line and the pump contains information on T_e and T_i independent of the information contained in the ion line. Asymmetries of this frequency difference between the up- and downshifted lines are a measure of the macroscopic ion velocity in the direction of $\underline{k}$, but the theory has yet to be worked out. The energetic electrons generated by the decay instability should, like photoelectrons, but unlike these at any time of day and year, excite plasma lines. The difference between these plasma lines and the decay lines is that the first occur at $\omega_1(\underline{k})$ as given by the Langmuir wave dispersion relation, while the latter occur at $\omega_o - \omega_2(\underline{k})$. Thus, the frequency difference between the up- and downshifted plasma line contains information on T_e and the macroscopic electron velocity.

3. LARGE SCALE FIELD ALIGNED IRREGULARITIES

The most obvious F region modification phenomenon observed in the Boulder experiments was spread F on ionograms (e.g. Utlaut 1973), occurring a few seconds after heater switch-on. Using a narrow beam HF phased array radar, Allen et al. (1974) could clearly show that spread F is caused by large scale field aligned irregularities.

Three theories have been developed to explain the generation of large scale field aligned irregularities by a strong eletromagnetic wave.

3.1 Thermal self-focusing instability (Perkins et al. 1974)

Assume a pre-existing infinitesimal field aligned irregularity. Where the electron density N is low, the refractive index is high, causing a concentration of electromagnetic wave energy. Thus, the electron gas is preferentially heated where the density is low, and N is further reduced by thermal expansion. The irregularity grows to finite amplitudes. Threshold is defined by the condition that the temperature changes imposed by the pump are compensated by heat conduction along $\underline{B}$. The theory of Perkins et al. (1974) yields the threshold and growth rate (defined with respect to irregularity amplitude, whereas in Section 2.1 it was defined with respect to energy density) given below:

$$E^2_{oth} \approx 12.6\lambda \frac{(KT_e)^2 N}{\varepsilon_o m_e \nu^2_{ei} k_o H} k_2^2 \cos^2\alpha \qquad (15)$$

with k_2 the irregularity wave number, α the angle between $\underline{k}_2$ and $\underline{B}$, and λ a number of order unity. Equ. (15) is valid if $k_2^2 \cos^2\alpha >> 16\, D^4/H^2$ with $D = k_2^2 H/2k_o$. Squeezing this condition to its limit, i.e. replacing >> by > , we obtain

$$E^2_{oth} \gtrsim 12.6\lambda \frac{(KT_e)^2 N}{\varepsilon_o m_e \nu^2_{ei}} \frac{k_2^8 H}{k_o^5} \qquad (15a)$$

In ionospheric modification experiments, threshold can be exceeded if $\lambda_2 \equiv 2\pi/k_2$ exceeds, say, 500 m. The growth rate corresponding to (15) is

$$\gamma \approx \frac{2KT_e}{m_i \nu_{in}} k_2^2 \cos^2\alpha \left(\frac{E_o^2}{E^2_{oth}} - 1\right) \qquad (16)$$

and gives typical rise times $\tau \equiv 1/\gamma$ of the order seconds.

3.2 Stimulated Brillouin scatter instability (Cragin et al., 1974)

The stimulated Brillouin scatter instability results in the excitation of a second electromagnetic wave (ω_1, $\underline{k}_1$) and an ion acoustic wave (ω_2, $\underline{k}_2$) or, in its purely growing mode, a spatially periodic structure (Re $\omega_2 = 0$, $\underline{k}_2$), by the pump (ω_o, $\underline{k}_o$). The selection rules (1) have to be satisfied. The nonlinear force driving the instability is the ponderomotive force discussed in Section 2.1. However, in order to be applied to large scale irregularities with considerable wave lengths along $\underline{B}$, another nonlinear force has to be incorporated, namely the pressure gradient force set up by differential heating in the interference field of waves 0 and 1. The expression derived by Fejer (1979) from the electron energy equation,

$$\underline{F}_{NL} = - \frac{K\nu\varepsilon_o}{k_{2\parallel}^2 \varkappa} \left(\frac{\omega_p}{\omega_o}\right)^2 \nabla \left\langle (\underline{E}_o + \underline{E}_1)^2 \right\rangle \qquad (17)$$

with $\varkappa$ the thermal conductivity of the electron gas along $\underline{B}$ and $k_{2\parallel}$ the component of $\underline{k}_2$ along $\underline{B}$, turns out to be larger than (2) if $\lambda_2 \equiv 2\pi/k_{2\parallel} > 4\lambda_{MFP}$ where λ_{MFP} is the electron mean free path which is of the order of a few hundred meters.

Cragin et al. (1974) discussed the case of irregularity generation by the purely growing version of this instability. The mechanism, in correspondence to the purely growing decay instability, involves a standing electromagnetic wave, which gives rise to differential heating and, thereby, irregularity amplification, provided the phase relation is appropriate and the selection rule $\underline{k}_o - \underline{k}_1 = \underline{k}_2$ is satisfied. The results are similar to those of the thermal self-focusing theory in that λ_2 should not be less than about 500 m. In a later paper (Cragin et at. 1977), the effect of $\underline{B}$ on the electromagnetic wave propagation was taken into account, and furthermore growth rates were calculated, yielding rise times of a few seconds.

3.3 Stimulated diffusion scatter instability (Berger et al. 1975)

This process resembles the stimulated Brillouin scatter instability. The difference lies in the nature of the low frequency perturbation, which is neither an ion acoustic wave nor a purely spatially periodic structure, but rather a so-called "diffusion quasi-mode". By this, the following is meant: Write down the coupled continuity, momentum, and energy equations for the electron gas. Incorporate in the energy equation a differential heating term, resulting from the interference field of the incoming and the scattered electromagnetic wave. Assume diffusion and heat conduction to be field aligned. Derive a dispersion relation for a low frequency mode (ω_2, $\underline{k}_2$) which satisfies $|\omega_2| << \nu_{ei}$, $k_{2\parallel} \cdot \lambda_{MFP} << 1$, and $k_{2\perp} >> k_{2\parallel}$. The results indicate that for $E^2 = 0$ the mode is purely decaying, with Im $\omega_2 = - \gamma_D$ ($\gamma_D = 2.19\, k_{2\parallel}^2 \cdot \lambda_{MFP}^2\, \nu_{ei}$ = damping rate due to field aligned diffusion), while at threshold it is purely propagating at the same rate, Re $\omega_2 = - \gamma_D$. In contrast to the theories outlined in 3.1 and 3.2, this theory explains the occurrence of field aligned irregularities with scale lengths down to less than 100 m across the magnetic field. The threshold is given by

$$E^2_{oth} = 29 \frac{m_e^3 \omega_L^2 \lambda_{MFP}^2 \nu_{ei} \omega_o}{m_i e^2} \cos^2 \alpha \qquad (18)$$

with ω_L the electron gyrofrequency. This value is easily exceeded in ionospheric modification experiments for α close to 90°. The corresponding growth rate reads

$$\gamma = \gamma_D \left(\frac{E_o^2}{E^2_{oth}} - 1 \right) \qquad (19)$$

Typical rise times are less than a second.

Although, as we have seen, ideas have been developed to understand the excitation of large scale field aligned irregularities, a unifying linear theory appears necessary. Saturation should also be considered. Furthermore, mutual interaction between these and the decay instabilities should be taken into account. On the experimental side, the temporal development of the irregularity spectrum after switch-on and its dependence on pump power should be determined, and drifts should be carefully measured (in particular, it should be found out if there exists a dependence of the drift velocity on λ_2). What we know from previous experiments, employing the satellite beacon scintillation technique (Bowhill

1974), is that the spectrum peaks at λ_2 around 300 - 400 m. The possibility of suppressing naturally occurring irregularities should also be investigated.

4. SHORT SCALE FIELD ALIGNED IRREGULARITIES

Short scale striations with scale lengths of the order meters across the magnetic field have been detected in the Boulder experiments by VHF and UHF radars (Minkoff et al. 1974). Two types of signals were received, a highly aspect sensitive center line at ω_R and two much less aspect sensitive plasma lines at $\omega_R \pm \omega_o$. At UHF, the different lines turned out to be about equally strong, while at VHF the center line was much stronger than the plasma lines.

4.1 Theoretical explanation

Das et al. (1979) formulated a theory in which certain shortcomings of previouns theories were removed. Assume a pre-existing infinitesimal stationary striation $N_2(\underline{k}_2)$. The electromagnetic wave $(\omega_o, \underline{k}_o)$ impinging upon the striation generates Langmuir waves $(\omega_1, \underline{k}_1)$ by linear mode conversion. These propagate in the direction of $\nabla N_2 \parallel \underline{k}_2 \perp \underline{B}$, and their frequency is $\omega_1 = \omega_o$ provided the striations do not move or propagate. Differential heating occurs in the interference field of waves 0 and 1, having a zero frequency component proportional to

$$Q(\underline{k}_2) \sim E_o^2 \, N_2 \, \frac{\omega_o^2-\omega_r^2}{(\omega_o^2-\omega_r^2)^2+\nu^2\omega_o^2} \qquad (20)$$

where ω_r is the Langmuir wave resonance frequency. Since Q should be positive where N_2 is negative in order to enhance the striation, $\omega_o < \omega_r$ is required. In the cold plasma approximation, $\omega_r = (\omega_p^2 + \omega_L^2)^{1/2}$ or $\omega_r = n\omega_L$ $(n = 2,3,...)$. In the first case, i.e. resonance at the upper hybrid frequency, ω_r increases with altitude. Thus, enhancement of the striation occurs above, damping below the resonance height. A net enhancement results if, at the resonance height, E_o^2 increases with altitude within its standing wave structure; otherwise, damping occurs. Since the phase of the standing wave in relation to the resonance height cannot be expected to remain constant in view of the variable nature of the ionosphere, amplification and damping follow each other. It is conceivable, therefore, that an almost constant time averaged striation level results without involving any saturation mechanism, where the averaging has to be performed over a few "amplification-damping-cycles". If this is a correct view, the striation level should decrease with increasing ionospheric variability which should be easy to test experimentally.

The case $\omega_r \approx n\omega_L$ is different from the previous case in that the resonance frequency changes much less with altitude. Thus, the height range in which resonance occurs is much wider, and this can result in a significantly reduced threshold. A further difference is that this instability can also be excited by an x-mode pump, while the first can only be excited by an o-mode pump, because the resonance height lies above the reflection height of the x-mode.

4.2 Airglow enhancement

Airglow enhancement due to energetic electrons generated by Langmuir waves, which are excited through the decay instability, was discussed in Section 3.6. The Langmuir waves generated in conjunction with striations can cause the same effects, and since they are expected to be particularly strong for $\omega_o \approx n\omega_L$, airglow should also be strong for heating at a multiple of the gyrofrequency. Strong airglow for $\omega_o = 2\omega_L$ has indeed been observed (Haslett et al. 1974). Theoretically, electron acceleration by Langmuir waves propagating perpendicular to $\underline{B}$ has been treated by Nicholson (1977) as a diffusion process in $v_\perp$-space.

4.3 Wide band absorption

Since mode conversion of an electromagnetic wave into a Langmuir wave by the striations is a linear process, it affects a weak diagnostic wave as much as it affects the strong heating wave. Thus, an electromagnetic wave reaching the altitude where the striations are exited suffers absorption through scattering into Langmuir waves. If the diagnostic wave with frequency ω has the same polarization as the heating wave, it reaches the excitation height provided $\omega > \omega_o$. Thus, a wide frequency band is affected by scattering losses. In the usual case, i.e. resonance at the upper hybrid frequency, only o-mode waves are affected by wide band absorption. Since Langmuir wave generation by mode conversion requires an electric field component parallel to ∇N_2, anomalous absorption should show a distinct dependence on the dip angle I and thus be larger at higher latitudes.

Theoretically, wide band absorption has been studied by Graham et al. (1976b). Measurements of wide band absorption with an ionosonde (Utlaut et al. 1971) showed an attenuation of about 10 dB which was reached within about 10 sec after heater switch-on. It thus appears that the rise time of the instability discussed above is of the order seconds, and that the Langmuir wave level could well exceed the level excited by the decay instability.

4.4. Plasma line overshoot

The reduction of the pump power a few seconds after heater switch-on due to the attenuation process described above should lead to a reduction of the decay line intensity in the backscatter spectrum. This phenomenon has indeed been observed (Showen et al. 1978). The plasma line builds up within milliseconds, but drops, within about 10 sec, to a new level which is typically a factor of 10 lower. Phenomenologically, this phenomenon may be described as overshoot, but physically it is rather a damping process, and a manifestation of the fact that the decay instability and the "striation instability" have different rise times.

Perkins (1974) described the generation of striations in terms of a parametric instability, excited by Langmuir waves which, in turn, were excited as a result of the decay instability. In view of the overshoot phenomenon, this process should probably be ruled out, but it may support the striation build-up in the initial phase.

5. F REGION HEATING BY RADIO WAVES

In the D region of the ionosphere, the energy flux of the heating wave and the ionospheric absorption coefficient are large, and consequently a large electron temperature increase will result. For the Tromsø heating experiments we expect T_e to be raised by up to a factor 15 around about 85 km (Stubbe at al. 1979). In the F region, the energy flux is at least a factor of 10 lower. A significant electron temperature enhancement is nonetheless expected. The reason is that, for a given flux, the ratio of heat input by the wave to heat loss by inelastic collisions is much larger in the F region. Furthermore, anomalous absorption can occur in the F region. Both factors together could easily overcompensate the smaller energy flux values in the F region.

5.1 Electron temperature enhancement in the F region

The electron temperature in the F region will be much more enhanced at night than at day because of the absence of D region absorption, and more by o-mode than by x-mode heating because of anomalous absorption (see Sections 2.5 and 4.3). Theoretical estimates (Stubbe et al. 1979) show that electron temperature enhancements of about 2000 - 3000 K are to be expected for the conditions of the Tromsø experiments, with rise and fall times of the order 30 - 60 sec. The ion temperature may be enhanced by 100 - 200 K.

There are several mechanisms for N_e to be changed when T_e is enhanced. The recombination coefficients for O_2^+ + e and NO^+ + e decrease with increasing T_e. Thus, N_e will be enhanced in the lower F region. In the O^+ dominated portion of the F region, enhanced vibrational excitation of N_2 (due to increased T_e or energetic electrons) will give rise to an enhanced reaction rate of $O^+ + N_2 \rightarrow NO^+ + N$ and, thus, to a reduction of N_e. Furthermore, the ambipolar diffusion coefficient will be enhanced, leading to a depression of the layer height and electron density.

By measuring $T_e(t)$ after heater switch-on, one can deduce the anomalous absorption coefficient. From $T_e(t)$ after switch-off, electron cooling rates can be directly inferred. From $N_e(t)$ and $T_e(t)$, the reaction rates for O_2^+ + e, NO^+ + e and $O^+ + N_2$ can be derived as a function of T_e.

In the F region, the ratio N_e/N_n exceeds 10^{-3}. Thus, a slight neutral gas heating becomes possible by energy transfer from the plasma to the neutrals. Attempts will be made to excite internal gravity waves and infrasonic waves by periodic heating. For $T_i - T_n$ = 100 degs, $N_e = 10^6$ cm^{-3}, the heat input to the neutral gas amounts to a temperature change of $\Delta T_n \approx 10^{-1}$ degs/sec or $\Delta T_n \approx 60$ degs/10 min. Although this value will be decreased by thermal conduction, it sould be possible to excite easily detectable gravity waves.

5.2 Airglow reduction

If the only heating effect is an increase of T_e, and no energetic electrons are generated, as can be realized by x-mode heating, the airglow intensity will be suppressed (Haslett et al. 1974). This can be easily understood. In the absence of energetic electrons, the red oxygen line is excited by dissociative recombination $O_2^+ + e \rightarrow O(^1D) + O(^3P)$. With increasing T_e, the recombination coefficient α decreases. Thus, the number density $n(O(^1D))$ decreases, too. Later in the process, when $n(O_2^+)$ begins to increase due to reduced α, the $n(O(^1D))$ decrease will be slowed down and finally be turned into an increase. From the resulting temporal variation of the red line intensity, on may again determine α as a function of T_e.

REFERENCES

Aliev, Yu.M., Silin, V.P., and Watson, C. 1966, Sov. Phys. JETP, 23, 626.
Allen, E.M., Thome, G.D., and Rao, P.B. 1974, Radio Sci., 9, 905.
Arnush, D., Fried, B.D., and Kennel, D.F. 1974, J. Geophys. Res., 79, 1885.

Berger, R.L., Goldman, M.V., and DuBois, D.F. 1975, Phys. Fluids, 18, 207.
Bowhill, S.A. 1974, Radio Sci., 9, 975.
Chen, H.C., and Fejer, J.A. 1975, Phys. Fluids, 18, 1809.
Cragin, B.L., and Fejer, J.A. 1974, Radio Sci., 9, 1071.
Cragin, B.L., Fejer, J.A. and Leer, E. 1977, Radio Sci., 12, 273.
Das, A.C., and Fejer, J.A. 1979, J. Geophys. Res., 84, 6701.
DuBois, D.F., Goldman, M.V., and McKinnis, D. 1973, Phys. Fluids, 16, 2257.
Fejer, J.A., and Leer, E. 1972a, Radio Sci., 7, 481.
Fejer, J.A., and Leer, E. 1972b, J. Geophys. Res., 77, 700.
Fejer, J.A. 1979, Rev. Geophys. and Space Phys., 17,135.
Fejer, J.A., and Kuo, Y.Y. 1973, Phys. Fluids, 16, 1490.
Fejer, J.A., and Graham, K.N. 1974, Radio Sci., 9, 1081.
Gordon, W.E., and Carlson, H.C. 1973, p. 4/1 in AGARD Conference Proceedings No. 138. Edinburgh.
Graham, K.N., and Fejer, J.A. 1976a, Phys. Fluids, 19, 1054.
Graham, K.N., and Fejer, J.A. 1976b, Radio Sci., 11, 1057.
Haslett, J.C., and Megill, L.R. 1974, Radio Sci., 9, 1005.
Kantor, I.J. 1974, J. Geophys. Res., 79, 199.
Kaw, P.K., and Dawson, J.M. 1969, Phys. Fluids, 12, 2586.
Kruer, W.L., and Dawson, J.M. 1972, Phys. Fluids, 15, 446.
Minkoff, J., Kugelman, P., and Weissman, I. 1974, Radio Sci., 9, 941.
Nicholson, D.R. 1977, J. Geophys. Res., 82, 1839.
Perkins, F.W., Oberman, C., and Valeo, E.J. 1974, J. Geophys. Res., 79, 1478.
Perkins, F.W., and Valeo, E.J. 1974, Phys. Rev. Lett., 32, 1234.
Perkins, F.W. 1974, Radio Sci., 9, 1065.
Porkolab, M. 1972, Nucl. Fusion, 12, 329.
Showen, R.L., and Kim, D.M. 1978, J. Geophys. Res., 83, 623.
Stubbe, P., and Kopka, H. 1979, Report MPAE-W-02-79-04. Max-Planck-Institut für Aeronomie, 3411 Katlenburg-Lindau 3.
Utlaut, W.F., and Cohen, R. 1971, Science, 174, 245.
Utlaut, W.F. 1973, p. 3/1 in AGARD Conference Proceedings No. 138. Edinburgh.
Weinstock, J., and Bezzerides, B. 1973, Phys. Fluids, 16, 2287.
Weinstock, J. 1975, J. Geophys. Res., 80, 4331.
Yngveson, K.O., and Perkins, F.W. 1968, J. Geophys. Res., 73, 97.

THE LOWER IONOSPHERE AT HIGH LATITUDES

L. Thomas

SRC, Rutherford & Appleton Laboratories, Slough, UK.

ABSTRACT

A review is given of the present knowledge and understanding of the E and D regions at high latitudes, and comparisons are made with current ideas concerning the formation of these regions at lower latitudes. Processes operating at E-region heights during auroral events and at D-region heights during solar particle and electron precipitation events are considered. In addition, the growth of water cluster ions, $H^+.(H_2O)_n$, during conditions of low mesospheric temperatures in summer is considered in relation to the formation of noctilucent clouds.

1. INTRODUCTION

Radio-propagation effects first showed that the polar ionosphere displays a very different behaviour from that at middle and low latitudes. The majority of the high-latitude effects arise from the increase in ionization caused by energetic charged particles, particularly below 100 km, a very pronounced variability often being observed. The particle energies responsible vary from about 2 keV for electrons and 10 keV for protons for the ionization changes at 130 km to 80 keV for electrons and 2 MeV for protons for those at 85 km.

Mass-spectrometer measurements of the positive-ion composition associated with particle sources of ionization have been interpreted to provide a good understanding of the chemistry involved, and have also been used to make deductions concerning certain neutral constituents at high latitudes. For negative ions in the

C. S. Deehr and J. A. Holtet (eds.), Exploration of the Polar Upper Atmosphere, 99–112.

lower D region, our ideas are chiefly derived from theoretical models based on laboratory measurements, as for lower latitudes.

A particular feature of the quiet high-latitude mesosphere is the cold and well defined mesopause (Theon et al. 1967). The ion-composition measurements carried out at high latitudes during summer are of particular interest in view of the marked temperature dependence of the D-region positive-ion chemistry (Thomas 1976; Reid 1977; Arnold and Krankowsky 1977), and their likely relevance to mechanisms for the formation of noctilucent clouds.

The purpose of this paper is to review our present knowledge and understanding of the lower ionosphere at high latitudes, and to highlight differences from lower latitudes.

2. THE E REGION

The sources of ionization in the normal daytime E region are known to be the photoionization of the major gases by solar radiations, chiefly of wavelengths 80.0-102.7 nm and 5.2-30.0 nm. This has been illustrated by solar-flux measurements for a moderately quiet sun by Heroux et al. (1974) which indicated an E-region peak ion pair production rate of 4×10^3 cm^{-3} s^{-1}; O_2^+ and N_2^+ represented the major primary ions below about 120 km with O^+ superseding O_2^+ at greater heights. For nighttime conditions, Lyman-α and Lyman-β radiations in the night sky result in production rates of about 10 cm^{-3} s^{-1} or less. The ionization rates associated with precipitating electrons at high latitudes can be substantially larger and can result in major perturbations, especially at night. In Fig. 1 are shown the ion pair production rates computed by Kondo and Ogawa (1976) for an electron flux representative of an IBC III class aurora and for the associated Bremsstrahlung radiation. The relative rates for different ions would be dependent on neutral composition but approximate to N_2^+ (62%), O_2^+ (17%), N^+ (14%) and O^+ (7%) (Swider 1969); the O^+ and O_2^+ rates can be further subdivided by assuming certain proportions to be formed in the $O^+(^2P)$, $O^+(^2D)$ and $O_2^+(a^4\pi)$ excited states, respectively. The reaction scheme adopted in an analysis of ion composition data by Swider and Narcisi (1977) is then shown in Table 1.

It is evident that the ion composition of the E region is controlled by the concentrations of nitric oxide and atomic nitrogen, as well as those of the major constituents O, O_2, and N_2. Indeed, any theoretical model of the E region needs to include information on the odd nitrogen constituents NO, $N(^4S)$, $N(^2D)$ and $N(^2P)$; it is customary to ignore the distinction between $N(^2D)$ and $N(^2P)$ since it is considered that similar rate coefficients apply to most reactions involving these excited states.

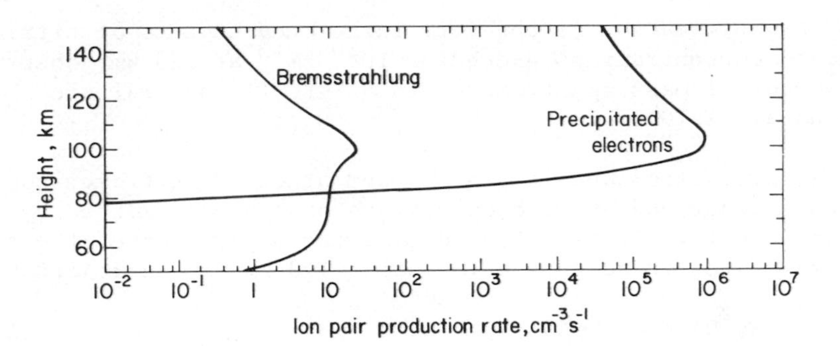

Fig. 1. Ion pair production rates derived for an IBC III aurora by Kondo and Ogawa (1976).

One feature of the ion mass-spectrometer measurements during auroral activity is the relatively large values observed for the ratio of NO^+ to O_2^+ concentrations at E-region heights compared with mid-latitudes (Donahue et al. 1970; Swider and Narcisi 1970, 1977).

1.	$O_2^+ + e \rightarrow O + O$	8.	$N_2^+ + O_2 \rightarrow O_2^+ + N_2$
2.	$NO^+ + e \rightarrow N + O$	9.	$N^+ + O_2 \rightarrow NO^+ + O$
3.	$O^+ + O_2 \rightarrow O_2^+ + O$	10.	$N^+ + O_2 \rightarrow O_2^+ + N$
4.	$O^+ + N_2 \rightarrow NO^+ + N$	11.	$O_2^+(a^4\pi) + O \rightarrow O^+ + O_2$
5.	$O^+(^2P, ^2D) + O_2 \rightarrow O_2^+ + O$	12.	$O_2^+(a^4\pi) + O_2 \rightarrow O_2^+ + O_2$
6.	$O^+(^2P, ^2D) + N_2 \rightarrow N_2^+ + O$	13.	$O_2^+(a^4\pi) + N_2 \rightarrow N_2^+ + O_2$
7.	$N_2^+ + O \rightarrow NO^+ + N$	14.	$O_2^+ + NO \rightarrow NO^+ + O_2$

Table 1 : Ionic processes of auroral E region

From an analysis of eight rocket measurements of ion composition, Swider and Narcisi (1977) deduced NO concentrations for the E region which ranged from 10^8 cm^{-3}, representative of normal values, to 1.9×10^9 cm^{-3} which corresponded to an IBC III aurora. Satellite measurements of the fluorescence of the NO γ bands in the dayglow at latitudes greater than 60° by Rusch and Barth (1975) indicated nitric oxide concentrations larger by a factor of about four than middle-latitude abundances, and similar measurements by Gérard and Barth (1977) suggested enhancements by a factor of eight. It is to be noted that these estimates provided no confirmation of the very large enhancements of nitric oxide, to concentrations exceeding 10^{10} cm^{-3} at 120 km, observed with a neutral mass spectrometer in an IBC II^+ auroral arc (Zipf et al. 1970).

The production and loss of NO and other odd nitrogen constituents under undisturbed conditions have been examined by several workers. It is believed that the major source of nitric oxide molecules at heights below about 140 km is the reaction

$$N(^2D) + O_2 \rightarrow NO + O \qquad (1)$$

The corresponding reaction involving $N(^4S)$ atoms is considerably slower at E-region temperatures.

Consequently, attention has recently been concentrated on the production mechanisms for the $N(^2D)$ state. Analyses for middle-latitude conditions by Strobel et al. (1976) and Kondo and Ogawa (1977) have demonstrated that the sources dominant at E-region heights are reactions (2) and (7) of Table 1, and dissociation of N_2 molecules by photoelectron impact and predissociation in the absorption band between 80 and 100 nm. The total production rates, represented by the broken curve in Fig. 2, were deduced by Kondo and Ogawa (1977) assuming a quantum efficiency of 90% for the production of $N(^2D)$ atoms by the various reactions.

Under auroral conditions, together with reaction (2) and (7) of Table 1, the dissociation of N_2 molecules by energetic electrons accounts for the major production of $N(^2D)$ atoms :

$$N_2 + e_f \rightarrow N + N(^2P, {}^2D, {}^4S) + e \qquad (2)$$

$$N_2 + e_f \rightarrow N^+ + N(^2P, {}^2D, {}^4S) + 2e \qquad (3)$$

From an analysis of auroral data and complementary laboratory measurements, Zipf et al. (1980) have concluded that about 50% of the N atoms are produced in either the $N(^2D)$ or $N(^2P)$ states. The total production rate of the $N(^2D)$ atoms for an IBC III aurora, corresponding to the ion pair production rates in Fig. 1, have been derived from the results of Kondo and Ogawa (1976) by assuming a 50% yield of these atoms in (2) and (3), and are represented by the full curve in Fig. 2. It is evident that the difference in nitric oxide concentrations between middle and high latitudes arise directly from differences in ionization rates.

The processes controlling the odd nitrogen chemistry under auroral conditions are represented in Fig. 3. At E-region heights the loss of NO is predominantly through its reaction with ground-state nitrogen atoms :

$$N(^4S) + NO \rightarrow N_2 + O \qquad (4)$$

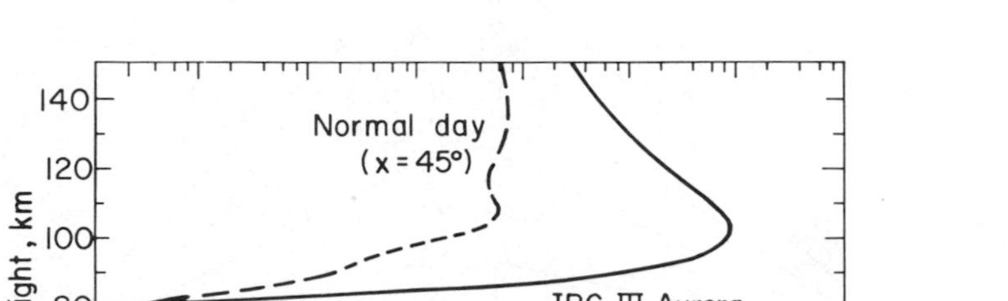
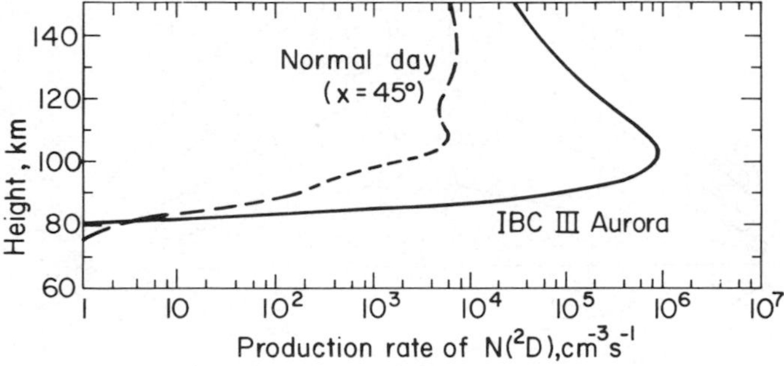

Fig. 2 The total production rates of $N(^2D)$ atoms computed for a solar zenith angle of 45° under normal daytime conditions by Kondo and Ogawa (1977), and the production rates for fast-electron dissociation and dissociative ionization of molecular nitrogen as derived from the results of Kondo and Ogawa (1976) for an IBC III aurora.

The relative production rates of $N(^2D)$ and $N(^4S)$ states are then of particular importance in controlling the morphology of nitric oxide, as illustrated in model studies by Rees and Roble (1979). Their results indicate that if the dissociations of N_2 molecules by energetic electrons, reactions (2) and (3), correspond to a 50% or greater yield of $N(^2D)$, the nitric oxide is enhanced and an auroral storm of a few hours duration could influence the E-region ion composition long after the cessation of ion production. In contrast to the relatively small yield claimed by Zipf et al. (1980), Rusch and Gérard (1980) have suggested an 80% efficiency for these two reactions, the overall auroral efficiency for these and reactions (2) and (7) of Table 1 being required to be as high as 83% in order to explain the 520 nm, $NI(^4S-^2D)$, auroral emission observed on Atmospheric D satellite.

It is evident from Fig. 3 that since the rate coefficients for reactions of $N(^2D)$ atoms with O_2 and NO molecules are 5×10^{-12} cm^3 s^{-1} (Lin and Kaufman 1971) and 7×10^{-11} cm^3 s^{-1} (Black et al. 1969), respectively, a limit is imposed on the growth of NO production from $N(^2D)$ atoms. Hyman et al. (1976) have argued on this basis that the NO concentration cannot exceed 10% of the corresponding O_2 value.

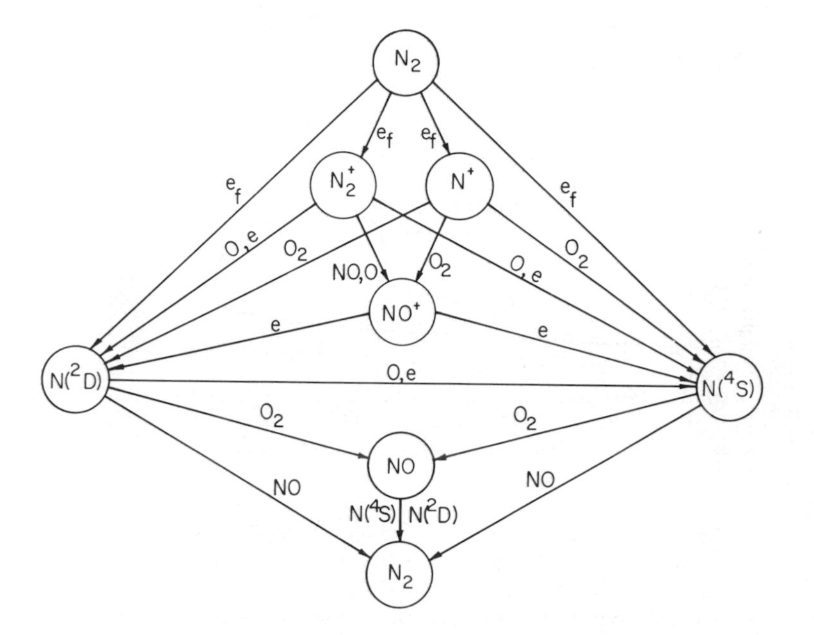

Fig. 3 Odd nitrogen reactions during auroral conditions.

In their analysis of mass-spectrometer data Swider and Narcisi (1977) paid particular attention to the interpretation of O^+ data since early studies (Swider and Narcisi 1970; Donahue et al. 1970) suggested that atomic oxygen was also enhanced during aurorae. Their analysis revealed that the measured O^+ concentrations were consistent with computations based on the CIRA (1972) atmospheric model thus indicating no substantial increase in atomic oxygen concentrations. This is consistent with the results of a study by Maeda and Aikin (1968) which showed that atomic oxygen enhancements above 90 km due to dissociation of molecular oxygen by auroral electrons would be important only for events persisting for several hours.

3. THE D REGION

3.1 The production of ionization

For heights between about 70 and 90 km, the photoionization of nitric oxide is believed to be the major daytime source of ionization for quiet conditions. Nitric oxide distributions deduced by Baker et al. (1977) and Witt et al. (1976) for quiet conditions at middle and high latitudes, respectively, are shown in Fig. 4. The ion pair production rates deduced for these distributions, with a solar angle of 60° and a Lyman-α flux of 6.3 ergs cm^{-2} s^{-1}

corresponding to solar-maximum conditions (Weeks 1967), are shown as the boundaries of the shaded area in Fig. 5. The corresponding production by Lyman-α radiation in the nightglow and by precipitated electrons are believed to provide the greatest contributions to the nighttime D region but the production rates are about two orders of magnitude smaller than the daytime values. Also included in Fig. 5 are the production rates due to galactic cosmic rays for solar minimum conditions at 70° magnetic latitude.

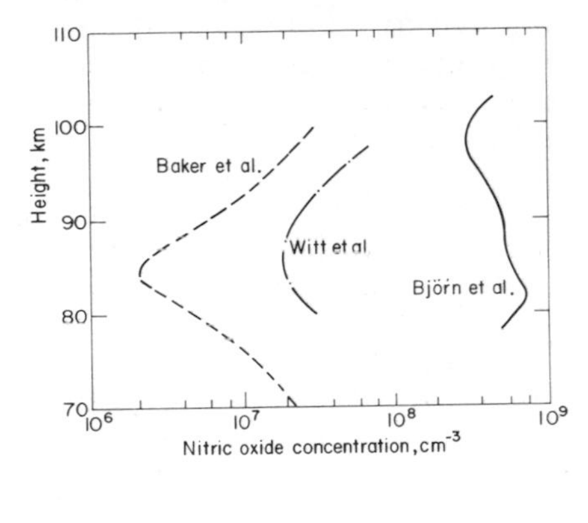

Fig. 4

Height distribution of nitric oxide derived from rocket-borne observations of γ-band fluorescence in the dayglow under quiet conditions at mid-latitudes (Baker et al. 1977) and high latitudes (Witt et al. 1976), and of ion composition under auroral conditions (Björn et al. 1979).

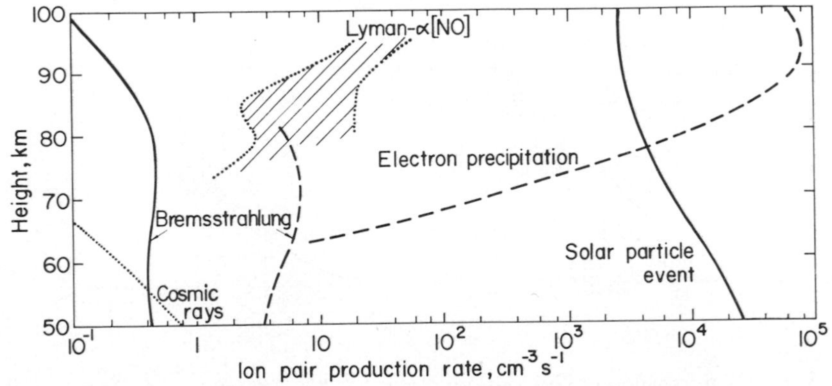

Fig. 5 Ion pair production rates derived by Reagan (1977) for the solar particle event of 4 August 1972 (full curves), and for an electron precipitation event on 1 November 1972 (broken curves). Also shown are the photoionization rates for nitric oxide concentrations intermediate between those of Baker et al. (1977) and Witt et al. (1976) in Fig. 4 for a solar zenith angle of 60° near solar maximum, and the production rate by galactic cosmic rays at 70° magnetic latitude near solar minimum.

For high latitudes, the high-energy part of the auroral particle spectrum will produce D-region ionization, as shown in Fig. 1. However, more spectacular changes are observed following major solar disturbances. Solar particle events are associated with the arrival of energetic protons, alpha particles and electrons at geomagnetic latitudes greater than 60°; electron precipitation events arise from the loss of trapped electrons in the outer radiation belt at the time of major magnetic storms, the D-region effects being confined to middle and auroral latitudes. The ion pair production rates representative of these two types of events have been considered by Reagan (1977) and are illustrated in Fig. 5. It is to be noted that the ionization rates at middle D-region heights resulting from these particle events will exceed those due to photoionization, even if the NO concentrations are enhanced to the values deduced for an auroral event by Björn et al. (1979), Fig. 4. Furthermore, the primary ions will be predominantly N_2^+ (62%) and O_2^+ (17%) rather than NO^+ as in the normal mid-latitude situation. Although fast reactions with atomic oxygen and nitric oxide to produce NO^+ are possible, the main loss of N_2^+ at D-region heights is by charge transfer with molecular oxygen because of the greater concentrations of this constituent.

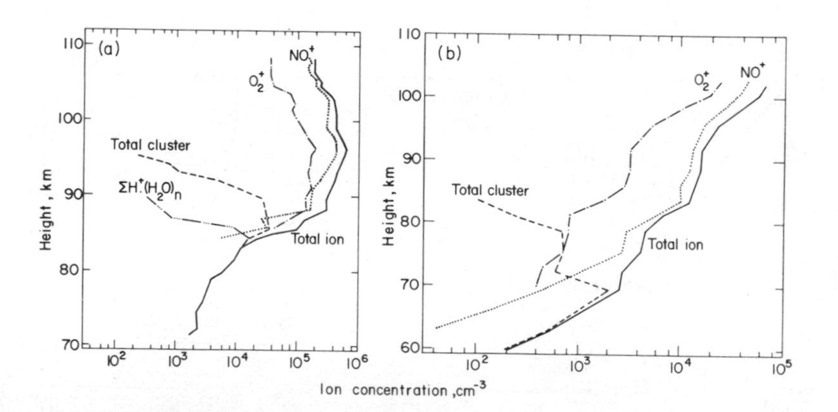

Fig. 6 Ion composition measurements by Arnold and Krankowsky (1977) during electron precipitation events; (a) 0757 LT, 31 May 1972, Andoya (69°N) (b) 2106 LT 21 February 1976, Kiruna (68°N).

3.2 Positive-ion composition

As in the normal and middle-latitude D region, mass-spectrometer measurements of ion composition during both solar particle and electron precipitation events have shown transitions from predominantly NO^+ and O_2^+ at the greater heights to water cluster ions, $H^+.(H_2O)_n$, below. Results for two electron precipitation events are shown in Fig. 6 (Arnold and Krankowsky 1977); for the daytime measurements in May 1972 both the total water cluster ions and all cluster ions are shown, whereas for the nighttime data in February 1976 only the total cluster ions are presented. It can be seen that whilst the May results show a transition height between O_2^+/NO^+ ions and water cluster ions rather similar to that for normal daytime conditions (82 km), that for February must have been considerably lower than the normal nighttime values (86 km). Measurements by Narcisi et al. (1972b) at Churchill during the solar particle event of November 1969 showed that during both daytime and nighttime conditions the transition heights were several kilometres lower than the corresponding levels for normal conditions.

The reaction scheme for the formation of water cluster ions with O_2^+ as the main precursor ion was proposed by Fehsenfeld and Ferguson (1969), Ferguson and Fehsenfeld (1969) and Good et al. (1970b). It is based on the three-body formation of O_4^+ and subsequent two-body reactions to form H_3O^+ and $H^+(H_2O)_2$, with successive three-body reactions leading to higher-order ions :

$$O_2^+ + O_2 + M \rightarrow O_4^+ + M \tag{5}$$

$$O_4^+ + H_2O \rightarrow O_2^+ . H_2O + O_2 \tag{6}$$

$$O_2^+ . H_2O + H_2O \rightarrow H_3O^+ + OH + O_2 \tag{7}$$

or

$$\rightarrow H_3O^+ . OH + O_2 \tag{8}$$

$$H_3O^+ . OH + H_2O \rightarrow H^+.(H_2O)_2 + OH \tag{9}$$

and

$$H^+.(H_2O)_n + H_2O + M \rightarrow H^+.(H_2O)_{n+1} + M \tag{10}$$

The upper boundary of the water cluster ions was associated by Ferguson (1971) with an increase in the concentration of atomic oxygen with height and the corresponding interruption of the scheme by the reaction :

$$O_4^+ + O \rightarrow O_2^+ + O_3 \tag{11}$$

Competition for the O_2^+ ions is provided by reactions with molecular nitrogen and oxygen. The NO^+ ions resulting from these reactions, together with the small production resulting from the interaction with mesospheric gases of N_2^+, N^+ and O^+ produced during the initial ionization process, imply that a comprehensive treatment also needs to take account of the formation of cluster ions from NO^+. Swider and Narcisi (1975) did include a simplified treatment of the formation of water cluster ions from NO^+ ions with the main O_2^+ scheme in their reproduction of the nighttime ion composition measurements of Narcisi et al. (1972b) for the November 1969 solar particle event. In their analysis, Swider and Narcisi took account of collisional dissociation of O_4^+. This process for which a rate of coefficient of 3.3×10^{-6} (300/T) exp (- 5030/T) has been reported by Durden et al. (1969), implies a temperature dependence analogous to that arising from collisional dissociation of intermediate complexes with N_2 and CO_2 in the NO^+ scheme (Thomas 1976; Reid 1977).

3.3 Negative ions

It has long been realised that the relatively large ambient pressures encountered at D-regions height permit the formation of negative ions by three-body attachment of electrons to neutral atoms and molecules. However, there has been little agreement on the negative-ion composition from the two sets of mass-spectrometer measurements (Arnold et al. 1971; Narcisi et al. 1971, 1972a). Consequently, an understanding of the negative-ion chemistry for both middle and high latitudes has been derived largely from laboratory measurements. It is believed that the O_2^- ions formed by the rapid three-body attachment of electrons to O_2 molecules (Chanin et al. 1959) can initiate a complicated sequence of reactions involving neutral constituents to produce a range of negative-ion species, such as O_3^-, O_4^-, CO_3^-, CO_4^-, HCO_3^-, NO_2^- and NO_3^-. Negative-ion schemes have been devised to incorporate these reactions (e.g. Ferguson 1974) and a number of theoretical models have been based on these schemes (e.g. Reid 1970; Thomas et al. 1973; and Swider et al. 1978).

Measurements at Ft. Churchill, Canada, under quiescent conditions (Narcisi et al. 1971), near midday and midnight during a solar particle event (Narcisi et al. 1972c), and near totality of a solar eclipse (Narcisi et al. 1972a) all showed evidence of $NO_3^- \cdot (H_2O)_n$ ions with some possible admixture of $CO_3^- \cdot (H_2O)_n$, where $n = 0$ to 5. To date, no account has been taken in theoretical models of hydration processes. Laboratory measurements have indicated that for certain reactions hydration of the negative ion involved has little effect (Fehsenfeld and Ferguson 1974) whereas for others such hydration renders the reaction endothermic (Ferguson 1975).

Important effects for sunlit periods might arise from the influence of hydration on the photodetachment rate of electrons from negative ions. The electron affinities of non-hydrated ions considered in theoretical models range from 0.43 eV for O_2^- to 3.9 eV for NO_3^-. However, for an hydrated ion the energy supplied must be sufficient to overcome the energy in the hydration bond as well as the electron affinity of the ion; it has been estimated that the effective electron affinity increases by 0.5 eV for each water molecule added to the negative ion (Swider 1977). This would be important only during summer and equinox periods at high latitudes. During such periods, photodissociation of both hydrated and non-hydrated ions could occur, leading to ions of lower electron affinities (e.g. Peterson 1976; Vestal and Mauclaire 1977; Smith et al. 1978).

$$CO_3^-.H_2O + h\nu\ (\lambda < 742\ \text{nm}) \rightarrow CO_3^- + H_2O \tag{12}$$

$$CO_3^- + h\nu\ (\lambda < 652\ \text{nm}) \rightarrow O^- + CO_2 \tag{13}$$

$$O^- + h\nu\ (\lambda < 843\ \text{nm}) \rightarrow O + e \tag{14}$$

Since the electron affinities of $CO_3^-.H_2O$ (3.3 eV) and CO_3^- (2.8 eV) correspond to radiations of wavelengths 376 nm and 443 nm, respectively, photo-destruction of these ions and subsequent release of electrons from O^- ions will precede photodetachment from the molecular ions during sunrise periods. The combination of these photodissociation and photodetachment processes and the onset of associative detachment of electrons from O_2^- and other reactions involving atomic oxygen will result in a very complicated sequence of negative-ion changes during sunrise periods.

For disturbed high-latitude conditions, the measurements during the solar particle event of November 1969 (Narcisi et al. 1972c) have attracted the greatest attention. The dominance of O^- and its relatively large concentrations above 75-80 km at nighttime and the dominance of O_2^- from 76 to 94 km in daytime were surprising features in view of the rapid associative detachment of these ions by atomic oxygen.

4. THE CHARACTERISTICS OF THE D REGION DURING CONDITIONS OF LOW MESOSPHERIC TEMPERATURES IN SUMMER

Current ideas concerning the formation of noctilucent clouds in the high-latitude summer mesosphere are based on the condensation of water on nucleation centres to form ice particles. Mesospheric temperatures as low as 140 K were recorded during a cloud display by Theon et al. (1967) but the presence of sufficient water vapour for supersaturation to occur has not been confirmed. Instead, Witt (1969) has suggested that nucleation might occur about the water cluster ions $H^+.(H_2O)_n$ discussed in section 3.2 or about metallic ions. Mass-spectrometer measurements of ion composition at high latitudes during summer are relevant to these ideas.

Johannessen and Krankowsky (1972) showed from measurements carried out near noon at Andoya (69°N) on 8 August 1971 that water cluster ions $H^+.(H_2O)_n$ dominated the composition below 80 km. The order of clustering, n, increased with height up to 84.5-86 km where the cluster ions were replaced by metal ions, and NO^+ and O_2^+ which became the dominant species; the largest clusters observed corresponded to $H^+.(H_2O)_6$ at 84.5 km on rocket ascent and $H^+.(H_2O)_5$ between 82 and 85 km on descent. Estimates of temperature derived from a simultaneous measurement of Lyman-α extinction indicated a value of 150 K near 85 km. A subsequent morning measurement by Goldberg and Witt (1977) at Kiruna (68°N) in August 1973, following the sighting of a noctilucent cloud display on the previous evening, showed that cluster ions were dominant up to 85 km, the higher order ions up to n = 7 appearing above 80 km. In addition, a series of ions interpreted as Fe^+, its oxides, and their hydrates were also observed at these heights. A third measurement, by Arnold and Joos (1979) during an early morning in June 1975 at Andoya, again showed evidence of the order of cluster ions increasing with height up to the transition to NO^+ and O_2^+ ions near 85 km.

The increasing order of hydration with increasing height and the presence of large clusters serve to focus attenuation on the hydration processes for cluster ions. Most positive-ion reaction schemes have considered that the growth of cluster ions arises through direct hydration

$$H^+.(H_2O)_n + H_2O + M \rightarrow H^+.(H_2O)_{n+1} + M \quad . \tag{15}$$

Rate coefficients of these reactions for room temperatures have been reported by Good et al. (1970a).

From comparisons of observed concentrations of ions $H^+.(H_2O)_n$ and $H^+.(H_2O)_{n+1}$, with allowance being made for loss of the higher-order ion by dissociative recombination, Arnold and Joos (1979) deduced hydration times for the cases n = 3 and n = 7, as functions of mesospheric temperatures measured simultaneously. The results indicated that cluster-ion hydration occurred much more rapidly than was expected for the room-temperature values of reaction (15). Furthermore, a marked temperature dependence was found, the hydration time for $H^+.(H_2O)_3$ at 85 km increasing by almost three orders of magnitude for a temperature change of 120 to 220 K. An attempt to reproduce this change on the basis of reaction (15) alone requires an unlikely temperature (T^{-8}) dependence for the rate coefficient. In order to explain the rapid hydration, Arnold and Joos (1979) invoked indirect processes analogous to those considered for NO^+ :

$$H^+.(H_2O)_n + N_2 + M \rightarrow H^+.(H_2O)_n.N_2 + M \tag{16}$$

$$H^+.(H_2O)_n.\ N_2 + CO_2 \rightarrow H^+.(H_2O)_n.CO_2 + N_2 \tag{17}$$

$$H^+ \cdot (H_2O)_n \cdot CO_2 + H_2O \rightarrow H^+ \cdot (H_2O)_{n+1} + CO_2 \qquad (18)$$

Arnold and Joos (1979) proposed that the association of N_2 with $H^+ \cdot (H_2O)_3$ proceeds at a similar rate to that for NO^+ at low temperatures but that thermal dissociation of the intermediate complexes with N_2 and CO_2 occurs at lower temperatures than in the NO^+ case. The increase in dissociation with increasing temperature is then responsible for the increase in hydration time constant observed between 120 and 220 K. On this basis, an increase in the order of hydration of cluster ions as the low-temperature mesopause is approached would be expected. However, observations of positive-ion composition carried out on 31 July 1978 by Kopp (private communication, 1980) do not seem to be consistent with these ideas. The increasing order of water cluster ions with increasing height was again shown very clearly, extending up to 90 km, but the temperature profile deduced by Witt (private communication, 1980) from Rayleigh scattering measurements indicated a minimum near 86 km.

5. CONCLUSIONS

The formation of the disturbed E region at high latitudes is reasonably well understood but, as for normal conditions at lower latitudes, quantitative treatments require a better knowledge of the odd nitrogen constituents. In the chemistry of such constituents, the relative production rates of $N(^2D)$ and $N(^4S)$ atoms in reactions (2) and (7) of Table 1 and in the dissociation of molecular nitrogen by fast electrons, are of particular important.

The major D-region disturbances are associated with solar particle and electron precipitation events. Measurements of positive-ion composition during such events have been successfully interpreted on the basis of a scheme for the formation of water cluster ions originating with O_2^+, reactions (5) - (11).

The negative-ion composition of the D region has not been established at either low or high latitudes, and studies have been confined to theoretical models devised from laboratory measurements. The observation of O_2^- and O^- as predominant ions in daytime and nighttime, respectively, during a solar particle event cannot be understood on the basis of such models.

Measurements of positive-ion composition at high latitudes during summer show increases in the order of hydration of water cluster ions, $H^+ \cdot (H_2O)_n$, as the low-temperature mesopause is approached. These results are understandable in terms of rapid hydration processes involving intermediate complexes with N_2 and CO_2, reactions (16) - (18), and are relevant to the formation of noctilucent clouds by nucleation about water cluster ions.

REFERENCES:

Arnold, F. and Joos, W, 1979, Geophys. Res. Lett. 6, 763.
Arnold, F., Kissel, J., Krankowsky, D., Wieder, H., and Zahringer, J. 1971, J. Atmos. Terr. Phys., 33, 1169.
Arnold, F. and Krankowsky, D. 1977, p. 93 in Dynamical and Chemical Coupling Between the Neutral and Ionized Atmosphere (Eds. B. Grandal and J.A. Holtet). Reidel Dordrecht-Holland.
Baker, K.D., Nagy, A.F., Olsen, R.O., Oran, E.S, Randhawa, J., Strobel, D.F., and Tohmatsu, T. 1977, J. Geophys. Res., 82, 3281.
Björn, L.G., Arnold, F., Krankowsky. D., Grandal, B., Hagen, E., and Thrane, E.V. 1979, J. Atmos. Terr. Phys., 41, 1184.
Black, G., Slanger, T.G., St. John, G.A., and Young, R.A. 1969, J. Chem. Phys., 51, 116.
Chanin, L.M., Phelps, A.V., and Biondi, M.A. 1959, Phys. Rev. Lett., 2, 344.
Donahue, T.M., Zipf, E.C., and Parkinson, T.D. 1970, Planet. Space Sci., 18, 171.
Durden, D.A., Kebarle, P., and Good, A. 1969, J. Chem. Phys. 50, 805.
Fehsenfeld, F.C. and Ferguson, E.E. 1969, J. Geophys. Res., 74, 2217.
Fehsenfeld, F.C. and Ferguson. E.E. 1974, J. Chem. Phys., 61, 3181.
Ferguson, E.E. 1971, p. 188 in Mesospheric Models and Related Experiments (Ed. G. Fiocco), Reidel, Dordrecht-Holland.
Ferguson, E.E. 1974, Rev. Geophys. Space Phys., 12, 703.
Ferguson, E.E. 1975, p. 197 in Atmospheres of Earth and the Planets (Ed. B.M. McCormac) Reidel, Dordrecht-Holland.
Ferguson, E.E. and Fehsenfeld, F.C. 1969, J. Geophys. Res. 74, 5743.
Gérard, J-C. and Barth, C.A. 1977, J. Geophys. Res., 82, 674.
Goldberg, R.A. and Witt, G. 1977, J. Geophys. Res., 82, 2619.
Good, A., Durden, D.A., and Kebarle, P. 1970a, J. Chem. Phys., 52, 212.
Good, A., Durden, D.A., and Kebarle, P. 1970b, J. Chem. Phys., 52, 222.
Heroux, L., Cohen, M., and Higgins, J.E. 1974, J. Geophys. Res., 79, 5237.
Hyman, E., Strickland, D.J., Julienne, P.S., and Strobel, D.F. 1976, J. Geophys. Res., 81, 4765.
Johannessen, A. and Krankowsky, D. 1972, J. Geophys. Res. 77, 2888.
Kondo, Y. and Ogawa, T. 1976, J. Geomag. Geoelectr., 28, 253.
Kondo, Y. and Ogawa, T. 1977, J. Geomag. Geoelectr., 29, 65.
Lin, C.L. and Kaufman, F. 1971, J. Chem. Phys. 55, 3760.
Maeda, K. and Aikin, A.C. 1968, Planet. Space Sci., 16, 371.
Narcisi, R.S., Bailey, A.D., Della Luca, L., Shearman, C., and Thomas, D.M. 1971, J. Atmos. Terr. Phys., 33, 1147.
Narcisi, R.S., Bailey, A.D., Wlodyka, L.E., and Philbrick, C.R. 1972a, J. Atmos. Terr. Phys., 34, 647.
Narcisi, R.S., Philbrick, C.R., Thomas, D.M., Bailey, A.D., Wlodyka, L.E., Baker, D., Federico, G., Wlodyka, R.A., and Gardner, M.E. 1972b, p. 421 in Proceedings of COSPAR Symposium on Solar Particle Event in November 1969 (Ed. J.C. Ulwick). AFCRL-72-0474.
Narcisi, R.S., Shearman, S., Philbrick, C.R., Thomas, D.M., Bailey, A.D., Wlodyka, L.E., Wlodyka, R.A., Baker, D., and Federico, G. 1972c, p. 441 in Proceedings of COSPAR Symposium on Solar Particle Event of November 1969, (Ed. J.C. Ulwick). AFCRL-72-0474.
Peterson, J.R. 1976, J. Geophys. Res. 81, 1433.
Reagan, J.B. 1977, p. 145 in Dynamical and Chemical Coupling Between the Neutral and Ionized Atmosphere (Eds. B. Grandal and J.A. Holtet). Reidel, Dordrech-Holland.
Rees, M.H. and Roble, R.G. 1979, Planet. Space Sci., 27, 453.
Reid, G.C. 1970, J. Geophys. Res. 75, 2551.
Reid, G.C. 1977, Planet. Space Sci., 25, 275.
Rusch, D.W. and Barth, C.A. 1975, J. Geophys. Res., 80, 3719.
Rusch, D.W. and Gérard, J-C. 1980, J. Geophys. Res., 85, 1285.
Smith, G.P., Lee, L.C., Cosby, P.C., Peterson, J.R., and Moseley, J.T. 1978, J. Chem. Phys. 68. 3818 .
Strobel, D.F., Oran, E.S., and Feldman, P.D. 1976, J. Geophys. Res., 81, 3745.
Swider, E. 1969, Rev. Geophys., 7, 573.
Swider, W. 1977, Space Sci. Rev., 20, 69.
Swider, W., Keneshea, T.J., and Foley, C.I. 1978, Planet. Space Sci., 26, 883.
Swider, E. and Narcisi, R.S. 1970, Planet. Space Sci., 18, 379.
Swider, W. and Narcisi, R.S. 1975, J. Geophys. Res., 80, 655.
Swider, W. and Narcisi, R.S. 1977, Planet. Space Sci., 25. 103.
Theon, J.S., Nordberg, W., Katchen, L.B., and Horvath, J.J. 1967, J. Atmos. Sci., 24, 428.
Thomas, L. 1976, J. Atmos. Terr. Phys. 38, 1345.
Thomas, L., Gondhalekar, P.M., and Bowman, M.R. 1973, J. Atmos. Terr. Phys. 35, 397.
Vestal, M.L. and Mauclaire, G.H. 1977, J. Chem. Phys., 67, 3758
Weeks, L.H. 1967, Astrophys. J. 147, 1203.
Witt. G., p. 157 in Space Research IX (Eds. K.S.W. Champion, P.A. Smith, and R.L. Smith-Rose). North-Holland, Amsterdam, Holland.
Witt, G., Dye, J.E., and Wilhelm, N. 1976, J. Atmos. Terr. Phys. 38, 223.
Zipf, E.C., Borst, W.L., and Donahue, T.M. 1970, J. Geophys. Res., 75, 6371.
Zipf, E.C., Espy, P.J., and Boyle, C.F. 1980, J. Geophys. Res. 85, 687.

ENERGY SOURCES OF THE HIGH LATITUDE UPPER ATMOSPHERE

P. M. Banks

Physics Department and
Center for Atmospheric & Space Sciences
Utah State University
Logan, Utah 84322

ABSTRACT

A short review is given of electrodynamic and plasma wave heating processes as they occur in the high latitude upper atmosphere.

1. INTRODUCTION

At the present time it is possible to identify five principal mechanisms which provide energy to the upper atmosphere at high latitudes. These include:

1. The absorption of solar extreme ultraviolet radiation
2. The transport of wave energy from the lower atmosphere
3. Electrodynamic (Joule) dissipation
4. Auroral particle energy loss
5. Plasma wave heating

This review is directed towards the description of electrodynamic (or Joule) dissipation and plasma wave heating, processes which assume particular importance at high latitudes where magnetospheric electrical effects are present. These sources supplement the heating arising from the absorption of solar EUV radiation and the energy loss of auroral particles, both of which involve complicated channels of ionization and internal excitation of atmospheric particles. Information about these latter modes of atmospheric heating can be found in the literature [1,2].

C. S. Deehr and J. A. Holtet (eds.), Exploration of the Polar Upper Atmosphere, 113–127.

$$\frac{D}{Dt}\left(\frac{3}{2}p_j\right) + \frac{5}{2}p_j\,\nabla.\bar{u}_j + \nabla.\bar{q}_j + \bar{\tau}_j : \nabla\,\bar{u}_j \qquad (1)$$

$$= Q_j - L_j + n_j \sum_s \frac{m_j \nu_{js}}{m_j + m_s}\left[3\,k(T_s - T_j) + m_s(\bar{u}_j - \bar{u}_s)^2\right]$$

with the subscript j representing electrons, ions, or neutrals and the subscript s is summed over the other species present. The term Q represents a variety of inelastic processes including ion-electron recombination, de-excitation and chemical heating, while the term L includes j-th species energy losses such as radiative cooling.

The final term of (1), which is of principal interest for electrodynamic heating, expresses the energy consequences of ordinary elastic collisional processes. It involves both thermal transfer proportional to $(T_s - T_j)$ and frictional heating due to differences in bulk flow velocity $(\bar{u}_j - \bar{u}_s)^2$. The quantity ν_{js} in this term is an effective collision frequency between particles of types j and s and is related to the usual momentum transfer collision frequency by the expression

$$\nu_{js} = \frac{m_s\,\nu^M_{js}}{m_j + m_s} \qquad (2)$$

Further details concerning (1) and (2) can be found in reference [8]. For later purposes we note the equality of collision rates is given by $n_s\,\nu^M_{sj} = n_j\,\nu^M_{js}$.

When the plasma and neutral gas are stationary with respect to each other, the temperatures of each species can be determined by using energy budgets of the form of (1) omitting the effect of frictional heating. Such is the case, for example, at middle geomagnetic latitudes. Consider, however, the situation where plasma motions occur in the upper atmosphere as a consequence of a substantial electric field $\bar{E}$. In most instances it is found that the ion gas energy balance is dominated by the elastic collisonal terms of (1); i.e., one can ignore transport effects, inelastic collisions, and the material derivative in comparison with the energy gained and lost via elastic collisions. In this situation, energy gained by the ions as a result of bulk motion through the neutral gas is dissipated back to the neutral gas through ordinary thermal transfer. In the lower thermosphere the rate at which this occurs is very rapid and other energy processes, such as electron to ion thermal transfer and dissociative recombination, usually cannot significantly influence the ion temperature. Thus, from (1), we obtain the basic relation

$$T_i = T_n + \frac{m_n}{3K} (\bar{u}_i - \bar{u}_n)^2 \qquad (3)$$

This result shows that ion motions induced by electromagnetic or other forces, as well as neutral winds, result in ion temperature increases above the local neutral gas temperature. For ions moving through an N_2 atmosphere, (3) becomes

$$T_i = T_n + 1129.\ U_*^2 \quad {}^\circ K \qquad (4)$$

where U_* is the ion-neutral relative velocity expressed in (km/sec). Since ion convective drift velocities of 1 to 2 km/sec are found regularly at high latitudes, it can be seen that the ion gas will often be several thousand degrees hotter than the background neutrals.

A separate feature of (3) is the absence of ion or neutral gas densities. Examination of (1) shows that these quantities determine the total rate at which energy is exchanged between ions and neutrals. The result given by (3) is accurate only so long as competitive processes, such as transport or other terms in (1), remain as minor contributors to the overall ion energy balance. As one goes to the F_2 - region and above, for example, the electron to ion energy transfer rate frequently dominates the ion to neutral rate and equations (3) and (4) become increasingly incorrect.

Observational evidence for enhanced ion temperatures resulting from convection electric fields is well documented from a variety of techniques, including incoherent scatter radar Atmosphere Explorer results, and rocket probes. In fact, the simultaneous observation of ion temperature and ion drift velocity can be used to estimate the neutral gas velocity of the thermosphere.

The effect of ion collisions upon the heating of the neutral atmosphere is two-fold. First, because $T_i > T_n$ there will be an ordinary thermal transfer term proportional to $(T_i - T_n)$. Secondly, as with the ions, there is the frictional dissipation proportional to $(\bar{u}_i - \bar{u}_n)^2$. In contrast to the ions, however, a description of the thermal structure of the neutral atmosphere involves all of the terms of equation (1); i.e., transport, EUV heating, particle heating, and temporal changes all enter into the upper atmospheric heat balance. The final term representing heating from elastic ion collisions can be simplified, however, if we use the special result given by (3) for the ion temperature. Thus, if we let Q_J represent this source of electrodynamic heating, we have from (1) the simple expression:

$$Q_J = n_n m_n \nu_{ni} (\bar{u}_n - \bar{u}_i)^2. \quad (5)$$

In terms of the momentum transfer collision frequency ν^M_{ni}, this result becomes

$$Q_J = n_n \mu \nu^M_{ni} U_r^2 \quad (6)$$

where μ is the ion-neutral reduced mass and U_r is the relative speed of ion-neutral motion. Thus, we see that atmospheric heating due to ion-neutral relative motions will be proportional to U_r^2, to the neutral gas density, and to the ion density, the latter quantity being hidden within the neutral-ion collision frequency.

In order for (6) to be useful, it is necessary to know the relation between the ion-neutral relative velocity and the ionospheric electric field. When an electric field is present within the lower thermosphere, ions and electrons are driven in directions perpendicular to $\bar{B}$ through the neutral gas, which may have its own characteristic velocity arising from a variety of different processes. In this situation the ion-neutral relative velocity can be written from the ion momentum equation as

$$U_r^2 = \frac{\Omega_i^2 \nu_{in}}{(\Omega_i^2 + \nu_{in}^2)} \frac{E^{*2}}{B^2} \quad (7)$$

where $\Omega_i = eB/m_i$ is the ion cyclotron frequency, ν_{in} is the effective ion-neutral collision frequency and $\bar{E}^* = \bar{E} + \bar{U}_n \times \bar{B}$. The effects of pressure gradients, gravity, and other terms are neglected. From this result we see that in the presence of $\bar{E}^*$, U_r will change with altitude, being $\bar{E}^*/B$ above about 150 km altitude, $1/2\ E^*/B$ at 125 km, and tending towards zero below 100 km. Values for ν_{in} can be found in [8].

Combining (6) and (7), and again noting that $n_i \nu^M_{in} = n_n \nu^M_{ni}$, we find that the atmospheric heating rate given by (6) becomes, with $n_i = n_e$,

$$Q_J = \frac{en_e}{B} \frac{\Omega_i \nu_{in}}{\Omega_i^2 + \nu_{in}^2} E^{*2} \tag{8}$$

or $$Q_J = \sigma_p E^{*2} \tag{9}$$

where σ_p is the Pedersen conductivity [9].

This last expression brings us back to an alternate mode of presenting atmospheric energy dissipation arising from electric currents. Within the ionosphere the current density, $\bar{j}$, is related to the effective electric field, $\bar{E}^*$ by the relation

$$\bar{j} = \sigma_p E^* + \sigma_H \hat{b} \times \bar{E}^* \tag{10}$$

where $\bar{E}^* = \bar{E} + \bar{U}_n \times \bar{B}$, $\hat{b} = \bar{B}/B$ and σ_H is the Hall conductivity. Using (9), we see that

$$Q_J = \bar{j} \cdot \bar{E}^* \tag{11}$$

in accord with expectations based on electromagnetic theory.

Up to this point the contribution to atmospheric heating made by the thermal electron population has been neglected. If one assumes an energy balance for electrons which insures that all energy given to the electron gas is deposited locally (no transport) and that radiative processes which lead to the net loss of energy from the atmosphere are absent, then we can follow the same method used above for the ions to show that the neutral gas heating arising from the electrons and ions together is

$$Q_J = \left[\frac{en_e}{B} \quad \frac{\nu_{in} \Omega_i}{\nu_{in}^2 + \Omega_i^2} + \frac{\nu_{en} \Omega_e}{\nu_{en}^2 + \Omega_e^2} \right] E^{*2} \tag{12}$$

where the electron collision and cyclotron frequencies have been introduced.

As shown in Figure 1, the collision frequency terms of (12) give two maxima; one at 125 km where $\nu_{in} = \Omega_i$ and heating due to ion-neutral collisions occurs and one at 70 Km where $\nu_{en} = \Omega_e$ and electron heating dominates. However, since the electron density below 100 km is usually much smaller than at higher altitudes, it is the ions which contribute most effectively to

Electrodynamic dissipation and plasma wave heating of the atmosphere arise as a consequence of electrical coupling between the magnetosphere and the ionosphere. As described elsewhere in this volume [3,4], global scale currents penetrate the ionosphere along magnetic field lines at high latitudes, driving horizontal currents which close in patterns determined by the ionospheric electrical conductivity and the associated effective electric field. Energy dissipation in this coupled system of plasma and neutral gas occurs through plasma-neutral gas elastic collisions, with the result that the ionospheric electrical circuit exhibits a net resistivity and there is a net energy transfer to the neutral gas. From past studies, energy dissipation in the thermosphere by electric currents is widely accepted as an important source of upper atmospheric heating [5,6]. The added heating of the neutral atmosphere by means of plasma waves which maintain elevated electron and ion temperatures, however, is a recent discovery [7]. The basic mechanism for plasma wave heating depends upon the generation of unstable waves in the electrojet region near $\sim$ 110 km altitude where there exist different velocities for the ions and electrons which are driven under the influence of the convection electric field. The scattering of radar signals from these waves provides important information about the auroral regions. The fact that these waves, via electron and ion collisions with neutral particles, can provide substantial heat to the lower thermosphere has been documented only recently in the incoherent scatter radar observations of Schlegel, et al. [7].

In Section 2 the salient features of electrodynamic dissipation are described, while the processes leading plasma wave heating are presented in Section 3. A summary of the results is given in Section 4.

2. ELECTRODYNAMIC HEATING

2.1 Basic Equations

To describe electrodynamic heating in the thermosphere we must consider the separate energy balance equations for electrons, ions and the neutral gas. For convenience, only one species of each is assumed to be present. Using standard notation, the generalized energy balance equation can be written as

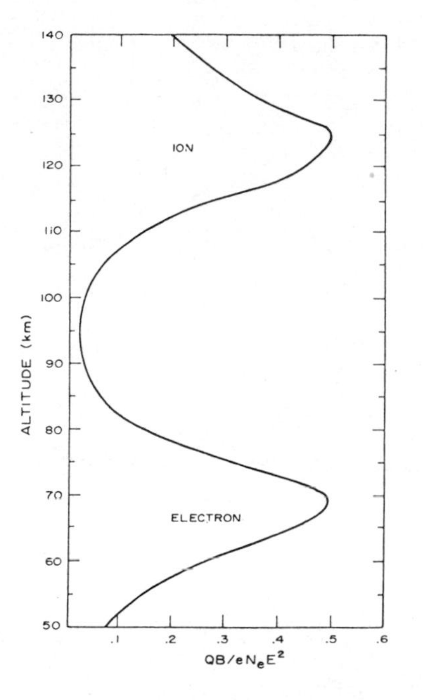

Figure 1. Altitude profile of the collisional factors which enter into the expression for electrodynamic dissipation. Ions dominate above 95 km while electrons are effective below. The effect of negative ions has been omitted. From Banks [10].

atmospheric heating, and this occurs in the lower thermosphere. For special circumstances, such as during a solar proton event or a relativistic electron precipitation event, mesospheric heating can become important [10]. It should also be noted that (12) ignores inelastic electron-neutral collisions and that when these become sufficiently frequent the simple result given by (11) will no longer be obtained.

2.2 Applications

Estimates of the importance of electrodynamic heating at high latitudes are frequently made on the basis of the height-integrated values. Since the electrostatic field, $\bar{E}$, is thought to vary only slightly with altitude throughout the E-region, integration of (8) with height gives

$$Q_J^T = \Sigma_p E^{*2} \tag{13}$$

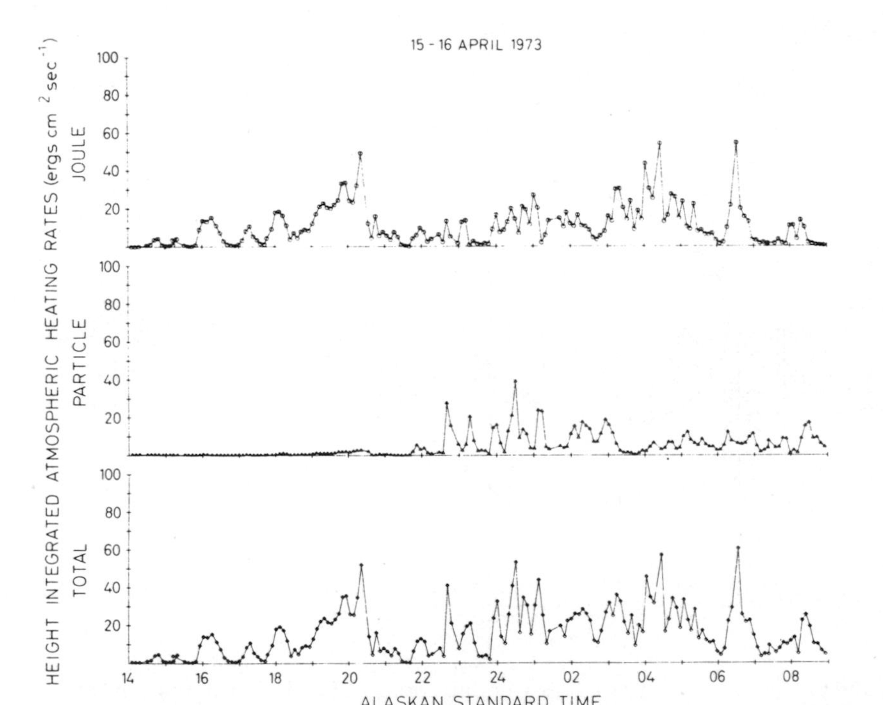

Figure 2. Height-integrated atmospheric heating rates for electron precipitation and electrodynamic dissipation. Results obtained from Chatanika, Alaska incoherent scatter radar data. From Banks [11].

where Σ_p is the Pedersen conductivity integrated over altitude. A typical auroral value for Σ_p is $\simeq$ 20 mho so that with $\bar{E}^* \simeq 40$ to 80 mV/m, we have $Q_J^T \simeq 0.032$ to .13 watt/m^2.

Unfortunately, height integration of the Joule heating rate hides two important aspects of atmospheric heating; namely, the important changes of $\bar{U}_n$ as one passes from the lower to upper thermosphere, and the fact that the atmospheric response to heating depends greatly upon the heating rate per neutral gas particle. With respect to the effects of neutral winds, since it is the relative velocity squared which enters into Q_J, winds on the order of 200-400 m/sec can appreciably influence the total heating rate, especially when there is a substantial variation with altitude, such as often occurs within the thermosphere. In this situation the height integration leading to (13) has no practical utility.

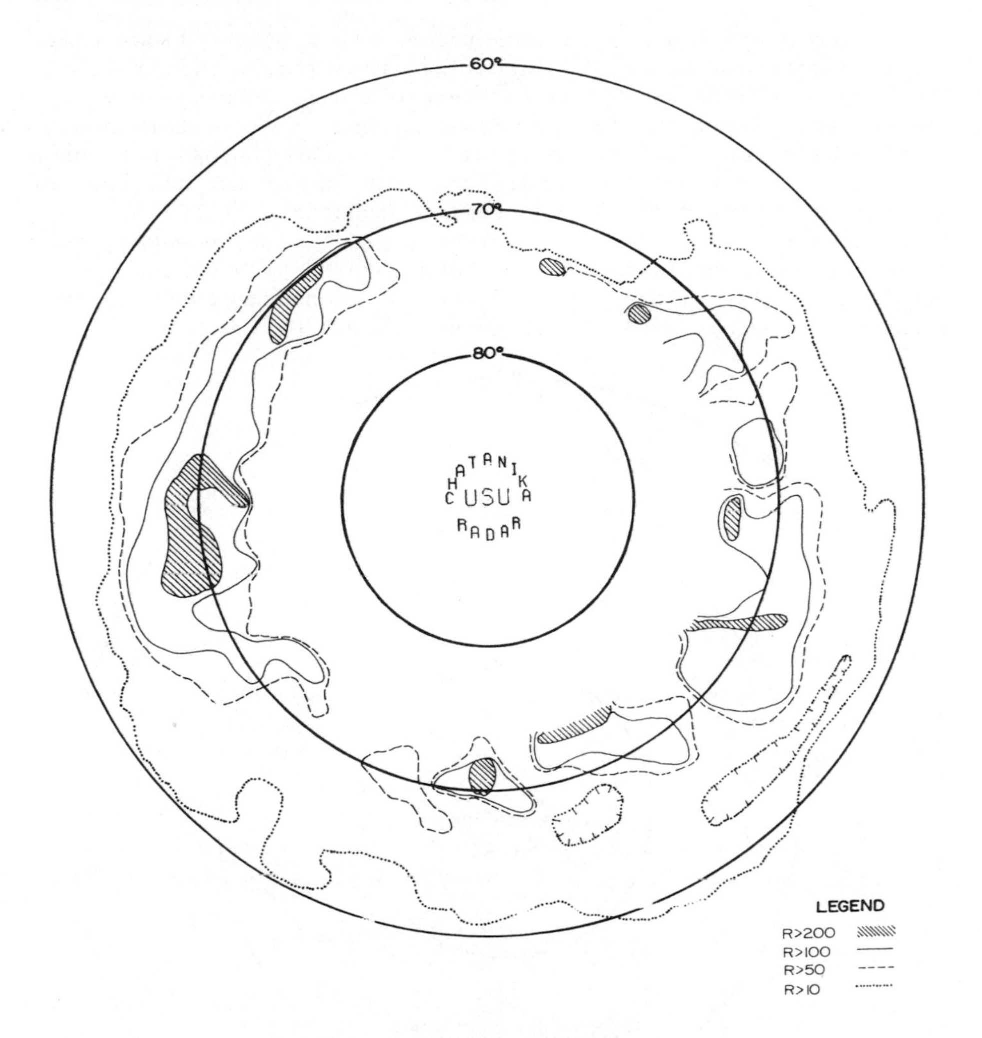

Figure 3. Electric field contribution to electrodynamic heating. Contours of the quantity R, defined through eqn. (15) are shown for data obtained from the Chatanika radar for four days in June, 1978.

The second point relating to power per particle follows from (1) where division by the neutral gas density, n_n, yields an expression of the form

$$\frac{3}{2} k \frac{DT}{Dt} + (\text{transport terms})/n = Q/n_n + L/n_n \qquad (14)$$

$$+ m_n \nu_{ni} U_r^2$$

Since Q and L are each proportional to n_n, it follows that the time rate of change of neutral gas temperature will reflect the local electrodynamic heating rate which is determined by n_e and the relative velocity U_r. Since $\bar{u}_i$ and $\bar{u}_n$ are determined by greatly different factors, effects seen in the thermosphere as a whole may not reflect the conditions deduced for the altitude of maximum energy production, the region emphasized by height integrated heating rates. Up to the present time, however, only barium release observations have been able to provide the simultaneous information about $\bar{u}_i$ and $\bar{u}_n$ needed to estimate the atmospheric per particle heating rate above 200 km.

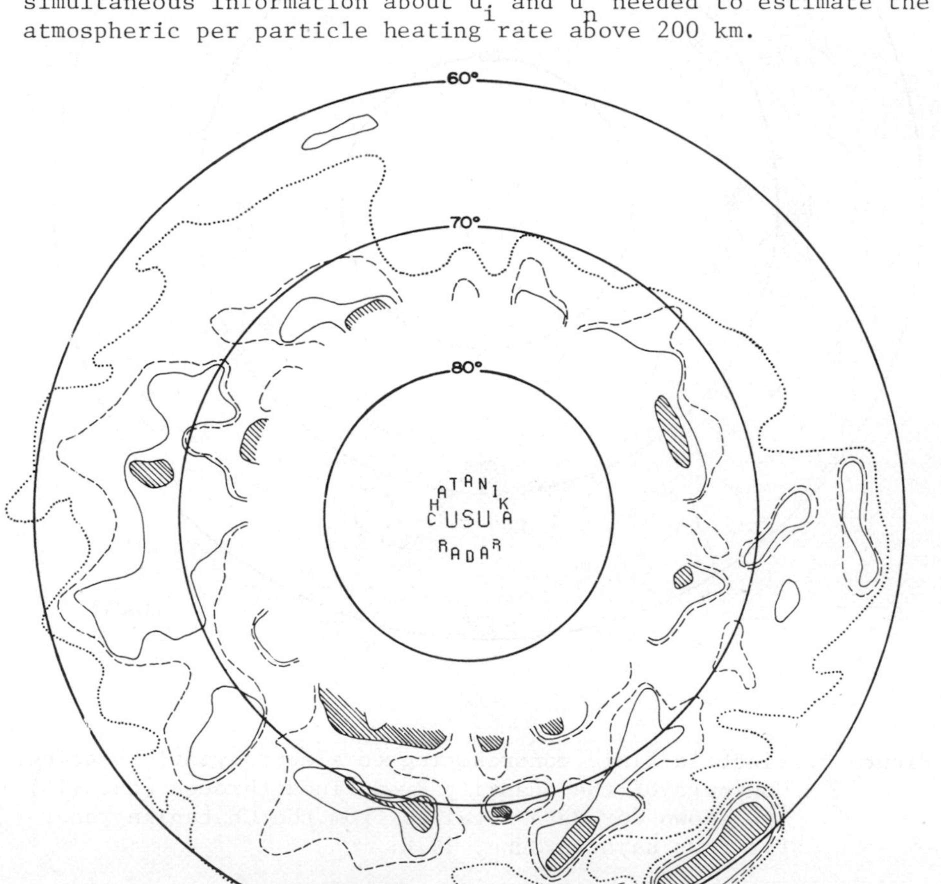

Figure 4. Electric field contribution to electrodynamic heating for July, 1979.

Within a geophysical context, the global pattern of high latitude electrodynamic heating is known only crudely. In terms of height integrated values, the results of Figure 2 show the diurnal variation of the height integrated Joule and energetic particle heating measured by incoherent scatter radar at Chatanika, Alaska [11]. The heating is clearly associated with the auroral regions and shows a diurnal variation linked to the passage of the measuring site through the auroral oval.

Although there is a wide variation of Σ_p in the auroral oval, Q_J^T is most strongly influenced by E^{*2}. In the absence of experimental results for neutral winds, only the electrostatic field, $\bar{E}$ can be reliably estimated. In Figure 3 we show the electric field contribution to the height-integrated joule heating by giving contours of the quantity R, which is related to Q_J^T by the expression

$$Q_J^T = 6.25\ \Sigma_p R\ \mu\text{watts/m}^2 \qquad (15)$$

where Σ_p is the height-integrated Pedersen conductivity given in mhos. The data of Figure 3, obtained by J.R. Doupnik and J.C. Foster with the Chatanika, Alaska incoherent scatter radar, represent a four day average of plasma drift observations in June, 1978. Strong heating tends to occur in isolated patches which often are at the edge of the radar's northern range. Furthermore, since the data used in compiling Figure 3 show R determined from averages of $\bar{U}$ rather than U^2, there may be a systematic bias towards lower values of joule heating than actually are present.

Figure 4 gives similar data for a four day period in July, 1979. Again, the peak heating rates seem to occur at the highest latitudes. To accompany Figures 3 and 4 it would be useful to have equivalent maps of Σ_p and $\bar{u}_n$. Without such data, however, one must rely the data of these figures to deduce the general features of electrodynamic heating.

3. PLASMA WAVE HEATING

In recent experiments with the Chatanika incoherent scatter radar, electrons between 95 and 115 km altitude have been observed to be up to 1000°K hotter than the neutral atmosphere [7]. Typical altitude profiles of the electron and ion temperatures are given in Figure 5. As shown in Figure 6, it was also found that T_e at 112 km is strongly correlated with the strength of the convection electric field. While such a behavior is expected on the basis of eqn. (4) to be true for

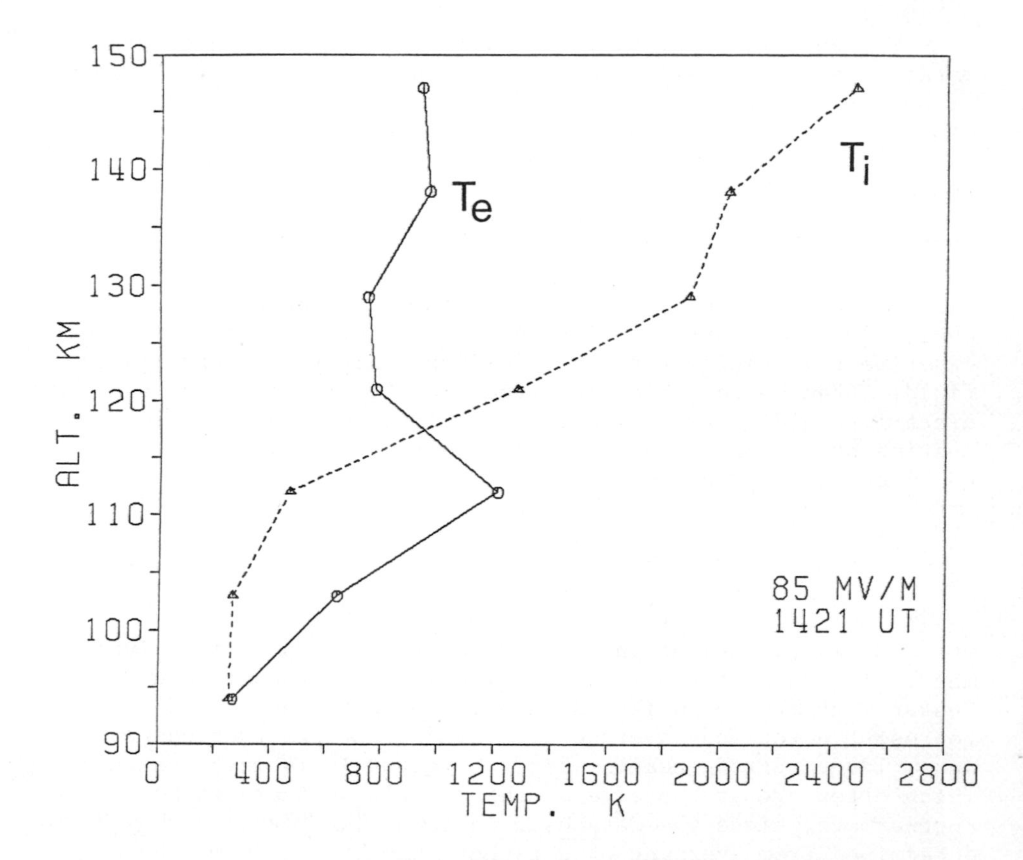

Figure 5. Electron and ion temperature profiles obtained with the Chatanika radar showing the electron temperature enhancement at 112 km associated with plasma wave heating. The high ion temperatures above 120 km are a direct consequence of ion motions through the neutral gas in response to the 85 mV/m electric field present at this time. From Schlegel and St. Maurice [7].

ions at higher altitudes (see the center panel), joule dissipation at 112 km is totally inadequate to explain the large excursions of T_e.

Following an analysis of the situation [12], it has been concluded that the enhanced electron temperatures result from unstable plasma waves arising from the Farley-Buneman modified two-stream instability. Heating of electrons and ion by these waves occurs primarily through wave trapping, while electron-neutral and ion-method collisions provide the mechanism of energy transfer to the neutral atmosphere.

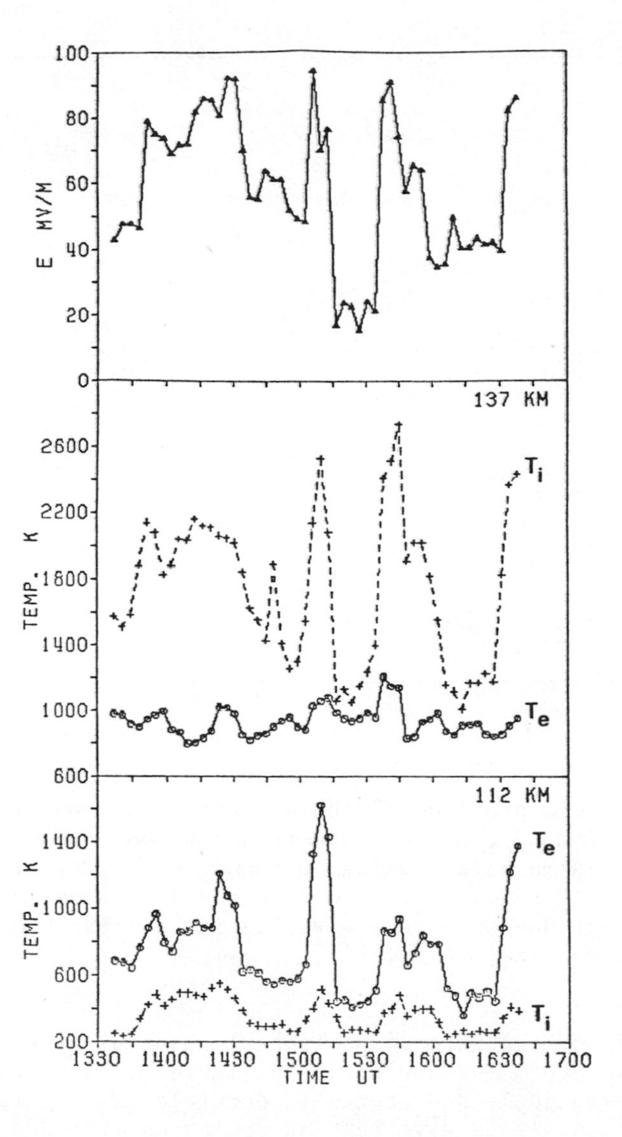

Figure 6. Observed variation of electric field and electron and ion temperatures at 112 km and 137 km on 13 November, 1979. The profiles of Figure 5 are found within this time span. Note the strong correlation of electron temperature at 112 km with the electric field. The similar variation of ion temperature at 137 km is a consequence of electrodynamic ion heating. From Schlegel and St. Maurice [7].

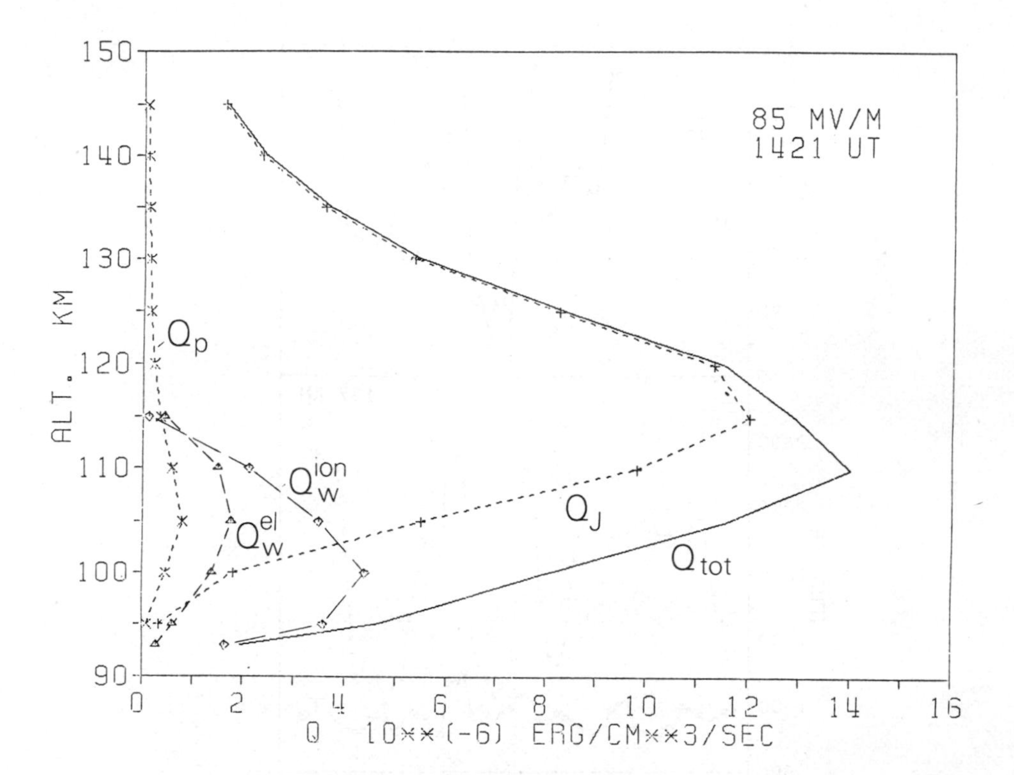

Figure 7. Altitude profiles of joule heating, Q_J, particle heating, Q_p, and the electron and ion components of plasma wave heating, Q_w^{el} and Q_w^{ion}. The overall heating rate, Q_{tot}, demonstrates a pronounced effect due to plasma wave heating at this time. From St. Maurice, et al. [12].

Estimates of the rate of atmospheric heating arising from this plasma wave process are shown in Figure 7, along with value for the observed joule and energetic particle heating rates. In general, the ion plasma wave heating dominates that due to electrons, with the ion peak lying somewhat lower than the electron peak. The example shown in Figure 7 corresponds to a large electric field. Theoretical studies [12] indicate that very little plasma wave heating occurs when the effective electric field is less than 45 mV/m. In turns of height-integrated quantities, plasma wave heating may approach 30 to 50% of the height-integrated joule dissipation rate. However, because it occurs at low altitudes, it may initiate different types of physical processes than are thought to occur in the regions of maximum joule heating.

The full range of values expected for plasma wave heating is not yet known, nor is there yet a complete understanding of the role such heating has in global scale problems.

ACKNOWLEDGEMENTS

This work was supported, in part, through NASA grant NSR-7289 and NSF grant ATM78-5797.

REFERENCES

[1] Stolarski, R.S., P.B. Hays and R.G. Roble, J. Geophys. Res., 80, 2266, 1975.

[2] Hays, P.B., R.A. Jones and M.H. Rees, Planet. Sp. Sci., 21, 559, 1973.

[3] Stockflet-Jorgensen, T., Models of high latitude electric fields, this volume.

[4] Potemra, R., Field-Aligned currents, this volume.

[5] Dickenson, R.E., E.C. Ridley and R.G. Roble, J. Atmos. Sci. 32, 1737, 1975.

[6] Straus, J.M., S.P. Creekmore, R.M. Harris and B.K. Ching, J. Atmos. Terr. Phys., 37, 1545, 1975.

[7] Schlegel, K. and J.P. St. Maurice, Anomalous Heating of the polar E region by unstable plasma waves: I. Observations, submitted to J. Geophys. Res., June, 1980.

[8] Banks, P.M. and G. Kockarts, Aeronomy, Academic Press, New York, 1973.

[9] Bostrom, R. in Cosmical Electrodynamics, edited by A. Egeland, O. Holter and A. Omholt, Universitetsforlaget, Oslo, 1973.

[10] Banks, P.M., J. Geophys. Res., 84, 6709, 1979.

[11] Banks, P.M., J. Atmos. Terr. Phys., 39, 179, 1977

[12] St. Maurice, J.-P., K. Schlegel and P.M. Banks, Anomalous heating of the polar E-region by unstable plasma waves: II. Theory, submitted to J. Geophys. Res., June, 1980.

TECHNIQUES FOR OBSERVING D REGION IONIZATION

Olav Holt

Institute of Mathematical and Physical Sciences
University of Tromsø

Abstract

Different methods for observing ionization in the D region are discussed. Radio wave experiments can be made with ground-based equipment, or between ground and rocket. Observations of differential absorption and phase are discussed in some more detail than other techniques. The possible importance of incoherent scatter observations also in the D region is pointed out. Better spatial resolution is obtained with probes carried on rockets. Some probes, which have been used successfully in the D region, are described.

1. INTRODUCTION AND OUTLINE

In the early days of radio communication, rather low frequencies were used to some extent, and the reflection of the waves in long distance communication must have been from D region heights. As radio communications turned to higher frequencies, the D region became a source of occasional trouble. Due to the high collision frequency of the electrons, the efficiency of each free electron in causing absorption of radio waves is also high in the D region. At times of increased electron density in this lower part of the ionosphere, HF communications may therefore be completely blocked - a phenomenon known as "radio blackouts". Many of the radio experiments used to study the D region, make use of this attenuation of the waves. A problem has been to obtain height resolution. The absorption of waves penetrating the region is an integrated effect. This is to some extent overcome in the study of weak

C. S. Deehr and J. A. Holtet (eds.), Exploration of the Polar Upper Atmosphere, 129–141.

partial reflections of HF waves from parts of the D region, and in recent studies of incoherent scattering of VHF waves. In transmissions between the ground and a rocket, the difficulty is removed altogether, the movement of the rocket providing the differentiation with altitude. Beside absorption, the refractive properties of the ionosphere are now frequently used also in the study of the D region, with rockets as well as in ground based experiments. The use of rockets also make more direct experiments in the D region possible. Different types of probes have been used to study electron and ion densities, and also temperatures. Mass spectrometers are difficult to operate in the D region, due to the relatively high pressure there, but interesting results on ion composition have been obtained. Rocket observations in general, and the probe measurements in particular, will give a much better height resolution than can be obtained with ground based techniques. Some radio experiments using rockets also have a high absolute accuracy in electron density measurements. Probe experiments often provide only relative values, and need to be calibrated. The advantage of groundbased observations is the possibility of obtaining long time series, and being (relatively) independent on weather conditions.

In the following, a number of experimental methods used in the study of D region ionization will be discussed. We shall attempt to emphasize the physical principles involved, omitting detailed mathematical and technical descriptions. First, we shall discuss radio wave experiments, both ground based and between ground and rocket. Then we shall continue with some examples of in situ measurements with instruments carried in rockets.

2. WAVE PROPAGATION EXPERIMENTS

2.1 Ionosonde and HF absorption measurements

The ionosonde is not particularly well suited for studying the D region. This is partly because of the technical difficulties in operating swept frequency equipment at such low frequencies that are reflected from the D region, but also because the region does not exhibit such clear layered structure as the E and F regions. Occasionally, partial reflections from the D region are seen on ionograms, but they yield little in the way of quantitative information. The parameter f_{min}, the lowest frequency on which ionospheric reflections are received, is a rough measure of the degree of ionization in the D region, since, in a simple theory, the absorption per free electron is inversely proportional to the square of the wave frequency. However, f_{min} will also depend upon equipment parameters: transmitted power, antenna gain, receiver sensitivity etc. Comparison of f_{min} from two different ionosondes, is therefore of little value. Such difficulties are partly overcome in more elaborate absorption measurements at fixed frequencies, where the transmitters and receivers are carefully calibrated, using multiple ionospheric returns. It is also possible to discriminate to a fair accuracy between the non-deviative absorption due to the high electron-neutral collision frequency in the D region, and the deviative absorption suffered by the wave near its reflection level, because the group velocity there is so low. The non-deviative absorption remains as a measure of the integrated absorption in the D region. Important results concerning the diurnal and seasonal variations in the D region were obtained with this method, but it is not much in use today.

2.2 Riometer measurements

The riometer (relative ionospheric opacity meter) developed by LITTLE and LEINBACH (1958), is an excellent instrument for routine observation of ionospheric absorption. The cosmic radio noise is recorded at a high (or very high) frequency (typically 30 MHz), within a narrow bandwidth. Normally, a relatively wide-angle vertically directed antenna is used. Narrow beam antennas, and obliquely directed antennas are sometimes used for special purposes. As the antenna, due to the rotation of the earth, sweeps over the sky, the strength of the source varies. This variation is repeated every day in sidereal time, since the cosmic radio sources do not show significant variation on such a short time scale. During normal, undisturbed ionospheric conditions, there will be some absorption of the cosmic radio noise in the ionosphere, but again there is little difference from one day to the next. Thus, from observations on selected days, it is possible to determine a "quiet day level" of cosmic radio noise recorded on a particular riometer. If the electron density in the lower ionosphere increases, for instance as a result of sudden increases in the intensity of ultraviolet or X-rays from the sun, or due to energetic particle precipitation, the cosmic radio noise level recorded by the riometer decreases. Figure 1 shows an auroral absorption event, recorded on a 30 and 40 MHz riometers at Tromsø, Norway. The quantity actually recorded with most riometers, is a current proportional to the received power. The absorption is measured as the power level P, relative to the "quiet day value" P_o, in units of dB.

$$L = 10 \log_{10}(P_o/P) \quad [\text{dB}]. \tag{1}$$

We shall consider how this is related to the electron density. We will assume that the ionosphere is horizontally stratified over the area covered by the antenna, i.e. that the electron density N, is a function of altitude, h, only. Then, for a plane wave at vertical incidence

$$P = K \exp\{-2 \frac{\omega}{c} \int_o^\infty \chi \, dh\} \; . \tag{2}$$

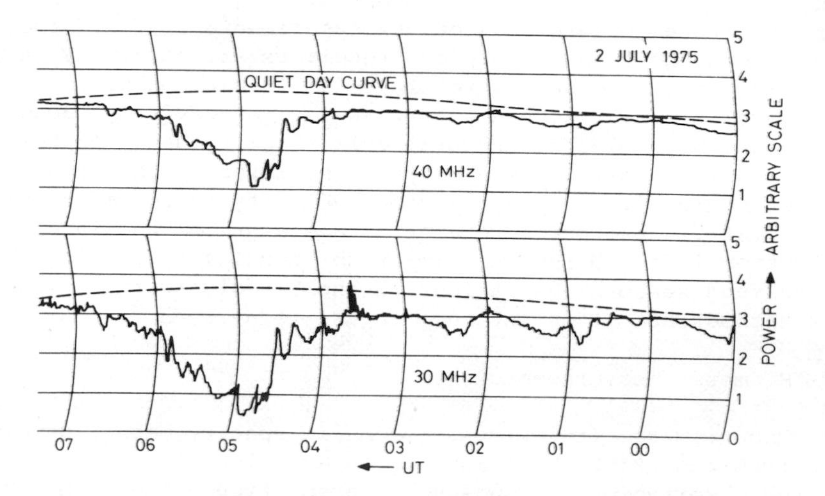

Figure 1. Example of riometer recording of an auroral absorption event at Ramfjordmoen, Tromsø.

where χ is the imaginary part of the refractive index. K is a proportionality factor, depending on equipment parameters and cosmic radio noise level outside the atmosphere. Since χ is proportional to electron density N, we can write

$$2 \frac{\omega}{c} \chi = f(\nu_M) \cdot N \tag{3}$$

where $f(\nu_M)$ is a known function of the collision frequency ν_M. Thus

$$L = 20 \log_{10} e \int_0^\infty (N_{disturbed} - N_{quiet}) f(\nu_M) dh \tag{4}$$

is a measure of the integrated absorption by the excess ionization.

Often $N_{disturbed}$ is orders of magnitude larger than N_{quiet}. Very often only part of the absorption measured by riometers, is due to increased electron density in the D region. This has become evident from comparison with electron density profiles measured by other techniques, for example using rockets. Only partly can this be explained by the fact that some of the absorption takes place at higher altitudes. LEINBACH (1962) considered a number of error sources, and found that for moderate absorption ($\lesssim$ 2 dB) they are probably less than the scaling accuracy in the recordings. For very large absorption (> 15 dB) the possible corrections would still be < 20 percent. A larger correction may come from a proper treatment of the direction sensitivity of the antenna. This was also treated by LEINBACH, who found that for small absorption (~ 1 dB) this could increase the apparent absorption by as much as 50 percent. For larger absorption, the correction would be smaller. In many cases, it still seems that the observed electron densities cannot account for the simultaneously observed absorption. It does not seem likely that an experimental error of this magnitude should have been overlooked, so it is reasonable to look for a source of additional attenuation of the cosmic noise. REID (1964) has studied the influence of a high energy tail of the electron energy distribution. (In the generalized magnetoionic theory a maxwellian distribution is assumed.) He finds that such "non-thermal" absorption may contribute significantly to auroral absorption observed with riometers. Scattering by irregularities in the F and E-regions is an effect that should be investigated (D'Angelo 1976).

It is possible to obtain some information on the altitude of the increased ionization from riometer observations at several frequencies simultaneously. This is due to the nonlinear variation of the absorption index per electron, with wave frequency and collision frequency. In the region above $\sim$ 70 km, an increase in electron density results in the same relative increase in k for all frequencies ($\gtrsim$ 15 MHz). Below this height, the observed absorption due to the same increase in electron density will be different for the different frequencies. Without going into the detailed procedure (PARTHASARATHY et al., 1963) it is clear that this information can be used to estimate N(h). The method is, however, not very accurate unless a large number of frequencies are used, and the altitude region for which the method is applicable, is rather limited. Anyway, there are other methods for determining electron densities which are in most respects, far better.

Despite all the uncertainties mentioned, the riometer has proved an invaluable tool in ionospheric research, particularly in the study of the morphology of absorption phenomena. The cosmic radio noise level is recorded continuously, with a minimum of inspection, so that even very short events will not escape observation. Most riometers are self-calibrating. At the relatively high frequencies used (20-50 MHz), the equipment is not "blacked out" even during very strong absorption.

2.3 LF and VLF observations

The reflections of VLF and LF waves from the ionospheric D region can be described by parameters such as apparent reflection heights, reflection coefficients and polarization. It does not seem possible to determine the ionization from the measurement of these parameters. An approach adopted by PITTAWAY (1965) is to use a full wave integration technique to calculate the wave-fields set up by a transmitter in a D region model (N and ν_M). An accuracy of ± 20 percent in electron density at any altitude between 55 and 90 km has been claimed.

2.4 Differential absorption and phase measurements

In these experiments the difference in the absorption and phase variation along the propagation path, between the ordinary and extraordinary wave modes, are used to deduce the electron density as a function of altitude. In some cases the two wave components are transmitted in alternating pulses, in other experiments a linearly polarized wave is transmitted and the two components are separated on the receiving side.

Assume that equal power is transmitted in each of the two modes. Let $E_{o,x} = A_{o,x} \exp\{i\, \varphi_{o,x}\}$, where indices refer to the ordinary and extraordinary components. Then

$$E_x/E_o = (A_x/A_o)\exp\{i(\varphi_x-\varphi_o\} = \exp\{i \int_T^R (k_x-k_o)\,dh\}. \qquad (5)$$

the integral being taken from transmitter to receiver along the propagation path where k_o and k_x are the complex wave numbers $k_{o,x} = n\cdot k^o = \mu k^o + i\chi k^o$, $k^o = \omega/c$ being the free space wave number and $n = \mu + i\chi$ the complex refractive index. Substituting this in (5) we obtain for the amplitude ratio and the Faraday rotation

$$\frac{A_x}{A_o} = \exp\{-\frac{\omega}{c}\int_T^R (\chi_x-\chi_o)\,dr\} \qquad (6)$$

$$\varphi_x - \varphi_o = \frac{\omega}{c}\int_T^R (\mu_x-\mu_o)\,dr \,. \qquad (7)$$

If the transmitter-receiver path is vertical, the integral is taken from zero to an altitude h, and by differentiation we obtain

$$\frac{d}{dh}\left(\ln \frac{A_x}{A_o}\right) = -\frac{\omega}{c}(\chi_x-\chi_o) = F(\nu_M)\cdot N \qquad (8)$$

$$\frac{d}{dh}(\varphi_x-\varphi_o) = \frac{\omega}{c}(\mu_x-\mu_o) = G(\nu_M)\cdot N \qquad (9)$$

Thus, if we can measure both the amplitude ratio and the phase difference as functions of altitude, both ν_M and N can be determined.

Using a rocketborne receiver, both A_x/A_o and $\varphi_x-\varphi_o$ can be measured in an elegant and quite simple experiment, making use of the rocket's spin around its own axis. A linearly polarized wave is transmitted from the ground, and received on a linear antenna in the rocket, perpendicular to the spin axis. Now, the received amplitude in the rocket will have a maximum when $\varphi_x-\varphi_o = 0$, since then the amplitudes of the two wave modes are added. There will be a minimum in the received signal when $\varphi_x-\varphi_o = \pi$, corresponding to the difference of the mode amplitudes.

A typical speed of the rocket is ~ 1 km/s. A normal spin rate is a few Hz, which is much more rapid than the change in $\varphi_x-\varphi_o$. Thus, for each rotation of the rocket, we obtain a value for A_o+A_x and A_o-A_x, which can then be reduced to give $A_x/A_o(h)$.

The rocket spin can be measured for instance by recording the magnetic field component in one fixed radial direction. The phase difference between the rocket spin and the variation of the received amplitude of the wave, will then be a measure of the Faraday rotation, $\varphi_x-\varphi_o$. This is illustrated in figure 2.

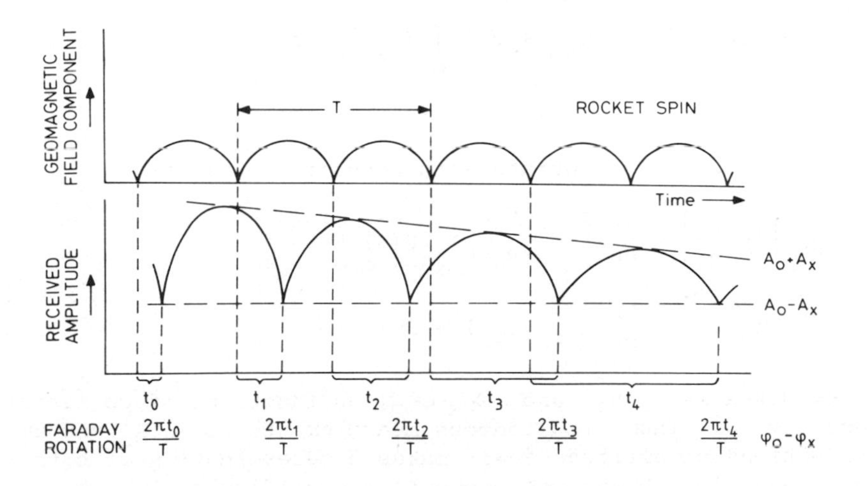

Figure 2. Principle of obtaining differential absorption and phase from simple rocket experiment.

In an entirely ground based experiment it would seem difficult to obtain A_x/A_o and $\varphi_x-\varphi_o$ as a function of altitude. However, in the frequency range of say 2-8 MHz, partial reflections of pulsed transmissions are often observed from the D region, sometimes over a rather wide range of altitudes.

The mechanism of reflection is not well known. Small irregulatities in the electron density, distributed throughout the D region, or large parts of the region, could produce the observed reflections. There is actually considerable evidence for this mechanism, known as volume scattering. In some cases it seems as if other mechanisms must be sought. For the present, we shall not dwell upon this point, but be content with describing the reflection by a reflection coefficient, R, known as a function of altitude. In general R must describe a change in both amplitude and phase upon reflection, i.e. $R_{o,x} = |R_{o,x}|\exp\{i\ \varphi_{R,o,x}\}$. Then, for a partial reflection from the height, h, the received field is given as

$$E_{o,x} = K \cdot R_{o,x} \cdot \exp\{2i \int_o^h k_{o,x}\, dh\} \tag{10}$$

K is a factor including equipment parameters and geometrical factors. We assume here that K is the same for both wave modes. Then, for the amplitude ratio and phase difference, we obtain

$$\frac{A_x}{A_o} = \frac{|R_x|}{|R_o|} \exp\{-2 \frac{\omega}{c} \int_o^h (\chi_x - \chi_o)\, dh\} \tag{11}$$

$$\varphi_x - \varphi_o = \phi_{Rx} - \phi_{Ro} + 2 \frac{\omega}{c} \int_o^h (\mu_x - \mu_o)\, dh \tag{12}$$

Again, we differentiate with respect to height:

$$\frac{d}{dh}\left[\ln \frac{A_x}{A_o}\right] = \frac{d}{dh}\left[\ln \frac{|R_x|}{|R_o|}\right] + 2F(\nu) \cdot N \tag{13}$$

$$\frac{d}{dh}(\varphi_x - \varphi_c) = \frac{d}{dh}(\phi_{Rx} - \phi_{Ro}) + 2G(\nu) \cdot N \tag{14}$$

Theory gives $|R_x|/|R_o|$ and $(\varphi_{Rx}-\varphi_{Ro})$ as functions of collision frequency ν_M. Thus simultaneous measurements of A_x/A_o and $\varphi_x-\varphi_o$ as functions of altitude again makes a determination of both ν_M and N possible. Phase measurements are difficult with the partial reflection method. In most experiments, therefore, only A_x/A_o has been measured. It is then necessary to know ν_M from some other source to obtain the electron density. Figure 3 shows an example of partial reflection amplitude observations, for an assumed model of $\nu_M(h)$. It should be noted that for the lowest altitudes, there is very little absorption below the reflection level. Then $A_x/A_o \sim |R_x|/|R_o|$, so that, at these altitudes, we obtain some information on ν_M from amplitude measurements alone.

The observed amplitudes A_o and A_x vary with time, i.e. they show what we call <u>fading</u>. This is connected with the reflection mechanism, and we shall not discuss it further here. To obtain electron densities from (13), we use average values from a large number of individual echoes. Observations show little difference between the averaged ratio, $\langle A_x/A_o \rangle$, or the ratio of averages $\langle A_x \rangle / \langle A_o \rangle$.

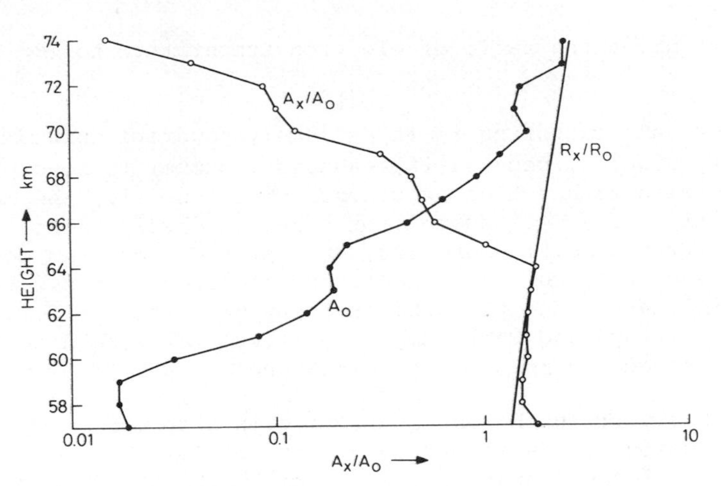

Figure 3. Example of observations of the ratio A_x/A_o, averaged over a 30 s period. Also shown is the amplitude A_o on a relative scale.

2.5 Incoherent scattering

At frequencies much higher than the plasma frequency (VHF and UHF), small fluctuations in electron density can lead to scattering of the incident wave, called incoherent scattering. The fluctuations are connected with the thermal motion of the ions, in combination with electrostatic forces between ions and electrons. The fluctuations can be described by a spatial spectrum, and the scattering is due to the components in this spectrum that has wave-numbers near twice the wave-number in the transmitted wave (Bragg condition). The ion motion leads to a doppler broadening of the received wave. The plasma can support ion-acoustic waves of the required wave-numbers. The velocity of these will give a doppler offset of the scattered wave, and lead to "shoulders" in the observed power frequency spectrum. Figure 4 shows the radar scattering cross section as function of frequency. The electron density is proportional to the total scattered power, which is given by the area beneath the curve of figure 4. The width of the spectrum is given by the ion temperature, whereas the position of the

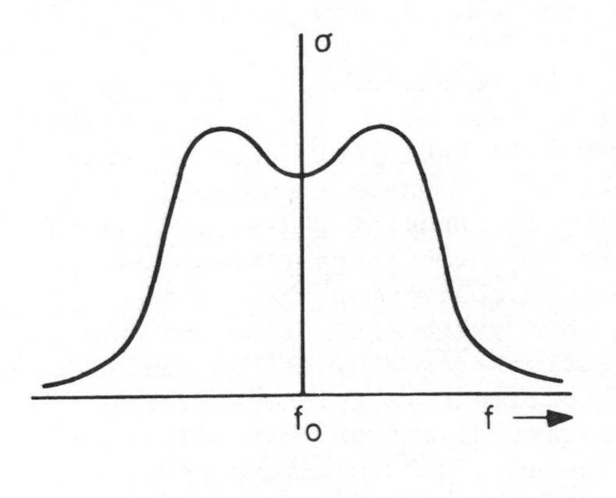

Figure 4.
The shape of the radar scattering cross section σ.

"shoulders" gives the ratio of electron temperature to ion temperature.

Incoherent scattering is an extremely powerful method for ionospheric studies, because it measures a number if ionospheric parameters with rather high accuracy. Unfortunately, the method is not particularly well suited for D region observations, because 1) the electron densities are low, so that the scattered power is low, and hence difficult to detect, and 2) irregularities in the D region ionization due to turbulence may give rise to relatively strong scattering, and there may be difficulties with discriminating between this and the incoherent scattering.

The theory of incoherent scattering is also more complicated for the D region. This is partly due to collisions between the charged particles and neutrals, which will make the observed spectrum more narrow, and partly to the presence of negative ions, which will have the opposite effect on the spectrum. If the collision frequencies can be measured in some way, or be reliably estimated, there is a possibility that negative ion densities can be measured. This seems to be the only possibility of obtaining this information with groundbased equipment, and would indeed be extremely interesting. Mainly for this reason we have included incoherent scatter here as a technique for studying D region ionization.

2.6 Modification experiments

The absorption of an HF radio wave in the ionosphere results in an increase in the temperature of the electron gas. Since the degree of ionization in the D region is so low, it is not possible to heat the neutral gas in this way; and the cooling rate of the electron gas when the HF wave is removed, is very rapid.

This effect has been used in the so-called cross modulation experiments. It is arranged that a pulse at a frequency reflected from the E or F layers, on its way down meets with another pulse on a separate frequency, going up. The down-going pulse then passes through a region "heated" by the upgoing pulse. Since the collision frequency increases with electron temperature, the down-going pulse will suffer a little more absorption than if the up-going pulse had not been there. This extra absorption depends on the electron density and the collision frequency at the altitude where the two pulses meet. The experiment is further arranged so that this meeting altitude can be varied, and so that only every second of the downgoing pulses meet with a "disturbing" pulse. Thus it is possible to measure the extra absorption continuously for the selected altitudes. With the power normally used in these experiments, this is a small effect, and the method is not very accurate for measuring N and ν_M.

A much stronger heating effect has been obtained by the use of very powerful transmitters and more directive antennas. Equivalent isotropic power with such transmitters is more than 100 MW. For technical reasons, and also from the point of view of frequency allocations, these transmissions can not be pulsed. The increased electron temperature will therefore occur throughout the D region simultaneously. Experiments like partial reflections or incoherent scatter must be used as diagnostic tools. The strong increase in electron temperature will affect the collision frequency as well as some of the chemical reaction rates that determine the electron density. Such modification experiments may therefore be a means of studying the reaction kinetics in the D region in a semi-controlled fashion.

3. PROBE EXPERIMENTS

Different types of probes carried on rockets have been used to study the electron and ion densities and temperatures in the ionosphere. The theory for most probes is complicated by the motion of the rocket through the ionosphere. This is particularly difficult for the D region. Still, valuable results have been obtained, and we shall discuss briefly two types of probes that have been used in D region studies.

The general remark should be made that absolute values of ionization density is hard to obtain by probes, and more so in the D region than higher in the ionosphere. Some form of calibration by comparison with other experiments must normally be used. The main advantage with the probes is the possibility of studying fine spatial structure that is not readily obtained with wave propagation experiments.

3.1 Electrostatic probes

By an electrostatic probe we mean a device where the current to a conductor immersed in a plasma, can be measured as a function of the potential of the conductor relative to the plasma. In its simplest form this is known as a Langmuir probe. Figure 5 shows the current-voltage characteristic of a Langmuir probe in a laboratory plasma.

The potential V_f at which the current is zero, is called the floating potential. This is the potential to which the probe would adjust itself if left alone in the plasma. The electron current then equals the ion current to the probe. For sufficiently high negative voltages all incident ions are collected and all electrons are rejected. For constant ion mobility, the ion saturation current should be proportional to ion density in the plasma. As the potential on the probe is increased above the floating potential,

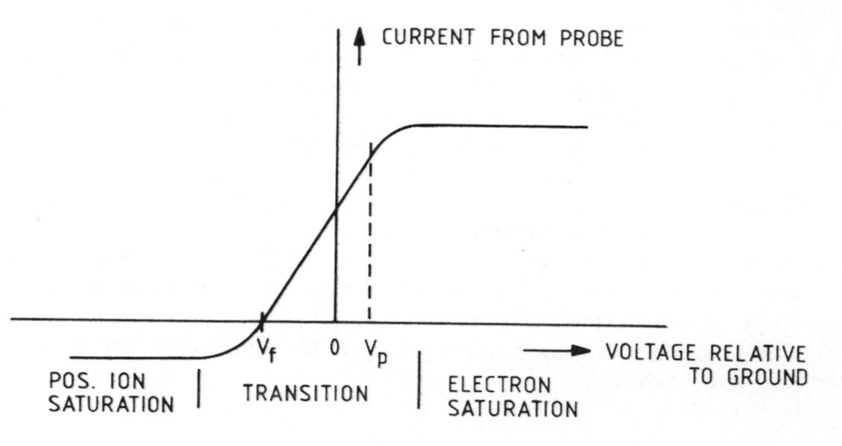

Figure 5. Simplified picture of the current-voltage characteristic of a Langmuir probe.

more and more electrons are collected, until we reach a potential, V_p, where all ions and electrons are collected. The potential on the probe is then the same as in the ambient plasma. If we increase the voltage further, ions are rejected, and for sufficiently high potential the current will again saturate, and be proportional to the electron density. Electron temperature can be obtained from the lower part of the transition region. This is to a certain extent valid even if the probe is put on a rocket, because the thermal velocities of the electrons is much higher than the rocket velocity. There are a number of theoretical and practical problems with probe measurements in the ionosphere. In the D region this is particularly so, since the plasma is collision-dominated, and the rocket is moving at supersonic speed. For a more extensive discussion of this, reference is made to FOLKESTAD (1970). For ion diagnostics in particular, it has been found advantageous to insert one or more grids between the collecting probe and the plasma. The grids may be kept at fixed or varying potentials relative to the collector, and may serve several purposes, such as to reduce the effect of photoelectric emissions from the probe surface or to reject unwanted ambient particles.

Recently, THRANE and GRANDAHL (1980) have obtained very interesting results on positive ion density fine structure in the D region with a screened spherical probe. The instrument has a large dynamic range, high accuracy and good time resolution. The fluctuations seen are of the order of 1 percent, but with considerable variation with altitude.

3.2 RF Capacitance Probe

At RF frequencies, a conductor immersed in a plasma, can be looked upon as a capacitor in a dielectricum. The impedance of this capacitor as a circuit element depends upon the plasma parameters: electron density, temperature and collision frequency. This is utilized in the RF capacitance probe, where the impedance is made part of the frequency-determining circuit of an oscillator, and the oscillator frequency is measured.

In rocket experiments, spherical probes have the advantage that their geometry relative to the gas will be independent of the orientation of the rocket. Performing the experiment at several frequencies, and with two spheres of different size, good measurements of electron densities in the D region can be obtained (HOLT and LERFALD 1967).

REFERENCES

D'Angelo, N.: 1976, J. Geophys. Res. 81, 5581.
Folkestad, K.: 1970, NDRE-Report No. 59, Norwegian Defence Research Establishement, N-2007 Kjeller, Norway.
Holt, O. and Lerfald, G.M.: 1967, Radio Science 2, 1283.
Leinbach, H.: 1962, Scientific Report 3, NSF Grant No. G 14133, Geophysical Institute, University of Alaska.
Little, C.G. and Leinbach, H.: 1958, Proc. IRE 46, 334.
Parthasarathy, R., Lerfald, G.M., and Little, C.G.: 1963, J. Geophys. Res. 68, 3581.
Pittaway, M.: 1965, Phil. Trans. Roy. Soc. 257, 219.
Reid, G.C.: 1964, J. Geophys. Res. 69, 3296.
Thrane, E.V. and Grandahl, B.: 1980, FFI-rapport 80/7002, Norwegian Defence Research Establishement, N-2007 Kjeller, Norway. (Also submitted to J. Atmosph. Terr. Phys.)

AURORAL RADIO ABSORPTION IN RELATION TO MAGNETOSPHERIC PARTICLES

P. N. Collis and J. K. Hargreaves

Department of Environmental Sciences,
University of Lancaster, Lancaster LA1 4YQ England.

A. Korth

Max Planck Institut für Aeronomie,
D-3411 Katlenberg - Lindau 3, W. Germany.

ABSTRACT

A comparison of auroral radio abosrption observed by riometers with simultaneous measurements of electrons in the 15 to 300 keV range at geosynchronous orbit has been made. The results indicate that for the events studied the effective loss rate in the D region is directly proportional to the electron density, rather than, as is conventionally assumed, to the square of the electron density, i.e. under steady conditions the production rate q is given by $q = \beta N_e$ rather than by $q = \alpha N_e^2$. This suggests a two step loss process rather than simple dissociative recombination.

The most important contribution to the total absorption during winter, night time events is shown to be due to electrons in the 40 to 80 keV range, which produce absorption in the 85 to 95 km altitude range.

The use of a riometer network to identify the "footprint" of the GEOS-2 satellite is illustrated by an example in which the footprint appears to be 30°W of its assumed position.

1. INTRODUCTION

Although auroral radio absorption measured by riometers is often used as a monitor of high energy electron precipitation, there is still little knowledge of the quantitative relationship between the ionisation source and the observed absorption. One

C. S. Deehr and J. A. Holtet (eds.), Exploration of the Polar Upper Atmosphere, 143–148.

of the greatest areas of uncertainty is in the height profile of the effective recombination coefficient and in the chemical processes governing this in the D region of the ionosphere. In order to study this relationship further, comparisons have been made between auroral absorption and simultaneous measurements of electron fluxes at geosynchronous orbit.

The riometer observations are from the University of Lancaster riometer network in northern Scandinavia which consists of five spaced stations operating at 30 MHz (Figure 1). The stations are between 100 and 200 km apart, allowing estimation of the velocity of events across the network, when the events can be timed at three or more stations.

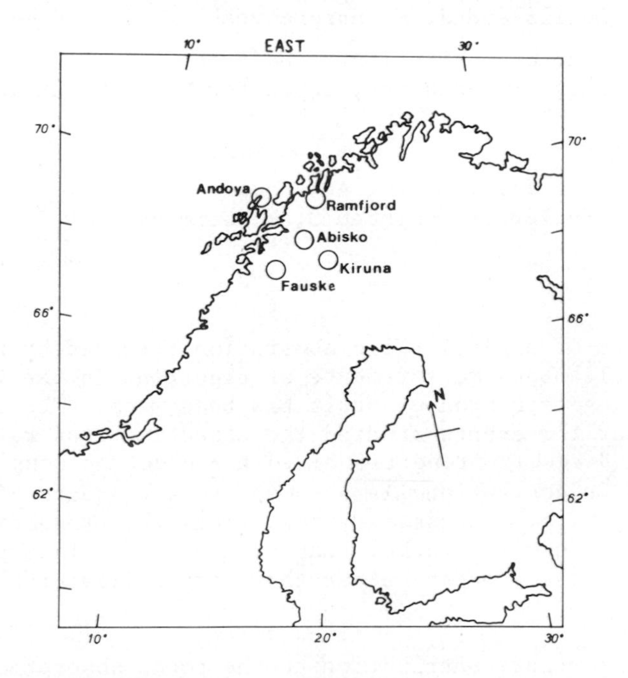

Figure 1 The University of Lancaster riometer network in Scandinavia.

The period covered by the present work is September to December 1978 when the satellite GOES-2 was in the Scandinavian sector. At this time the satellite footprint was estimated to be to the north east of the riometer network, some 200 km from Abisko. Measurements of electron fluxes in the 15 to 300 keV range are provided by the charged particle spectrometer on GEOS-2 (Korth and Wilken, 1978).

2. TEMPORAL VARIATIONS OF AURORAL ABSORPTION AND ELECTRONS AT SYNCHRONOUS ORBIT

In selecting intervals for the comparison of satellite and ground based data, it has to be borne in mind that the satellite footprint may not be located close to a riometer station and therefore there may be a time delay between an event at the ground and the similar feature observed at the satellite. For the predicted geometry and for typical event velocities, similar features in the absorption and in the electron flux are expected to be observed within a few minutes of each other, the sign of the time difference depending upon local time.

The data generally bear this out, but in one case the time delay seems anomalously large. Figure 2 shows that the enhancement of electrons occurs approximately 17 minutes after the onset of absorption at Abisko. Previous work (Hargreaves, 1967) has shown that in this time zone absorption events typically move westwards, and a time delay of 17 minutes is equivalent to an east-west separation of two hours local time, that is, 30° of longitude. This would indicate that the field line connected to the satellite was probably nearer to Iceland than to Scandinavia. Evidence supporting this is seen in the absorption record from the riometer at Siglufjördur in Iceland. Although the amplitude has decreased, this is presumably the same event as the one observed at Abisko but occurring 25 minutes later at Iceland. This is consistent with the westward propagation of the event, and with the footprint of the satellite field line lying between Iceland and Scandinavia.

This example illustrates that before comparing magnitudes it is necessary to first ascertain that the events at the satellite and on the ground are indeed simultaneous.

3. COMPARISON OF AURORAL ABSORPTION AND SIMULTANEOUS OBSERVATIONS OF ELECTRONS AT SYNCHRONOUS ORBIT

Early work has shown conclusively that a riometer responds to the higher energy particles of the auroral electron spectrum. Detailed measurements of the electron spectrum may be used to compute the absorption profiles and from that the expected absorption, but this procedure requires a knowledge of the effective recombination coefficient (α) as a function of height, which is a rather uncertain quantity.

Detailed spectra were selected for 9 separate intervals of winter, night time conditions during which auroral absorption ranging between 0.2 and 6 dB was observed, and two intervals were also included during which there was no detectable absorption. Ionospheric production rates (q) due to the precipitating

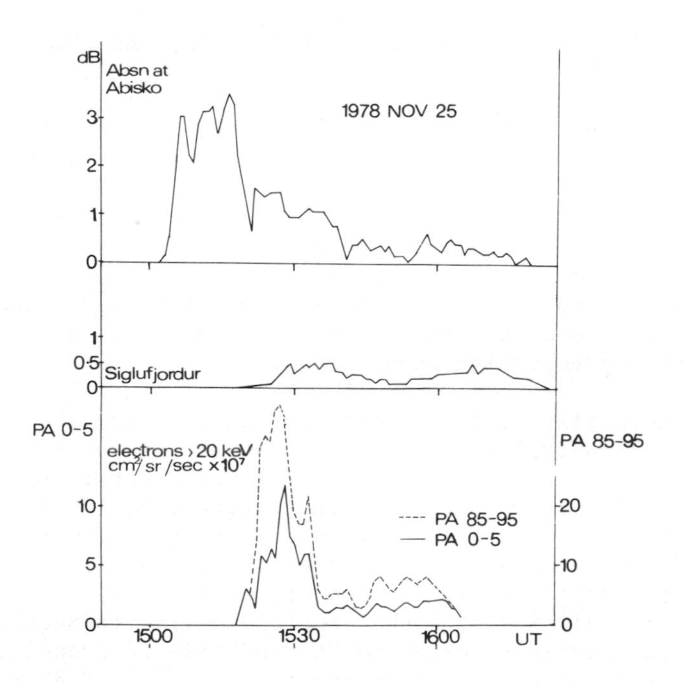

Figure 2 Simultaneous observations of integral electrons fluxes > 20 keV at GOES-2 and auroral absorption at Abisko (long. 18.5°E) and Siglufjordur (long. 18.5°W).

electron flux in the 15 to 300 keV range were computed as a function of height using the method of Rees (1963), as applied by Evans et. al. (1977).

The absorption suffered by a radio wave passing through the atmosphere is equal to the height-integrated product of electron density and specific absorption. Profiles of specific absorption for 30 MHz are available in the literature and the electron density is usually taken to be $\sqrt{\frac{q}{\alpha}}$. To get some idea of the heights important for the production of absorption we have therefore plotted $\sqrt{q}$ against the observed absorption over the height range of interest, which we take to be 70 to 120 km. It was found that the results did not support the square root relationship particularly well at any height.

The conventional assumption that the electron density depends upon $\sqrt{q}$ is a result of assuming that the loss rate is governed by simple dissociative recombination of electrons with molecular ions. However, suggestions have been made (Haug and Landmark, 1970) that a two-ion model may apply in the D region, where one species, namely cluster ions, has a much faster recombination rate than the simple ions. In this case the electron density depends linearly upon q, and we might expect a linear relationship between q and absorption. In fact, the present data do seem to support this idea, particularly at heights of 85 to 95 km.

If N_e depends linearly on q, we can determine an approximate value for β, the effective attachment coefficient, from the results (by summing the production rates for a chosen absorption), and this value appears to be within the range of strongly temperature dependent possibilities given by Arnold et. al. (1980), where the rate controlling reaction is the first step in the formation of cluster ions from NO^+. The present analysis gives $\beta \sim 1.1\ s^{-1}$ at 85 km. Since that first step is a 3-body reaction, β should vary as (air density)^2, and the overall process should go over to an α-type process via the simple ions at some altitude (e.g. above 95 km). Thus there does appear to be some chemical basis for the result that the radio absorption varies linearly with the production rate.

Besides indicating the possible loss mechanism, the results indicate the important heights for production of absorption at night as 85 to 95 km, and the heights where production of absorption is unimportant - below about 80 km. This is best seen in a plot of correlation coefficient between production rate and absorption as a function of height, Figure 3. The apparent high correlation above about 100 km is due to the restriction of the lower energy limit to 15 keV. If lower energy electrons were considered the correlation coefficient at these heights would decrease. The peak values between 85 and 95 km indicate the main region of absorption and correspond to the height of maximum production rate for 40 to 80 keV electrons. Below 85 km the correlation coefficient falls rapidly to insignificant values. Looking at these results empirically, we can give a linear relation between production rate at 90 km (or N_e at 90 km) and 30 MHz radio absorption. For this group of events:

$$\text{absorption} = \frac{q\ (90\ \text{km})}{4 \times 10^4}\ \text{dB}$$

where q is in units of $cm^{-3}\ s^{-1}$, or

$$\text{absorption} = \frac{N_e\ (90\ \text{km})}{10^5}\ \text{dB}$$

where N_e is the electron density in cm^{-3}.

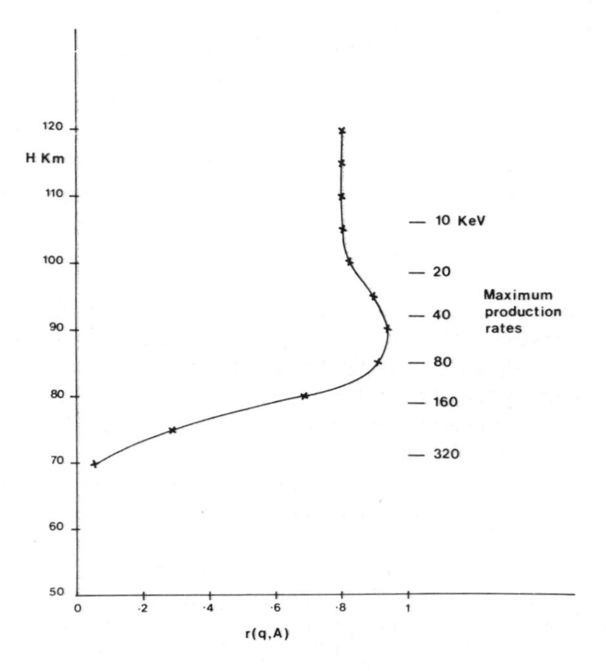

Figure 3 Height variation of the correlation coefficient between q and absorption.

ACKNOWLEDGEMENTS

Thanks are due to Dr. A. Brekke (University of Tromsø), Dr. G. Gustafsson (Kiruna Geophysical Institute), Mr. S. Kristensen (Fauske) and personnel at the Abisko Scientific Research Station and Andøya Rocket Range for their help with the riometer network in Scandinavia, and Mr. S. Kristinsson (Siglufjördur) in Iceland. The study has been supported by the U.K. Science Research Council.

REFERENCES

Arnold, F., Krankowsky, D., Zettwitz, E., and Joos, W. 1980, J. Atmos. Terr. Phys., 42, 249.

Evans, D.S., Maynard, N.C., Trøim, J., Jacobsen, T., and Egeland, A. 1977, J. Geophys. Res., 82, 2235.

Hargreaves, J.K. 1967, J. Atmos. Terr. Phys., 29, 1159.

Haug, A., and Landmark, B. 1970, J. Atmos. Terr. Phys., 32, 405.

Korth, A., and Wilken, B. 1978, Rev. Sci. Instrum., 49, 1435.

Rees, M.H. 1963, Planet. Space Sci., 11, 1209.

OPTICAL REMOTE SENSING OF THE POLAR UPPER ATMOSPHERE

Gordon G. Shepherd

Centre for Research in Experimental Space Science
York University, 4700 Keele Street,
Toronto, Canada M3J 1P3

INTRODUCTION

Optical observations from satellites and from the ground provide a useful way of studying the magnetosphere - atmosphere interface, where precipitating electrons give up their energy. Optical remote sensing of the upper atmosphere can be conducted by viewing upwards from the ground, or downwards from a satellite (Fig. 1(a)).

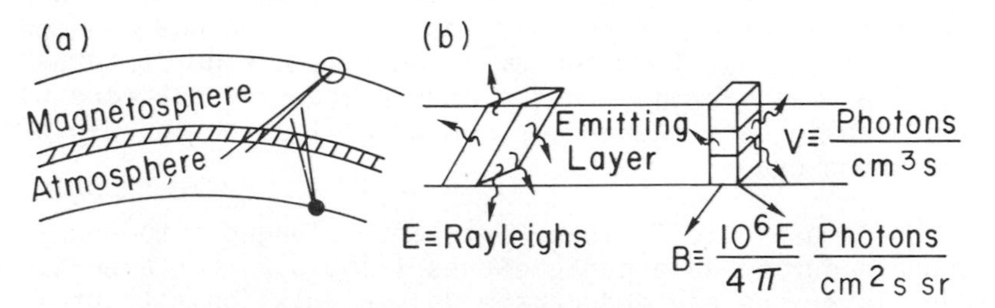

Fig. 1(a) The magnetosphere-atmosphere interface.
(b) Integrated emission rate E, volume emission rate V and surface brightness B.

Normally the thickness of the emitting layer is small compared to the observer's distance from it, so that one idealizes the situation as viewing at some angle through a thin layer. This layer may be considered the interface between the magnetosphere and the ionosphere, which is one of the reasons why its study is so useful and important. What is measured is the number of photons emitted per second from a 1 cm^2 column along the line of

C. S. Deehr and J. A. Holtet (eds.), *Exploration of the Polar Upper Atmosphere*, 149–158.

sight. (Fig 1 (b)). The emission of 10^6 photons/sec from such a column is defined to be one rayleigh. The surface brightness (radiance) of this layer for normal incidence (Fig 1(b)) is $10^6 E/4\pi$ photons cm^{-2} sec^{-1} sr^{-1} where E is the emission rate in rayleighs. Such integrated measurements are very useful but may be difficult to interpret, since one is implicitly trying to solve an integral equation. In a rocket flight one normally flies through the layer (Fig. 2).

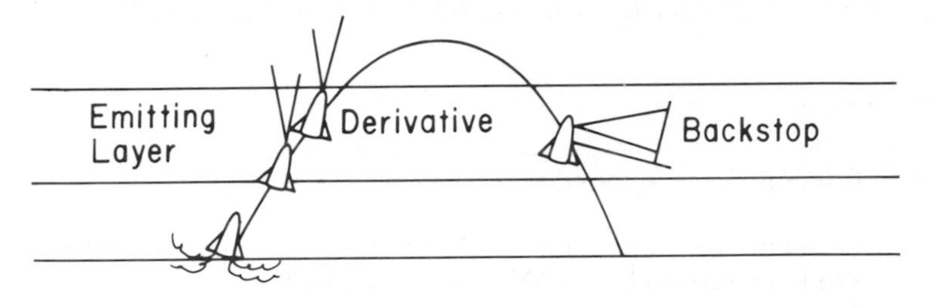

Fig. 2 Illustrating the backstop and derivative methods of determining volume emission rates from rockets.

If one measured the intensity against a dark backstop the optical measurement would become a direct one, giving the volume emission rate, the photons/sec emitted from one cm^3. Since a 10 km layer is 10^6 cm thick, a volume emission rate V of 1 photon cm^{-2} sec^{-1} corresponds to 1 rayleigh of emission rate for this layer. For reasonable sensitivity one would need a backstop about one km from the rocket and to my knowledge this has not been done. Instead one uses the derivative method, in which the difference between the intensities with the rocket at intervals of one km (say) is determined. With good instrumentation one can measure down to 1R, so that 1 photon cm^{-3} sec^{-1} is a kind of threshhold for optical measurements in the atmosphere. Recent careful work is now breaking the 1R barrier, which will allow the exploration of new phenomena.

The Atmospheric Explorer Satellite plunged into the emitting atmosphere during each perigee pass (Fig. 3). For a horizontally uniform layer one can deduce the volume emission rate profile from a single satellite rotation. This is a powerful method when the conditions are satisfied, and yields similar information to a rocket flight, once per pass. For more complex patterns vertical and horizontal volume emission rate profiles can be calculated by multiple "looks" through an emitting region using image reconstruction procedures sometimes called tomography. This can be done using a satellite, or multiple ground stations as shown in Fig. 4.

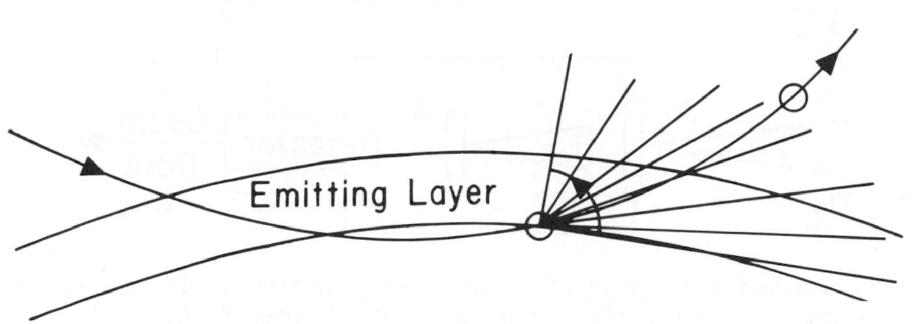

Fig. 3 Showing a schematic Atmospheric Explorer perigee pass through the atmosphere. The angular scan can be in verted to a vertical volume emission rate profile.

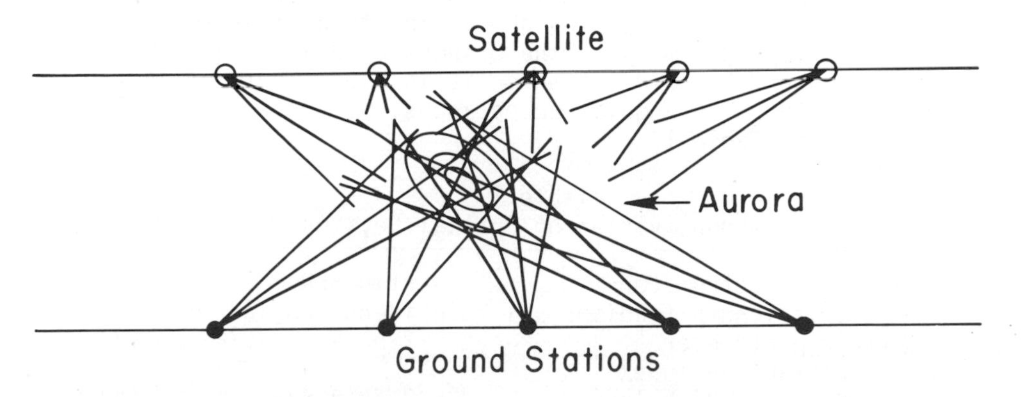

Fig. 4 Illustrating the image reconstruction (tomographic) analysis of an auroral form from above and below.

RESPONSIVITY OF OPTICAL INSTRUMENTS

We consider now a generalized instrument, shown in Fig. 5, of entrance area A_c, entrance solid angle Ω_c, optical transmission τ, and detection quantum efficiency p. The photon count rate c, for a source of brightness B is given by:

$$c = B\, A_c \Omega_c \tau p = \frac{E}{4\pi} \times 10^6\, A_c \Omega_c\, \tau p \text{ photoelectrons/sec} \quad (1)$$

where E is the emission rate in rayleighs. The responsivity can be defined as c/B or c/E. The characteristics of any spectroscopic device are determined by the dispersing element and what we really are concerned about is the Ω at this element. But since the $A\Omega$ product has the same value at all elements in an optical system we can consider $A\Omega$ at the dispersing element just as well as the $A_c\Omega_c$ in front. If one wants a small Ω_c in

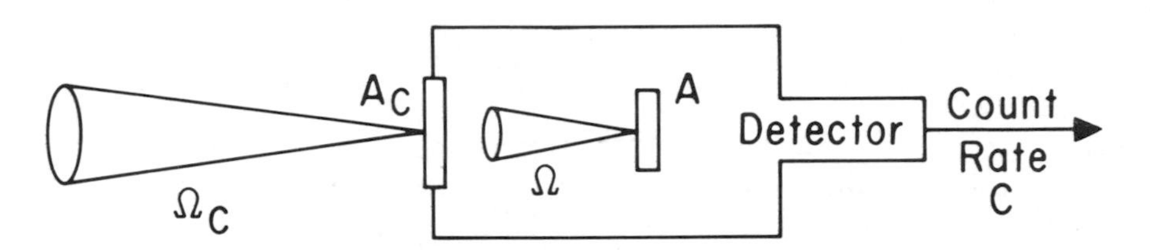

Fig. 5 Schematic drawing of an optical instrument with collecting optics $A_c\Omega_c$ and a dispersing element of area A and solid angle Ω.

front for a small field of view, then a large light collection system can be used, but it will not influence the photoelectron count rate of the system. τ and p are technical factors that are generally fixed and A is normally a matter of cost. Hence the critical quantity in comparing one device to another is Ω. The values of Ω depend simply on the dispersing device, and the resolving power, $R = \lambda/d\lambda$ at which it is used, where $d\lambda$ is the spectral width. The angular acceptances, Ω, for some useful devices are in Table 1.

Table 1
Responsivities of Optical Devices

Grating spectrometer	$\Omega = \beta \tan i/R \cong 0.1/R$	(2)
β = slit length (radians), i = incidence angle		
Filter photometer	$\Omega = 2\pi n^2/R \cong 15/R$	(3)
n = refractive index		
Fabry Perot, Michelson	$\Omega = 2\pi/R \cong 6/R$	(4)
Wide Angle Michelson	$\Omega = 4\pi/\sqrt{R} \cong 12\sqrt{R}/R$	(5)

Thus for these devices the responsivities are respectively in the ratio 0.1, 15, 6, $12\sqrt{R}$. In fact one does not use these devices at the same resolving power, but at roughly the same responsivity, so that the above ratios indicate the relative resolving powers that can be achieved. These concepts give only rules-of-thumb since other practical factors enter into particular observing situations, but they are still very useful.

OPTICAL ATMOSPHERIC MEASUREMENTS

From the conceptual instrument described, the emission rates can be determined from the count rates, normally using a calibration against a known optical source. However, consideration of the spectroscopy of the source often permits the measurement of other physical quantities.

From the relative intensities of vibrational bands the vibrational temperature can be deduced. From the shape of a molecular band, which can be determined from the relative intensities of rotational lines, or from the integrated intensities of passbands located at different places on the band, the rotational temperature can be determined. The rotational temperature normally corresponds to the kinetic temperature, since electron impact seems not to perturb the angular momentum of the relatively heavy molecule.

Atmospheric atomic oxygen has zero nuclear spin and no significant isotopic structure, so its emission lines have no hyperfine components and the line shape is determined purely by the Doppler motion. (At upper atmospheric temperatures the natural linewidth is negligible). Normally the Doppler profile yields the kinetic temperature of the neutral atmosphere, but some excitation processes give kinetic energy to the excited products which could be reflected in the Doppler profile. Emission line shifts can normally be interpreted as winds. By looking at O and O^+ emissions one can in principle determine both the neutral and ion winds. The effect of integrating in altitude must be considered, and the most advantageous method is to view from a satellite at the earth's limb. The measurement of neutral winds is described by Smith (1980).

THE OPTICAL ATMOSPHERIC SOURCE

We mentioned earlier the thinness of the emitting layer. When the energy source is precipitating electrons, the production rate of an emitting species y, at altitude h is given by:

$$P_y(h) = \int \phi(E,h)\ \sigma_{xy}(E)\ N_x(h)\ dE \qquad (6)$$

where $\phi(E,h)$ is the electron flux, $N_x(h)$ the number density of the target species x, and $\sigma_{xy}(e)$ the cross section for impact from the target to the emitting species. The upper altitude limit of the emission is determined by the level at which the production rate has a significant value. The bottom is determined by where the electrons run out of energy. The other major energy source is chemical reaction, through energy stored in ionization, or molecular dissociation. The recombination of molecular ions with electrons for example yields a production rate of:

$$P(h) = k\ N_+(h)\ N_e(h) \qquad (7)$$

where k is the rate constant and $N_e(h)$ is the electron density. The thickness of this emitting region is determined by the altitude range over which the $N_+(h)N_e(h)$ product is large enough to give P(h) a significant rate.

Loss processes that compete with radiative loss must be considered. For a steady state condition, which is a reasonable assumption for all sources except those like pulsating aurora, the production rate is equal to the loss rate by radiation of the emission we are interested in, equal to N_y/τ_y where τ_y is the lifetime of the state, plus the loss rate by collisional quenching $k_q N_y N_z$ where N_z is the number density of the quencher and k_q is the quenching rate constant. Solving the equation for the emission rate $E(h) = N_y/\tau_y$ (this assumes the emission of interest is the only one from N_y, otherwise a different τ-value is required) one obtains

$$E(h) = \frac{P(h)}{1 + k_q N_z \tau_y} \qquad (7)$$

When quenching is negligible the emission rate is clearly equal to the production rate, but this is never the case at all altitudes when N_y is a metastable species. For the 6300A emission the product $k_q N(N_2)\tau_y$ equals unity at about 250 km, so that quenching by N_2 of the $O(^1D)$ state is dominant below that altitude and auroral emission from 100 km, say, will not contain any 6300A emission. For the 5577A emission the corresponding level for quenching of $O(^1S)$ by O is about 110 km.

ATMOSPHERIC MODELS AND METHODS

A number of theoretical models exist for the calculation of emission rate profiles in aurora. One begins with an assumed or measured electron energy spectrum. Then one needs a model (or measured) atmosphere, (giving the ion and neutral composition and the temperature) and a catalog of ionization and excitation cross sections, transition probabilities and rate constants. This assumes that a particular set of reaction processes has been defined. As the primary beam enters the atmosphere, secondary electrons are produced and the changing electron spectrum of the combined primary and secondary beam must be kept track of. It is the secondary electrons that are responsible for most of the excitation.

Rather sophisticated models have been developed, to the point where time dependence, and horizontal winds in neutrals and ions have been introduced. However, much of the early development was in a period when few measurements were being made. So far, the record of agreement between observed and calculated emission rates, where the incident electron fluxes have been measured, is poor. The succession of failures has prompted new activity, and new proposed processes.

From time to time there have been suggestions of interaction processes between the energetic electrons and the ionosphere that are not of a collisional nature. This would make invalid most of the current models. However, since the prompt N_2^+ emission is reasonably well predicted, and most of the difficulties are with metastable states, these suggestions have been disregarded by the modellers.

Because of the integrated nature of most optical measurements, models are essential, not just to satisfy ourselves that we understand the atmosphere, but to solve the integral equations and make deductions about the atmosphere. Better success has been obtained with the solar electromagnetic source than with the aurora. Measurements made in the twilight by Meriwether et al. (1978) suggest that the atomic oxygen concentrations can be measured from the ground, from the intensity of the O+ 7319Å line. Much earlier, Noxon and Johansen (1972) inferred O_2/N_2 rates from the intensity of the 6300A emission in twilight. However their knowledge of the relevant processes was then not very accurate. We now know that photoelectron impact on O is the dominant process, rather than the photodissociation of O_2 in the Schumann-Runge continuum.

Because of difficulties in understanding the nightside auroral emissions, Shepherd, et al. (1980) developed an empirical method. The ISIS-II 6300A emission rates were compared with electron fluxes measured by the Soft Particle Spectrometer on the same spacecraft, in the following way. The energy fluxes were divided into four bands, E_1 to E_4, corresponding to the energies 5-60 eV, 60-300 eV, 300 eV - 1 keV, and 1-15 keV. It was assumed that the 6300A emission rate could be written as a linear superposition of the energy in these bands.

$$I(6300) = C_1E_1 + C_2E_2 + C_3E_3 + C_4E_4 \qquad (8)$$

By a regression analysis on the four variables the coefficients C_1-C_4 were determined. This gives the 6300A production rate as a function of energy. The values of the coefficients are given in Table 2.

Such measurements extended to other emissions would permit an accurate assessment of the accuracy with which the energy

Table 2
6300Å production coefficients and efficiencies

Electron eneryg	5-60 eV	60-300eV	.3-1 keV	1-15 keV
Production coefficient R erg^{-1} cm^2 sec	1100	1500	280	15
Production efficiency photons/electron	.05	0.5	0.2	0.1

spectrum can be determined. However, the collection of the data would be very difficult.

Much of the optical data presented at this meeting comes from observations of the dayside aurora, and these are important for a number of reasons. The dramatic differences between the optical spectra of the dayside and nightside aurora affirms the sensitivity of the optical spectrum to the electron energy spectrum and gives us hope that accurate estimates of electron energy spectra may yet be made from the ground, or high altitude satellites. The quiet diffuse nature of the dayside cleft aurora is very amenable to modelling, unlike the nightside phenomenon. Its stability from day to day may even make possible some monitoring of upper atmospheric composition and temperature.

OPTICAL MEASUREMENTS OF THE MAGNETOSPHERE

It was stated earlier that in auroral observations we see the particles as they exit from the magnetosphere and into the ionosphere. Awareness of the auroral zone has been around for a long time but the concept of an auroral oval fixed to the sun came much later, and the observation of an instantaneous auroral oval came only with the ISIS-II, DMSP and now the KYOKKO satellite. Satellite observations show the auroral region to have a greater spatial integrity, and systematic behavior than was evident from ground-based observations. This is because a ground-based observer has a limited view from one site. The dynamical changes of the aurora (diurnal and substorm) are great enough to cause sweeps in and out of the viewed region, whereas viewed on a global scale, the changes appear relatively small and systematic. Put differently, at a single observatory, auroras come and go, but from a satellite there is always aurora somewhere, and in fact always a defined auroral oval.

The differences in behavior for different wavelengths are particularly striking. The 6300A emission forms an oval with the brightest portion at noon, with intensity weakening on either side of noon, as shown in Fig. 6. This is a polar projection of 6300A isointensity contours, in invariant coordinates with noon at the top. It is provided by the Red Line Photometer on ISIS-II as described by Shepherd (1979). The gradual decrease persists around to the nightside where there may be a secondary maximum, or where the emission may virtually disappear.

In Fig. 7 are shown ISIS-II polar projected images in three wavelengths - 6300A (top row), 5577A (middle row) and 3414A (bottom row). In the left-most image in each row the invariant coordinates are superimposed - noon is at the bottom and the triangles indicate the satellite track. The other pictures in the same row are for different contrast levels with only the invariant pole marked with a cross. The reader should now

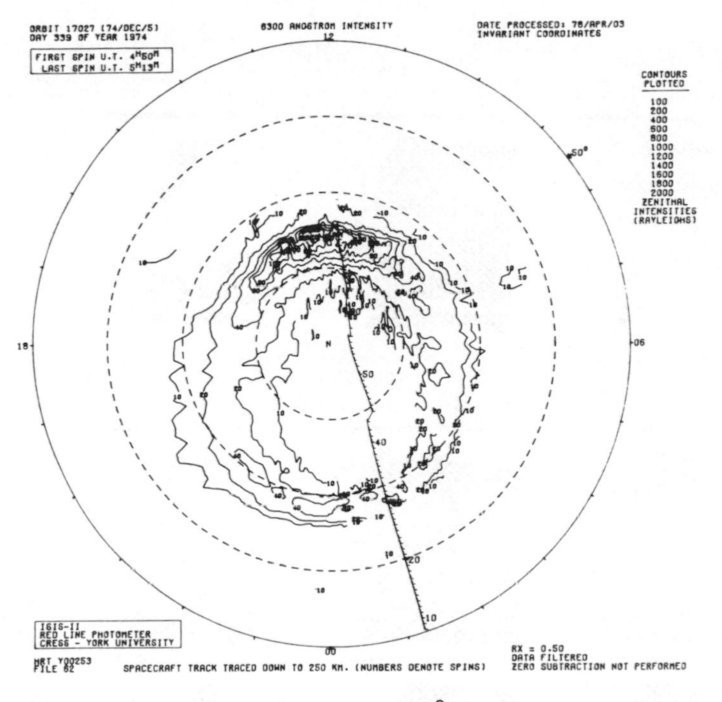

Fig. 6. Isointensity contours of 6300Å emission plotted in a polar invariant projection, with noon at the top. These data were taken with the Red Line Photometer on the ISIS-II spacecraft, and were acquired during a pass with quiet magnetic conditions. The brightest region of emission is at noon.

concentrate on the second column from the right. The upper image (6300A) shows a uniform intensity across the noon section with some small localized bright spots and a brighter region at 14 hr. The evening arc continues upward and into the night sector, where there are some bright 6300A regions. The middle image shows three bright 5577A arcs in the pre-noon region and a noon gap in intensity before the 14 hr maximum. (The faint strip upwards from this region arises from sunlight scattered in the baffles). The nightside region is bright. For the 3914A (bottom) image the situation is similar to the 5577A image except that the intensity does not drop out at the noon sector. As shown by Sivjee (1980), this is probably due to resonance fluorescence of N_2 ions. The dayside "gap" in 5577A arcs was first observed by Cogger et al. (1977) and identifies a narrow region of purely soft precipitation. It is a cusp-like feature of the broader 6300A region that has become known as the dayside cleft.

More detailed observations of this cleft region and of auroras in the polar cap, will be shown during this session. When properly measured and interpreted, optical remote sensing can provide a wealth of information about the polar atmosphere.

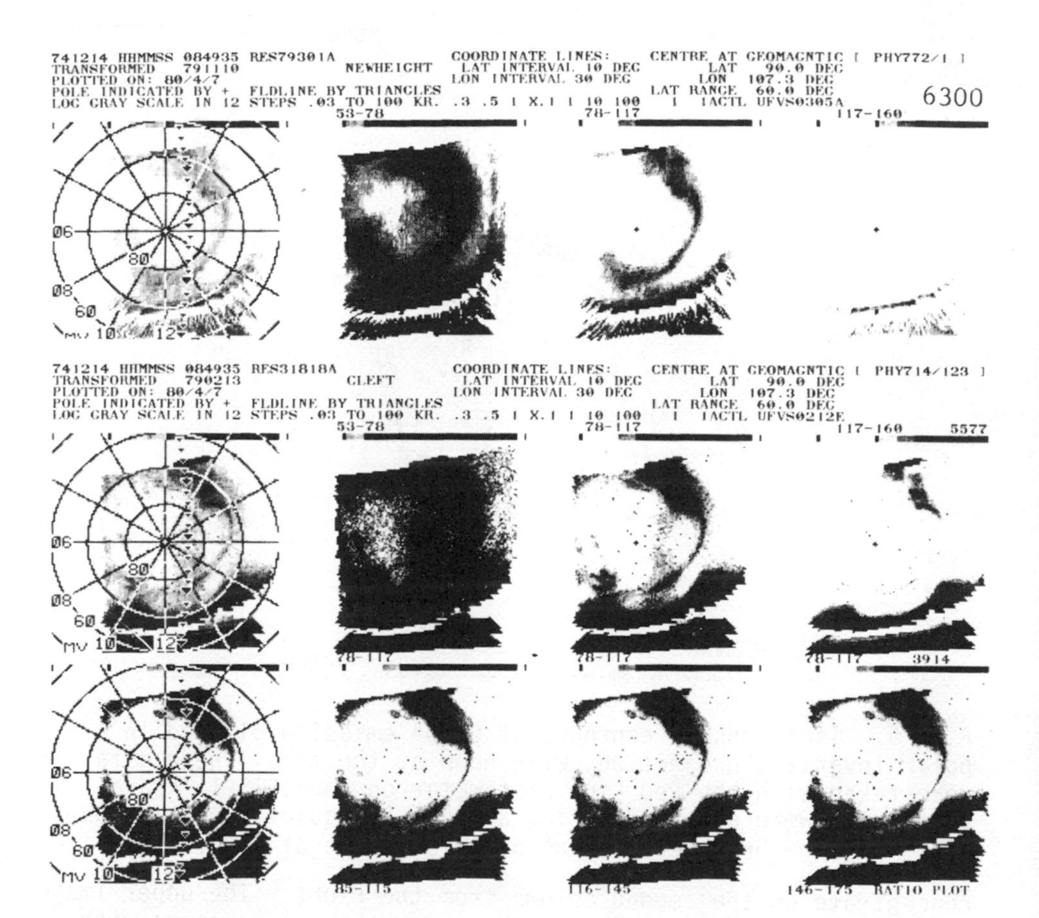

Fig. 7 Composite of ISIS images in three wavelengths, in polar invariant projections with noon at the bottom. The top row of pictures is for 6300 A emission, the second row is for 5577A emission and the bottom for 3914A emission. For further details see text.

REFERENCES

Cogger, L. L., Murphree, J. S., Ismail, S. and Anger, C. D., 1977. Geophys. Res. Lett., 4, 413.
Noxon, J. R. and Johanson, A. E., 1972. Planet. Space Sci. 20, 2125.
Meriwether, J. W., Torr, D. G. and Walker, J. C. G., 1978. J. Geophys. Res., 83, 3311.
Shepherd, G. G., 1979. Res. Geophys. Space Phys. 17, 2017.
Shepherd, G. G., Winningham, J. D., Bunn, F. E., and Thirkettle, F. W., 1980. J. Geophys. Res., 85, 715.
Sivjee, G. G., 1980. This volume.
Smith, R. W., 1980. This volume.

POLAR CAP OPTICAL EMISSIONS OBSERVED FROM THE ISIS2 SATELLITE

L.L. Cogger

Department of Physics, The University of Calgary
Calgary, Alberta, Canada

INTRODUCTION

The application of polar orbiting scientific satellites to the study of polar cap phenomena has led to significant advances in our understanding of polar cap processes. The ISIS2 satellite, which is in a circular orbit 1400 km above the earth's surface, has provided monochromatic optical images of the high latitude region since April 1971. The 5577Å emission from atomic oxygen and the 3914Å emission from the First Negative bands of N_2^+ have been monitored by the auroral scanning photometer (Anger *et al*. 1973), and the 6300Å emission from atomic oxygen has been obtained by means of the red line photometer (Shepherd *et al*. 1973). In order to view the entire polar cap during a single pass of the satellite, several restrictive conditions have to be met which limit the quantity of useful data. Nevertheless, there have been a sufficient number of satisfactory observations to allow the characteristic features of high latitude emissions to be identified. In the sections that follow, a summary of results of observations of dayside aurora and polar cap arcs is given.

It is important to remember that polar cap optical emissions are extremely variable in space, time and intensity, and that as a consequence of this any summary fails to represent faithfully the complexity of the phenomena. Some idea of the variable nature of the emissions can be obtained from Fig. 1 which contains images at 3914Å transformed onto a corrected geomagnetic latitude - geomagnetic local time grid. The images were acquired from successive passes over the northern polar cap on December 3, 1975. Note especially the presence of polar cap arcs and the protrusion of the auroral oval emissions to high latitudes.

C. S. Deehr and J. A. Holtet (eds.), Exploration of the Polar Upper Atmosphere, 159–164.

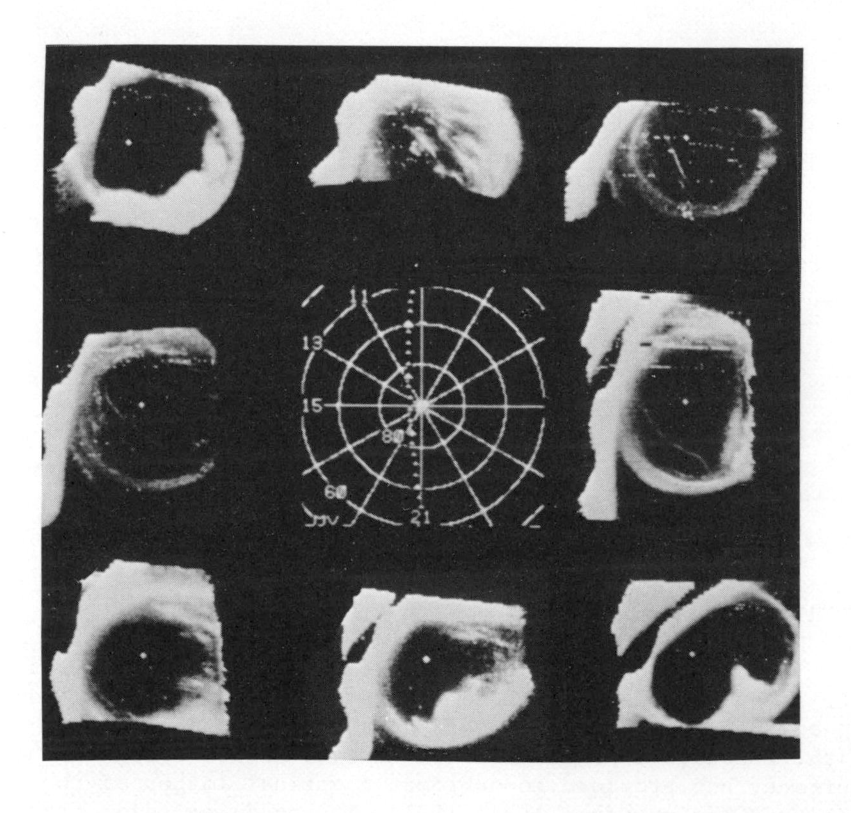

Fig. 1 Eight successive passes over the northern polar cap, 00 to 13 UT on December 3, 1975. The 3914Å auroral intensities range from 0.5 to 3.9 kR. Coordinates are corrected geomagnetic latitude and local magnetic time. Scattered sunlight limits the coverage on the dayside.

DAYSIDE AURORA

Observations from ISIS2 of optical emissions in the dayside portion of the auroral oval have been reported by Shepherd and Thirkettle (1973), Shepherd *et al*. (1976), Cogger *et al*. (1977), and Murphree *et al*. (1980). On the morning side, the discrete auroral forms are irregular, often resembling short arcs or patches especially near the poleward boundary of the oval. By contrast, the afternoon oval is more regular at 5577Å and 3914Å. Spectral ratios are consistent with higher characteristic energies of precipitating electrons in the morning.

Other characteristics of the dayside aurora are illustrated by the example shown in Fig. 2. A persistent feature in 3914Å and 5577Å images is the presence of an enhancement in both intensity and width near 14 MLT. This enhancement is probably related to a statistically deduced maximum in particle fluxes observed in this sector (McDiarmid et al. 1976). Another characteristic is the minimum in 5577Å and 3914Å emissions between 10 and 12 MLT. This "gap" in discrete aurora has also been identified in images obtained from the DMSP satellites (Dandekar and Pike 1978). The auroral emissions on either side of this minimum occur at approximately 76° CGL with a width of 1 to 3° latitude.

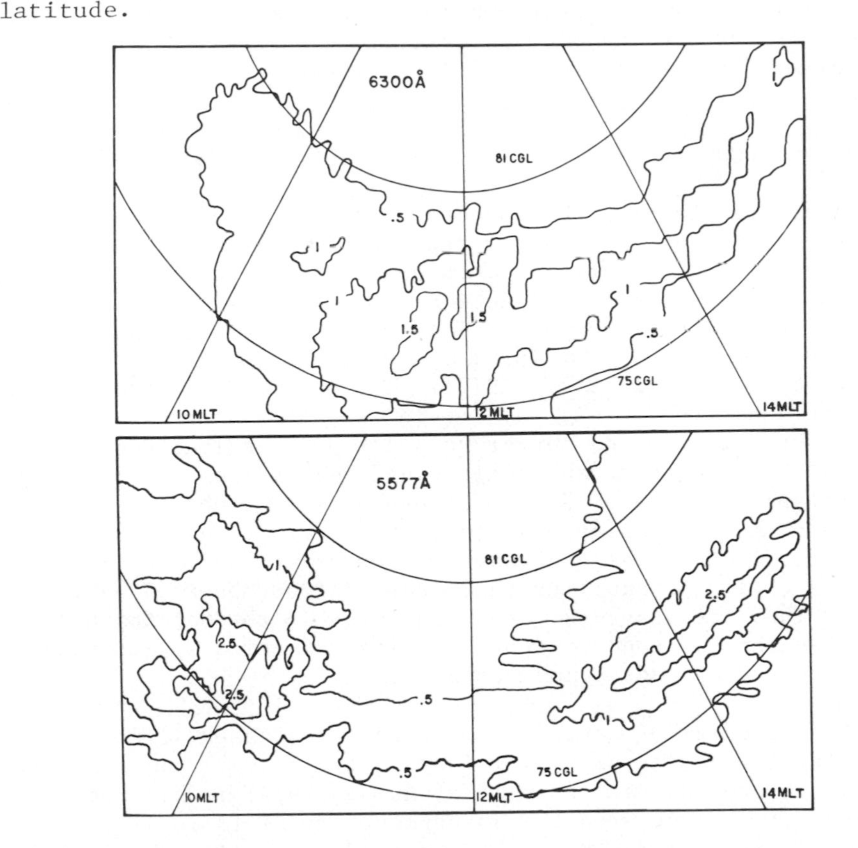

Fig. 2 Dayside 5577Å and 6300Å aurora at 0855 UT on December 14, 1974. The contour interval is 0.5 kR.

The description of dayside 6300Å is very different. The 6300Å emission region is much broader than for the other wavelengths, being typically 5° in latitude. It has been found that large scale discrete features are often not apparent in the 6300Å data (e.g. the aurora in 5577Å near 9 MLT in Fig. 2). The 6300Å emission maximizes in the noon sector, but not necessarily at the time of minimum 5577Å and 3914Å intensity. Although the 6300Å emission is preferentially excited by low energy electrons, the spatial distribution of this emission does not appear to provide an obvious signature of the dayside cusp region or the boundary between open and closed field lines.

Dayside auroral motions are correlated with substorm activity (Eather *et al*. 1979), and polar cap emissions in general respond to the direction and magnitude of the B_y and B_z components of the interplanetary magnetic field (IMF). However, at the present time relatively little is known regarding the causes of the persistent features of dayside aurora, and even less is known about mechanisms for the diversity of auroral variations.

POLAR CAP ARCS

A relatively rare, but spectacular feature of the polar cap is the presence of long narrow regions of auroral emission which are often called sun-aligned arcs due to their tendency to be aligned in the sun-earth direction. Characteristics of arcs of length greater than 400 km and intensity greater than 0.5 kR have been identified from ISIS2 observations during the period 1971 to 1975 (Ismail *et al*. 1977). The observed positions of the arcs are indicated in Fig. 3. It is seen that arcs can vary in length, orientation and curvature. The average length of these arcs is about 1400 km, and their width varies from 30 to 200 km with a typical value of 70 km. These arcs were observed on approximately 6% of the passes examined. However, it was found that when arcs do occur, several may be present at the same time and there is a high probability that arcs will be observed over several hours. They are observed more frequently in the morning half of the polar cap and tend to appear during magnetically quiet periods when both the Kp and AE indices are small.

From studies of northern polar cap arcs observed from the satellite (Berkey *et al*. 1976; Ismail *et al*. 1977) and from the ground (Lassen and Danielson 1978; Lassen 1979) it has been found that the arcs occur when the interplanetary field is directed northward (ie positive B_z). Lassen (1979) has also shown that the pattern of arcs is dependent on the sign of B_y, the east-west component of the IMF, and has concluded that these quiet-time patterns are optical signatures of shears and flow irregularities in the magnetospheric convection.

Although the intensity along an arc may vary over a large range, the spectral ratio 5577Å/3914Å remains relatively constant. For the arcs observed this ratio ranged from 1 to 3 and the ratio 6300Å/3914Å was about 7 for the few cases in which the 6300Å and 3914Å data were both available. These ratios are consistent with the precipitation of low-energy electrons of characteristic energy in the range 0.2 to 1 keV which would produce the emission in the

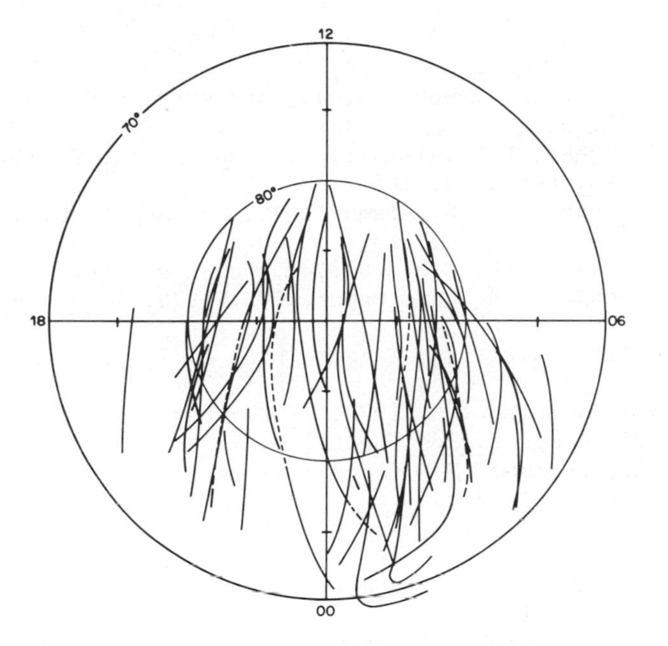

Fig. 3 Observed positions of polar cap arcs in the northern hemisphere. Arcs on the dayside at latitudes less than 80° CGL, and arcs of length less than 400 km have been omitted.

lower F region. Polar cap arcs appear to be the optical signature of polar shower precipitation identified by Winningham and Heikkila (1974). Except for their position and somewhat lower associated average electron energy, polar cap arcs bear a strong resemblance to oval aligned arcs, thus supporting the existence of a single electron energization process for all auroral arcs.

CONCLUSION

In addition to their immediate scientific value, the ISIS2 observations of the aurora have demonstrated the general usefulness of instantaneous monochromatic imaging of the polar caps. Future space missions will undoubtedly incorporate optical instruments to provide two-dimensional images of the pattern of auroral emissions created by precipitating energetic particles.

ACKNOWLEDGMENTS

Drs. C.D. Anger, S. Ismail and J.S. Murphree have contributed extensively to the processing and interpretation of ISIS2 data. Dr. G.G. Shepherd provided the 6300Å data used in Fig. 2.

REFERENCES

Anger, C.D., Fancott, T., McNally, J., and Kerr, H.S. 1973, Appl. Opt., 12, 1753.

Berkey, F.T., Cogger, L.L., Ismail, S., and Kamide, Y. 1976, Geophys. Res. Lett., 3, 145.

Cogger, L.L., Murphree, J.S., Ismail, S., and Anger, C.D. 1977, Geophys. Res. Lett., 4, 413.

Dandekar, B.S., and Pike, C.P. 1978, J. Geophys. Res., 83, 4227.

Eather, R.H., Mende, S.B., and Weber, E.J. 1979, J. Geophys. Res., 84, 3339.

Ismail, S., Wallis, D.D., and Cogger, L.L. 1977, J. Geophys, Res., 82, 4741.

Lassen, K., and Danielson, C. 1978, J. Geophys. Res., 83, 5277.

Lassen, K. 1979, Geophys. Res. Lett., 6, 777.

McDiarmid, I.B., Burrows, J.R., and Budzinski, E.E. 1976, J. Geophys. Res., 81, 221.

Murphree, J.S., Cogger, L.L., Anger, C.D., and Ismail, S. 1980, Geophys, Res. Lett., 7, 239.

Shepherd, G.G., Fancott, T., McNally, J., and Kerr, H.S. 1973, Appl. Opt., 12, 1767.

Shepherd, G.G., and Thirkettle, F.W. 1973, Science, 180, 737.

Shepherd, G.G., Thirkettle, F.W., and Anger, C.D. 1976, Planet. Space Sci., 24, 937.

Winningham, J.D., and Heikkila, W.J. 1974, J. Geophys, Res., 79, 949.

OPTICAL EMISSIONS IN THE POLAR AURORAL E-REGION

E.J. Llewellyn and B.H. Solheim

Institute of Space and Atmospheric Studies
University of Saskatchewan
Saskatoon, Canada S7N OWO

Abstract. The use of optical measurements of the aurora and airglow to remotely sense the atmosphere are briefly described. It is shown that observations of the molecular oxygen emissions in the airglow require a peak atomic oxygen concentration of about 1.E12 cm-3*. It is suggested that these emissions are excited in a two step process similar to the Barth mechanism for the green line. It is proposed that monitoring of the green line and the atmospheric bands will allow the atomic oxygen concentration in the lower thermosphere to be determined under both auroral and non-auroral conditions.

1. INTRODUCTION

At high latitudes the natural atmospheric radiation, the airglow, is dramatically modified by the effects of extensive particle precipitation into dynamic and visible forms, the aurora, which have long held man in awe. The spectacular nature of the aurora has also encouraged the efforts of aeronomers and auroral physicists to understand the phenomena which comprise it. For the optical aurora at least each step in our understanding has seemingly generated new problems and we are finding that the aurora itself is often the best laboratory for the study of these problems.

The visual aurora may be as low as 80 km or as

C. S. Deehr and J. A. Holtet (eds.), Exploration of the Polar Upper Atmosphere, 165–174.

high as 400 km so that atomic and molecular processes may be investigated over a range of pressure and mean free path which it is difficult to reproduce in ground based laboratories. Auroras commonly have lower borders in the region 100-120 km and a vertical visual extent of some 50 km so that frequent observation of aurora at a single location can provide information on the temporal variation of the atmosphere at that location. It should not be presumed the auroral emission is restricted to this vertical extent as some spectral features will occur at much higher altitudes due to either quenching or excitation processes. However, in the present paper we are concerned with optical emissions associated with the main auroral form of normal aurora, the E-region emissions. The spectrum of the aurora is briefly reviewed and the various excitation mechanisms for some features are discussed. The possible use of the auroral emissions to remotely sense the state of the atmosphere and some restrictions on the neutral atmosphere, particularly the concentration of atomic oxygen, are also considered.

2. THE AURORAL SPECTRUM

The early studies of the auroral spectrum were limited by both atmospheric transmission and the availability of appropriate detectors, however, with the advent of rocket and spacecraft instrumentation it has been possible to extend spectral measurements throughout the entire wavelength range from x-rays to the far infrared. These measured spectra have indicated a wealth of atomic and molecular features. In some wavelength regions it is a simple matter to identify the different features while in others it is possible to completely neglect, or miss, some feature which may be present. Indeed in the region near 3500A, where the Herzberg I bands of O_2 are expected, the spectrum is so rich that the presence of these bands is in some doubt (Yau and Shepherd 1979). Thus the study of auroral features using low resolution instrumentation (e.g. filter photometers) must frequently be assessed quite carefully if the results are not to be misinterpreted. However, the use of synthetic spectra (Vallance Jones and Gattinger 1975) provides a simple way to allow for the effects of contaminating emissions and affords some confidence in the predicted maximum intensities for as yet undetected emission features.

For some features the synthetic spectra are incapable of providing the necessary information. The NO γ-bands have been identified as being present, although recent observations (Beiting and Feldman 1978) have concluded that the 2150A feature is not due to NO. The precise identification of this spectral feature will probably require a very high resolution spectrum.

In the visible region it is readily apparent that apart from the strong feature at 5577A, the oxygen green line, the spectrum is dominated by features due to molecular nitrogen. This exactly reflects the situation with the afterglows of oxygen and nitrogen in which the latter is much stronger as many more bound states are available through atomic recombination. Further, for oxygen the stable states have long radiative lifetimes so that quenching mav be quite significant. It was this dominance of the nitrogen emissions in the auroral spectrum that persuaded Vegard (1939) that the excitation of the oxygen green line must involve nitrogen (see Section 4). However, the observed spectrum must reflect the energy input to the atmosphere from the precipitating particles, the excitation mechanisms for the various emissions and the atmospheric composition. Thus it should be possible to monitor different spectral features to provide measurements of these different parameters, e.g. N_2^+(1NG) to measure total energy flux, and Doppler line widths to measure temperature (Hilliard and Shepherd 1966). However, it is frequently found that optical measurements can only provide unique information when they are used in conjunction with other measured parameters.

Optical sensing is an obvious method for the determination of atomic oxygen concentrations in the lower thermosphere although any feature monitored must provide information both in the presence, and absence, of particle precipitation. This restricts the choice to the molecular oxygen features, the atmospheric bands ($b^1\Sigma_g^+ - X^3\Sigma_g^-$), the Herzberg I bands ($A^3\Sigma_u^+ - X^3\Sigma_g^-$) or the infrared atmospheric bands ($a^1\Delta_g - X^3\Sigma_g^-$) and the oxygen green line at 5577A and it is of value to consider the information that is presently available from airglow studies.

3. THE OXYGEN EMISSIONS IN THE AIRGLOW

These emissions have been studied by many investigators, using both ground based and rocket borne techniques, and typical results for the nightglow are summarized in Table 1. It is immediately apparent that the different emission features are not concentrated at a single altitude, but rather throughout an altitude range of some 10 km, which suggests that either different excitation mechanisms exist for the different emissions or that the effects of quenching are very important. The identification of the excitation mechanism for these emissions has been the subject of much recent effort (Slanger 1978; Kenner et al. 1980;Llewellyn et al. 1980) but it is agreed that the airglow energy is supplied by the recombination of oxygen.

TABLE 1

Feature	Intensity	Emission Altitude
$O(^1S)$	150 R	98 km
$O_2(a^1\Delta_g)$	50 kR	98 km
$O_2(A^3\Sigma_u^+)$	1 kR	96 km
$O_2(b^1\Sigma_g^+)$	6 kR	94 km
$O_2(a^1\Delta_g)$	50 kR	87 km
OH(Meinel)	1 MR	86 km

3.1 The oxygen green line

The oxygen green line in the nightglow is believed to be excited in either the single step Chapman mechanism,

$$O + O + O \rightarrow O_2 + O(^1S) \qquad (1)$$

or the two step Barth mechanism,

$$O + O + M \rightarrow O_2^* + M \qquad (2a)$$
$$O_2^* + O \rightarrow O_2 + O(^1S) \qquad (2b)$$

however, in each case the source of the excitation is the three-body recombination of atomic oxygen. The exact path has been extensively discussed since the Barth (1964) mechanism was first proposed but recent laboratory measurements (Slanger and Black 1977) suggest that the transfer mechanism is the major excitation path. Some support for this conclusion has been provided by the rocket observations of Thomas et al. (1979) and Witt et al. (1979) although the interpretation of these measurements is not straightforward.

These green line profiles can be inverted to give the atomic oxygen concentration profile in the emission region. For the Chapman mechanism the peak concentration is typically 1.E11 cm-3 at 98 km while for the Barth mechanism the concentration is approximately 7.E11 cm-3 at the same altitude. Obviously such a large difference in the concentrations is unacceptable so that other measurements which may indicate the concentrations and, therefore, the green line excitation mechanism should be considered.

3.2 The molecular emissions

If these emissions are excited in the direct recombination of atomic oxygen, step 1 in the Barth mechanism, then the volume emission rate is given by,

$$V(h) = \frac{\eta k_{2a}[O]^2[M] A}{\Sigma A_{v''} + \Sigma k_i[X_i]} \qquad (3)$$

where η is the fractional yield of the excited state in the recombination and the other terms have their accepted meanings. Equation (3) may be inverted to give the atomic oxygen profile from observations of the emission height profiles in the atmosphere. For the atmospheric bands the derived concentration is 1.E11 cm-3 (Deans et al. 1976) in agreement with the measurements of Howlett et al. (1980). However, if these concentrations are applied to the infrared atmospheric bands the derived intensity is approximately a factor 10 less than that observed even for a recombination yield of unity. As neither the a nor the b state is quenched in the airglow the direct recombination mechanism also fails to explain the differences in the emission altitudes. Thus if the measurements are accepted then the postulated direct recombination mechanism must be incorrect for at least one of these emissions.

Laboratory measurements (Clyne et al. 1965) did not detect either the $a'\Delta_u$ or the $b'\Sigma_u^+$ states in the direct recombination and suggested that at least the b state is excited through a precursor state. This idea has been studied by Greer et al. (1980) who have shown that the height profile of the atmospheric bands can be explained by transfer with O_2 as the transfer agent. This excitation mechanism makes the effective yield for the $b'\Sigma_g^+$ state in the direct recombination height dependent in agreement with the observations of Witt et al. (1979).

For the O $(a'\Delta_g)$ emission to have a different height profile from that for the atmospheric bands a different transfer species must act, and a comparison with the green line suggests that the transfer agent is atomic oxygen. That part of the emission centered on 87km is probably due to an OH reaction and is not considered here. The derived atomic oxygen concentrations, 7.E11 cm-3 at the peak, are in good agreement with that obtained from an analysis of the green line. Thus the airglow observations indicate that the peak atomic oxygen concentration is large, typically 7.E11 cm-3, in agreement with the measurements of Thomas et al. (1979) but in contradiction to those of Howlett et al. (1980). These airglow measurements also suggest that the observed singlet oxygen states are excited through energy transfer mechanisms which can be represented by the following reactions:

$$O + O + M \; - \; O_2{*} + M \qquad (2a)$$

$$O_2{*} + O \; - \; O + O('S) \qquad (4a)$$

$$- \; O\,(a'\Delta_g) + O \qquad (4b)$$

$$O_2{*} + O_2 \; - \; O\,(b'\Sigma_g^+) + O \qquad (5)$$

Obviously it is important to identify the probable intermediate state in the proposed excitation system. As the transfer of vibrational energy to electronic energy to yield O('S), O $(b'\Sigma_g^+)$ and O $(a'\Delta_g)$ is unlikely only the states $A^3\Sigma_u^-$, $C^3\Delta_u$ and $c'\Sigma_u^-$ need be considered. The Herzberg I bands (A-X) are present in the airglow but the measured rate constants do not support the A state as the precursor. For the C state the long radiative lifetime for the observed Chamberlain bands suggests that quenching cannot be particularly effective. Thus we conclude that O $(c'\Sigma_u^-)$ is the most probable precursor; emission from this state has not yet been observed in the terrestrial airglow although Slanger (1979) has suggested that the Herzberg II bands (c-X) may be present. These bands have in fact been observed in the laboratory (Slanger 1978) and in the Venus airglow (Krasnopolsky et al. 1976). The atmospheric bands were also observed in the laboratory experiment, but no attempt was made to observe the infrared atmospheric bands. These latter bands have also been detected (Connes et al. 1979) in the Venus airglow and it has been suggested that essentially all oxygen recombinations on Venus must yield the $a'\Delta_g$ state. If the only reaction path available is that leading to O $(a'\Delta_g)$ then essentially all recombinations will apparently yield this state; this can occur if the O_2 concentration is so small that the transfer path through O_2 is a negligible loss and the required energy to convert from $O_2(c'\Sigma_u^-)$ to O('S) is unavailable. If this energy is supplied from vibrationally hot $c'\Sigma_u^-$ then the rapid relaxation in CO_2 would preclude the formation of O('S). Thus the Venus results are in agreement with the concept of a Barth mechanism for the singlet oxygen emissions and the identification of the intermediate as the $c'\Sigma_u^-$ state.

4. OXYGEN EMISSIONS IN THE AURORA

The measurement of atomic oxygen concentrations in an aurora have been attempted but with apparently contradictory results, low concentrations, 1.E11 cm-3, have been reported by Deans et al. (1976) and Sharp (1978) and concentrations near 1.E12 cm-3 derived by us. These low concentrations are based in part on measurements of the atmospheric bands in the airglow (Section 3.2), on quenching of the V-K bands of N_2 in the aurora and the excitation of the green line in the aurora. Thus there is a need to obtain unambiguous measurements of the concentration in aurora.

For a determination from optical measurements only emissions which directly involve atomic oxygen, either in the excitation process or in quenching, can be used. This precludes both the oxygen atmospheric bands and the infrared atmospheric bands in the aurora, although the long radiative lifetime of the $a'\Delta_g$ state will allow a significant nightglow emission to be present under auroral conditions. However, the identification of the nightglow component and the determination of the atomic oxygen concentration is difficult. The obvious emission is the green line although, as with the nightglow, the excitation mechanism is uncertain. Various mechanisms have been proposed and two of these are discussed below.

(i) Energy transfer from nitrogen

$$N_2(A^3\Sigma_u^+) + O \rightarrow N_2 + O(^1S)$$

This mechanism was originally considered by Vegard (1939) and interest was revived following the laboratory studies of Meyer et al. (1970) and the analysis of Parkinson (1971) which showed that auroral fluctuation results did not preclude a transfer mechanism. However, the importance of this mechanism depends on the quenching coefficient for the N $(A^3\Sigma_u^+)$ state and the $O(^1S)$ yield, both of which are uncertain. Further the mechanism would have a height profile which is N_2 dependent rather than the observed O_2 dependence (Feldman 1978).

(ii) Energy transfer from oxygen

$$O_2^* + O \rightarrow O_2 + O(^1S)$$

It was noted in the discussion of the airglow emissions that the oxygen emissions could be excited in a two step process and Solheim and Llewellyn (1979) have proposed that a similar process applies to the excitation of the auroral green line. These authors suggested that electron impact on O_2 would lead to the $c^1\Sigma_u^-$ state which subsequently transfers to the $O_2(a^1\Delta_g)$, O $(b^1\Sigma_g^+)$ and $O(^1S)$ states as with the airglow. Thus the volume emission rate of the green line in the aurora, under steady state conditions, is given by

$$V(h) = \frac{A_{5577}\,[O_2]\,k_{4a}\,[O]\int \mathcal{F}(\varepsilon)\,\sigma_{O_2}(\varepsilon)\,d\varepsilon}{\{A_{^1S} + k_6[O] + k_7[O_2]\}\,\{A^* + k_4[O] + k_5[O_2]\}}$$

and if the dominant term in the first denominator is $k_6[O]$ then the volume emission height profile will have the O_2 dependence observed. It should also be noted that under these conditions the effect of a decreased atomic oxygen concentration is to increase the green line volume emission rate. There is some support for this mechanism from the observations of the vibrational development of the atmospheric bands in aurora (Vallance Jones and Gattinger 1976). A vertical transition from the $c^1\Sigma_u^-$, v=0 level, would give v=5 in the b-state in agreement with the observations.

5. CONCLUSION

If the proposed excitation mechanisms for the oxygen emissions are correct it is possible to use optical remote sensing of the green line and the atmospheric bands to determine the atmospheric composition and temperature. Under non-auroral conditions the volume emission profiles of the green line and the atmospheric bands may be used to determine the atomic oxygen concentration profile in the emitting region; if the infrared atmospheric bands are also observed in twilight it is possible to determine the mesospheric ozone profile. The measurement of the line width for the green line will also provide the kinetic temperature. Under auroral conditions the ratio of the green line to the oxygen atmospheric bands is given by

$$\frac{O(^1S)}{O_2(b'\Sigma_g^+, v=5)} = \frac{k_{4a}[O] \quad \text{Loss } O_2 \text{ b, v = 5}}{k_5[O_2] \quad \text{Loss } O\ (^1S)}$$

which may be inverted to derive the atomic oxygen concentration. In this case the v = 0 and 1 levels are not used as the contributions to the auroral emission from other mechanisms are not negligible.

Although many of the processes discussed in this brief review represent a different approach to auroral and airglow excitation they still show that optical remote sensing can provide valuable knowledge of many required parameters. It is probable that a comparison of these quantities with those derived from other techniques will provide an improved understanding of atomic and molecular interactions.

Acknowledgement

Much of this work has resulted from our collaboration with Dr.G.Witt and has been supported by grants from the Natural Sciences and Engineering Research Council of Canada.

REFERENCES

Barth, C.A. 1964, Ann. Geophys., 20, 182.
Beiting, E.J., and Feldman, P.D. 1978, Geophys. Res. Lett., 5, 51.
Broadfoot, L., and Kendall, K.R. 1968, J. Geophys. Res., 73, 426.
Clyne, M.A.A., Thrush, B.A., and Wayne, R.P. 1965, Photochem. Photobiol., 4, 957.

Connes, P., Noxon, J.F., Traub, W.A., and Carleton, N.P. 1979, Astrophys. J., 233, L29.
Deans, A.J., Shepherd, G.G., and Evans, W.F.J. 1976, Geophys. Res. Lett., 3, 441.
Feldman, P.D. 1978, J. Geophys. Res., 83, 2511.
Gault, W.A., Koehler, R.A., Link, R., and Shepherd, G.G. 1980, Planet. Space Sci. (in press).
Greer, R.G.H., Llewellyn, E.J., Solheim, B.H., and Witt, G. 1980, Planet. Space. Sci. (paper submitted).
Hilliard, R.L., and Shepherd, G.G. 1966. Planet. Space Sci., 14, 383.
Howlett, L.C., Baker, K.D., Megill, L.R., Shaw, A.W., and Pendleton, W.R. 1980, J. Geophys. Res., 85, 1291.
Kenner, R.D., Ogryzlo, E.A. and Turley, S. 1979, J. Photochem., 10, 199.
Krasnopolsky, V.A., Krys'ko, A.A., Rogachev, V.N., and Parchev, V.A. 1976, Cosmic Research, 14, 789.
Llewellyn, E.J., and Solheim, B.H. 1978, Planet. Space Sci., 26, 533.
Llewellyn, E.J., Solheim, B.H., Witt, G., Stegman, J., and Greer, R.G.H. 1980. J. Photochem., 12, 179.
Meyer, J.A., Stetser, D.W., and Stedman, D.H. 1970, J. Phys. Chem., 74, 2238.
Parkinson, T., Zipf, E.C., and Donahue, T.M. 1970, Planet. Space Sci., 18, 187.
Parkinson, T. 1971. Planet. Space Sci., 19, 251.
Sharp, W.E. 1978, J. Geophys. Res., 83, 4373.
Slanger, T.G. and Black, G. 1977, Planet. Space Sci., 25, 79.
Slanger, T.G. 1978, J. Chem. Phys., 69, 4779.
Slanger, T.G. 1979, private communication.
Solheim, B.H., and Llewellyn, E.J. 1979, Planet. Space Sci., 27, 473.
Thomas, L., Greer, R.G.H., and Dickinson, P.H.G. 1979, Planet. Space Sci., 27, 925.
Vallance Jones, A., and Gattinger, R.L. 1974, J. Geophys. Res., 79, 4821.
Vallance Jones, A., and Gattinger, R.L. 1975, Can. J. Phys., 53, 1806.
Vegard, L. 1939, p. 573 in 'Physics of the Earth' (Ed. J.A. Fleming), Vol. VIII. McGraw-Hill, New York.
Witt, G., Stegman, J., Solheim, B.H., and Llewellyn, E.J. 1979, Planet. Space Sci., 27, 341.
Yau, A.W. and Shepherd, G.G. 1979, Planet. Space Sci., 27, 481.

* In this paper concentrations $1x10^{12}$ cm^{-3} are written as 1.E12 cm-3.

A REVIEW OF OPTICAL F-REGION PROCESSES IN THE POLAR ATMOSPHERE

Jean-Claude Gérard

Institut d'Astrophysique - Université de Liège
B-4200 Liège, Belgium.

ABSTRACT

The importance of optical emissions is described in the context of the magnetosphere-ionosphere-neutral atmosphere interactions. Excitation processes of some important F-region emissions in the polar regions are reviewed. The importance of optical emissions for the determination of the quantum yields and quenching coefficients is illustrated for the $O^+(^2P)$, $O^+(^2D)$, $N(^2D)$, $N(^2P)$ and $O(^1D)$ states and a set of values deduced from recent experiments such as the Atmosphere Explorer satellites is given. The effects of transport by neutral winds and electromagnetic drifts on the morphology are also discussed and illustrated by recent model calculations for $N(^2D)$ horizontal transport.

1. INTRODUCTION

High latitude optical emissions result from the interaction of the magnetospheric plasma with the earth's neutral and ionized atmosphere. Although the initial energy source always originates in the precipitated energetic particles and E-field, many indirect excitation mechanisms enhanced by the primary precipitation make substantial contribution and occasionally dominate the production of optical radiation. Consequently, the task of untangling the excitation processes of a given emission is essential when the emission is used as a tool to probe the polar atmosphere.

The interaction of the primary particles with the atmosphere

C. S. Deehr and J. A. Holtet (eds.), Exploration of the Polar Upper Atmosphere, 175–187.

produces secondary electrons, enhances the local ion and electron densities and temperatures and generates neutral winds and composition changes in the thermosphere (figure 1) in a way formally similar to the EUV radiation. The optical emissions (box "airglow excitation rates") constitute one of the products of this complex system and result from 1) the electron ionization and excitation of neutral constituents, 2) the excitation of low energy levels by the high energy tail of thermal electrons, 3) the dissociative recombination of molecular ions and 4) energy-transfer processes from metastable species carrying excitation energy. The total energy released as optical radiation in a typical magnetospheric substorm is about 10% of the total energy deposited by particle precipitation (Rees, 1975).

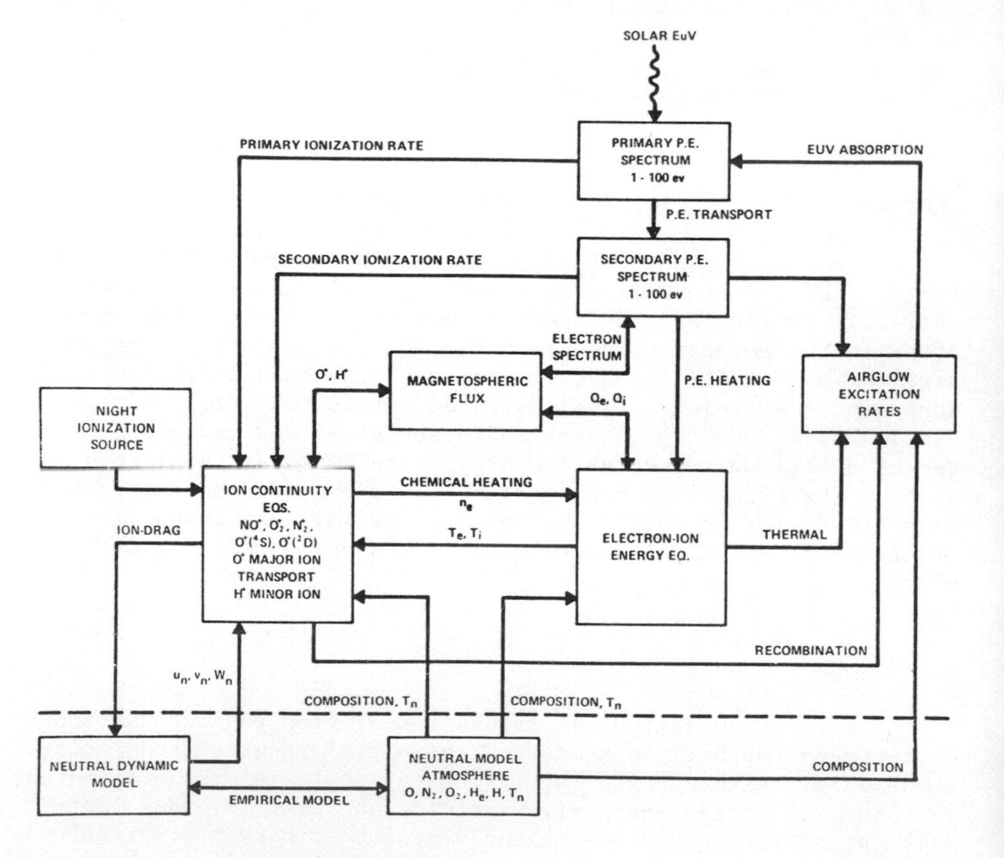

Figure 1 : Block diagram illustrating the magnetosphere-ionosphere-atmosphere interactions (from Roble 1975).

2. PRODUCTION AND QUENCHING OF F-REGION EMISSION

Spectroscopically, optical emissions fall into two categories: allowed transition having transition probabilities of the order of $10^8 s^{-1}$ and forbidden emissions arising from transition violating selection rules for electric dipoles. Corresponding transition probabilities range from more than 1 to less than $10^{-5} s^{-1}$. The importance of forbidden transitions in this context is due to the fact that energy levels of atoms (or ions) lying a few electric volts above the ground state belong to the fundamental electronic configuration. Laporte's selection rule states that the parity of the total electronic angular momentum $l = \Sigma l_i$ must change during an electronic transition. Consequently, transition such as NI $^4S^o - {}^2D^o$ between states of same parity are forbidden. Other selection rules such as $\Delta S = 0$ and $\Delta L = \pm 1$ may also decrease the transition probability.

Quenching of metastable species X^n in the n state may proceed through one of the two paths:

$$X^n + M \rightarrow X^l + M \qquad \text{(quenching)}$$

or

$$X^n + M \rightarrow Y + Z \qquad \text{(chemical reaction)}$$

In the former case, M is a second body not altered by the process, whereas in the latter, an actual chemical reaction involves the second body. In a steady state, the production and loss rates of X^n are equal and given by:

$$P_n = (\sum_l A_{n,l} + \sum_i k_i[M_i])\,[X^n] = \tau_e^{-1}[X^n],$$

where P_n is the volume production rate of X^n, $A_{n,l}$ the probability of the $n \rightarrow l$ transition, k_i the deactivation coefficients and $\tau_e = (\sum_l A_{n,l} + \sum_i k_i[M_i])^{-1}$ is the effective lifetime. The volume emission rate $n \rightarrow m$ is thus given by:

$$\eta_{n,m} = [X^n]\,A_{n,m} = \frac{P_n}{B^{-1} + A_{n,m}^{-1} \sum_i k_i[M_i]}$$

where $B = \dfrac{A_{n,m}}{\sum_l A_{n,l}}$ is the radiative branching ratio.

The effective lifetime at the peak of the emission rate profile may be considerably smaller than the radiative lifetime as will be shown below. Recent measurements (Deans and Shepherd 1978, Rees *et al.*, 1977, Rusch and Gérard, 1980) have indicated that the neutral composition at high latitudes may be significantly different from that given by standard models such as MSIS or Jacchia. Until self-consistent models are available, we recommend the use of a neutral atmosphere where O is roughly half and O_2 three times the values given by MSIS for moderate magnetic activity in the nighttime auroral zone.

3. REVIEW OF F-REGION EMISSIONS

This section gives a brief review of current knowledge of sources and quenching of a few important F-region high latitude emissions. Much of the progress during recent years has been made by the Atmosphere Explorer (AE) mission and particularly with the Visible Airglow Experiment (VAE) (Hays et al. 1973).

$O^+(^2P)$. In particle bombardment conditions, the $O^+(^2P)$ states arise from electron collisions with $O(^3P)$. The quenching is due to atomic oxygen and N_2 (Rusch et al., 1977). Deactivation by thermal electrons deduced from Henry et al. (1969) is also compatible with the V.A.E. observations. So far, a major discrepancy existed between the auroral measurements and the model calculations of 7320-30 A using this set of deactivation coefficients coupled to the quantum yield for $O^+(^2P)$ calculated by Dalgarno and Lejeune (1971). Calculated intensities tended to exceed the ground-based as well as the VAE measurements. This problem seems to be solved at the present time by the adoption of a neutral atmosphere model with reduced O/N_2 ratio. (Rees, pers. comm.). We shall return to this point in section 4.

$O^+(^2D)$. This emission at 3729 A was not observed by the AE satellites. However, since $O^+(^2D)$ plays an important role in the ion chemistry, its production and loss was investigated in depth using the ion composition measurements and reanalysed recently by Torr and Torr (1980). The reaction scheme for $O^+(^2D)$ and $O^+(^2P)$ is shown in Figure 2. Since quenching by N_2 dominates the loss of ^2D ions, this reaction is a significant source of N_2^+ ions at high altitudes and competes only with quenching for electrons. It is clear that in the daytime polar cusp, where $O^+(^2D)$ is enhanced by large fluxes of soft electrons, this process may become a major source of N_2^+ ions in the F-region. In the nighttime high altitude aurora, the 3729 A line is hardly identified in the emission spectrum (Vallance-Jones and Gattinger, 1975).

$N(^2P)$. Recent observations have given fairly consistent sets of production and quenching rates of $N(^2P)$ (Young and Dunn, 1975, Gérard and Harang, 1980, Zipf et al., 1980). It appears that atomic oxygen dominates $N(^2P)$ quenching in the F-region. The main source is undoubtedly dissociative excitation of N_2 but

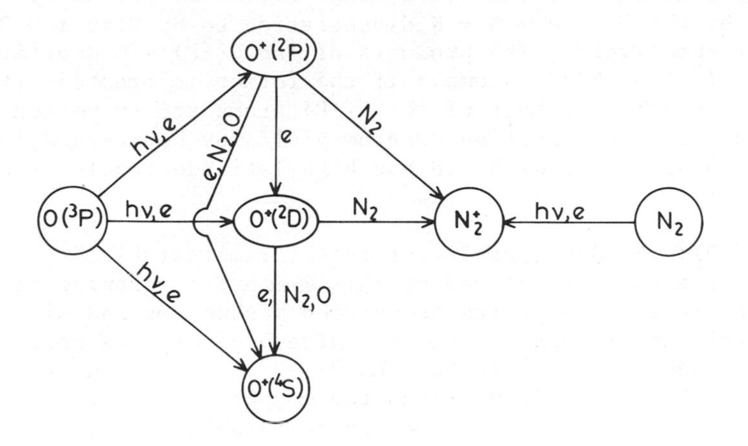

Figure 2 : Block diagram showing the role of $O^+(^2P)$ and $O^+(^2D)$ in the ionosphere.

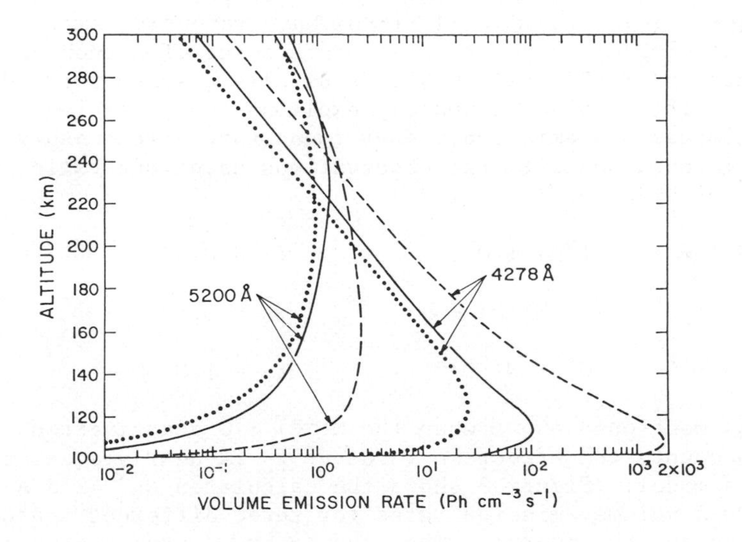

Figure 3 : Vertical distribution of the N_2^+ 4278 A and NI 5200 A emission rates for three different electron energy spectra (from Rusch and Gérard, 1980).

the quantum yield for $N(^2P)$ has not been measured. Zipf et al. 1980 have calculated the yield of $N(^4S)$, $N(^2D)$ and $N(^2P)$ atoms formed by the $N_2 + e \rightarrow N + N$ dissociation to be 0.46 : 0.35 : 0.19, respectively. The products of the $N(^2P) + O$ deactivation may be $N(^2D)$ or $N(^4S)$ atoms. If the former is created efficiently by this process, most of the $N(^2P)$ atoms are converted to $N(^2D)$ and a high effective quantum yield may be reached for $N(^2D)$, in agreement with mid and high latitude models of nitric oxide.

<u>$N(^2D)$</u>. A considerable interest for metastable $N(^2D)$ atoms arises from the role played by this species as a source of thermospheric nitric oxide. Its production mechanisms and sinks seem to be well understood due to the large body of data collected by the AE-C and D satellites for the ^4S - ^2D transition at 5200 A. Figure 3 shows the odd nitrogen thermospheric cycle. $N(^2D)$ has three sources of magnitude roughly comparable to the N_2 ionization rate, which explains why, in spite of its extremely long radiative lifetime (26 hrs), this doublet easily reaches measurable emission rates. Quenching rates have been deduced from the analysis of the VAE dayglow measurements (Fredrick and Rusch, 1977). The latter value is about one order of magnitude larger than the laboratory determination by Davenport et al. (1976) when the probable increase of k_O with temperature is taken into accounts. Rusch and Gérard (1980) have recently examined on the basis of AE-D measurements whether this set of quenching coefficients and $N(^2D)$ quantum yields usually assumed for mid-latitude conditions also adequately explains the particle-excited nighttime 5200 A emission. They found that a reasonably good match is obtained with the observations using the following values of $N(^2D)$ efficiencies :

$$NO^+ + e \rightarrow N(^2D) + O\ , \qquad f_1 = 0.80$$

$$N_2 + e \rightarrow N + N(^2D) + e \qquad f_2 = 0.60$$

and $$N_2^+ + O \rightarrow NO^+ + N(^2D)\ , \qquad f_3 = 1.0$$

As mentioned above when the $N(^2P) + O$ deactivation is taken into account, the effective value of f_2 is nearly equal to 0.8 in this model. Figure 3 shows the calculated N_2^+ 4278 A and NI 5200 A volume emission rates for three different measured electron energy spectra. They show a broad peak of $N(^2D)$ roughly extending from 140 to 300 km and reaching about 1×10^5 cm^{-3}. The comparison between mid-latitude and high latitude sources is also of interest. When the production results from electron impact on N_2, the three sources listed above are of nearly equal magnitude, whereas at mid-latitudes the second one reaches only 10% of the total production. In the aurora, roughly four N atoms are created for each N_2 ionization and with the efficiences adopted in this mode, the global efficiency for $N(^2D)$ production is 83%. Consequently, an empirical ratio of ∿ 3.3 $N(^2D)$ produced/N_2 ionization may be adopted. The wide range of altitudes where $N(^2D)$ atoms are found makes it unrealistic to define single effective lifetimes for this state but the large values of τ_e make it a very good candidate for horizontal transport by winds. This point will be illustrated in section 4.

$O(^1D)$. The OI ^3P - ^1D doublet at 6300 - 6363 A is by far the most prominent feature amongst the F-region emissions. Understanding of its altitude distribution and excitation mechanisms is essential since ground-based measurements of its intensity and Doppler shift are used extensively to determine the electron energy spectrum and the neutral winds respectively.

The classical sources of $O(^1D)$ due to particle precipitation are :

$$O(^3P) + e_{s,th} \rightarrow O(^1D) + e_{s,th}$$

and $$O_2^+ + e \rightarrow O + O(^1D) ,$$

where the electrons exciting $O(^3P)$ may be secondaries or fast thermal electrons and the quantum yield of $O(^1D)$ for dissociative recombination of O_2^+ is 0.66. This state is mainly quenched by N_2 (Hays et al. 1978).

This theory could hardly be tested against ground-based observations due to the diffuse character of the 6300 A morphology. However, analysis of a coordinated AE-rocket measurement over Fort-Churchill showed that another source of $O(^1D)$ was needed to explain the observed emission rate. Figure 4 shows the measurements as well as the $O(^1D)$ production rate by the two processes described above, where the curve marked "Harp" is the contribution of the first process calculated using the measured electron flux. The comparison clearly shows that an extra source of ^1D is needed to account for the observations. Rusch et al. (1978) have proposed that the energy transfer process :

$$N(^2D) + O \rightarrow NO + O(^1D) + 3.8 \text{ eV} ,$$

may be the missing source. They showed that the efficiency should be near 100% to explain the ^1D vertical profile. Laboratory spectra of the infrared emission from nitric oxide formed by this process show that the vibrational distribution of NO breaks up at v' = 7 which corresponds to the threshold of formation of $O(^1D)$ instead of $O(^3P)$. The close temporal correlation between the 5200 A and the 6300 A intensities observed during time variations is also in agreement with this hypothesis.

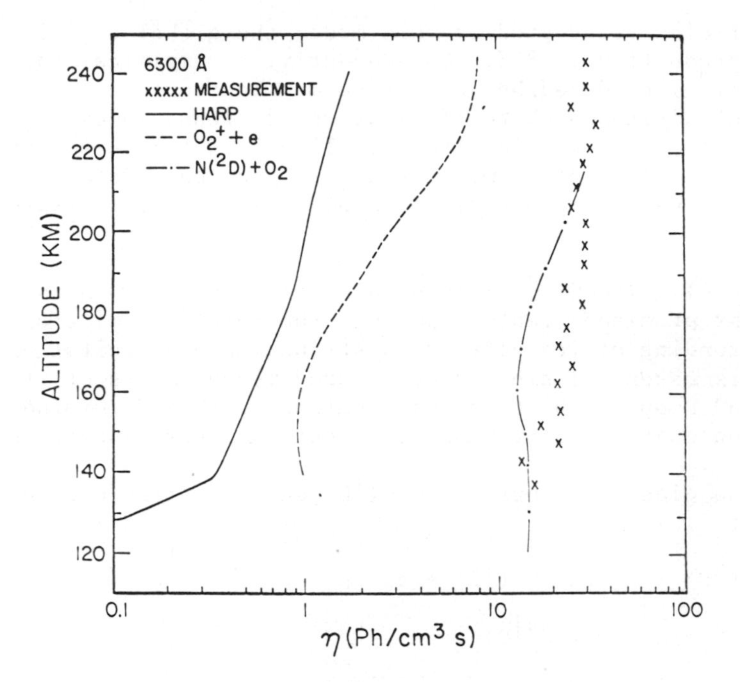

Figure 4 : Measured and calculated volume emission rates of $O(^1D)$ for various excitation processes (from Rusch et al. 1978).

Recently, efficiencies for the production of $O(^1D)$ by primary electrons in various energy channels have been deduced by Shepherd et al. (1980) by comparing the measured 6300 A emission rate with the ISIS II low energy electron spectrometer.

Table I summarizes our recommended set of values concerning the sources and quenching of the five important metastable states described above. The last two columns list the altitude of the peak and the calculated effective lifetime at these altitudes. These figures are only approximate since the altitude of the maximum is sometimes loosely defined and depends strongly on the primary electron spectrum.

4. REMOTE SENSING BY F-REGION OPTICAL EMISSIONS

The use of optical emissions as a remote sensing technique of the polar atmosphere and ionosphere has developed along three major axes :

Table 1 : F-region metastable atom sources and quenching

transition	λ(A)	sources	quantum yield	loss coefficients ($cm^3.s^{-1}$)	peak altitude (km)	effective lifetime (s)
[OI] $^3P - {}^1D$	6300-64	$O + e \rightarrow O(^1D) + e$ $O_2^+ + e \rightarrow O(^1D) + O$ $N(^2D) + O_2 \rightarrow O(^1D) + NO$	0.66	$A = 9.1 \times 10^{-3} s^{-1}$ $k_{N_2} = 3 \times 10^{-11}$	300	20
[OII] $^4S - {}^2P$ and $^2D - {}^2P$	2470 7320-30	$O + e \rightarrow O^+(^2P) + 2e$	0.22	$A = 0.22\ s^{-1}$ $k_O = 5.2 \times 10^{-10}$ $k_{N_2} = 4.8 \times 10^{-10}$ $k_e = 1.9 \times 10^{-7}(300/T_e)^{1/2}$	200	0.4
[OII] $^4S - {}^2D$	3729	$O + e \rightarrow O^+(^2D) + 2e$ $O^+(^2P) \rightarrow O^+(^2D) + h\nu$ $O^+(^2P) + M \rightarrow O^+(^2D) + M$	0.32	$A = 7.7 \times 10^{-5} s^{-1}$ $k_O < 1 \times 10^{-11}$ $k_{N_2} = 1 \times 10^{-10}$ $k_e = 1.5 \times 10^{-7}(300/T_e)^{1/2}$	300(?)	15
[NI] $^4S - {}^2P$ and $^2D - {}^2P$	3466 10,395-404	$N_2 + e \rightarrow N(^2P) + N + e$ $N_2^+ + e \rightarrow N(^2P) + N$ $(N_2^+)^* + O \rightarrow NO^+ + N(^2P)$	0.05 to 0.20 0.50	$A = 8.3 \times 10^{-2}\ s^{-1}$ $k_O = 1 \times 10^{-11}$ to 3×10^{-11} $k_{O_2} = 2.6 \times 10^{-12}$ $k_e = 4 \times 10^{-10}(300/T_e)^{-1/2}$	120	0.4
[NI] $^4S - {}^2D$	5199-5201	$N_2 + e \rightarrow N(^2D) + N^{(+)} + (2)e$ $NO^+ + e \rightarrow N(^2D) + O$ $N_2^+ + O \rightarrow NO^+ + N(^2D)$	0.75 0.8 1.0	$A = 1.1 \times 10^{-5}\ s^{-1}$ $k_O = 4 \times 10^{-13}$ $k_{O_2} = 6 \times 10^{-12}$ $k_e = 4 \times 10^{-10}(300/T_e)^{-1/2}$	250	1000

i) determination of the characteristic energy of the precipitated electron energy.

ii) evaluation of composition changes caused by vertical upwelling due to local Joule heating.

iii) measurement of neutral winds and electromagnetic drifts by observation of the Doppler width and shift of F-region emissions. Developments of the three techniques are briefly described in the following paragraphs.

4.1. Determination of the precipitation characteristic energy

The emission rate ratio of a forbidden to an allowed emission may, once the excitation mechanisms are known quantitatively, be used to evaluate the spectral index of the flux of precipitated electrons. For example, Rees and Luckey (1974), assuming a maxwellian distribution of energies have calculated the 6300 A/4278 A ratio versus I(4278 A) for various values of the characteristic energy using the "classical" production mechanisms for $O(^1D)$. Conversely, the observed ratio and absolute 4278 A intensity may be used to determine the hardness of the precipitation. However, as mentioned in the previous section, a new source of $O(^1D)$ atoms needs to be added to the model and until revised tables including this extra production term are available (Rees, pers. comm.), the figures of Rees and Luckey should be used with caution.

4.2. Neutral composition changes

Observations of forbidden lines whose relative intensities are sensitive to the N_2(or O_2)/O ratio may be used to monitor the neutral composition in the polar thermosphere. For example, the (OII 7320 A)/(N_2^+ 4278 A) ratio is given by :

$$\frac{\eta(7320\ A)}{\eta(4278\ A)} = C \frac{[O]}{[N_2]} \quad Q \ ,$$

where Q is the quenching factor described in section 2. Consequantly, this ratio may be used to determine changes of the O/N_2 ratio due to vertical upwelling generated by Joule heating.

4.3. Wind and drift measurements

The technique using the 6300 A line shift and Doppler width measurements to determine F-region winds has been used extensively and will be employed for the first time on a satellite in the Dynamics Explorer mission. The method and new results for the polar regions are described by Smith (this volume) and will not be repeated here. Although it has never been tested, progress in sensitivity of Fabry-Perot and Michelson interferometers should permit the method to be extended to weaker features such as NI 5200 A.

It is clear that morphology of metastable emission can be greatly influenced by the presence of horizontal neutral winds. Table I shows that effective lifetimes may reach several seconds near the peak of the emission profile. To illustrate this point, figures 5 and 6 show the effect on the morphology of the $N(^2D)$ distribution of a horizontal wind blowing through an auroral arc. The particle precipitation is similar to that used by Roble and Gary (1979). The altitude dependence of the wind velocity is given by :

$$v(z) = 50 + 0.75(Z - 75) \quad m/S.$$

The contours are shown in units of $\log(N^2D)$ in figure 5, 15 and 60 min after the wind is turned on. A downwind plume of metastable atoms is formed, similar to the result for NO described by Roble and Gary. Consequently, the 5200 A intensity is enhanced by a large factor on the downwind side of the arc. The particle flux is turned off after 60 min and figure 6 illustrates the morphology at t = 75 min and 90 min. The enhanced region of $N(^2D)$ gradually disappears under the effects of the wind and quenching by O and O_2. Evidence for such transport effects was presented by Shepherd et al. (1970) and Frederick and Hays (1978).

Measurements of Doppler shifts of emissions from metastable ions such as $O^+(^2P)$ have also been used to determine ion drift velocities in the high latitude thermosphere. Preliminary observations were reported by Meriwether et al. (1974) but with the improved sensitivity of recent instruments, better accuracies could be obtained and this technique should receive greater attention in the future.

Finally, another optical technique for remotely sensing drift velocities has been suggested by Haerendel (1976). A balloon-borne all-sky camera is used to observe the resonance scattering of sunlight by Mg^+ ions at 2800 A. These ions are produced in the E-region by meteor ablation and transported upwards by winds and drifts. Their presence was observed optically in the F-region by Gérard (1976) and by Grebowsky and Brinton (1978) using a mass spectrometer. Striations similar to those observed after Ba rocket releases should be observed in the presence of field-aligned currents.

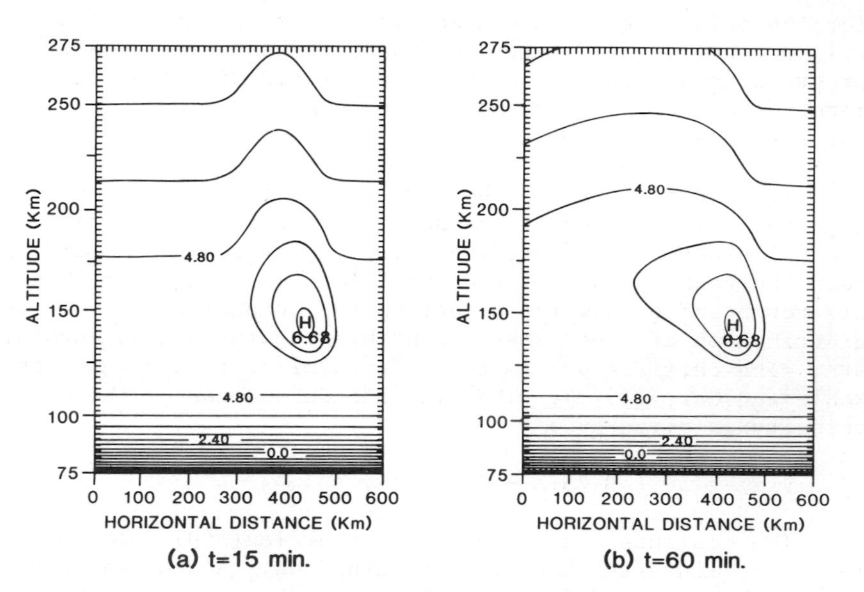

Figure 5 : Contour plots of $\log_{10}(N^2D)$ for an auroral arc with neutral wind. Contours are separated by 0.6 unit. (Courtesy of R.G. Roble).

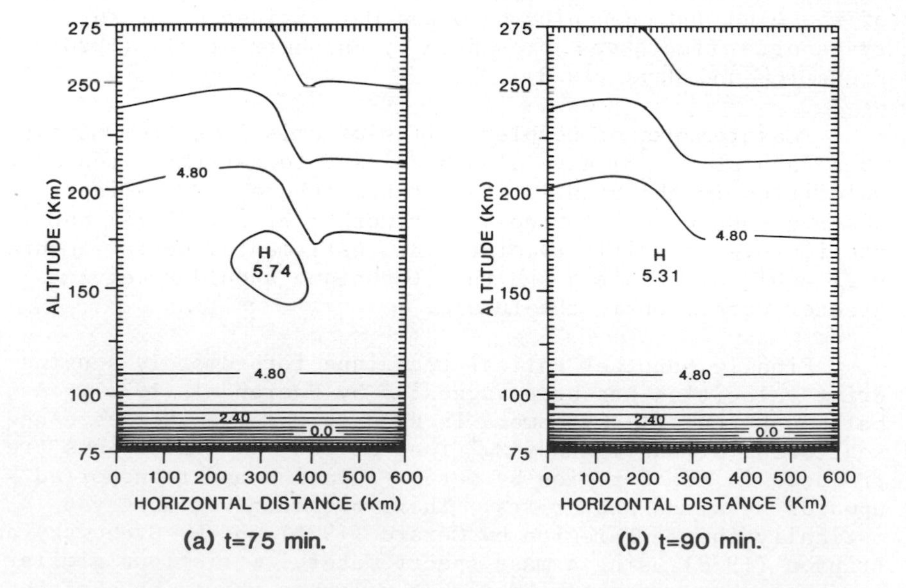

Figure 6 : Same as figure 5, except that the source has been turned off at t = 60 min.

ACKNOWLEDGMENTS

The author is supported by the Belgian Foundation for Scientific Research (FNRS).

REFERENCES

Dalgarno A.L., and Lejeune G. 1971, Planet. Space Sci., 19, 1653.
Davenport J.E., Slanger T.E., and Black G. 1976, J. Geophys. Res. 81, 12.
Deans A.J., and Shepherd G.G. 1978, Planet. Space Sci., 26, 319.
Frederick J.E., and Rusch D.W. 1977, J. Geophys. Res., 82, 3509.
Frederick J.E., and Hays, P.B. 1978, Planet. Space Sci., 26, 339.
Gérard J.C. 1976, J. Geophys. Res., 81, 83.
Gérard J.C., and Harang O.E. 1980, J. Geophys. Res., 85 Accepted for publication.
Grebowsky J.M., and Brinton H.C., 1978, Geophys. Res. Lett., 5 791.
Haerendel G. 1976, Proceeding ESA Symposium, ESA/SP115.
Hays, P.B., Carignan G., Kennedy B.C., Shepherd G.G., and Walker J.C.G. 1973, Radio Sci., 8, 369.
Hays P.B., Rusch D.W., Roble R.G., and Walker J.C.G. 1978, Rev. Geophys. Space Phys., 16, 225.
Henry R.J., Burke P.G., and Sinfailam A.L. 1969, Phys. Rev., 178, 218.
Meriwether J.W., Hays P.B., McWatters H.D., and Nagy A.F.V. 1974, Planet. Space Sci., 22, 636.
Rees M.H. 1975, Planet. Space Sci., 23, 1589
Rees M.H., and Luckey D. 1974, J. Geophys. Res., 79, 5181.
Rees M.H., Stewart A.I., Sharp W.E., Hays, P.B., Hoffman R.A., Brace L.H., Doering J.P., and Peterson W.K. 1977, J. Geophys. Res., 82, 2250.
Roble R.G. 1975, Planet. Space Sci., 23, 1017.
Roble R.G., and Gary J.M. 1979, Geophys. Res. Lett., 6, 703.
Rusch D.W., Torr D.G., Hays P.B., and Walker J.C.G. 1977, J. Geophys. Res., 82, 719
Rusch D.W., Gérard J.C., and Sharp W.E. 1978, Geophys. Res. Lett., 5, 1043.
Rusch D.W., and Gérard J.C. 1980, J. Geophys. Res., 85, 1285.
Shepherd G.G., Pieau J.F., Creuzberg F., McNamara A.G., Gérard J.C., McEwen D.J., Delana B., and Whitteker J.H. 1976, Geophys. Res. Lett., 3, 69.
Shepherd G.G., Winningham J.D., Bunn F.E., and Thirkettle K.W. 1980, J. Geophys. Res., 85, 715.
Torr D.G., and Torr M.R. 1980, J. Geophys. Res., 85, 783.
Vallance-Jones A., and Gattinger R.L. 1975, Can. J. Phys., 53 1806.
Young R.A., and Dunn O.J. 1975, J. Chem. Phys., 63, 1150.
Zipf E.C., Espy P.J., and Boyle C.F. 1980, J. Geophys. Res., 85, 687.

NEUTRAL WINDS IN THE POLAR CAP

R. W. Smith

Ulster Polytechnic, Co. Antrim, U.K.

ABSTRACT The optical Doppler technique of the ground-based remote sensing of the thermospheric wind is reviewed with special consideration given to the admissibility of the assumptions made in the reduction of data from a single station. Examples of observations made in the Auroral Zone indicating substantial wind gradients and vertical winds are given. Such observations indicate the importance of small scale disturbances in the Polar thermosphere. Data from three stations in the Northern Auroral Zone and Polar Cap show how the flow pattern, the wind speed and its variation with magnetic activity strongly suggest that the main source of the momentum of the Polar Thermosphere is the convecting ions in the ionosphere.

1. INTRODUCTION

This paper concerns the remote sensing of the neutral wind in the Polar Thermosphere by a ground-based method which is based on the Doppler principle. The radiation from the discrete and diffuse aurora, and also from the airglow, is emitted by gas particles which, in many cases, are in equilibrium with the ambient gas. The distributions of random and bulk velocities of such particles are representative of the population as a whole and may be studied as tracers which can yield both the temperature of the gas and the component of the neutral wind along the line of sight of the detector. Reliable tracer measurements may be made using any spectral feature of absorption or emission provided that the initial state for the optical transition is long lived compared to the interval between collisions and the altitude range over which the emission or absorption takes place is sufficiently narrow.

C. S. Deehr and J. A. Holtet (eds.), Exploration of the Polar Upper Atmosphere, 189–198.

Owing to the existence of wind shears in the 90-150 km altitude region (see for example Rees (1971)), it is difficult to make use of many of the auroral spectral features in this range since their altitude profiles are not sufficiently narrow. Most measurements, to date, have been made using the 630 nm emission of atomic oxygen whose initial state is $O(^1D)$ having a radiative lifetime of 110 seconds which is adequately long for the 200-300 km altitude region in which it is produced.

2. INSTRUMENTATION

Hays and Roble (1971) and Nagy et al (1974) have published observations of neutral thermospheric winds made at College, Alaska, in the Auroral Zone using the 630 nm oxygen line. Their results demonstrated the optical technique in the aurora. It has also been in use for many years in midlatitudes at Fritz Peak (Hernandez 1976). The instrumentation is required to measure the Doppler shift in wavelength to a precision of 3 parts in 10^8 in order to achieve 10 ms^{-1} resolution in the wind component along the line of sight. In addition, this measurement needs to be made in less than 5 minutes if reasonable time resolution is to be achieved. The normal approach to this high specification is to use an interferometer in order to obtain the necessary resolving power and luminosity (or êtendue). A typical instrumental specification is given below in Table I.

TABLE I
SPECIFICATION OF A SUITABLE DOPPLER WIND DETECTOR

1. Spectroscopic resolving power: $R = \frac{\lambda}{\Delta\lambda} = 3 \times 10^5$

   Peak wavelength can be measured to 0.01 of the spectroscopic resolution element.

2. Overall sensitivity: $C = BA\Omega\tau pf$

   Where B = surface brightness of the source; A - area of the critical aperture; Ω = solid angle of acceptance at the critical aperture; τ = transmission factor; p = quantum efficiency of the detector; and f = scanning factor, the loss due to scanning either side of the spectral feature. The minimum value of C is about 0.1 recorded photons per second per Rayleigh.

3. Stability: in short term, it must be to 0.001 of the spectroscopic resolution element; long term drift, it may be 0.01 if sufficiently frequent calibrations are made.

Conditions 1 and 2 in Table I can be fulfilled by suitable versions of either the Michelson or Fabry-Perot interferometers (for example, see Despain et al (1973) and Hernandez and Mills (1973)). The wide angle Michelson interferometer (WAMI) has a superior ΩR product compared to the plane Fabry-Perot interferometer (PFPI). If $R = 3 \times 10^5$, then for the WAMI; $\Omega R = 12\sqrt{R} = 6.6 \times 10^3$ and for the PFPI; $\Omega R = 2\pi = 6.3$. However, when the WAMI is used with a small modulation of optical path, only a single Fourier component of the profile of the spectral line is obtained whereas the PFPI typically yields 8 components which is useful for studies and checks on the line profile. Also, with a suitable scanning mask (Hirschberg 1971) or an imaging photon detector (Rees et al 1980a), considerable improvements can be made in the effective ΩR product. More importantly, post hoc spatial scanning can be used which allows the full multiplex advantage to be obtained. In the opinion of the author, each type of interferometer has its merits and deserves consideration for the particular application in view.

The stability requirements affect both types of interferometer in that the optical path difference must be held constant within the required limits. These are: for pressure, ±10 mB (air); and for temperature, ±0.01°C assuming the expansivity of the optical path difference is $10^{-6}\ {}^{\circ}C^{-1}$.

3. DATA REDUCTION

The determination of wavelength from the interference pattern, or interferogram, is carried out by finding the position of a fringe maximum, or peak of photocurrent as a function of position in the scan and its separation from the position in the scan whose wavelength is well known. The peak position may be found to 0.01 of a resolution element by fitting the observed profile to a suitable nominal shape. In many cases, an inverted parabola fits the upper part of the profile sufficiently well. An example of this is shown in figure 1.

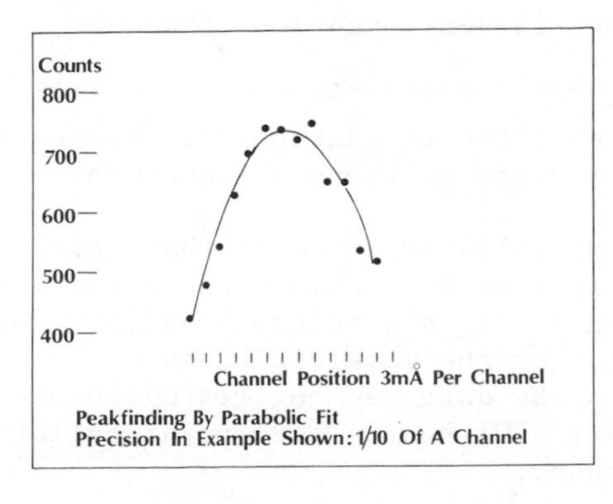

Figure 1. An example of the use of a parabolic fit to the peak of a PFPI fringe. The random error in the determination of the peak position is 0.1 channel widths.

For instruments which record the image of the interferogram and employ spatial scanning, the interference ring pattern must be reduced to a one dimensional representation of photocurrent versus the square of the radius and then treated in a similar fashion.

The most suitable source for wavelength calibration is a laboratory lamp giving the spectral line which is being measured in the aurora. However, when using a transition with a metastable initial state, it is often difficult to obtain a good source for use in the field. A satisfactory substitute is to use an rf excited, water cooled single isotope ^{198}Hg discharge lamp, although this requires an extra calibration. Measurements of the displacement of the 546 nm line of this lamp from the observed position of the auroral line when looking in the zenith, averaged over a long period of time (say 24 hours), provide the necessary data for an immediate wavelength calibration using the lamp alone.

Table II summarises the assumptions used in the interpretation of data from observations of the 630 nm line of oxygen observed at 30° elevation.

TABLE II
ASSUMPTIONS USED IN THE INTERPRETATION OF DATA

1. $O(^1D)$ atoms in the emission layer have the same bulk and random speed distribution as the ambient gas.
2. The wind vector is constant over the height range of emission (200-300 km).
3. The horizontal scale size of irregularities in the wind is greater than the baseline of the observations, 800 km in this case.
4. Discrete auroral structures do not cause local effects in the neutral wind.
5. The time scale of irregularities is greater than the cycle time of the measurements.
6. The vertical wind component is very much less than the horizontal component.

When measurements are presented later in this paper, reference will be made to the validity of some of these assumptions.

Observation schemes are normally arranged so that measurements of the Doppler shift are obtained at some chosen elevation angle looking N, S, E or W in sequence. If a wavelength calibration has been made giving the zero point on the wavelength scale for nil bulk speed of the source, then the data may be treated in one of the two ways shown in Table III. Without a zero point calibration only the first option can be used.

TABLE III
DEDUCTION OF THE HORIZONTAL COMPONENTS OF THE NEUTRAL WIND

OPTION	COMMENT
1. Subtract the wavelength in one direction from that at its opposite and hence obtain a single meridional and zonal component of the vector.	A single average wind vector is found over the region of atmosphere sampled.
2. Find the difference of wavelengths in each direction from the zero point and hence obtain two values of each component.	Two independent vectors are obtained. Spatial gradients and vertical winds may be studied.

Note that option 2 is more vulnerable to failure of assumptions 3-6 than option 1.

4. WIND MEASUREMENTS IN THE AURORAL ZONE AND POLAR CAP

Data treated according to option 1 has been reported by Nagy et al (1974) and is shown in figure 2. To make comparisons of these wind measurements with published models of the neutral wind, two examples are taken using the studies of Kohl and King (1967) and Fedder and Banks (1972). The value in this particular choice lies in the fact that the Kohl and King model is dominated by the influence of the Solar EUV heating whilst that due to Fedder and Banks demonstrates the expected thermospheric circulation at high latitudes when the main driving agent is momentum transfer from ion convection. It is clear that there is a striking difference between the expectations of the two models in the nighttime sector of the Auroral Zone. The measurements in figure 2 favour the ion convection source.

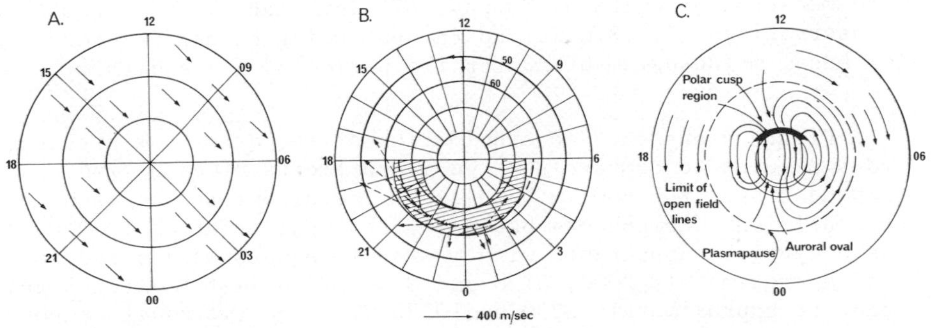

Figure 2. Polar plot in magnetic co-ordinates of the thermospheric neutral wind: (a) measured at College, Alaska February 27, 1973 (Nagy et al 1974), (b) predicted by Kohl and King (1967), and (c) flow pattern calculated by Fedder and Banks (1972).

Data obtained jointly by the author and J R Whiteford at Skibotn, Norway (Invariant latitude 67°), is shown in figure 3 and confirms the conclusion from the data in figure 2. The Skibotn measurements were reduced using option 2 giving two sequences of vectors which were obtained by combining the N and W observations and the E and S pair. Although there are clearly identifiable differences, the general character is the same in both cases. A westward wind developed prior to the substorm (onset determined by the trace of the magnetometer H component) followed by a weaker south-eastward wind after the substorm break-up. This is interpreted in terms of momentum transfer from the westward ion convection which is expected during the early evening and is characteristically found in the evening diffuse aurora followed by a reversal after passage under the Harang discontinuity. The development of the equatorward component in the post midnight sector was probably due to a combination of the antisolar cross-Polar flow combined with the Joule heating effect in the auroral zone during the substorm.

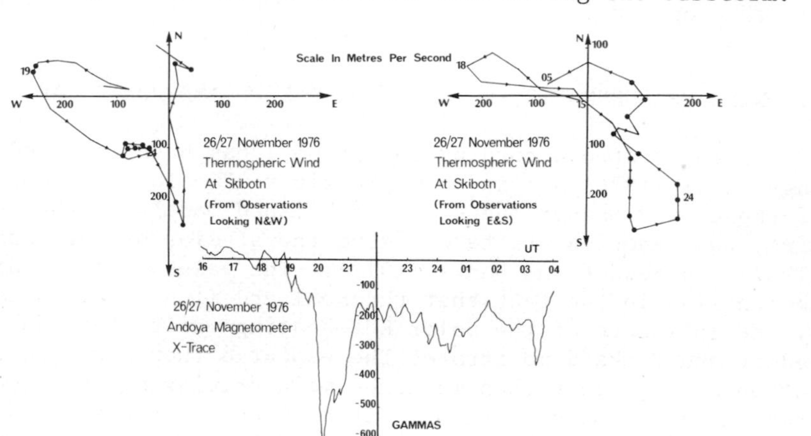

Figure 3. (a) Vector Hodograph plot of the thermospheric neutral wind measured at Skibotn, Norway, 67° Inv. lat. Vector from N, and W observations. (b) Similar to (a) but using E, and S observations. (c) Local magnetogram of ΔH for the period of (a) and (b).

Figure 4 presents an examination of the data from figure 3 redrawn so as to compare the two measurements of each wind component over the 800 km baseline. The difference between the two meridional components becomes most apparent after 1900 UT and has a wavelike appearance with maxima in equatorward speed looking N at approximately 2000, 0130 and 0430 UT, whilst looking S they occur at approximately 2230 and 0230 UT. The two zonal components agree very closely during the westward acceleration but show evidence of a local time effect at 1800 UT when the component observed looking E changed first, about 1.5 hours before the other. This is consistent with the later passage of the atmosphere at the observation point looking W under the Harang discontinuity. Hence the effect of ion velocity reversal was seen later. Clear differences in excess of 200 ms^{-1} occurred after 2200 UT. The occurrence of such large apparent wind gradients must call into question the validity of assumption 3 in Table III.

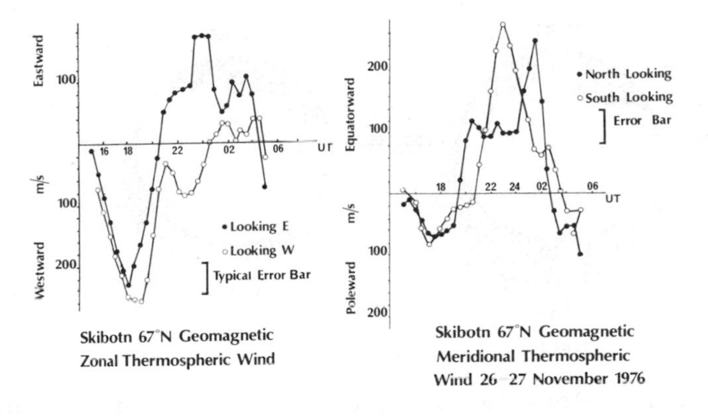

Figure 4. (a) Meridional components of the thermospheric neutral wind measured at 67° Invariant latitude at Skibotn, Norway. (b) Similar to (a) but showing the zonal components.

The local time coverage which can be easily achieved using this ground-based technique is limited to the nighttime period between civil twilight in the evening and morning. Successful daytime measurements of midlatitudes have been reported by Cocks and Jacka (1979) using a multiple etalon system, but very strict criteria have to be imposed on the sky conditions in daytime before the data can be interpreted reliably. However, by taking advantage of the continual darkness during the period near winter solstice in Polar latitudes, full coverage can be obtained for all local times with a single etalon system. Observations of this type have been reported by Smith and Sweeney (1980) who were collaborators in the Multinational Auroral Expedition to Spitzbergen in 1978/9 (Deehr et al 1980). Figure 5 shows some of the neutral wind data from that campaign. The data was reduced using option 1 which had the effect of suppressing local irregularities so as to provide the average wind over a sample of thermosphere 800 km in diameter centred on 75° invariant latitude. Comparison of figure 5 with the models in figure 2 shows that the directions of the wind are generally compatible with either model. However, there are strong suggestions of a vortex on the morning and evening sides of the pattern, and the wind speeds on 27 January 1979 are generally greater than would be expected for the Solar EUV effect alone.

Once again, the data strongly suggests that the neutral wind correlates with ion convection. The 27 January was a day of moderate magnetic activity (A_p = 23) where a strong twin cell of thermospheric circulation is suggested, whereas on the 21 January, a less active day with A_p = 11, the evening cell seems to have been enlarged and the morning cell diminished.

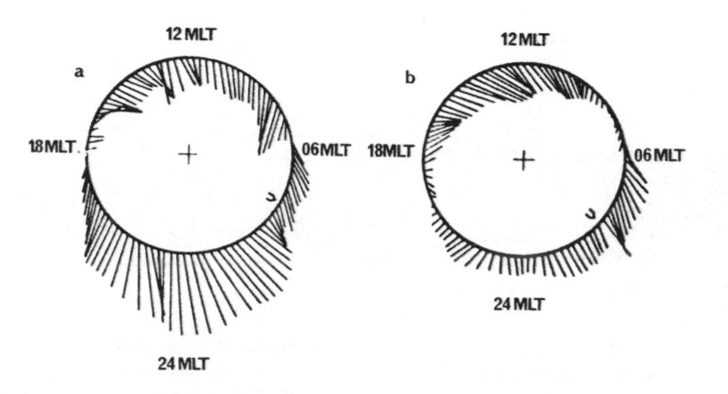

Figure 5. Polar plots in magnetic co-ordinates of the thermospheric neutral winds at Invariant latitude 75° in the Polar Cap at Longyearbyen, Spitzbergen. (a) For 27 January, 1979, when Ap was 23. (b) For 21 January, 1979, when Ap was 11.

This is consistent with the patterns of equivalent ionospheric currents observed by Friis Christensen (1980) for periods of high and low magnetic activity in which a twin cell pattern is observed in which the evening cell becomes dominant in weak magnetic activity. Since F region ions have the same flow pattern as the ionospheric current in the E region, this is further evidence for the correlation of ion and neutral flow at F region heights. The data in figure 5 for 27 January 1979 has been shown to be compatible with the UCL Global Dynamic Model of the Thermosphere (Rees et al 1980b), in which momentum transfer from convecting ions is the most important wind driving effect over the Polar Cap.

It is notable in this respect that the wind speed in the midnight sector on 27 January 1979 was about four times that on 21 January 1979 which indicates that the antisolar wind speed in the Polar Cap increases with magnetic activity. A study of this effect using data from nine 24 hour periods of observation at Spitzbergen shows that the wind speed V at midnight can be expressed as $V = 50K_p^* - 52\ ms^{-1}$ with a correlation coefficient of 0.75. K_p^* is the sum of K_p values for the previous 6 hours.

The Doppler shift measured when observing in the zenith was studied in order to test assumption number 6 in Table III and to investigate its relationship to the occurrence of discrete auroral forms. Results from a sequence of data interpreted in terms of a vertical wind are shown in figure 6.

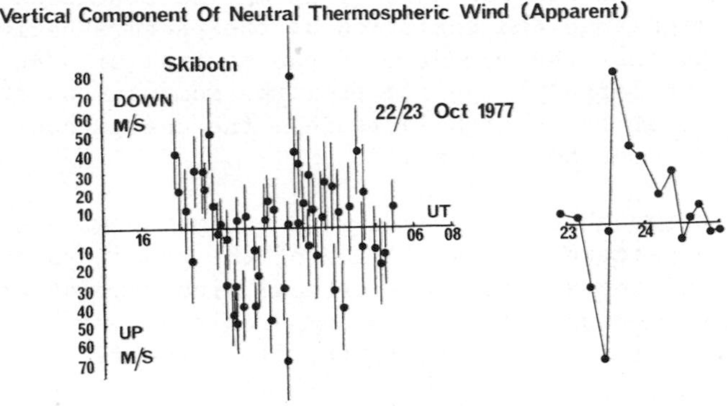

Figure 6. Observations of the vertical component of the thermospheric neutral wind made at 67° Invariant latitude at Skibotn, Norway on October 1977.

The evening began with a general downward wind of 30 ms^{-1} which steadily changed direction to upward at 50 ms^{-1} by 2200 UT and then returned to zero. In the period 2145-2215 UT there was a break-up in progress with negative ΔH at 300γ. After the recovery the auroral display died away apart from a single stable arc which steadily approached the zenith from the poleward side. It arrived overhead at 2250 UT, died away, but reappeared in the same place for some minutes at 2315 UT with about half its former intensity. At this time, it was the only form in the sky apart from some weak unstructured activity at some distance on the poleward side. The observed response of the neutral atmosphere at this time was sharper than measured during break-up. There was a rapid onset of downward motion peaking at 70 ms^{-1} followed by a smart reversal to upward motion of about 80 ms^{-1} with a subsequent recovery lasting approximately one hour. The most likely explanation of this event is that it was due to a gravity wave passing overhead since it is similar to the results of the model of Richmond and Matsushita (1975).

5. CONCLUSIONS

The optical Doppler technique is based on a set of assumptions which are generally valid at midlatitudes, but some of which may be called into question in the active auroral zone. In particular, the possibility of small scale sizes of irregularity in the wind, less than 800 km across, and the substantial apparent vertical winds, giving unexpectedly large systematic errors in the winds deduced in the manner indicated. Uncertainty in the wind speed could be as high as 50 ms^{-1} due to the failure of these assumptions. The problem may be overcome by using three spaced interferometers pointing at the same small region of thermosphere so as to obtain an unambiguous wind vector for that restricted region.

Despite these considerations, it may be concluded from the direction, magnitude and variation of the thermospheric wind in the Polar Cap that the momentum of the neutral atmosphere above 200 km is very largely supplied from the momentum of the convecting ions. Hence, at these high latitudes, the energy source for the circulation is the Solar wind.

Also, although the averaged wind vector shows a steady and recognisable pattern, investigation of the gradients in the wind components and the substantial vertical wind demonstrates that during active periods there is a comparable amount of energy in the irregularities of flow as in the flow itself.

REFERENCES

Cocks, T.D., and Jacka, F. 1979, J. Atmos. Terr. Phys., 41, 409.
Deehr, C.S., Sivjee, G.G., Henriksen, K., Egeland, A., Sandholt, P.E., Smith, R.W., Sweeney, P., Duncan, C., and Gilmer, J., 1980, J. Geophys. Res., (in press).
Despain, A.M., Baker, D.J., Steed, A.J., Tohmatsu, T., 1973, Appl. Opt., 12, 126.
Fedder, J.A., and Banks, P.M., 1972, J. Geophys. Res., 77, 2328.
Friis-Christensen, E. 1980, This volume.
Hays, P.B., and Roble, R.G., 1971, J. Geophys. Res., 74, 4162.
Hernandez, G., and Mills, G.A., 1973, Appl. Opt., 12, 126.
Hernandez, G., and Roble, R.G., 1976, J. Geophys. Res., 81, 2065.
Hirschberg, J,G., Fried, W.I., Hazelton, L.Jr., and Wouters, A., 1971, Appl. Opt., 10, 1979.
Kohl, H., and King, J.W., 1967, J. Atmos. Terr. Phys., 29, 1045.
Nagy, A.F., Cicerone, R.J., Hays, P.B., McWatters, K.D., Meriwether, J.W., Belon, A.E., Rino, C.L., 1974, Rad. Sci., 9, 315.
Rees, D., 1971, J. Brit. Interplan. Soc., 24, 233.
Rees, D., Fuller-Rowell, T., Smith, R.W., 1980b, Planet. Space Sci., (in press).
Rees, D., McWhirter, I., Rounce, P.A., Barlow, F.E., and Kellock, S.J., 1980a, J. Phys. E., (in press).
Richmond, A.D., and Matsushita, S., 1975, J. Geophys. Res., 80, 2839.
Smith. R.W., and Sweeney, P.J., 1980, Nature, 284, 437.

DIFFERENCE IN POLAR ATMOSPHERIC OPTICAL EMISSIONS BETWEEN MID-DAY AND NIGHT-TIME AURORAS

G. G. Sivjee and C. S. Deehr

Geophysical Institute, University of Alaska
Fairbanks, Alaska 99701

ABSTRACT

Atomic emissions, especially from metastable atmospheric species, dominate the mid-day auroral optical spectrum. In addition to [OI]5577A and 6300 A lines, the [OII] lines at 3727-29A and 7320-30A as well as [NI] lines at 3466A and 5199-5201A are prominent. Unlike the night-time auroras, the cusp auroras are almost devoid of $N_2$1P and $N_2$2P bands. N_2^+1NG bands in mid-day auroras are probably enhanced through resonant scattering of sunlight by N_2^+ ions formed, in almost equal amounts, by particle impact on N_2 and charge exchange of [OII] (^{2}D) with N_2.

INTRODUCTION

Various rocket and satellite measurements have shown that auroral electrons precipitating during mid-day are about an order of magnitude less energetic than in night-time auroras. Hence, while most of night-time auroras are formed around 110-120 km in the atmosphere, the mid-day auroral optical emissions emanate above 150 km height. Differences in the relative abundance of various atmospheric species and their mean free paths at these heights should be reflected in the spectral distribution of optical emissions from the day and night-time auroras. Since night-time auroras are formed at a height where molecular species (in particular N_2) are the main atmospheric constituents, molecular bands (e.g. $N_2$2P, $N_2$1P, N_2^+1NG, N_2^+M and O_2At.) should form the strong auroral emissions. On the other hand, daytime auroral emissions come from an altitude region where atomic species (in particular O) form a significant portion of the atmosphere and

C. S. Deehr and J. A. Holtet (eds.), Exploration of the Polar Upper Atmosphere, 199–207.

their collision frequency is relatively low. Consequently, atomic emissions, including those from long-lived metastable atmospheric species, can be expected to dominate the daytime auroral optical spectrum. Additionally, the N_2^+1NG emissions are the only molecular bands observed in daytime aurora. Their detection is aided by intensity enhancement from resonant scattering of sunlight by N_2^+ ions formed in equal parts by particle impact and [OII] (^2D) charge exchanging with N_2. Spectroscopic observations of night and mid-day auroral optical emissions between ∿3300A and ∿8600A from Svalbard and Alaska may be analysed to show that the expected differences between day- and night-time aurorae do indeed exist.

INSTRUMENTATION

The measurements reported here were made with two large throughput Ebert-Fastie spectrophotometers. One of the spectrometers has a focal length of 1 meter and employs a 25.6 x 15.4 cm^2 grating and 15 cm long curved slits. The second spectrometer has a focal length of 1/2 meter and is fitted with a 12 x 7 cm^2 grating and curved slits, each 7 cm long. Both spectrometers, together with an auroral camera and a 4278A photometer, are mounted in an insulated spherical shell and aligned to view the same region of the sky. Remotely controlled elevation-azimuth drive of the housing facilitates the pointing of the spectrometer-pointing in any desired direction.

The two spectrometers use thermoelectrically cooled extended S-20 photomultiplier tubes as detectors. They are both operated in photon-counting mode and are coupled to a real-time digital data handling system consisting of a 64 K byte mini-computer, a 9-track, digital magnetic tape recorder-reproducer, an interactive graphics unit and an electrostatic copier. Spectroscopic measurements are made in the second or third order using broadband, glass filters to isolate the desired wavelength region.

Depending on the free spectral range covered in each scan, the measurements presented here were made at resolution in the range of 0.5 to 7A, with most of the data gathered at 2.5A resolution.

Night-time auroral spectroscopic observations were performed in Fairbanks, Alaska from 1975 to 1978. Daytime auroral emissions were observed from Svalvard during the winter solstice in 1978, 1979 and 1980.

AURORAL SPECTRA

Figure 1 shows auroral emissions in the wavelength region

~3400-3900A, both from night and daytime auroras. Note that while the night-time spectrum is replete with $N_2$2P bands and shows some N_2VK bands, the day-time spectrum is completely devoid of these features. Differences in rotational and vibrational distributions of N_2^+1NG bands in the two spectra are due to resonance scattering of sunlight in the cusp region. Another difference is the presence of relatively intense [OII]3727-29A emissions in the day-time aurora. The emission at 3889A has been shown to be due to incoming neutral reliance atoms in a number of cases (Sivjee, et al., 1980). In this particular series, however, it is probably Hg emission from the lights of Longyearbyen.

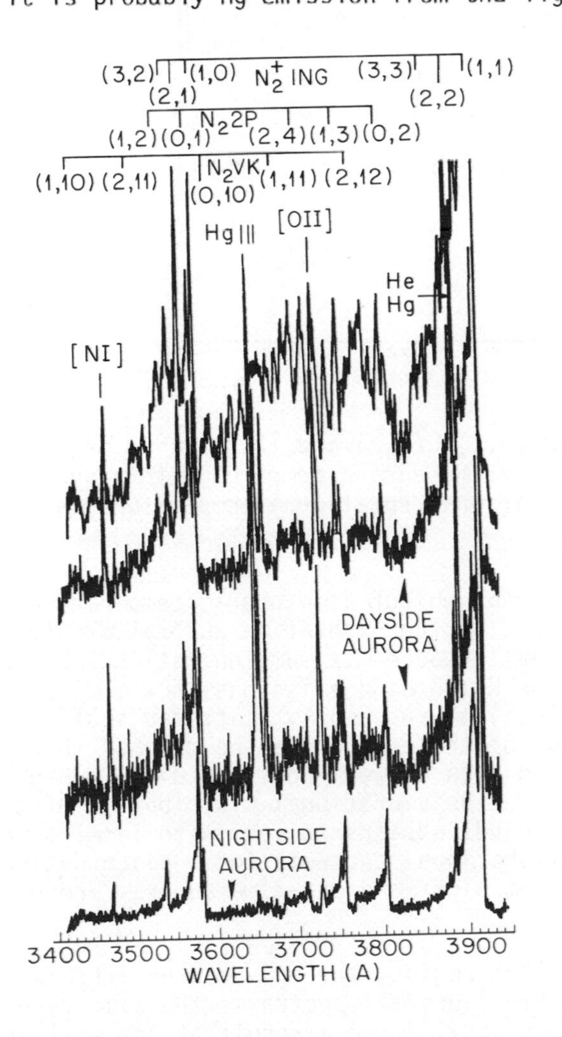

Figure 1. A comparison between auroral emission spectra in the wavelength region 3400A-3900A between night- and day-side aurorae.

Relatively high resolution spectra of auroral N_2^+1NG (0,1), (1,2) and (2,3) bands are displayed in Figure 2. Both vibrational and rotational distributions of these bands in night-time

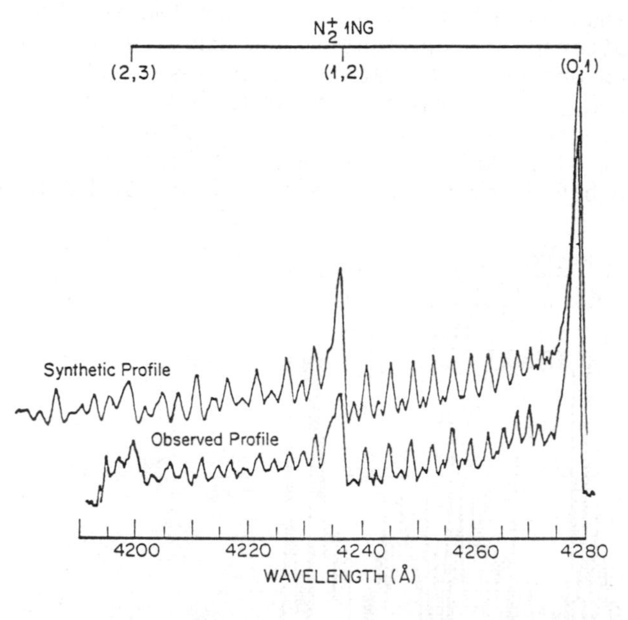

Figure 2. Showing the (0,1), (1,2) and (2,3) bands of N_2^+1NG observed in the daytime aurora compared with a synthetic spectrum where the vibrational rotational temperature is 2500°K.

auroras are thermalized to ambient atmospheric temperature at ∿115 km height where most of the night-time auroral electrons (average energy $\tilde{\sim}$3-5 keV) produce maximum ionization. On the other hand, the day-time N_2^+1NG band emissions show a clear signature (Swing's effect)[2] of resonantly scattered sunlight. The rotational distribution of these bands can be matched to a synthetic spectrum based on a Maxwell-Boltzman distribution of $N_2(X\ ^1\Sigma,v)$ for $T \tilde{\sim} 2500$°K. The vibrational distribution of the day-time auroral N_2^+1NG bands appears to vary with local time. These differences in vibrational and rotational distributions of N_2^+1NG band emissions from night and day-time auroras are observed at all wavelengths.

In the 4500A to 4900A region, the only marked difference between day and night-time auroral spectra is the paucity of Hβ in the cusp. This is probably more a result of the smaller excitation cross sections for incoming protons below 1 kev than it is a reduction in number flux from the night-side.

From 5000A to 6000A the spectra exhibit two main differences. In addition to very low level of band emissions in the day-time aurora, the ratio of [NI] 5198.5A to 5200.7A lines seem to vary from ~1.18 to 1.56 in the cusp (Figure 3a) while in bright night-time auroras this ratio has a constant value of 1.56 ± 0.06 (Figure 3b). The latter is close to the theoretical value of the ratio of the statistical weights and transition probabilities of [NI] $^2D_{3/2}$ and $^2D_{5/2}$ levels. The variations in this ratio observed in day-time auroras can be explained as a consequence of the conservation of total angular momentum, of N and O in a quenching reaction, and differences in temperatures at atmospheric heights where day-side auroral particles dissipate their energy (Sivjee, et al. 1980).

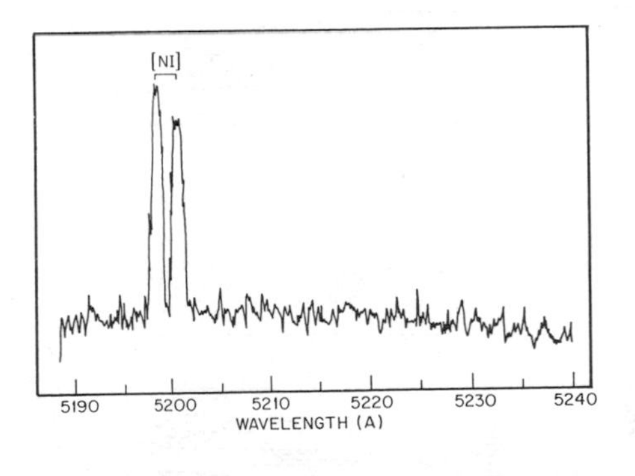

Figure 3a. Dayside auroral emission spectrum from Svalbard in the wavelength region 5190A-5230A. Note the lack of band emissions.

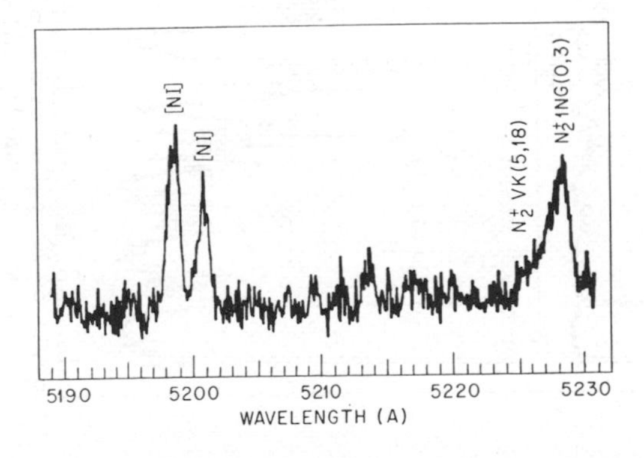

Figure 3b. Typical nightside auroral emission spectrum over the wavelength region 5190A-5230A. Note the presence of N_2^+1NG (0,3) and N_2VK (5,18) bands as well as unidentified weak emissions at 5190A, 5195A, 5209A, 5211A, 5213A and 5217A.

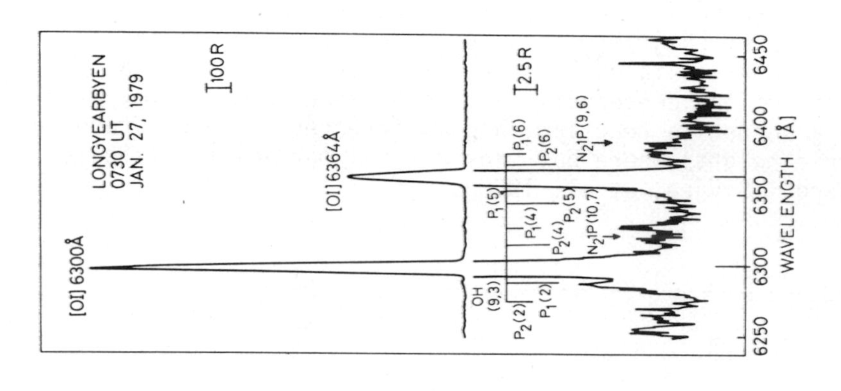

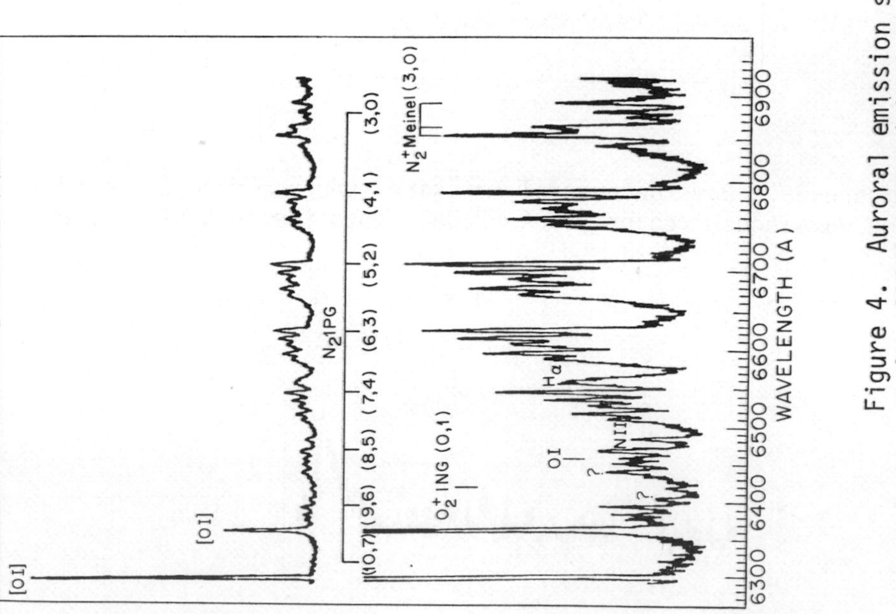

Figure 4. Auroral emission spectra in the wavelength region 6000A-7000A (a) nightside (b) dayside.

The red spectra (6000A-8500A) of day and night-time auroras displayed in Figures 4, 5 and 6, are strikingly different. While the $N_2$1P, N_2^+ Meinel and O_2 Atmospheric Bands dominate the night-time auroral emissions, in the cusp aurora [OI] 6300-64A, [OII] 7320-30A and OI 7774-8446A lines are the dominant emissions.

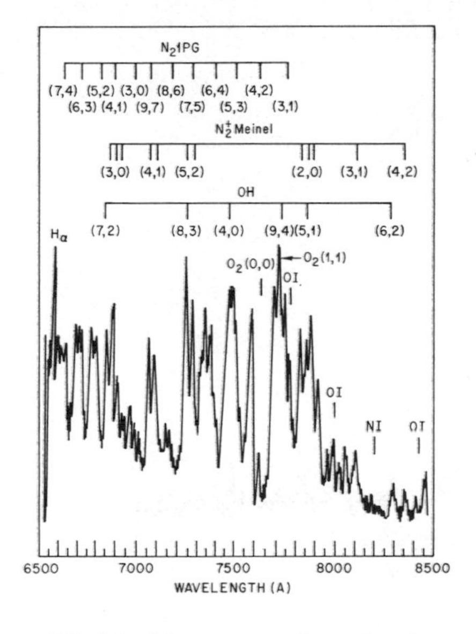

Figure 5a. Nightside auroral emission spectra in the wavelength region 6500A-8500A.

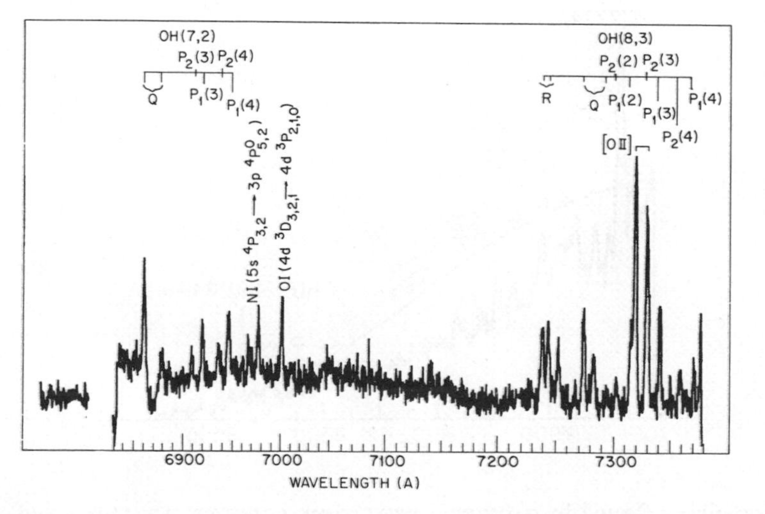

Figure 5b. Dayside auroral emission in the wavelength region 6500A-8500A.

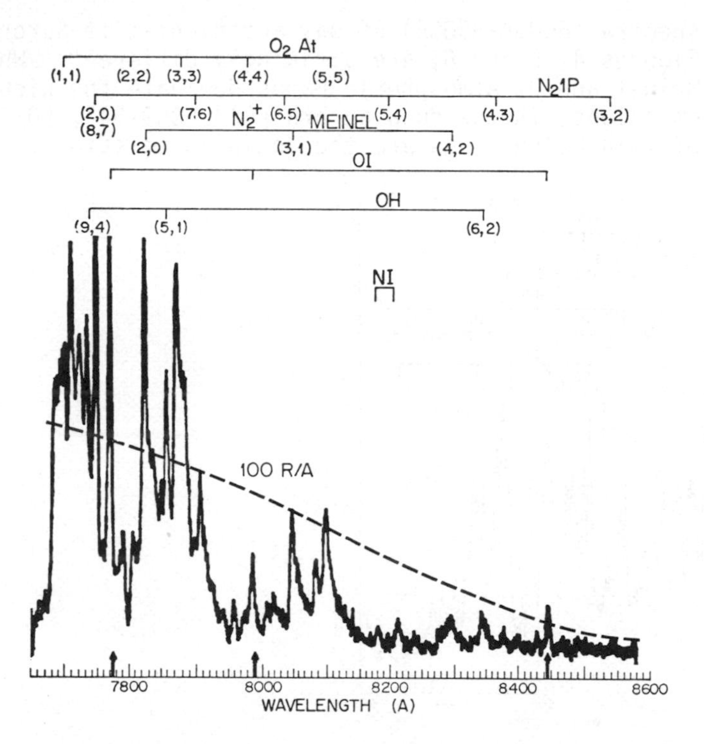

Figure 6a. Nightside auroral emission in the wavelength region 7000A-8400A.

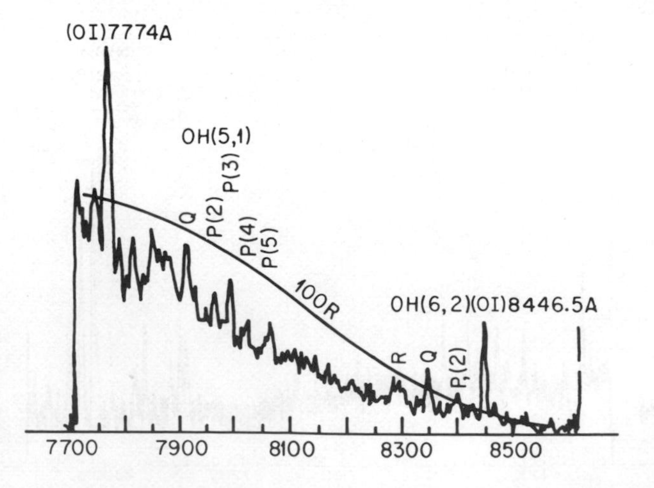

Figure 6b. Dayside auroral emission spectra in the wavelength region 7700A-8500A.

DISCUSSION

Figures 1 through 6 clearly illustrate that band emissions contribute most of the light from night-time auroras, while line emissions from [OI], [OII], [NI] and He form the bulk of day-time auroral spectral features. The He 3889A line has not been observed in the auroral zone, but it is a prominent feature of both night- and dayside aurora in the polar cap, although its intensity is variable. The [OII] 3727-29A lines, which have been observed only once before (in the great red aurora of 1958, Wallace 1950) are continuously present on the dayside for about eight hours around local magnetic noon. Similarly, the [OII] 7320-30A lines which are relatively weak (<5R) in most of the night-time auroras have almost a constant intensity of about 100R in the day aurora for the same period. The only band system with an individual band intensity >10R observed in the quiet and diffuse cusp around magnetic local noon is the N_2^+1NG. The particle excited portion of this emission accounts for slightly less than half of the total band intensity; resonant scattering of sunlight by $N_2^+(X, {}^1\Sigma, v)$ (formed both by auroral particle impacting on, and [OII] (^2D) charge exchanging with N_2 (Broadfoot 1967)) contributes the other half of the total intensity (Deehr, et al. 1980).

Because of the enhanced rotational distribution of N_2^+1NG band in day-time auroras, photometers calibrated for night-time auroral measurements of say the (0,1) band around 4278A will underestimate the intensity of this band in day-time aurora. In calibrating the photometer for cusp measurements, the extended rotational distribution of N_2^+1NG bands must be considered in evaluating the overlap integral of the band intensity distribution and transmission profile of the photometer's interference filter.

ACKNOWLEDGEMENTS

This work was funded by the National Science Foundation, Atmospheric Sciences section, Grants ATM77-24837 and ATM77-24838.

REFERENCES

Broadfoot, A. L., 1967. Planet. Space Sci. 15, 1801.
Deehr, C. S., Sivjee, G. G., Egeland, A. Henriksen, K., Sandholt, P.E., Smith, R., Sweeney, P., Duncan, D., and Gilmer, W. J., J. Geophys. Res., 85, 2185.
Sivjee, G. G., Henriksen, K., and Deehr, C. S., 1980. J. Geophys. Res., (in press).
Wallace, L., 1959. J. Atmos. Terr. Phys., 17, 46.

THE SOLAR WIND-MAGNETOSPHERE-IONOSPHERE SYSTEM: AN OVERVIEW

Juan G. Roederer

Geophysical Institute, University of Alaska,
Fairbanks, Alaska 99701

The upper atmosphere at high geomagnetic latitudes is an active component of the solar-terrestrial system. This system, comprising a chain of several distinct plasma regions, features two important characteristics: (1) As one proceeds from the main particle, energy and momentum source on the sun through the solar wind, the magnetosheath, magnetosphere and ionosphere to the sinks in the neutral atmosphere, one observes an increasingly important feedback coupling between adjacent regions; (2) The time scales of response to perturbations originating on the sun are increasingly determined by the properties of the local medium in each region. Being the last link in the solar-terrestrial chain, the upper atmosphere exerts an important feedback effect on the preceding region, the magnetosphere; this in turn influences the form and rate of solar energy delivery to the upper atmosphere. The magnetosphere thus plays an important role as a non-linear transducer of solar energy. For a general review of the solar-terrestrial system see Parker et al. (1979).

Figure 1 depicts in schematic form three main channels of energy flow from the sun to the earth. The main power delivery to the atmosphere occurs via the practically constant solar black-body radiation absorbed on the earth's surface and in the troposphere. Of the variable components of solar emissions, the solar wind exerts the principal control on the magnetosphere, the ionosphere and the upper atmosphere. This control depends on several solar wind parameters which regulate the efficiency of energy and momentum transfer to the magnetosphere. The quantitative under-

C. S. Deehr and J. A. Holtet (eds.), Exploration of the Polar Upper Atmosphere, 209–218.

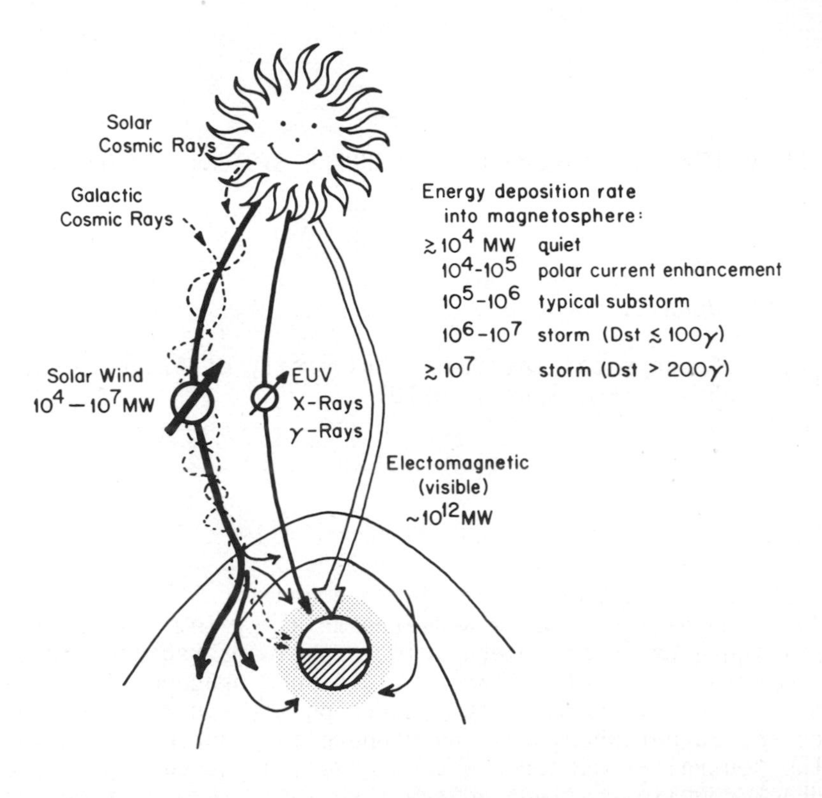

Figure 1. The three main channels of energy flow from the sun to the earth and the range of power transfer and corresponding magnetospheric response.

standing of the physical mechanisms responsible for this transfer is one of the most important current goals of solar-terrestrial physics. The table in Figure 1 indicates typical ranges of solar wind power transfer and related magnetospheric responses.

The interplanetary magnetic field (IMF) embedded in the solar wind modulates galactic cosmic rays and guides, traps and modulates energetic particles emitted by solar flares. Since these particles affect the ionization in the stratosphere and mesosphere, the polar wind thus exerts yet another, indirect kind of influence on the terrestrial upper atmosphere. The channel marked EUV and X-rays in Figure 1 represents sporadic electromagnetic emissions from active centers on the sun that also affect the ionization in the upper atmosphere.

In this chapter we are concerned with the key mechanisms operating in the terrestrial magnetosphere (for general and specific references, see for instance Parker et al. op. cit. Vols. 2 and 3.

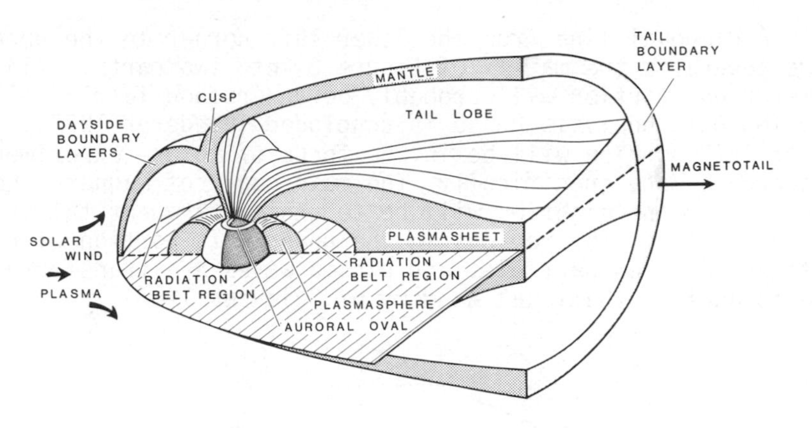

Figure 2. A sketch of fundamental plasma regions in the magnetosphere.

Figure 2 shows the principal magnetospheric plasma regions and how they are connected via magnetic field lines with the upper atmosphere. The same regions are displayed in Figure 3 in a matrix representation. Rows represent principal spatial domains; columns represent regions of specific magnetic field topology or morphology. This field configuration determines charged particle behavior (from left to right: field-aligned streaming, convection, transient trapping, stable trapping). It should be noted that cusp field lines probably are closed on the equatorward portion and open on the poleward side of the cusp. In the rows, resistivity is a distinctive parameter (zero in the plasma regions at the top; perhaps finite in the $E_{\parallel}$ region; finite in the bottom row).

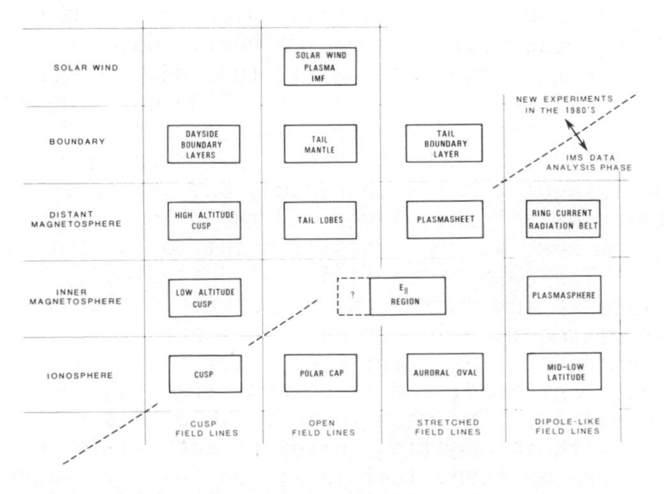

Figure 3. Plasma regions, spatial domains and magnetic field morphology, in matrix form.

A diagonal line from the lower left corner to the upper right side separates the matrix of Figure 3 into two parts: (1) the lower right portion will probably be understood fairly well when the IMS Data Analysis Phase is concluded (Roederer 1977): (2) the upper left portion will be in the focus of future experimental research. More specifically, the main goals of magnetospheric research in the eighties will be to achieve a quantitative understanding of how the regions in the upper left portion of Figure 3 interact and how particles and plasma waves are transferred from one to another (National Academy of Sciences 1980).

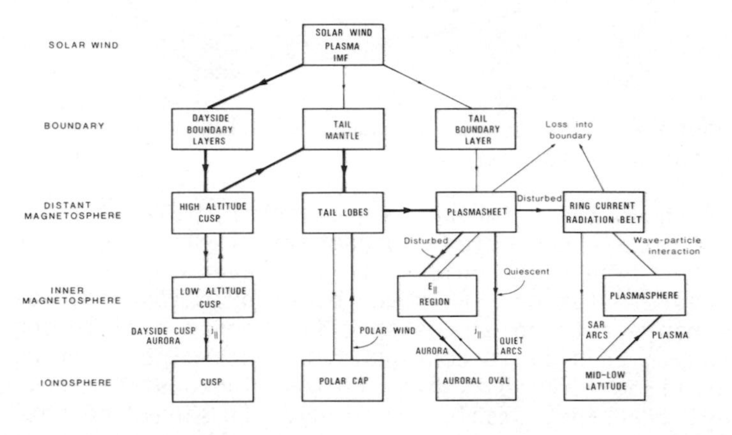

Figure 4. Magnetospheric particle transport routes.

Figure 4 shows the main particle transport routes in the magnetosphere, according to current (not necessarily universally accepted) thinking. The heavy lines represent what may be the principal entry route from the solar wind to the main magnetospheric plasma reservoir, the plasmasheet. Note the role of the entry layer and the high and low altitude cusp regions in these initial stages of particle transport. Figure 5 schematically illustrates this role. The entry layer consists of field-aligned streaming of plasma of density and temperature nearly the same as in the magnetosheath, but with reduced and irregular flow speed. Both the thickness of this layer and the plasma content increase gradually from zero at the subsolar point to the cusp latitude (about 78°). The cusp or cleft is a longitudinally elongated region of intense low-energy charged particle fluxes, composed of entry layer plasma that is approaching the earth's ionosphere; of return streaming of these particles after they have mirrored or have been backscattered; and of upward streaming particles of ionospheric origin. The magnetopause exhibits a clear indentation in this area with interesting vortex signatures in the magnetosheath flow. It is now believed that localized "patchy" magnetic field reconnection events in the region of the indentation play an important role in particle access to the magnetosphere. The high-latitude large-scale electric field, generally directed from dawn-to-dusk in the vicinity of the noon meridian, imposes a tailward drift on the down-streaming entry layer particles. Although an independent particle description may seem rather unrealistic under cleft conditions, it does explain qualitatively many observed features: the softening of proton spectra when the cleft is traversed in poleward direction; the flow and temperature profile of incoming particles that mirror and then "expand" along the open field line into the mantle (Figure 5); and the observed correlation of mantle thickness with the southward component of the IMF.

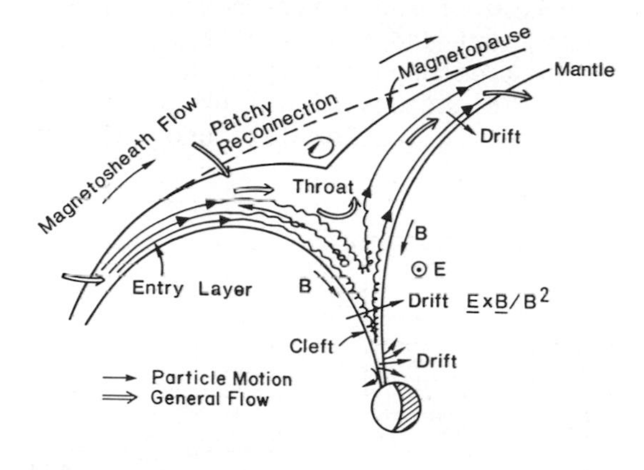

Figure 5. Sketch of particle motion and plasma flow into and out of the cleft region (not to scale).

Little is known about the specific transfer routes of solar wind plasma into the plasmasheet (Figure 4); the convection electric field is believed to be responsible for a general drift of streaming mantle plasma toward the center of the tail where particles are transiently captured to form the plasmasheet and, therein, the cross-tail current sheet.

During magnetospheric substorms, plasma is convected from the plasmasheet earthwards and betatron accelerated, feeding into the ring current, which is the most important energy reservoir of the magnetosphere. Further acceleration via radial diffussion feeds particles into the Van Allen radiation belts. During substorm events, a certain fraction of plasmasheet electrons is accelerated along the magnetic field by a transient parallel electric field; this process is responsible for the formation of substorm-associated auroral arcs.

Figure 4 also shows the main routes of particle transport from the ionosphere to the magnetosphere. The plasmasphere is a reservoir of ionospheric plasma corotating with the earth. The polar wind and upward acceleration of ionospheric ions by the $E_{\parallel}$

field during substorms represent the main particle routes linking the ionosphere with the plasmasheet.

Particle transport in the magnetosphere-ionosphere system is actually governed by electromagnetic coupling between the constituent plasma regions. This coupling is depicted schematically in Figure 6. We should point out five principal chains of electromagnetic linkage: (1) the "vertical" chain from the solar wind to the polar cap (and cusp); (2) the "horizontal" links between the ionospheric regions; (3) the "closed" chain involving the tail lobes, polar cap, auroral oval and plasmasheet; (4) the transient links between the plasmasheet and the ring current and $E_{\parallel}$ regions, respectively; (5) the link between the ionosphere and the plasmasphere. There are also indications of a link between the solar wind and the magnetosphere via the low-latitude boundary layer.

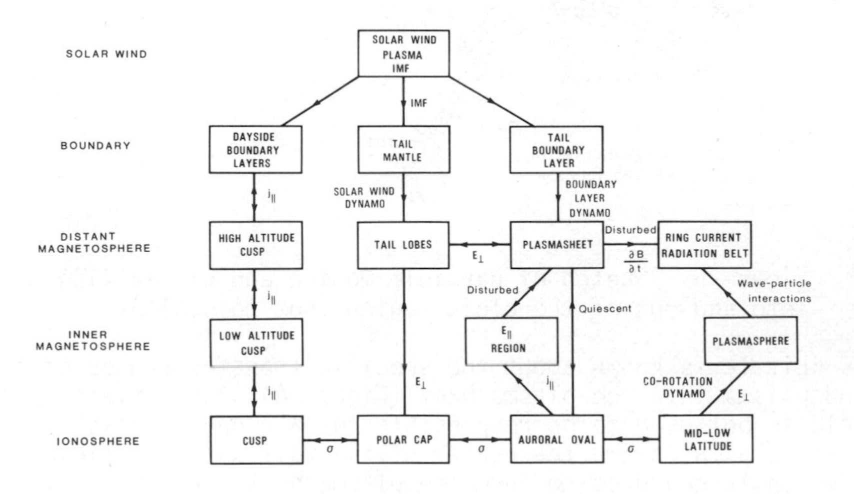

Figure 6. Principal electrical couplings.

A simple-minded first order description of chain (1) invokes the mapping of the $-\underline{V}\times\underline{B}$ electric field in the solar wind along the (equipotential) open field lines onto the polar cap ionosphere. This yields a convection electric field whose geometry is strongly dependent on the interplanetary magnetic field and the topology of its interconnection with the terrestrial field, but which bears some invariant features such as general direction from dawn to dusk for a wide range of interplanetary field configurations (except due northward IMF). Once the electric field is impressed on the open field lines of the polar cap, the distribution of ionospheric conductivity determines the electrostatic field in the remaining portion of the ionosphere (chain 2). Although this simple-minded picture of the solar wind acting as a "voltage source" of magnetospheric behavior yields some quantitative predictions in fairly good agreement with observations, a full quantitative understanding of the electrodynamic coupling between the solar wind, the magnetosphere and the ionosphere, requires a self-consistent solution of the closed system of interacting plasmas represented by the above mentioned chain (3), shown in Figure 6.

An important feature of electrodynamic coupling in the magnetosphere is the system of field-aligned currents (Birkeland currents) that appears in connection with the relevant dynamo processes. Consider a blob of plasma (Figure 7) under the action of an external force $\underline{F}$ (assumed applied with uniform force density). This force causes electrons and ions to drift in mutually opposite directions; the plasma is polarized in such a way as to generate an electric field that imparts a bulk drift $\underline{V}$ whose acceleration satisfies exactly $\dot{\underline{V}} = \underline{F}/m$ (i.e., the plasma follows the dictates of the external force). If this plasma is now coupled to a resistive region via magnetic field lines that join both regions (Figure 7), the polarization charges will be drained into (or draw neutralizing charges from) the resistive medium in the form of field-aligned currents $j_{\parallel}$: the polarization of the plasma will be altered, and so will the electric field configuration and associated plasma drift (bulk motion). The currents $j_{\parallel}$ will be closed by perpendicular currents $j_{\perp}$; the corresponding Lorentz forces $\underline{j} \times \underline{B}$ play the role of "viscous" forces. Energy is dissipated in the "load" of the resistive medium, and the plasma blob will be forced into co-motion with the former. If that state is achieved, all field-aligned currents disappear, and the field lines will appear as "frozen" into both co-moving media.

The most important magnetospheric dynamos are pointed out in Figure 6. In the solar wind dynamo, the plasma "blob" of Figure 7 is the solar wind, driven by coronal expansion, connected to the resistive polar cap ionosphere by the open magnetic field lines. The associated system of Birkeland currents will flow along the boundary of closed field lines and be directed into the ionosphere on the postmidnight side, and out of it at premidnight. This is

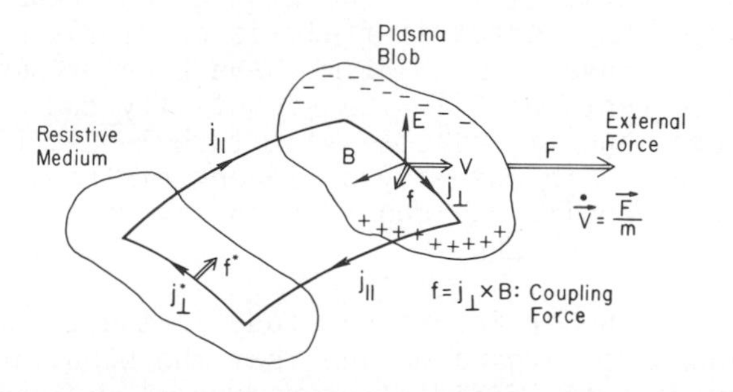

Figure 7. Schematic diagram of the solar wind dynamo.

indeed observed. Although the geometry and intensity of this system of field-aligned currents will depend on the actual configuration of the field interconnection, the direction of the currents and some other general features will remain the same as long as there are open field lines. The cusp regions feature a specific field-aligned current system (Figure 6) that is also controlled by the solar wind dynamo; the dominating solar wind parameter here is the B_y component of the IMF.

The corotation dynamo indicated in Figure 6 causes the plasmasphere to co-move with the low-latitude ionosphere, driven by the interaction with the neutral atmosphere. Since co-motion exists in this case, there is no field-aligned current system associated with the plasmasphere-ionosphere interaction in a steady-state situation.

The boundary layer dynamo marked in Figure 6 is assumed to be yet another form of solar wind dynamo, driven by plasma of solar wind origin flowing in the low latitude boundary layer on closed field lines in the antisunward direction, along the flanks of the magnetopause. This latter dynamo should be independent of whether or not there is magnetic interconnection between the solar wind and the magnetosphere. The associated field-aligned current system should lie along the auroral oval sunward of the Birkeland currents associated to the first solar wind dynamo. Its configuration should depend on the particular plasma flow in the low-latitude boundary layer.

A complex and highly variable field-aligned current system is driven by the bulk motions in the plasmasheet. This system is part of the closed chain of electrodynamic coupling shown in Figure 6. A complicating factor is that plasmasheet-ionosphere coupling depends on ionospheric conductivity, which in turn depends on auroral precipitation from the plasmasheet, which in turn depends on plasmasheet-ionosphere coupling!

Auroral precipitation during substorms is controlled by the field-aligned electric field in the $E_{\parallel}$ region (Figures 4 and 6). The generation of this electric field is currently a subject of intensive study. Several mechanisms have been proposed: (1) field-aligned currents with finite resistivity caused by electrostatic or electromagnetic plasma- wave turbulence; (2) trapped electrons and ions with mutually different distribution functions causing a charge density buildup (double layers); (3) current-driven electrostatic shocks.

Another topic under intensive study is the cause of substorms. It has been known for over a decade that the behavior of the interplanetary magnetic field plays a crucial role in causing, or creating conditions favorable for, substorm events. A commonly cited sequence of events is the following: (i) A southward turning of the IMF increases the rate at which plasma flux tubes are transferred from the solar wind to the tail; (ii) as a result, the cross-tail current intensity increases, tail field lines are increasingly stretched, and the intensity of the mutually opposing fields in the tail lobes increases; (iii) if the southward-directed IMF persists, the thickness of the plasmasheet decreases, an X-type neutral line (or several of them) is formed in the cross-tail current sheet, and a magnetic merging process is eventually triggered along the neutral line leading to an enhanced plasma flow away from the neutral line; (iv) the plasma stream flowing toward the Earth is being betatron accelerated, feeding ring current and particle precipitation; (v) as this process continues, the neutral line migrates down the tail until the magnetic field relaxes back to a less stretched, more dipole-like configuration.

This picture is presently being challenged on several fronts. (1) The existence of, and need for, a large-scale neutral line and associated magnetic merging process in the tail has been questioned (e.g., Heikkila and Pellinen 1977; Perrault and Akasofu 1978). The observed earthward convection and acceleration of plasma could be accomplished by an electric field induced by a time-dependent magnetic field in the tail caused by a sudden decrease of the cross-tail current density. (2) The picture of gradual energy storage and subsequent triggered release as a basic, perhaps even cyclic, feature of substorm dynamics has been questioned and replaced by a model of "real time" control of substorm activity by solar energy transfer to the magnetosphere. This transfer would be governed by a quantity ε that is a function of solar wind parameters: $\varepsilon \sim V B^2 \sin^4(\theta/2)$ (V: solar wind velocity; B: IMF magnitude; θ projection of the polar angle of the IMF on the y-z plane in magnetospheric coordinates).

Many of these questions will hopefully be clarified during the next few years; the answers may well be contained in the impressive data base acquired during the IMS. What will not be solved are problems pertaining to the <u>early</u> stages of entry of solar wind plasma into the magnetosphere and its transfer to the plasmasheet reservoir. The proposed OPEN mission and other high latitude, high altitude satellite programs such as PROGNOZ will be necessary to provide the needed information.

Since all solar wind plasma entry and early-stage transfer processes occur on magnetic field lines connected to the polar regions of the earth, many of their manifestations have effects on the upper atmosphere in the cusp, the polar cap and the auroral oval. This is why I like to tell the "man in the street" that the high latitude upper atmosphere is "where outer space meets planet earth"!

REFERENCES

Heikkila, W.J. and R.J. Pellinen 1977, J. Geophys. Res. 82, 1610.

National Academy of Sciences 1980, Solar-Terrestrial Research for the 1980's, in preparation.

Parker, E.N., C.F. Kennel and L.J. Lanzerotti 1979, (editors), Solar System Plasma Physics, North Holland Publ. Co., Amsterdam-New York-Oxford.

Perreault, P. and S.-I. Akasofu 1978, Geophys. J. Roy. Astron. Soc., 54, 547.

Roederer, J.G. 1977, Space Sci. Rev. 21, 23.

THE DISTANT MAGNETOSPHERE: RECONNECTION IN THE BOUNDARY LAYERS, CUSPS AND TAIL LOBES

Gerhard Haerendel

Max-Planck-Institut für Physick und Astrophysik
Institute für extraterrestrische Physik
8046 Garching, W-Germany

INTRODUCTION

There are three ways in which a cosmical plasma can interact with a self-gravitating system. They depend on the existence of an extended atmosphere and/or an appreciable intrinsic magnetic moment. If the system is lacking both of these qualities completely, the plasma will impinge directly on it, as, for instance, the solar wind does on the moon. In case of a zero or weak magnetic moment, but with the existence of an extended atmosphere, the plasma will interact with the latter via its ionized component. In the solar system, examples of this class are comets and the planet Venus.

The existence of a strong magnetic field shifts the location of interaction outside the atmosphere. It is typically collision-free in nature, although the magnetic field can transport the forces towards the central body where collisions are important. A *magnetosphere* is being created. It is the region in which the magnetic field linked to the self-gravitating system dominates the forces acting on the plasma component. Examples in the solar system are provided by the Earth, Mercury, perhaps Mars, Jupiter, Saturn, possibly Neptune and Uranus.

In this contribution, I want to deal with the detailed plasma and field configurations in the particle interaction regions, and among all possible acceleration processes only with reconnection. Although reconnection has been known conceptually for more than twenty years and has been studied extensively, we are only now beginning to collect conclusive direct observations

C. S. Deehr and J. A. Holtet (eds.), Exploration of the Polar Upper Atmosphere, 219–228.

of this process in the magnetosphere. There are several unexpected features emerging that should be taken into consideration when studying other reconnection situations.

RECONNECTION

Reconnection or merging of two oppositely directed magnetic fields in a highly conducting fluid was first studied in the astrophysical context by Sweet (1958) and Parker (1957). Dungey (1961) applied it to the magnetosphere and proposed the general scheme of magnetic flux transport between the dayside and nightside magnetosphere. Although it has undergone substantial modifications, this scheme is still believed to be valid. Depending on their orientation, interplanetary magnetic field lines carried past the magnetopause by the solar wind can connect temporarily with the Earth's field. Such field lines are then called "open". They give the solar wind an easy means to transfer flow momentum to the magnetospheric plasma via magnetic shear stresses and drag it into the antisolar direction. Thereby, the field is being stretched and a magnetic tail is formed. The central region of the tail which contains a layer of hot plasma, the so-called plasma-sheet, separates stretched fields of anti-parallel direction. They are subject to the merging or reconnection process under circumstances which are not yet fully understood. This occurs in events of typically 1 hour duration which are called "substorms". As a result, open field lines are transformed into closed ones (i.e. dipole-like field lines) plus field lines that are completely disconnected from the Earth and are carried away by the solar wind flow. Internal convection motions carry the closed magnetic flux-tubes back towards the dayside where the process can eventually start again.

The physics of reconnection events in the tail, the substorms, is subject to another contribution in this volume (Vasyliunas 1980). A good overview of substorms is contained in a recent textbook by Akasofu (1977). However, we are still lacking fully convincing data sets on the very process; much of the inferences are indirect.

On the front-side of the magnetopause, we are on somewhat safer grounds. Data have become available which allow quantitative checks of the predictions of the macroscopic theory of reconnection. Temporal and spatial scales of the process can as well be inferred. Therefore, and because our laboratory has been strongly involved in this area of research, I will deal in detail with reconnection at the front-side of the Earth's magnetopause.

In view of the great interest in that process and of the many satellites launched with the aim of investigating the magnetosphere, it is rather surprising that it took so long to identify it unambiguously by in situ measurements. All that is needed is the combination of a magnetometer and a plasma detector capable to establish the flow of the dominant plasma component in three dimensions. Because the process is rather short-lived, both measurements must be made sufficiently fast and (this requires) a high telemetry rate. The International Sun-Earth Explorers, ISEE 1 and 2, were the first satellites with the right orbits and instruments to meet these conditions. Paschmann *et al.* (1979) report an event of a few minutes duration at the dayside magnetopause near noon in which the observed changes of flow momentum at the magnetopause are consistent with the magnetic stresses that would exist if the internal and external fields were connected through a rotational discontinuity.

Whatever the detailed structure of a reconnection region may be (Petschek, 1964; Vasyliunas, 1975), it should contain a rotational discontinuity in which the magnetic field changes direction by a large angle and has a finite component normal to the plane of the discontinuity. This is sketched in Figure 1 for the simple case of almost anti-parallel fields. The plasma that

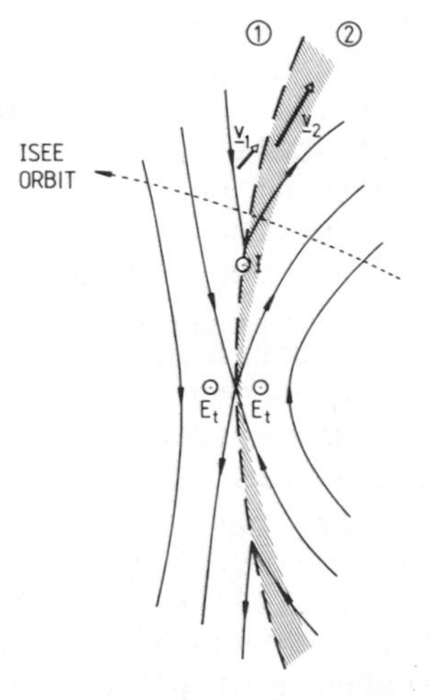

Figure 1. Reconnection situation at the dayside magnetopause with southward pointing interplanetary field. The shaded area shows a layer of accelerated plasma flow after transition of the solar wind through the rotational discontinuity (Paschmann *et al.*, 1979).

transits through the discontinuity should undergo little change of its thermodynamic properties, but be accelerated by an amount that is related to the balance of magnetic and mechanical stresses which (in an isotropic plasma) reads:

$$[\rho \, \underline{v}_t \, v_n] = \left[\frac{\underline{B}_t \, B_n}{4\pi}\right] \quad . \qquad (1)$$

Index "n" and "t" refer to the normal and tangential components, respectively. Square brackets indicate the jump across the discontinuity. Since B_n = const. and $\rho \cong$ const., we have the simple relations:

$$[\underline{v}_t] = \frac{[\underline{B}_t]}{\sqrt{4\pi\rho}} \qquad (2)$$

and

$$v_n = \frac{B_n}{\sqrt{4\pi\rho}} , \qquad (3)$$

which the plasma should obey. The jump in tangential flow velocity (Equation 2) is easily observable, in contrast to the quantities of Equation 3. The reason is that on the one hand most theories predict small normal components, and on the other hand one or even two satellites are not sufficient to establish with sufficient confidence the normal vector of an observed discontinuity.

In earlier studies, of the plasma flow at the magnetopause (Heikkila, 1975; Paschmann *et al.*, 1976; Haerendel *et al.*, (1978), it was quite disturbing that the predictions of Equation 2 were not encountered even when the magnetic fields inside and outside the magnetopause were almost oppositely directed. It became evident that if reconnection occurred at all in the explored regions it should be transient and small-scale in nature and thus escape detection. The recent measurements of Paschmann *et al.* (1979) were, however, sufficiently fast to cope with this difficulty. Figure 2 shows a set of data on three subsequent transitions through the magnetopause along a pass as sketched by the dashed line in Figure 1. The displayed data are total plasma density, N_p, magnitude of the flow velocity, v_p, the component of $\underline{B}$ perpendicular to the ecliptic, B_z, the plasma pressure P and magnetic pressure, B, and finally, the sum of gas and magnetic pressures, P_T. The units are respectively: cm^{-3}, km/sec, nT, and 10^{-7} N m^{-2}. The magnetopause (MP) usually undergoes radial oscillations; hence there are three transitions as revealed most clearly by the jumps of B_z from positive to negative values. The most important feature is the large increase of plasma flow velocity by several 100 km/sec just inside the magnetopause. Via the total density measurements which the same instrument yields and the magnetometer data, Equation 2 can be checked. Agreement is found within 10%. This must be considered as rather good in view of several sources of experimental error.

A special technique developed by Sonnerup (1971) and co-workers, which is called "minimum variance technique", allows the derivation of B_n if the orientation of the discontinuity is sufficiently stable during the transit time of the satellite. For the event contained in Figure 2 an inward-pointing normal component of 5.4 nT was found. An inward flow component of

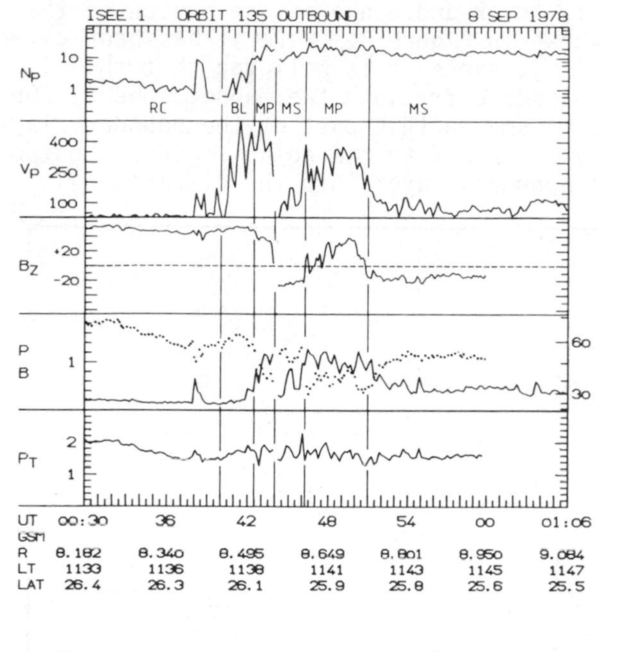

Figure 2. Plasma and magnetic field data from a transition of ISEE 1 through the magnetopause near local noon. N_p, V_p, B_z, P, B, and P_T are plasma density, flow speed, component of $\underline{B}$ normal to the ecliptic, gas and magnetic pressures and the total pressure, respectively. The units are cm^{-3}, km/sec, nT, and 10^{-7} N m^{-2}. The symbols RC, BL, MP, MS designate the different plasma regimes encountered, namely ring current, boundary layer, magnetopause layer, magnetosheath (i.e., shocked solar wind) (Paschmann *et al.*, 1979).

28 km/sec would go along with this value (Equation 3). The existence of two closely-spaced spacecraft (ISEE 1 and 2) allows the determination of the speed of normal motion of the magnetopause. This is needed to correct the value found for v_n by a similar minimum variance technique. Though affected by large error bars both values are in good agreement.

I have chosen to discuss this particular measurement in detail in order to give the reader some feeling for the difficulties involved in establishing with great confidence the existence of a fundamental plasma process in space, even under rather favorable circumstances. Meanwhile about 10 events of this kind have been identified, approximately 30% of the total number of cases in which the orientation of the interplanetary field was favorable for reconnection. Strong anti-parallel field components are apparently not sufficient for the process to occur. What a sufficient condition could be, is not known at this moment.

Earlier studies of the plasma near the magnetopause (Hones *et al.*, 1972; Akasofu *et al.*, 1973; Rosenbauer *et al.*, 1975; Paschmann *et al.*, 1976; Haerendel *et al.*, 1978) had revealed an important feature of the magnetopause. It is covered on its inside by a boundary layer which covers it down to the distant tail. It is particularly thick and dense in the region of the polar cusps, which is shown in Figure 3. Here it has been given the name "entry layer" (EL), since it is believed that this is the region of dominant plasma entry into the magnetosphere. Only a small fraction of the plasma in this part of the boundary layer penetrates along the field lines into the polar region. Most of it flows along the tail boundary layer into the distant tail.

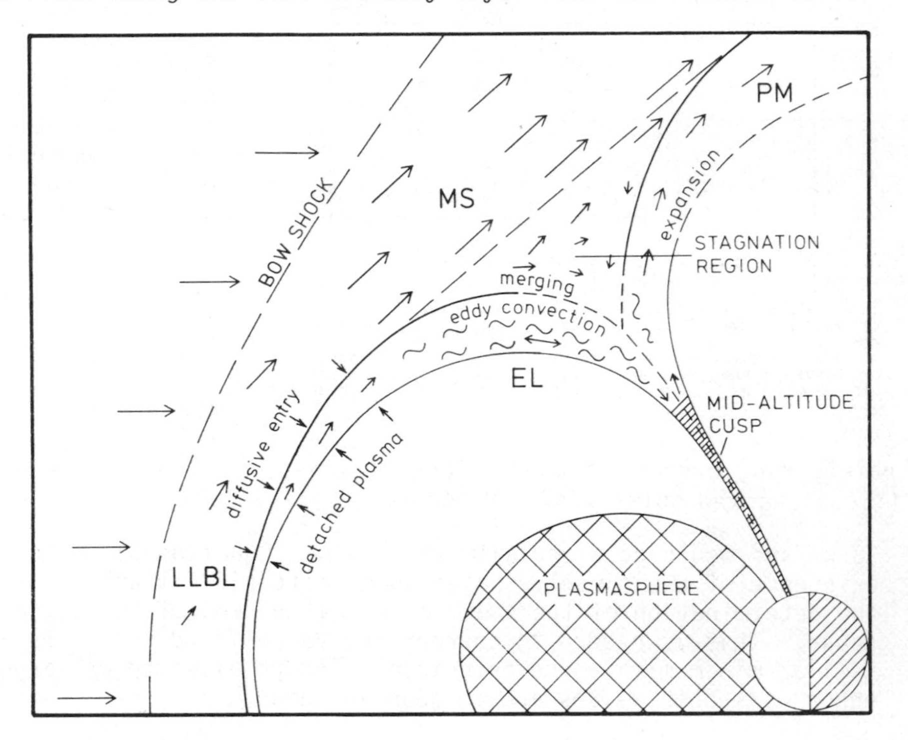

Figure 3. Meridian cut through the frontside boundary layers with indication of the dominant processes (MS = magnetosheath, LLBL = low latitude boundary layer, EL = entry layer, PM = plasma mantle) (Haerendel *et al.*, 1978).

This part has been called "plasma mantle" (PM). The boundary layer on the low-latitude dayside (LLBL) is rather thin, of lower density than in the adjacent solar wind (magnetosheath (MS)) and exhibits strong temporal modulations.

If the cusp regions play a dominant role in plasma entry, they should also be the location of frequent reconnection events (Haerendel, 1978). From the observation of very irregular flows in the entry layer it was concluded that these events should be transient and small-scale in nature. Scales of only 1000 km and 20 sec have been deduced from measurements with insufficient temporal resolution on the ESA satellite HEOS 2. This feature was probably also responsible for our inability to identify, in the same manner as discussed before, the signature of reconnection. However, the irregularity of the flow gives ground to the hypothesis that the mass transport inside the boundary layer is a kind of eddy convection process. An order of magnitude estimate of its efficiency is consistent with the implications of the drainage of this region by the mantle flow (Haerendel, 1978).

A more recent study of the low latitude part of the dayside boundary layer by Sckopke *et al.* (1980) has revealed its transient nature rather clearly. Here the time-scales are, however, much longer. They are of the order of a few minutes and the spatial scales of several Earth's radii. Of the three possible interpretations of the observations which are shown in Figure 4,

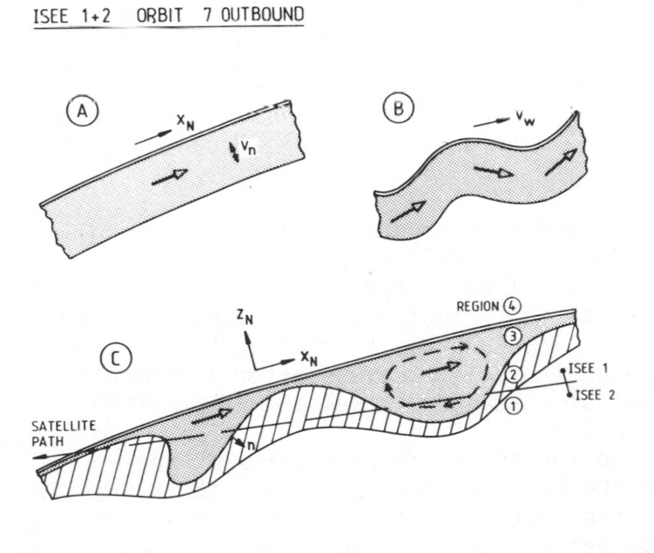

Figure 4. Possible interpretations of transient encounters of the low latitude boundary layer with ISEE 1 and 2. The circulation indicated in (c) is to be seen in a frame moving with the overall structure (Sckopke *et al.*, 1980).

case (c) appears as the most likely. This means that plasma is carried in form of separate "blobs" along the inside of the magnetopause in the downstream direction. The expected rotational motion of the plasma inside and outside the boundary layer has been confirmed observationally. In the event that has been extensively studied by Sckopke *et al.* (1980), the bending of the magnetic field inside the boundary layer is in a sense as if the plasma were pulled from above, i.e. the cusp region. This is again consistent with dominant reconnection in the cusps. The periodicity of the events on the low latitude side may be related to the separation of "vortices" from the stagnation region (shown in Figure 3) outside the cusp magnetosphere, as it would happen in an ordinary fluid streaming around a corner (Haerendel, 1978). It could also be the consequence of a Kelvin-Helmholtz instability of the boundary layer flow (Sckopke *et al.* 1980).

The consequences of reconnection in the cusps on the gross topology of the field is shown in Figure 5 taken from Haerendel *et al.* (1978). Magnetic flux is being eroded from the front-side of the magnetosphere. On the low latitude side, the magnetic field becomes stretched, i.e. its magnitude increases, as it is often observed. This means that reconnection does not necessarily imply a release of magnetic energy everywhere. Part of the space involved may experience a growth of magnetic energy at the expense [of kinetic energy of flow. The erosion of magnetic flux occurs in short-lived events of tens of seconds to a few minutes. Sometimes such flux-tubes can be identified outside the magnetopause in the solar wind plasma (Russell and Elphic, 1979). In addition to the magnetic signature (increase of $|B|$ one finds also hot electrons streaming away from the magnetopause. They may have been accelerated at the magnetopause or released from the interior. Whatever the origin of these energetic electrons is, we seem to observe an important step in the production of energetic particles by a magnetosphere.

A more important source of energetic particles is usually the geomagnetic tail (Anderson, 1965; Baker and Stone, 1976; Hones *et al.*, Sarris *et al.*, 1976). Again it seems that part of the energization is due directly to the tail reconnection process and part to a leakage of the energetic trapped particles from the outer radiation belt when the tail recovers after a reconnection event ("recovery phase" of a substorm) (Belian *et al.*, 1980).

In summary, we find the following properties of the reconnection process in the Earth's magnetosphere. It is transient and small-scale, i.e. the spatial scale is much smaller than the size of the overall magnetic configuration. These scales may be connected to the hydrodynamic properties of the plasma flow around the object (turbulence). The cusp regions seem to be the primary site of the reconnection process. Boundary layers inside the magnetopause are set up as a consequence. The short duration of the reconnection events leads to the erosion of magnetic flux in the form of rather discrete flux-tubes, which provide paths for the escape of energetic particles from the interior. At the same time, direct acceleration of energetic particles is observed.

When applying theoretical models of reconnection to an astrophysical system, we should be warned that it may be dangerous to use a stationary picture and scales of the size of the overall system as suggested by the well-known model of Petschek (1964) and its successors. As observable in so many phenomena, the plasma seems to "like" the formation of small-scale structure,

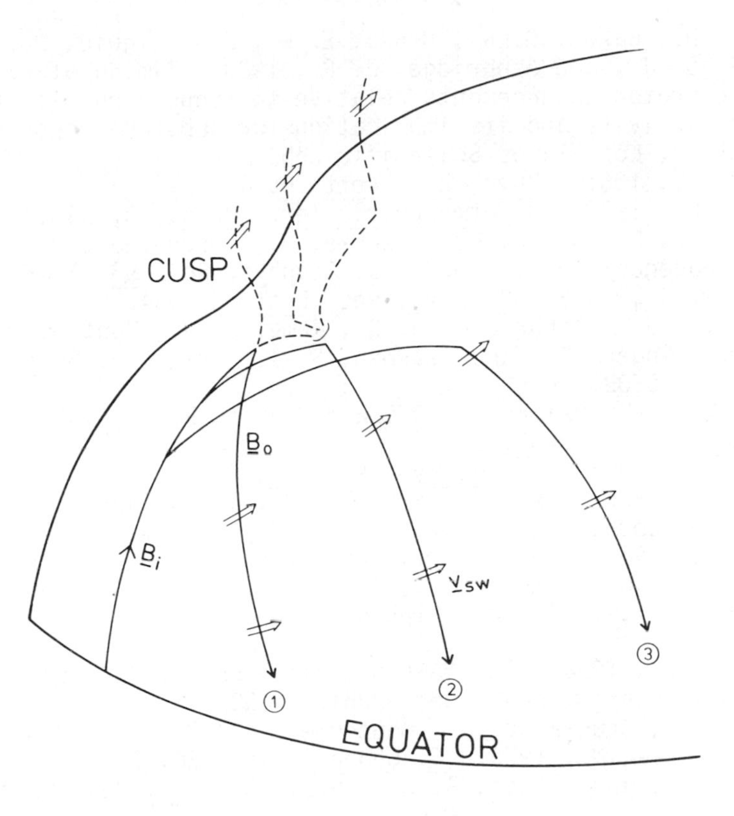

Figure 5. Sketch of the erosion of magnetic flux from the front-side of the magnetosphere initiated by reconnection in the cust regions (Haerendel *et al.*, 1978).

filaments. We must learn to understand the causes of this behavior in order to be able to make predictions for other situations. It is quite clear that the efficiency of a process is quite different when it is small-scale and transient from what it would be when it is large scale and stationary.

REFERENCES

Akasofu, S.-I., Hones, Jr., D. W., Bame, S. J., Asbridge, J. R., and Lui, Y., 1973. J. Geophys. Res. 78, 7257.
Akasofu, S.-I., 1977. Physics of Magnetospheric Substorms, Reidel Publ. Co., Dordrecht-Holland.
Anderson, K. A., 1965. J. Geophys. Res. 70, 4741.
Baker, D. N., and Stone, E. C., 1976. Geophys. Res. Lett. 3, 557.

Belian, R. D., Baker, D. N., Hones, E. W., Jr., Higbie, P. R., Bame, S. J., and Ashbridge, J. R., 1980. Timing of energetic proton enhancements relative to magnetospheric substorm activity and its implication for substorm theories, preprint, Los Alamos Scientific Lab.
Dungey, J. W., 1961. Phys. Rev. Lett. 6, 47.
Haerendel, G., 1978. J. Atmospheric Terr. Phys. 40, 343.
Haerendel, G., Paschmann, G., Sckopke, N., Rosenbauer, H., and Hedgecock, P.C., 1978. J. Geophys. Res. 83, 3216.
Heikkila, W. J., 1975. Geophys. Res. Lett. 2, 154.
Hones, E. W., Jr., Ashbridge, J. R., Bame, S. J., Montgomery, M. D., Singer, S., and Akasofu, S.-I., 1972. J. Geophys. Res. 77, 5503.
Hones, E. W., Jr., Palmer, I. D., and Higbie, P. R., 1976. J. Geophys. Res. 81, 3866.
Parker, E. N., 1957. J. Geophys. Res. 62, 509.
Paschmann, G., Haerendel, G., Sckopke, N., Rosenbauer, H., and Hedgecock, P. C., 1976. J. Geophys. Res. 81, 2883.
Paschmann, G., Sonnerup, B.U.O., Papmastorakis, I., Sckopke, N., Haerendel, G., Bame, S. J., Ashbridge, J. R., Gosling, J. T., Russell, C. T., and Elphic, R. C., 1979. Nature 282, 243.
Petschek, H. E., 1964. AAS-NASA Symposium on the Physics of Solar Flares, NASA Spec. Publ. SP-50, 425.
Rosenbauer, H., Grünwaldt, H., Montgomery, M. D., Paschmann, G., and Sckopke, N., 1975. J. Geophys. Res. 80, 272.
Russell, C. T., and Elphic, R. C., 1979. Geophys. Res. Lett., 6, 33.
Sarris, E. T., Krimigis, S. M., and Armstrong, T. P., 1976. J. Geophys. Res. 81, 2341.
Sckopke, N., Paschmann, G., Haerendel, G., Sonnerup, B. U. O., Bame, S. J., Forbes, T. G., Hones, E. W., Jr., Russell, C. T., 1980. Structure of the Low Latitude Boundary Layer, submitted to J.. Geophys. Res.
Sonnerup, B. U. O., 1971. J. Geophys. Res. 76, 6717.
Sweet, P. A., 1958. Electromagnetic Phenomena in Cosmical Physics, ed. by B. Lehnert, Cambridge Univ. Press, 123.
Vasyliunas, V. M., 1975. Rev. Geophys. Space Phys. 13, 303.
Vasyliunas, V. M., 1980. This volume.

PLASMA SHEET DYNAMICS: EFFECTS ON, AND FEEDBACK FROM, THE POLAR IONOSPHERE

Vytenis M. Vasyliunas

Max-Planck-Institut für Aeronomie
D-3411 Katlenburg-Lindau 3
Federal Republic of Germany

1. INTRODUCTION

1.1. What Is the Plasma Sheet

The plasma sheet as a distinct region of the terrestrial magnetosphere historically was first identified in the observations of plasma and energetic particles; in a systematic approach appropriate to a tutorial lecture, it is more conveniently described by reference to the configuration of the magnetic field. The magnetic field lines within the magnetosphere can be divided into two broad morphological classes: the basically dipolar (although sometimes significantly distorted) field lines of the inner magnetosphere, and the highly stretched-out field lines of the magnetotail. The plasma sheet lies in between and occupies the region of transition between these two classes of field lines; it may be viewed as a very broad boundary layer separating the inner magnetosphere from the geomagnetic tail and at the same time separating the two lobes of the magnetotail from each other. Figure 1 schematically illustrates the configuration of the plasma sheet in the noon-midnight meridian plane.

The plasma contained within the plasma sheet region is hot and relatively dense; the number density is considerably higher than that within the adjacent tail lobe regions, although still very much lower than the density in the magnetosheath or the solar wind, and the mean thermal energy exceeds the bulk flow energy of solar wind protons. This region of hot and dense plasma begins on highly stretched-out field lines next to the lobes of the magnetotail and extends inward well into the dipolar field region; the plasma sheet merges, in general continuously and

C. S. Deehr and J. A. Holtet (eds.), Exploration of the Polar Upper Atmosphere, 229–244.

without an obvious sharp discontinuity, into the ring current ion population. The so-called inner edge of the plasma sheet is a sharp decrease of electron temperature; the ion temperature and

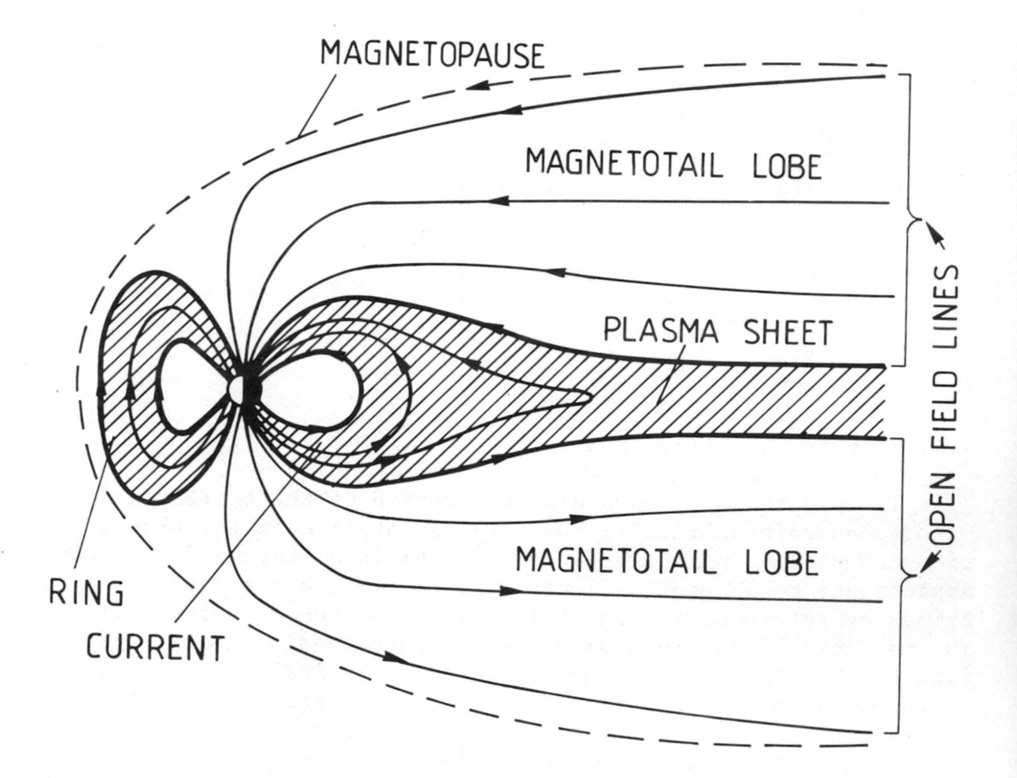

Fig. 1 Schematic configuration of the plasma sheet within the magnetosphere.

density undergo no sharp change across it but continue to increase with decreasing distance until the termination of the ring current, at or near the plasmapause, is reached (e.g. Vasyliunas 1972a).

Related to (and indeed generally viewed as the ultimate reason for) the morphological distinction of dipolar and stretched-out magnetic field lines is a topological distinction: closed field lines connected to the earth in both directions, and open field lines connecting in one direction to the earth and extending in the other direction out into the solar wind. Whereas the morphological distinction is not precisely defined (the field lines threading the plasma sheet are to a large extent intermediate between the two classes), the topological distinction is sharp and admits no intermediate cases: a particular magnetic field line from the earth at a specified time is either open or closed. Almost certainly, the field lines through the lobes of the magnetotail are open, and many of the field lines through the plasma sheet are closed, so that the separatrix surface between open and closed field lines must lie somewhere not far from the outer boundary of the plasma sheet. That the separatrix actually coincides with the boundary between the plasma sheet and the tail lobes (and that consequently all the field lines threading the plasma sheet are closed) is an assumption that is frequently made but up to now has not been firmly established.

1.2. Significance of the Plasma Sheet for the Polar Ionosphere

The open field line region, at the heights of the ionosphere and upper atmosphere, is generally associated with the polar cap. The projection of the plasma sheet along magnetic field lines down to ionospheric levels accordingly forms a ring-shaped region lying just equatorward of the polar cap and thus coinciding or at least nearly coinciding with the auroral oval.

The influence of the plasma sheet upon the polar ionosphere is two-fold; it has a particle aspect and an electrodynamic aspect. In the first place, the plasma sheet defines the spatial extent of charged particle precipitation into the atmosphere at high latitudes. The lobes of the magnetotail contain too little plasma to sustain significant precipitation; the high intensities of precipitating electrons associated with the aurora can only occur in regions connected by magnetic field lines to the plasma sheet (or its earthward extension, the ring current). It is thus no coincidence that the ionospheric projection of the plasma sheet is coextensive with the auroral oval. However, while the plasma sheet constitutes the primary source of precipitating particles for the aurora, it is well established that the particles are not simply "dumped" unchanged from the plasma sheet into the upper atmosphere but that they are subject to important acceleration processes in-between that determine their final, highly complex spatial and energy distribution.

In the second place, the plasma sheet largely determines the Birkeland currents (electrical currents flowing parallel to the magnetic field) that flow into and out of the polar ionosphere, and it exerts a major influence upon the ionospheric electric fields at high latitudes. Conversely, through electrodynamic coupling the polar ionosphere can also influence the dynamical processes occurring within the plasma sheet.

2. MAGNETOSPHERE-IONOSPHERE INTERACTION

2.1. Electric Field Mapping

The description of the interaction between the plasma sheet and the polar ionosphere is a particular case of the general theory of magnetosphere-ionosphere interaction, the basic principles of which I briefly review here. Since we are dealing with two distinct regions, the magnetosphere and the ionosphere, connected by magnetic field lines, an important first question is, what determines the mapping of electric fields along the magnetic field lines? This question has been the center of a long controversy, whose general course can be described in Hegelian terms: thesis-antithesis-synthesis.

The thesis, a first approach to the question which stems to an appreciable extent from the work of Alfvén and received its classical exposition in his monograph (Alfvén 1950), incorporates the basic insight that in space and astrophysical systems the behavior of electric fields is very different from that in a near vacuum. Although the charged particle number densities of 1 or 0.01 cm^{-3} may seem extremely tenuous by ordinary laboratory standards, they are offset by the large spatial scale of the systems, so that the number of particles in any volume element of interest is very large; if scaled down to laboratory dimensions, the cosmical systems would be characterized by very high densities as well as extremely high magnetic fields. The electric field is then strongly constrained: in a frame of reference where the plasma is locally at rest, any electric field is (very nearly) shorted out by the motion of charged particles along the magnetic field in response to it. Transformed to a general frame of reference, this implies that the electric field $\underline{E}$, the magnetic field $\underline{B}$, and the plasma bulk flow velocity V are related by the familiar magnetohydrodynamic (MHD) or "frozen flux" approximation

$$c\underline{E} + \underline{V} \times \underline{B} \approx 0 \qquad (1)$$

whose consequences are well known: the electric field component parallel to $\underline{B}$ is vanishingly small, the magnetic flux through any loop moving with the plasma remains constant, and the variation along the field lines of either $\underline{E}$ or the perpendicular component of $\underline{V}$ can be calculated from the magnetic field geometry.

The antithesis, a subsequent second approach also championed by Alfvén (1968), points to the observational fact that, in contradiction to the MHD approximation, the parallel component of the electric field is in many cases non-zero and non-negligible. The observational evidence comes from laboratory experiments and (on the basis of indirect but fairly persuasive inferences) from auroral particle observations, and it has led to the development of several theoretical models for the formation of large parallel electric fields (e.g. Fälthammar 1977; Goertz 1979).

To reconcile the two approaches into a synthesis, it is necessary to note under what conditions large parallel electric fields occur. Quite generally, (1) is an approximation to the generalized Ohm's law, which can be written in the form

$$\underline{E} + \frac{1}{c}\underline{V} \times \underline{B} = \eta\underline{J} - \frac{1}{ne}\nabla.P^{(e)} + \frac{1}{nec}\underline{J} \times \underline{B} + \frac{m_e}{ne^2}\left[\frac{\partial \underline{J}}{\partial t} + \nabla.(\underline{JV} + \underline{VJ})\right] \quad (2)$$

in an obvious notation (see, e.g. Vasyliunas 1975a); η is the effective resistivity, including any plasma turbulence or particle-wave interaction effects that may be present. All the terms of the right-hand side of (2) are proportional to spatial gradients (remember that $\underline{J} = (c/4\pi)\ \nabla \times B$). For many cosmical systems, all these terms become negligibly small as a consequence of the large spatial scales; (2) then reduces to (1) and the parallel electric field $E_{\parallel}$ becomes nearly zero. Exceptions are small-scale structures such as shocks, discontinuities, or neighbourhoods of topologically singular magnetic lines. A variety of specific situations where non-negligible $E_{\parallel}$ occurs have been identified (e.g. Fälthammar 1977; Goertz 1979). They may be classified into two general groups:

(a) $E_{\parallel}$ associated with pressure effects (anisotropies or gradients), i.e. with the parallel component of the $\nabla.P^{(e)}$ term of (2). Its occurrence need not be connected with a Birkeland current $J_{\parallel}$. Usually $E_{\parallel}$ has a specific value such that its line integral along the magnetic field line (the field-aligned potential difference) is proportional to, and generally of similar order of magnitude as, the mean particle thermal energy per charge. The additional energy that a particle can gain by acceleration through the field-aligned potential drop is thus somewhat limited. The deviations from the "frozen flux" condition (1) produced by $E_{\parallel}$ of this group are of the same order as those due to gradient, curvature, and magnetization drifts, which appear in (2) via the perpendicular components of the same $\nabla.P^{(e)}$ term, together with $\underline{J} \times \underline{B}/nec$.

(b) $E_{\parallel}$ associated with Birkeland currents. Included are various types of double layers and so-called "electrostatic shocks" (Goertz 1979 and references therein), which involve the inertial terms of (2), as well as $E_{\parallel}$ due to large "anomalous" resistivity η produced as a consequence of one of several possible current-driven instabilities (e.g. Papadopoulos 1977). The common feature of all these mechanisms is that the Birkeland current density $J_{\parallel}$ is required to reach or exceed a threshold value

$$J_{\parallel} \gtrsim n e w_c \qquad (3)$$

where n is the ambient number density and w_c a critical speed that depends on the specific mechanism or instability; for double layers, w_c is typically the electron thermal speed. The value of $E_{\parallel}$ or of the associated potential difference is, in contrast to case (a), not fixed but may become very large if that is necessary to prevent $J_{\parallel}$ from exceeding too much the threshold (3). The local deviations from (1) can thus be enormous, but the global applicability of the frozen flux concept is little affected since the large $E_{\parallel}$ are confined to narrow regions. The restriction to a narrow spatial space is apparent if the order-of-magnitude relation given by Ampère's law $J_{\parallel} \approx (c/4\pi)\, \Delta B/L$ is combined with (3):

$$L \lesssim \lambda_i \frac{V_A}{w_c} \frac{\Delta B}{B} \qquad (4)$$

where V_A is the Alfvén speed and $\lambda_i = (m_i c^2/4\pi n e^2)^{1/2}$ the ion inertial length.

To restate the synthesis: the basic insight of the first approach remains valid - in cosmical systems the electric field is strongly constrained, the plasma flow and the magnetic field are connected. However, as the second approach correctly indicates, the constraint and the connection are not always given by the simple MHD "frozen flux" concept - under specific conditions they become much more complicated. The specific conditions can take many forms but ultimately reduce to questions of scale. For sufficiently large spatial scales, MHD is an adequate approximation and electric fields may be simply mapped. For smaller scales, parallel electric fields may develop and must be taken into account in the mapping.

The determining factor in (4) is the value of λ_i; the remaining factor, in magnetospheric and ionospheric applications, is generally of order 1 or less. Within the plasma sheet, where $n \approx 0.1$ to $1\ cm^{-3}$, λ_i is of the order of a few hundred km and thus much smaller than the characteristic dimensions and gradient scales of the plasma sheet. Above the auroral ionosphere, $n \sim O(10^4)\ cm^{-3}$ and the ions are mainly O^+ rather than protons; then λ_i is near 10 km, interestingly (and perhaps not by mere coincidence) similar to the scale of many auroral structures. Global problems of the plasma sheet and the magnetotail, then, can for most purposes be adequately treated within a MHD framework; near the earth, however, the convergence of the magnetic field lines leads to strong reduction of scale and increase of Birkeland current density, so that at some distance above the auroral ionosphere non-MHD effects become important and significant parallel electric fields, generally viewed as the main accelerating mechanism for auroral particles, can develop.

2.2 Self-consistent Electrodynamic Coupling

With this background, I now review the general theory of magnetosphere-ionosphere coupling, developed quantitatively during the 1960's (Fejer 1964; Swift 1967) and formulated as a closed self-consistent chain of equations (Vasyliunas 1970b) shown schematically in Figure 2. Begin with the distribution of plasma

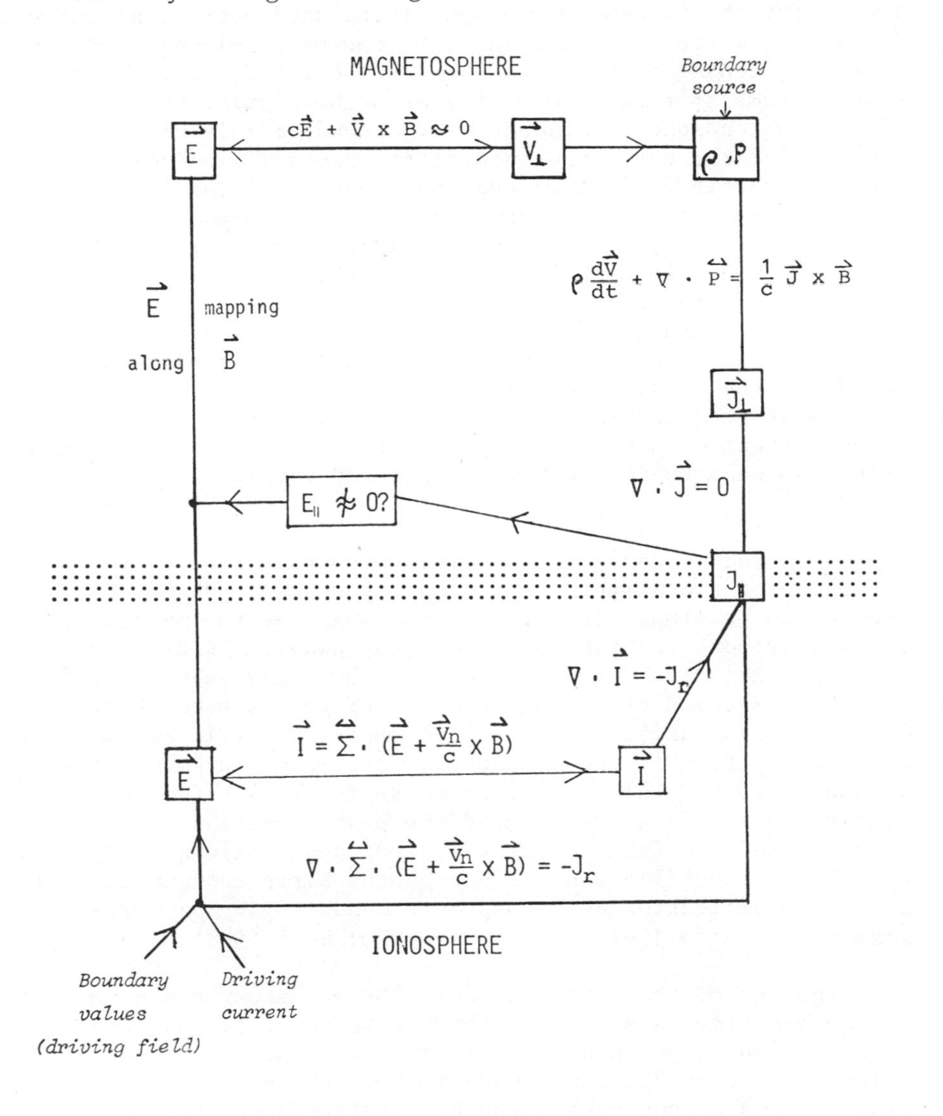

Fig. 2 Calculation scheme of magnetosphere-ionosphere coupling (from Vasyliunas 1979). Quantities being calculated are enclosed in boxes and the equations relating them are indicated along the lines joining the boxes. Italics denote boundary conditions. The dotted band in the middle separates magnetosphere and ionosphere.

density and pressure in the magnetosphere (the upper right hand corner of the diagram in Figure 2). From the momentum equation, calculate the $\underline{J} \times \underline{B}$ force density needed to balance the mechanical plasma stresses, and from it the perpendicular component of $\underline{J}$, whose divergence then equals the parallel divergence of $-J_{\parallel}$. Integrated along the field line, this becomes the Birkeland current flowing into or out of the ionosphere and acting as the source for the ionospheric height-integrated horizontal current density $\underline{I}$. In turn, $\underline{I}$ is related by the height-integrated conductivity tensor to the ionospheric electric field $\underline{E}_i$ (viewed from a frame of reference moving with the neutral atmosphere, since the ionospheric conductivity depends mainly on neutral-ion collisions; the appearance of the neutral wind velocity in the ionospheric Ohm's law leads to a possible dynamical coupling between the neutral upper atmosphere and the magnetosphere-ionosphere system). $\underline{E}_i$ is related to the magnetospheric electric field $\underline{E}_m$, which in turn is related to the magnetospheric plasma flow; it is only at this point in the calculation that the previously discussed questions concerning the mapping of the electric fields and the validity of the MHD approximation must be faced and the possible development of $E_{\parallel}$ when $J_{\parallel}$ is sufficiently large must be taken into account. Finally, the plasma flow must be consistent with the density and pressure distribution; this brings us back to the starting point and completes the chain of equations.

To develop quantitative models with this scheme of calculation, one of two limiting approaches has been applied. One approach can be simply described by saying that the magnetosphere is considered a voltage source and the diagram is traversed counterlockwise: the plasma flow, or the equivalent electric field, is assumed given in some part of the magnetosphere (typically at high latitudes); the ionospheric electric field is then calculated, from it the ionospheric currents and thence the Birkeland currents, calculation of whose feedback effect on the initially assumed flow constitutes the next iteration of the diagram (usually left to future work). Typical applications of this approach are modelling the basic magnetospheric convection pattern and its modification by ionospheric conductivity patterns (Iwasaki and Nishida 1967; Vasyliunas 1970; Wolf 1970).

The second approach considers the magnetosphere a current source and traverses the diagram clockwise: a plasma pressure distribution (e.g. an asymmetric ring current) is assumed, $\underline{J}_{\perp}$ is calculated, then $J_{\parallel}$, whose closure through the ionosphere determines $\underline{E}_i$ and thence by mapping $\underline{E}_m$ (its feedback effect on the assumed pressure distribution is usually included at least in a simplified fashion). Typically this approach is applied to modelling the electric field and plasma flow effects of the ring current (Swift 1967) and its interaction with magnetospheric convection (Vasyliunas 1972b; Wolf 1975 and references therein), the latter application requiring a partial combination of both approaches.

In the case of the interaction between the plasma sheet and the polar ionosphere, neither approach is particularly suitable. Effects where the plasma sheet is most simply viewed as a voltage source occur together with others where it is more conveniently considered a current source. A further complication is that the magnetic field is now significantly affected by currents in the magnetosphere-ionosphere loop; the validity of assuming an *a priori* given field model, as in the previously mentioned calculation, is questionable. In the rest of this paper I explore qualitatively some aspects of plasma sheet dynamics in relation to the polar ionosphere; quantitative modelling here, in contrast to the case of the inner ring current region, remains a task for the future.

3. THE STEADY-STATE PLASMA SHEET

It is a truism to say that the magnetosphere, or any other system, is never in a steady state. Nevertheless, steady-state models are potentially applicable in at least two cases. One is a quasi-stationary situation, where the system varies on a time scale long compared to its characteristic dynamical time scales, so that its evolution can be considered a succession of nearly steady-state configurations. The other is the time-averaged system, which by definition is a steady-state configuration.

In discussing the dynamics of the plasma sheet in the steady state, whether quasi-stationary or time-averaged, the starting point is the electric field directed dawn-to-dusk across the polar cap. This electric field has a well-understood physical explanation in terms of the MHD-dynamo-like solar wind interaction with the magnetosphere (e.g. Vasyliunas 1975b, 1979 and references therein), but for our purposes here it may be taken simply as an observational fact. From the polar cap the electric field is mapped out to the magnetotail, where it still has a dawn-to-dusk component (although it is easy to show that it cannot be uniform in the magnetotail if it was uniform over the polar cap). Therefore, according to (1), there is a component of plasma flow downward in the northern lobe of the tail and upward in the southern lobe; by continuity, this flow extends into the plasma sheet, where there is thus a flow pattern with components toward the center from both above and below. The two flows meet as they approach the magnetic field reversal region (the so-called magnetic neutral sheet) near the center of the plasma sheet; by continuity again, they must both be deflected into some direction parallel to the plane of the neutral sheet.

One possibility is for the flow near the neutral sheet to turn toward the earth. This requires an earthward-directed stress, which can only be provided by the magnetic $\underline{J} \times \underline{B}$ force (not by the plasma pressure gradient, which exerts an opposite stress, the pressure within the plasma sheet generally decreasing with increasing distance down the tail). Since $\underline{J}$ is directed dawn-to-dusk, the $\underline{J} \times \underline{B}$ force will be earthward if B_n, the component of $\underline{B}$ and normal to the plane of the neutral sheet, points northward. Within and near the field reversal region, the association of northward B_n with earthward flow is also required by (1), given the dawn-to-dusk component of the electric field. Thus the MHD connection between plasma flow and magnetic field is here consistent with stress balance. The configuration of flow and field is stretched in Figure 3a; it applies at least within the plasma sheet regions not too distant from the earth, since B_n there must be northward in order to agree with the dipole field direction. (It should be noted that the illustrated pattern is a projection on the noon-midnight meridian and does not show the dawn-to-dusk components, which in general do exist.)

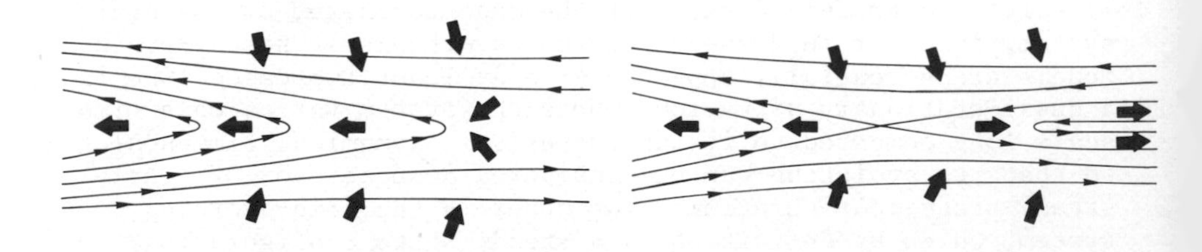

Fig. 3 Schematic configuration of the magnetic field lines and plasma flow (heavy arrows) near the center of the plasma sheet. (a) (left): region of earthward flow. (b) (right): region of both earthward and tailward flows.

The other possibility which must be considered, especially for the more distant regions of the plasma sheet, is that the flow, as it approaches the neutral sheet, turns anti-earthward or tailward. The MHD condition (1) requires in this case that B_n point southward, a conclusion that can also be reached if one imagines a loop moving with the plasma, aligned with the streamlines of the flow, and then requires that the magnetic flux through it be conserved as it is turned by the flow to become parallel to the neutral sheet. With a southward B_n, $\underline{J} \times \underline{B}$ points tailward and provides, together with the pressure gradient, the stress needed for the tailward turning of the flow; here, too, MHD and stress balance are consistent. The configuration of flow and field near the transition from earthward to tailward flow is sketched in Figure 3b. The reversal of B_n from northward to southward occurs at the magnetic X line (sometimes called, less precisely, the magnetic neutral line); magnetic field lines crossing the X line form the two branches of the separatrix surface that separates open and closed field lines.

Observation of a persistent tailward flow component throughout the thickness of the plasma sheet constitutes strong evidence for the topological configuration of Figure 3b, i.e. a reversal of B_n from northward to southward and hence a magnetic X line somewhere between the point of observation and the earth. (If $\underline{V}$ is aligned with $\underline{B}$, a tailward flow may occur without a southward B_n, but it can persist for an appreciable time only on the open field lines of the magnetotail and not near the middle of the plasma sheet. The earthward and tailward flows of Figure 3 are nearly field-aligned, except at the center of the neutral sheet, but even a small $V_\perp$ makes a crucial difference in (1).) The existence of an X line can often be more reliably inferred from the flow than from an attempted direct observation of a southward B_n: usually $B_n \ll B$ and a slight error in the orientation of the neutral sheet can lead to a large error in the measured B_n, whereas the earthward or tailward flow velocity components are expected to be much larger than the up and down components (e.g. Vasyliunas 1975a and references therein) and hence easily identifiable, even with fairly inaccurate observations.

The consequences of plasma sheet dynamics for the ionosphere can be summarized in three points:

(a) The electric field associated according to (1) with the flow within the plasma sheet is (at least partially) mapped to the ionosphere, where it constitutes the extension of the polar cap electric field equatorward into the auroral oval. The plasma sheet thus acts as a voltage source across the magnetic field lines connected with the aurora.

(b) Several mechanisms give rise to Birkeland currents flowing between the plasma sheet and the ionosphere. If a velocity shear develops as a result of stresses internal to the plasma sheet, magnetic field lines crossing the shear region themselves become sheared, implying that $\underline{B} \cdot \nabla \times \underline{B} \neq 0$, so that the plasma sheet becomes a current source for the auroral ionosphere. In addition, the ionospheric response to the plasma sheet as a voltage source discussed under (a) produces Birkeland currents whose closure within the plasma sheet modifies the flow there. In both cases there is essentially the same relation between the Birkeland current density and the flow acceleration, discussed by Rostoker and Boström (1976) without distinguishing the two mechanisms. (The distinction becomes apparent when the Birkeland currents are viewed as standing Alfvén waves in a flowing plasma: the wave is initiated within the plasma sheet in the first case and at the ionosphere in the second, although the subsequent multiple reflections render the complete configuration practically the same in both cases.)

(c) The spatial extent of the region of auroral particle precipitation on the ionosphere is strongly influenced by the location of the X line and the amount of magnetic flux crossing the neutral sheet earthward of it.

4. TIME-VARYING PLASMA SHEET: SUBSTORMS

Major variations of the flow pattern within the plasma sheet are observed to occur during magnetospheric substorms. Typically, there is an onset of tailward flow at the beginning of the substorm with a subsequent change to strong earthward flow (see Hones and Schindler 1979 and references therein). In view of the discussion in section 3, the natural interpretation is that a magnetic X line forms in the magnetotail during a substorm, with its location initially earthward of the observing spacecraft (which generally is at a distance of 15-20 earth radii) and afterwards changing to a much greater distance. In some cases only the earthward flow is observed, and it is assumed that the initial location of the X line lay already tailward of the spacecraft. This model, relating substorm dynamics to the formation of an X line, has been the subject of considerable controversy, but the arguments of its opponents (e.g. Lui et al. 1976, 1977a, b) can largely be reduced to two major points: lack of unambiguous evidence for the X line from magnetic field observations, and failure to observe a tailward flow during every substorm. The first merely reflects the previously mentioned inaccuracy of B_n determinations, and the second need imply, as already noted, no more than an initial location of the X line tailward of the observing spacecraft during some substorms.

The postulated topological development of the magnetotail during substorms is schematically illustrated in Figure 4, which shows a time sequence of magnetic field configurations. Panel a is the quiet-time magnetotail before the substorm; there is a distant X line associated with the separatrix between open and closed field lines and mapped to the polar cap boundary on the ionosphere. At the beginning of the substorm, a new X line appears relatively near the earth (b), associated with a magnetic "island" of field lines that do not connect either to the earth or to the solar wind. The island expands (c) and eventually reaches the boundary of closed field lines (d). Thereafter it is on open field lines (e) and is carried tailward by the plasma flow; the near-earth X line is also assumed to move tailward (not shown), with an increasing flux of closed field lines. It should be noted that Figure 4 is meant only to represent the topology; more "realistic"

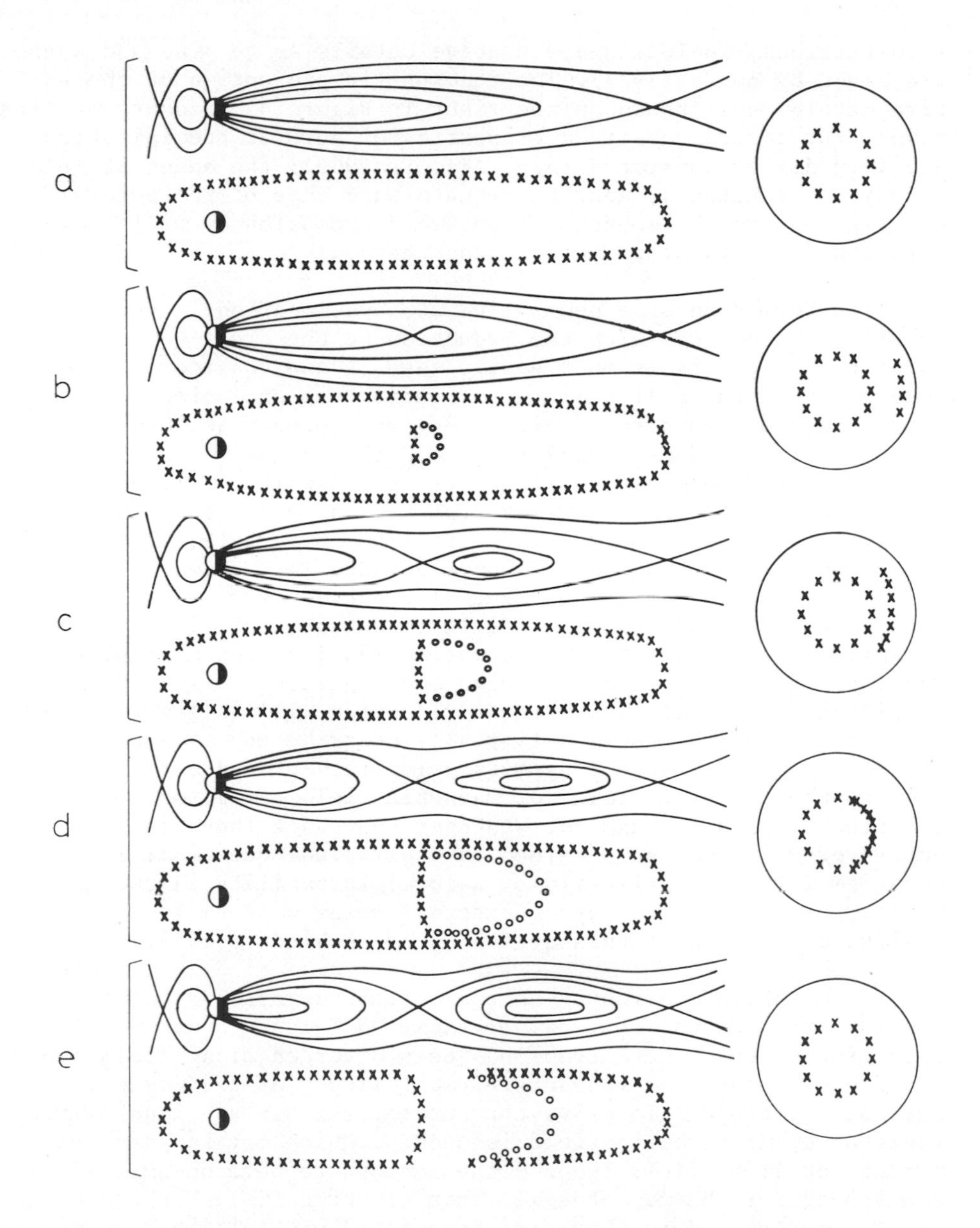

Fig. 4 Possible changes of magnetic field topology during substorms (from Vasyliunas 1976). Within each panel, at the top is the magnetic field line configuration in the noon-midnight meridian, at the bottom is the equatorial plane with the magnetic singular lines of the X or O type shown by the respective symbols, and at the right is the projection along magnetic field lines to the ionosphere.

illustrations, including speculative details as to size and shape, are given by Hones (1977). The ionospheric projection of the entire development (shown on the right in Figure 4) accounts qualitatively for several observed properties of auroral precipitation activity during substorms (e.g. Akasofu 1968): the onset at relatively low latitudes, near the equatorward edge of the auroral arc region, and the subsequent poleward expansion as well as westward and eastward propagation along the oval.

The formation of a new X line at some location in the magnetotail is associated with the reduction of the cross-tail electric current there; this is most easily seen if one considers the Ampèrean integral $\oint d\underline{l}.\underline{B}$ around a square loop enclosing the X line (e.g. the boundary of Figure 3b) and notes that the formation of the X line produces non-zero B_n and thus negative contributions from both side segments, while to first order it has little effect on the (positive) contributions from the top and bottom segments. Sometimes "current interruption" (really a hyperbole for "current reduction" since the cross-tail current does not actually disappear) is presented as an opposing alternative model to X line formation (e.g. Akasofu 1979). To a large extent, this is a purely semantic controversy: X line formation, as just noted, implies a current reduction, and conversely, the magnetic configuration resulting from an assumed current reduction will in general contain an X line. The two models really describe the same phenomenon viewed from two different aspects: electric current and its relation to the electric field, or magnetic field variation and its relation to plasma dynamics. Whatever substance there is in the controversy is really contained in a different question: is the substorm initiated primarily by a local instability in the magnetotail, or is it driven by an external solar wind variation? (For a wider discussion of this question, see Schindler 1979.)

To maintain continuity as the current across part of the magnetotail is reduced, it is widely assumed (although not yet firmly established) that the current has been diverted along field lines and flows through the ionosphere, most likely as the westward auroral electrojet. To drive the current through the ionospheric resistance, an electric field is needed, which may in turn be mapped out along field lines - the ionosphere here becomes a voltage source for the magnetotail. This electric field has a dawn-to-dusk component and is therefore associated with plasma flow toward the X line from above and below. The possibility of a feedback mechanism thus becomes apparent, whereby the diversion of current from the plasma sheet to the ionosphere enhanced the plasma flow, leading to a larger B_n and a further reduction of cross-tail current; instabilities based on a similar feedback mechanism have been discussed by Atkinson (1967, 1979).

5. CONCLUSION

The plasma sheet plays a major role in determining the electric fields, currents, and particle precipitation regions in the polar ionosphere and upper atmosphere; in turn, ionospheric influences on electric fields and currents within the magnetotail are potentially important for plasma sheet dynamics. The whole is a particularly complex instance of magnetosphere-ionosphere coupling, not adequately tractable by either of the hitherto usual approximations. A reasonable qualitative understanding of several features of the magnetospheric substorm on the basis of dynamical processes within the plasma sheet exists, but quantitative models as well as an adequate knowledge of detailed sequencing and cause-effect chains remain to be developed.

REFERENCES

Akasofu, S.-I. 1968, Polar and Magnetospheric Substorms. D. Reidel, Dordrecht, Holland.
Akasofu, S.-I. 1979, p. 447 in Dynamics of the Magnetosphere (Ed. S.-I. Akasofu). D. Reidel, Dordrecht, Holland.
Alfvén, H. 1950, Cosmical Electrodynamics. Oxford University Press, London.
Alfvén, H. 1968, Ann. Géophys., 24, 341.
Atkinson, G. 1967, J. Geophys. Res., 72, 5373.
Atkinson, G. 1979, p. 461 in Dynamics of the Magnetosphere (Ed. S.-I. Akasofu). D. Reidel, Dordrecht, Holland.
Fälthammar, C.-G. 1977, Rev. Geophys. Space Phys., 15, 457.
Fejer, J.A. 1964, J. Geophys. Res., 69, 123.
Goertz, C.K. 1979, Rev. Geophys. Space Phys., 17, 418.
Hones, E.W., Jr. 1977, J. Geophys. Res., 82, 5633.
Hones, E.W., Jr., and Schindler, K. 1979, J. Geophys. Res., 84, 7155.
Iwasaki, N., and Nishida, A. 1967, Rept. Ionosph. Space Res. Japan, 21, 17.
Lui, A.T.Y., Meng, C.-I., and Akasofu, S.-I., 1976, J. Geophys. Res., 81, 5934.
Lui, A.T.Y., Meng, C.-I., and Akasofu, S.-I. 1977a, J. Geophys. Res., 82, 1547.
Lui, A.T.Y., Frank, L.A., Ackerson, K.L., Meng, C.-I., and Akasofu, S.-I. 1977b, J. Geophys. Res., 82, 4815.
Papadopoulos, K. 1977, Rev. Geophys. Space Phys., 15, 113.
Rostoker, G., and Boström, R, 1976, J. Geophys. Res., 81, 235.
Schindler, K. 1979, p. 311 in Dynamics of the Magnetosphere (Ed. S.-I. Akasofu). D. Reidel, Dordrecht, Holland.
Swift, D.W. 1967, Planet. Space Sci., 15, 835.
Vasyliunas, V.M. 1970, p. 60 in Particles and Fields in the Magnetosphere (Ed. B.M.McCormac). D. Reidel, Dordrecht, Holland.
Vasyliunas, V.M. 1972a, p. 192 in Solar Terrestrial Physics/1970: Part III (Ed. E.R. Dyer). D. Reidel, Dordrecht, Holland.

Vasyliunas, V.M. 1972b, p. 27 in Earth's Magnetospheric Processes (Ed. B.M. McCormac). D. Reidel, Dordrecht, Holland.

Vasyliunas, V.M. 1975a, Rev. Geophys. Space Phys., 13, 303.

Vasyliunas, V.M. 1975b, p. 179 in The Magnetospheres of the Earth and Jupiter (Ed. V. Formisano). D. Reidel, Dordrecht, Holland.

Vasyliunas, V.M. 1976, p. 99 in Magnetospheric Particles and Fields (Ed. B.M. McCormac). D. Reidel, Dordrecht, Holland.

Vasyliunas, V.M. 1979, p. 387 in Proceedings of Magnetospheric Boundary Layers Conference (Ed. B. Battrick) ESA SP-148, Noordwijk, The Netherlands.

Wolf, R.A. 1970, J. Geophys. Res., 75, 4677.

Wolf, R.A. 1975, Space Sci. Rev., 17, 537.

RELATIONSHIPS BETWEEN THE SOLAR WIND AND THE POLAR CAP MAGNETIC ACTIVITY

Annick Berthelier

LGE - CNRS, 4 Av. de Neptune, 94100 St-Maur-des-Fossés
France

ABSTRACT. The influence of solar wind conditions on magnetic activity is described in order to delineate the differences in the response of the magnetic activity to the arrival on the magnetopause of different typical solar wind variations. By determining a new index of local magnetic activity free from seasonal and diurnal effects we put in evidence the dependance of the various effects upon the invariant latitude. Most important results are : (1) the main increase of the magnetic activity does not occur at the same invariant latitude for different interplanetary variations, e.g. peaks of Bz tend to increase magnetic activity mainly in the auroral zones while peaks of B correspond to a uniform increase in magnetic activity over the polar cap and auroral zone ; (2) there is a two steps response of magnetic activity to the high speed plasma streams ; (3) an increase of magnetic activity is observed for large and northward Bz, which probably indicates that the solar wind - magnetosphere coupling is efficient under these circumstances. The specific influences of the IMF polarity are also briefly reviewed.

1. INTRODUCTION

Variations of the solar wind conditions induce a number of perturbations of the magnetosphere, manifested in a complex way in the physical quantities which can be measured from the ground. Among them magnetic disturbances, which are related to the ionospheric and magnetospheric currents, appear as the largest available data base offering excellent temporal and spatial coverages. Results from their analysis are presented in this work, in an attempt to better understand the coupling of the solar wind with the high latitude magnetosphere and ionosphere.

C. S. Deehr and J. A. Holtet (eds.), Exploration of the Polar Upper Atmosphere, 245–258.

Variations in the earth's magnetic field appear as small amplitude pulsations with periods ranging from $\sim$ 0.1 to 100's of seconds, which reflect mainly the magnetospheric wave-particle interactions, and by more long term fluctuations having a larger amplitude and no clear periodicity. The morphology and occurrence of these fluctuations are highly variable and depend upon the time, the season, and the geographic and magnetic location of the observatory, and also upon the solar wind conditions. They are analysed along two main different ways. The first one consists in calculating the components of the magnetic vector perturbations taken as the departures from estimated quiet conditions, and to interpret them in terms of equivalent ionospheric currents (see a review by Friis-Christensen in the same volume). In the second approach one takes only into account the amplitude of the fluctuations of the magnetic field components during a certain interval of time. A detailed description of the various magnetic activity indices describing these fluctuations is given in Mayaud (1980).

In the last decade numerous studies have shown that the magnetic activity increases with the solar wind speed and temperature, with the variance of the interplanetary magnetic field (IMF), and is higher when the north-south IMF component Bz is southward (see e.g. Hirshberg and Colburn 1969 ; Arnoldy 1971 ; Kane 1972 ; Burton et al 1975 ; Murayama and Hakamada 1975 ; Maezawa 1978). Moreover, several attempts have been made to improve the correlations between the magnetic activity and the solar wind parameters by replacing one of them by quantities such as VBz, V^2Bz, VBz^2 in an effort to link the observations to physical quantities which are known to influence the interaction between the solar wind and the magnetosphere (e.g. Garrett et al 1974 ; Crooker et al 1977 ; Akasofu 1979 ; Iyemori et al 1979). We will here not consider such global relations, but only describe the variations of the local magnetic activity at high latitudes and their relation to different variations of the solar wind parameters.

2. MAGNETIC ACTIVITY INDICES AT HIGH LATITUDES

The local magnetic activity is described by means of the K indices which characterize the amplitude of the irregular variations of the horizontal components of the terrestrial magnetic field over each three hour interval. From the K indices an estimation of the average amplitude a_K of the corresponding actual fluctuations can be calculated accordingly by means of conversion coefficients (Mayaud 1955 ; Lebeau 1965). The diurnal and seasonal variations of the local magnetic activity are mainly governed by local time and season, due to the influence of solar radiations on ionospheric conductivity, and by magnetic local time. The variations depend also on the invariant latitude. This prevents studies of the influence of the solar wind on local magnetic activity directly from K and a_K indices, except if we should have enough data for each hour and season. By defining a new corrected index of local magnetic activity, a_c, one can however avoid this difficulty.

Our basic hypothesis is that the interplanetary conditions can be considered as an excitating factor exerted on the magnetosphere, and that, during a sufficiently long period of time (about 2 months), one gets an homogeneous sample of them without in particular any bias due to a self diurnal variation. Furthermore, we have supposed that the seasonal and local time influences can be expressed as a multiplicative modulation factor, a_{base}. This can be calculated for a given observatory and three hour interval by averaging the diurnal variation of a_K over running periods lasting about two months.

The new index of local magnetic activity corrected from the statistical influences of the diurnal and seasonal variations is thus defined : $a_c = a_K/a_{base}$. Its mean value is uniform over all latitudes (and ~ 1). This allows studies of the latitudinal response of the magnetic activity to interplanetary parameters variations. However, it will be useful to distinguish between daytime and night-time activity, since the physical phenomena involved are of very different origins.

3. MAGNETIC ACTIVITY RESPONSE TO THE ARRIVAL OF STRUCTURED INTERPLANETARY VARIATIONS AT THE MAGNETOPAUSE

We have considered the variations of the main parameters describing the characteristics of the solar wind, namely the velocity V, the thermal speed V_T and the density N of the plasma in the interplanetary medium, and the components of the interplanetary magnetic field (IMF) taken in geocentric solar magnetospheric (GSM) coordinates (Bx, By, Bz). Data used are the hourly values for the years 1967-1968, kindly supplied by the Explorer 28-33-34-35, Vela 3 and Mariner 5 experimenters. One observes different types of structured variations occurring sufficiently often to allow a statistical analysis of their influences by means of a superposed epoch analysis. As typical examples we will here consider the southward turnings of the IMF Bz component, the peaks of the total IMF intensity B, and the large increases of the velocity V corresponding to high speed plasma streams.

3.1 Southward peaks of Bz

The southward turnings of the north-south component of IMF Bz are seen as negative peaks in Bz without any related large variation of the IMF intensity B. Figure 1-a (left panel) shows the variation of each parameter averaged over the 54 cases that have been sorted out of the 1967-1968 data, taken from 2 days before to 2 days after the time of the minimum of Bz. The potential imposed in the frontal region of the magnetosphere, denoted Φ, has been calculated from the

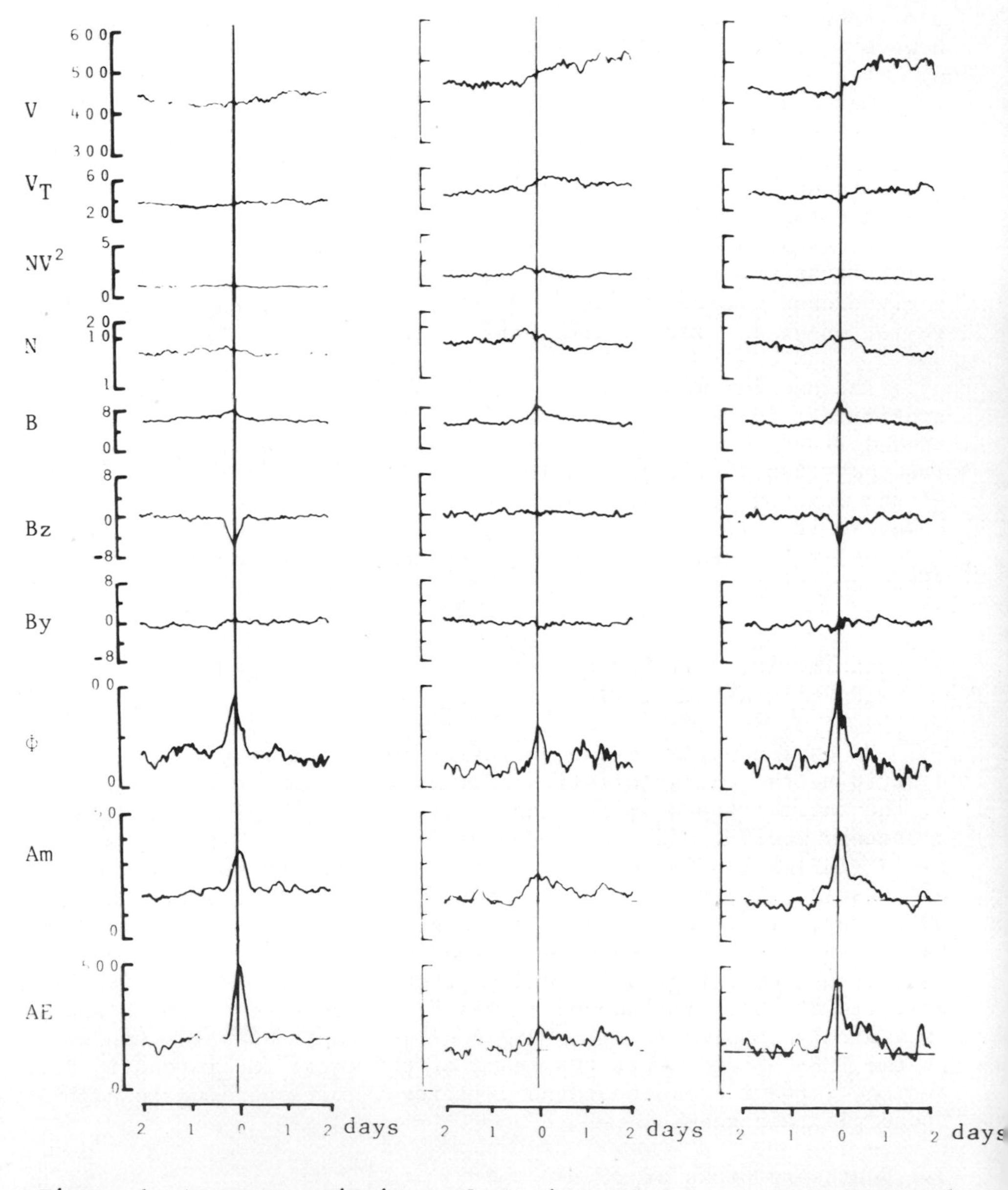

Figure 1. Average variations of the interplanetary parameters and magnetic activity indices versus time (days). From top to bottom : the velocity V and thermal speed V_T ($km.s^{-1}$), the quantity NV^2 which is proportional to the solar wind dynamic pressure (arbitrary unit), the density N, the IMF intensity B, and its north-south, Bz, and azimuthal, By, components (gammas), the potential Φ (see text), and the magnetic indices Am and AE (gammas). Panel (a) represents 54 southward peaks of Bz, (b) 35 peaks of B when Bz fluctuates, and (c) 15 peaks of B when Bz turns southward.

semi-quantitative model of reconnection between the interplanetary and terrestrial magnetic fields proposed by Gonzalez and Mozer (1974).

It appears that the average variation of the magnetic activity follows the variation of Bz, with a delay less than one hour, estimated from larger scale diagrams. The maximum of the auroral electrojet activity index AE is much larger than that of the Am index, calculated from K indices at subauroral observatories (Mayaud 1980). The relative increase of AE represented by the ratio of its maximum value to its mean value is 2.4, compared to 1.8 for the relative increase of Am.

The averaged variations of the a_c indices are calculated for the southern and northern high latitude stations (Figure 2). The relative values of the maxima (ratio of the maximum to the mean value) are shown for different invariant latitudes (dashed line, Figure 3-a). The relative maximum of a_c is the highest for invariant latitudes of about 65 to 62°. We believe that this indicates the position of the auroral zone which is about 5° lower than its average position under these circumstances.

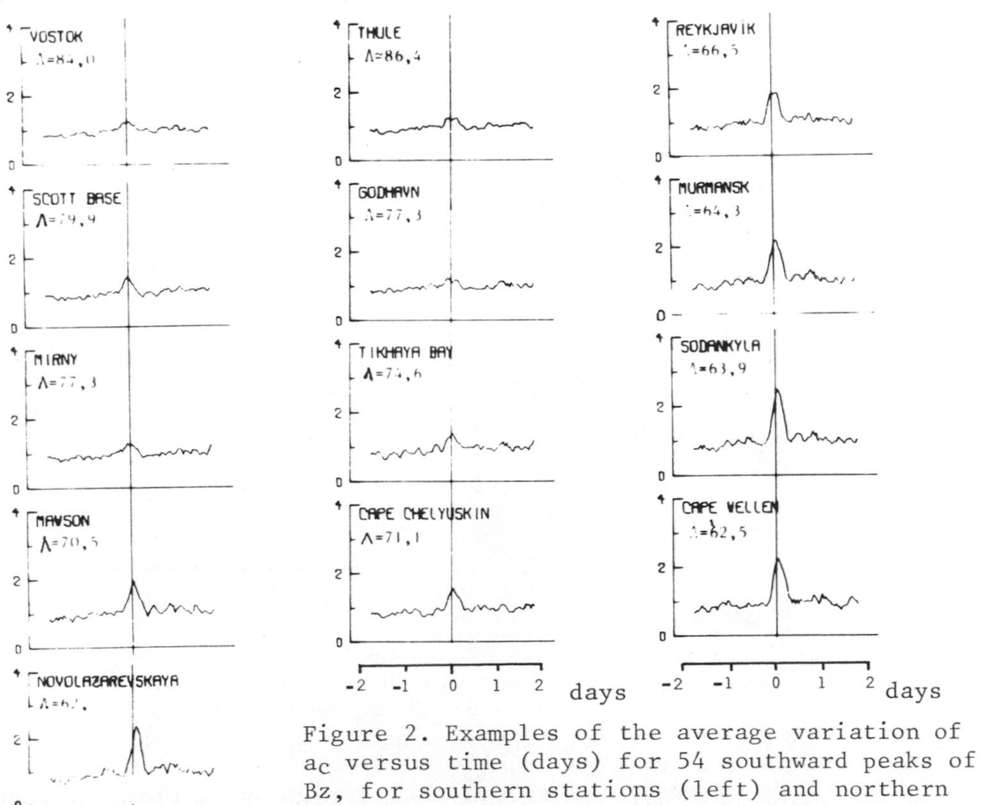

Figure 2. Examples of the average variation of a_c versus time (days) for 54 southward peaks of Bz, for southern stations (left) and northern stations (central and right columns). The invariant latitude Λ is given for each station (see also Figure 3).

3.2 Peaks of B

In the case of the peak-shaped variations of the IMF intensity B we have separated the cases when there are no large variations of Bz, but only fluctuations of rather low amplitude, and those when one observes a simultaneous increase of southward Bz. It should be noticed that these latter events are not identical to the peaks of southern Bz shown above, which do not involve a pronounced increase in B.

An increase of the magnetic activity is observed in both cases (Figure 1-b, -c), but in difference to the preceding case it is slightly higher for Am than for AE. The relative values of the maxima of a_c are plotted as function of invariant latitude (Figure 3-a). When Bz fluctuates (thin line) the relative increase of a_c is about 1.5 and does not vary with the latitude. This may indicate that the increase of B causes an energy input in the magnetosphere creating a global intensification of magnetospheric and ionospheric high latitude currents systems, but without any corresponding triggering of substorms. In the case of the B peaks related to Bz southward, the response of a_c (thick line) is higher at latitudes of 65 to 62°. This is similar to that found for southward turnings of Bz, although regularly larger. It is interpreted as a global intensification of the currents systems caused by the increase of B. This is added to the growing substorm activity related to the southward turning of Bz. Due to the substorm development the auroral zone is shifted to latitudes lower than its average position.

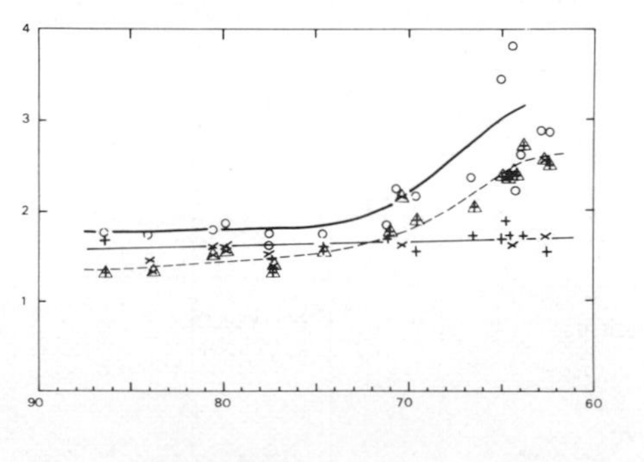

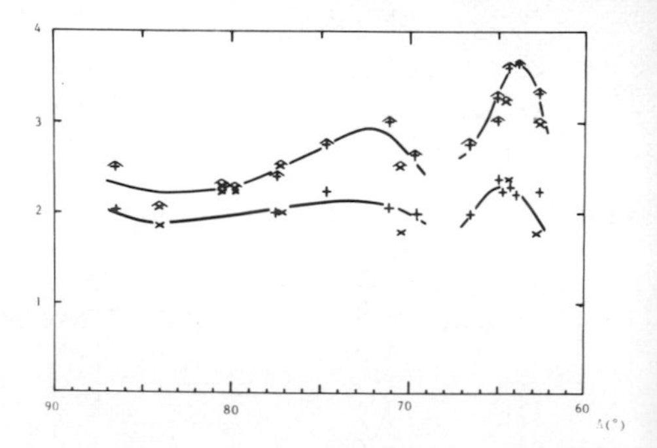

Figure 3. a. Relative values of the maxima of a_c shown as function of invariant latitude of the southern (x) and northern (+) stations, for (i) southward turnings of Bz (Δ, ▲, dashed line), (ii) peaks of B when Bz fluctuates (x, +, thin line), (iii) peaks of B when Bz is southward (o, thick line).

b. Same as (a) in the case of high speed plasma streams for the first quasi-instantaneous increase (x, +), and for the second increase, twelve hours later ($\hat{x}$, $\hat{+}$).

3.3 High speed plasma streams

The high speed plasma streams are seen as a large increase in V reaching its maximum after about one day (Figure 4). One observes on the average an increase in V_T, N and B lasting a few hours, and a sharp increase of the potential Φ, which recovers after about one day, while the recovery of V is slower. No clear associated varia-

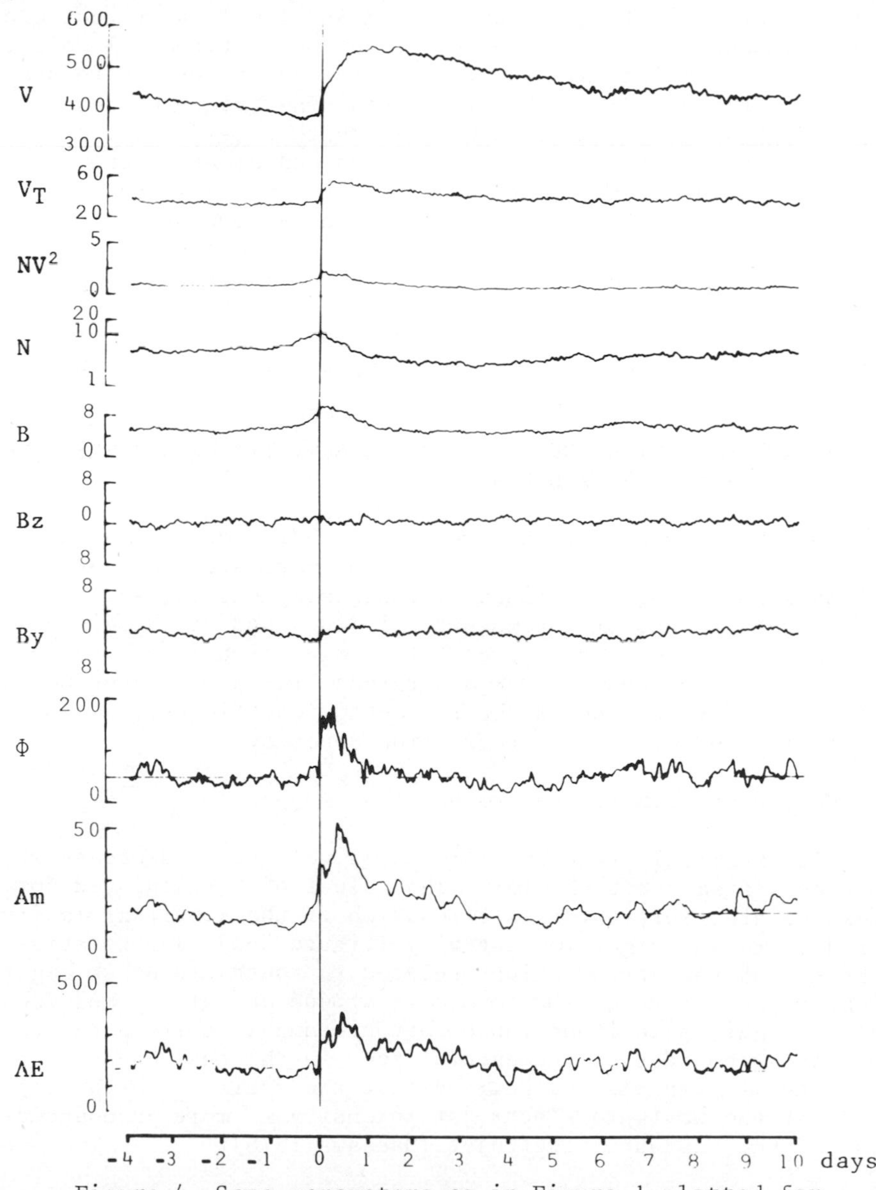

Figure 4. Same parameters as in Figure 1 plotted for the 34 high speed plasma stream events.

tions of By and Bz are seen from hourly averages, but it should be noticed that there could be changes in Bz on a shorter time scale (e.g. Rosenberg and Coleman 1978).

A first abrupt increase of Am and AE corresponds to the beginning of these events. It is followed by an enhancement of the magnetic activity, with a maximum about twelve hours later. The relative increase of Am is larger than that of AE. The local magnetic activity a_c also shows a two steps response, but the latitudinal variations of the relative intensities of the first and second increases are different (Figure 3-b). The first quasi-instantaneous response is about 2 at all latitudes. We believe that it reflects the influences of the compression of the magnetosphere and of an increase of the electric field in its frontal region. The secondary response is much larger than the first one and presents a maximum around 65° corresponding to the auroral region, and another one around 73°, thought to indicate the polar cusp. We believe that this corresponds both to an increase of the electric potential imposed in the frontal region of the magnetosphere causing in particular the polar cusp increase of activity, and to a development of auroral substorms.

4. AVERAGE VARIATIONS OF THE MAGNETIC ACTIVITY WITH RESPECT TO THE SOLAR WIND PARAMETERS

In order to derive the average variation of the magnetic activity indices as a function of solar wind parameters we have sorted the data into classes defined by increasing values, and calculated the average values of the magnetic indices and of the given solar wind parameter. One can approach the regression line by using the least-squares method to draw a straight line among these points. The slope of this line is the regression coefficient, k, which provides a measure of the regression tendency.

4.1 Variation with respect to Bz

It is firstly seen that Am, AE as well as a_c decrease when Bz increases from about $-5\,\gamma$ to 0. The values of k calculated for Bz negative are represented as a function of the invariant latitude for daytime and nighttime activity (Figure 5-a). The relative importance of the perturbations related to southward Bz is higher at night than at daytime. Maximum k is around 66° at day and lower than 63° at night, a latitude range that presumably corresponds to the equatorial boarder of the auroral zone in the case of substorm activity. It can also be noticed that the variation with respect to Bz of the westward electrojet intensity is more pronounced than that of the eastward electrojet (Maezawa 1978).

When Bz is northward and increases, a_c first slowly diminishes, then increases with Bz. Such an increase of the magnetic activity

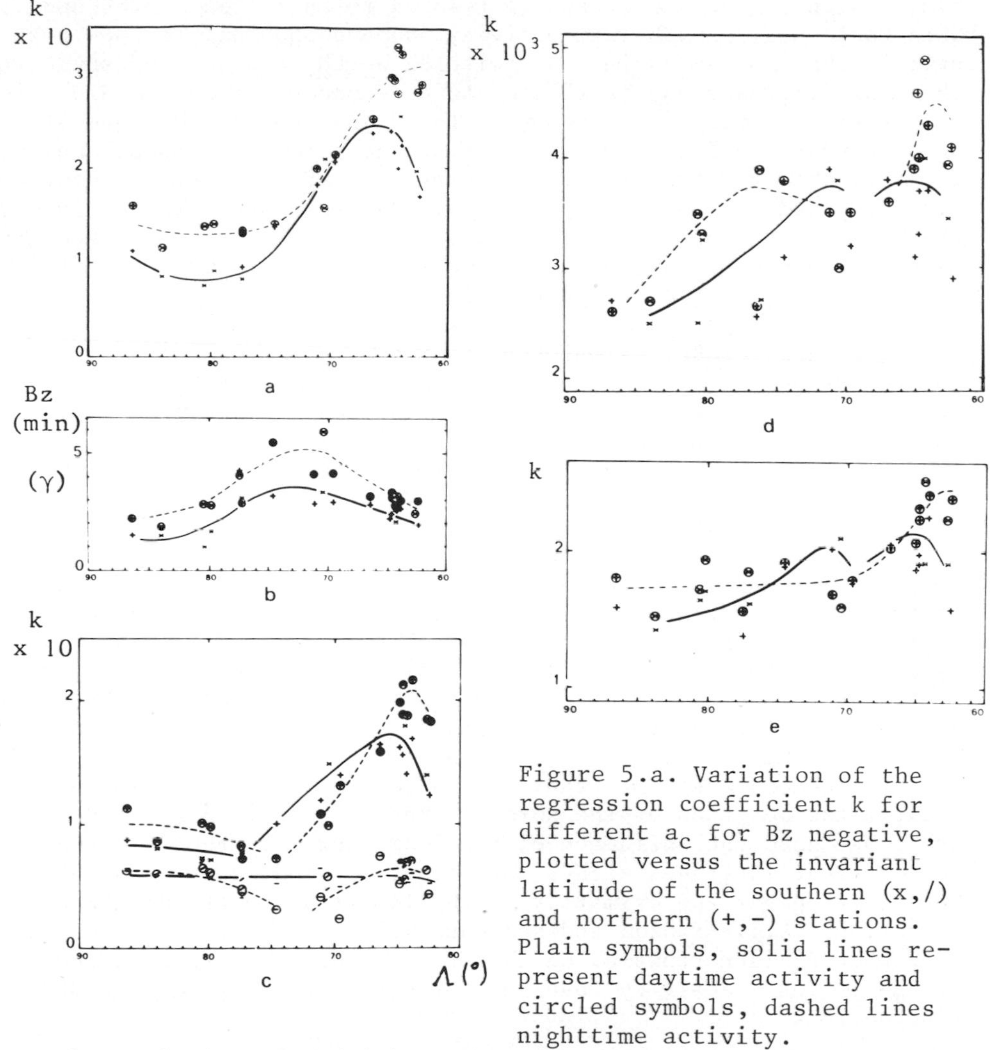

Figure 5.a. Variation of the regression coefficient k for different a_c for Bz negative, plotted versus the invariant latitude of the southern (x,/) and northern (+,-) stations. Plain symbols, solid lines represent daytime activity and circled symbols, dashed lines nighttime activity.

b. Variation of Bz(min) versus invariant latitude (same symbols as in (a)).

c. Same as (a) for the variation of a_c with respect to B in the cases of Bz south (x, +, upper curves) and Bz north (), /, lower curves).

d. Same as in (a) for the variation of a_c with respect to V.

e. Same as in (a) for the variation of a_c with respect to V_T.

with Bz north is also observed in the case of Am, especially when By is different from 0, but it does not appear in the case of AE. Maezawa (1978) showed that the ratios Am/AU and Am/AL increase with Bz north, and suggested that such a difference between the variation of Am and that of AU or AL could correspond to an influence of the compression of the magnetosphere related to an increase of B and V. This would not, as substorm activity, depend upon Bz. The new result here is that Am and a_c increase with Bz north, and we have suggested that this increase may be related to an enhancement of the electric potential in the frontal region of the magnetosphere. We have also noticed that the function representing the potential imposed on the front of the magnetosphere in the very simplified model of reconnection proposed by Gonzalez and Mozer (1974) increases when Bz increases to large northward values, a tendency which is more accentuated for higher values of By. (A more thorough discussion of this hypothesis may be found in Berthelier (1979)).

The value of Bz for which a_c is minimum, denoted Bz (min), gives an indication of the influence of Bz north : a low value of Bz (min) means that a_c increases with Bz north whereas a large value means that it does not increase insofar. Bz (min) is found to be higher for nighttime activity than for daytime, with maximum around 70 to 74° in the two cases (Figure 5-b). This has been interpreted by considering that there are two main causes of magnetic activity at high latitudes. One corresponds to global fluctuations of ionospheric currents which increase with southward Bz, and also with northward Bz under some conditions. This will be dominating when Bz (min) is lower. The other is the substorm activity characterized by a decrease when Bz south diminishes and a related contraction of the auroral zone toward higher latitudes. This dominates when Bz(min) is larger. Following this interpretation one can infer from the Figure 5-b that the substorm influence is higher at nighttime than at daytime, and most important around 70 to 74°, which also indicates the position of the auroral zone when Bz is northward. On the other hand the influence of the ionospheric current fluctuations dominates at latitudes higher than about 80° and lower than 68°. This corresponds approximately to the localisation of the DP2 perturbations (e.g. Nishida and Kokubun 1971 ; Mayaud 1978) which occur even when Bz is northward. We therefore think that at least one part of the magnetic activity observed for northward Bz is due to DP2 fluctuations.

4.2 Variation with respect to B

The magnetic activity is shown to increase with increasing values of the IMF intensity B, for northward and southward values of Bz. The variations of the regression coefficient with the invariant latitude are shown on Figure 5-c for daytime and nighttime activity. For Bz north k remains around 0.8, indicating a response of magnetic activity which is rather independant of the invariant latitude. This should correspond to the influences of the magnetospheric compression and of a global increase of the convection related to the increase of B. For southward Bz it is not possible to separate out the B-related component from the variation related to Bz.

4.3 Variation with respect to V and V_T

The magnetic activity is observed to increase with either V or V_T. The variation of the regression coefficient versus invariant latitude is shown in Figures 5-d, -e. The variation is interpreted as being mainly influenced by the high speed plasma streams. The maximum observed around 65° for the nighttime activity indicates growth of substorm activity with increasing V. The daytime maximum observed around 70° reflects an increase of the magnetic response in the polar cusp regions, then seen as the auroral zone at lower invariant latitudes than the average position.

The variation of the magnetic activity with respect to V_T is similar to that related to V, firstly because of the strong dependence between these two parameters. However, it should be noticed that the response of the magnetic activity to increases in V and in V_T is larger when both are associated, compared to the case of independent increases either of V, or of V_T (Berthelier 1979).

5. VARIATIONS OF MAGNETIC ACTIVITY RELATED TO THE IMF POLARITY

The polarity of the IMF is defined by the dominant direction of IMF in the ecliptic plane. For positive polarity this component is directed away from the sun, while negative polarity is defined by IMF directed toward the sun. The polarity is also related to the sign of the azimuthal IMF component By. Positive polarity corresponds generally to By positive and a negative polarity to By negative.

Effect from By has been recognized in dayside magnetic perturbations at high latitudes (Svalgaard 1968 ; Mansurov 1969), and has been interpreted by means of equivalent currents (e.g. Berthelier 1972 ; Svalgaard 1973 ; Berthelier et al 1974 ; Friis-Christensen et al 1975 ; Friis-Christensen 1979, and this volume). It is as yet considered as a signature of both Birkeland currents and convection on the dayside, which change drastically with By (Wilhjelm et al 1978).

For the magnetic activity indices there are two main changes related to the IMF polarity ; one concerns the diurnal and seasonal variations of the magnetic activity, the other is seen on the difference between the northern and southern magnetic activity.

5.1 Average magnetic activity

The annual variation of the magnetic activity presents generally two maxima around the equinoxes. This has been interpreted in different ways, for example by taking into account the seasonal variation of the angle between the geographic axis and the earth-sun line (Mc Intosh 1959). When the data are sorted according to the IMF polarity one observes a spring maximum significantly higher than the fall maximum for a negative polarity, and the opposite for a positive polarity. Russell and Mc Pherron (1973) have proposed that this effect stems from the seasonal variation of the north-south IMF component Bz taken in GSM coordinates. Two IMF vectors having the same Bz component in solar ecliptic coordinates and opposite polarities will have different Bz components in GSM coordinates, a difference which depends upon the time and the season. From a modeling of this effect and a detailed comparison to the seasonal and diurnal variations of Am and AE one can conclude that the influence of the polarity is well-described by the Russell - Mc Pherron hypothesis, but it can not account for the totality of the annual and diurnal variations. For Am there remains a semi-annual variation well-explained by the Mc Intosh effect, while the diurnal variation appears to be influenced both by the polarity and the Mc Intosh effects.

5.2 Hemispheric magnetic activity differences

The difference between the magnetic activity in the northern and southern hemispheres is slightly higher when the IMF polarity is positive than when it is negative (Siebert 1968 ; Wilcox 1968 ; Berthelier and Guérin 1973). The relative importance of this asymmetry has been estimated to be about 5 to 10% of the average magnetic activity (Berthelier 1979). This low value explains why this effect can be currently hidden by other variations.

It has been further shown that this hemispheric asymmetry also drastically depends upon the values of the north-south IMF component Bz (Berthelier 1979). The difference δA = (an -as) decreases when Bz increases in the case By > 0, and increases with Bz for By < 0 (Figure 6). It appears that the hemispheric asymmetry due to the IMF polarity, which has been previously observed without taking into account the value of Bz, i.e. δA higher for By > 0 than for By < 0, is observed for Bz southward, but not for Bz northward. The variations of δA are interpreted as being due firstly to the asymmetry

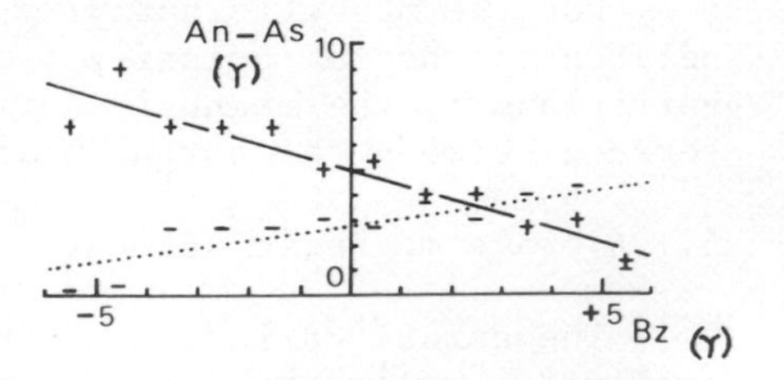

Figure 6. Variation with respect to Bz of the difference δA=An−As in the cases of By positive (+) and of By negative (−).

of the convection patterns at high latitude related bo By (Heppner 1972 ; Berthelier et al 1974). This leads to a change in the influence of the eastward and westward electrojets on subauroral magnetic activity according to the hemisphere and to the sign of By. Secondly the variations of the intensities of the eastward and westward electrojets with respect to Bz are different as it has been recently observed by Maezawa (1978). A detailed discussion may be found in papers by Berthelier (1979, 1980).

6. CONCLUSION

The aim of this paper was to delineate the differences in the response of the magnetic activity at high latitudes related to the different variations of the interplanetary medium parameters. The main conclusion that arises from this work is that the solar wind - magnetosphere coupling depends not only on the individual interplanetary parameters but also on the shape of their simultaneous variations, so that any law aiming to relate magnetospheric quantities to the values of the interplanetary parameters should be also dependent on the type of event considered.

REFERENCES

Akasofu, S.-I. 1979, Planet. Space Sci., 27, 4381.
Arnoldy, R.L. 1971, J. Geophys. Res., 76, 5189.
Bartels, J. 1938, Z. Geophys., 14, 68.
Berthelier, A. 1972, C.R.H. Acad. Sc. Paris, 275(B), 841.
Berthelier, A. 1976, J. Geophys. Res., 81, 4546.
Berthelier, A. 1979, Thèse de Doctorat d'Etat, Paris VI.
Berthelier, A., and Guérin, C. 1973, Space Res., 13, 661.
Berthelier, A., Berthelier, J.J., and Guérin, C. 1974, J. Geophys. Res., 79, 3187.
Berthelier, A., and Berthelier, J.J. 1979, Paper G3-18, IAGA Bull. 43, (Ed. N. Fukushima), IUGG Pub. Office, Paris.
Burton, R.K., Mc Pherron, R.L., and Russell, C.T. 1975, J. Geophys. Res., 80, 4204.
Crooker, N.U., Feynman, J., and Gosling, J.T. 1977, J. Geophys. Res., 82, 1933.
Friis-Christensen, E. 1979, p.280 in Magnetospheric Study 1979, (Ed. Japanese IMS Committee), Tokyo.
Friis-Christensen, E., and Wilhjelm, J. 1975, J. Geophys. Res., 80, 1248.
Garrett, H.B., Dessler, A.J., and Hill, T.W. 1974, J. Geophys. Res., 79, 4603.
Gonzalez, W.D., and Mozer, F.S. 1974, J. Geophys. Res., 79, 4186.
Heppner, J.P. 1972, J. Geophys. Res., 77, 4877.
Hirshberg, J., and Colburn, D.S. 1969, Planet. Space Sci., 17, 1183.

Iyemori, T., Maeda, H., and Kamei, T. 1979, J. Geomag. Geoelectr., 31, 1.
Kane, R.P. 1972, J. Atmos. Terr. Phys., 34, 1941.
Lebeau, A. 1965, Ann. Geophys., 21, 167.
Mac Intosh, D.H. 1959, Phil. Trans. Roy. Soc. London, Ser.A, 251, 525
Maezawa, K. 1978, Solar Terr. Env. Res. Japan, 2, 103.
Mansurov, S.M. 1969, Geomagn. Aeronomy, 9, 622.
Mayaud, P.N. 1955, Magnétisme terrestre, S.IV.2, Ed. by Expéditions Polaires Françaises, Paris.
Mayaud, P.N. 1968, Indices Kn, Ks et Km, 1964-1967, Editions du CNRS, Paris.
Mayaud, P.N. 1978, Ann. Geophys., 34, 243.
Mayaud, P.N. 1980, Derivation meaning and use of geomagnetic indices, Monograph n° 22, AGU, Washington.
Murayama, T., and Hakamada, K. 1975, Planet. Space Sci., 23, 75.
Nishida, A., and Kokubun, S. 1971, Rev. Geophys. Space Phys., 9, 417.
Rosenberg, R.L., and Coleman, P.J. Jr. 1978, IGPP Publication n° 1804 University of California, Los Angeles, California, USA.
Russell, C.T., and Mc Pherron, R.L. 1973, J. Geophys. Res., 78, 92.
Siebert, M. 1968, J. Geophys. Res., 73, 3049.
Svalgaard, L. 1968, Geophys. Pap. R-6, Dan. Meteorol. Inst., Charlottenlund, Denmark.
Svalgaard, L. 1973, J. Geophys. Res., 78, 2064.
Wilcox, J.M. 1968, J. Geophys. Res., 73, 6835.
Wilhjelm, J., Friis-Christensen, E., and Potemra, T.A. 1978, J. Geophys. Res., 83, 5586.

AURORAL MORPHOLOGY: A TELEVISION IMAGE OF SOLAR AND MAGNETOSPHERIC ACTIVITY

C. S. Deehr
G. J. Romick
G. G. Sivjee

The Geophysical Institute, Fairbanks, Alaska 99701

INTRODUCTION

The aurora is a radiant manifestation of solar particle emissions and their control by intervening electromagnetic fields. The analogy with a television system was first made, we believe, by Elvey, (1958). The latest concepts of solar-terrestrial control are included in description by Akasofu (1979) showing the phosphor screen as the upper atmosphere with an auroral image produced by particles from a source on the sun, modulated by electric and magnetic fields with the magnetohydrodynamic (MDH) generator formed by electrons and protons from the solar wind across the geomagnetic tail as the power supply. Thus, the size and shape of the aurora must reflect all the forces acting on the auroral particles on their way from the sun to the earth. Auroral morphology, therefore, is the study of the occurrence of aurora in space and time for the purpose of describing the origin of solar particles and the forces acting upon them between the time of their production on the sun and their loss in the atmosphere. The advantage of using the aurora as a television monitor of this process over any conceivable system of in situ measurements is obvious when one considers the large number of space vehicles which would be necessary to record the information concentrated in the auroral oval which differs in scale with the magnetosphere by perhaps 10^6.

MICROSCOPIC MORPHOLOGY

Microscopic morphology is the study of the temporal or spatial variations of individual auroral forms. Observations are

C. S. Deehr and J. A. Holtet (eds.), Exploration of the Polar Upper Atmosphere, 259–266.

carried out in total light or spectroscopically in order to map local electromagnetic field effects, chemical reactions and the interaction of energetic charges particles with the atmosphere. Of particular interest to studies of the aurora as an indicator of solar-terrestrial effects is the relationship between auroral height-luminosity and incoming particle energy. The energy of auroral primaries is determined from differences in the auroral emission spectrum due to various production and loss mechanisms associated with secondary electron impact on atmospheric constituents at different altitudes. Both the local volume emission rates and the height-integrated emission show the effects of differing incoming primary particle energy.

Of particular interest here, are two aspects of the height-luminosity distribution. The first is the visual observer's impression of different primary particle energy. Figure 1 shows how the ratios of the main visual emissions change relative to one another as a function of altitude. Specifically "blood-red" predominates over yellow-green above about 200 km. Yellow-green predominates in the middle region and magenta predominates below approximately 90 km. Blood-red is the [OI] emission at 6300-6364A, yellow-green is the [OI] emission at 5577A and magenta is a combination of N_2 and O_2^+ emissions near 6000A and N_2^+ emissions in the blue end of the spectrum. Thus, the observer reporting "blood-red" aurora is seeing aurora produced by a primary electron energy spectrum with a pronounced flux of low energy electrons (< 1 kev). Reports of "red lower borders" (usually fast-moving and therefore "crackling fire") indicate primaries of greater than 40 kev energy. The majority of observations refer to yellow-green or sometimes grey. The latter refers to observations below the color threshold of the eye and represent low particle flux at any altitude.

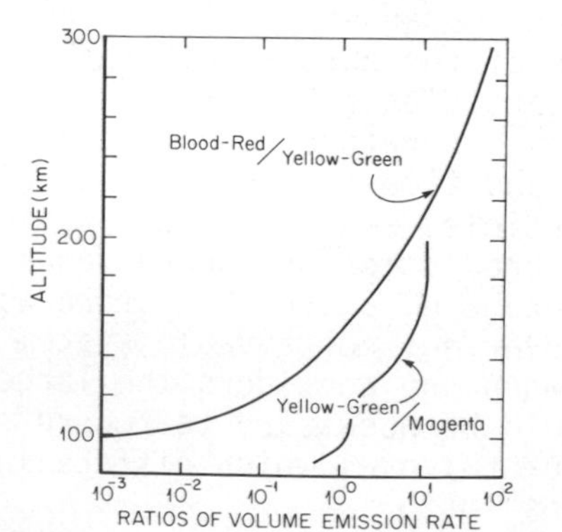

Figure 1. Showing the variation of the volume emission rate of the main visible auroral emissions as a function of altitude.

The second aspect of the height-luminosity distribution to be dealt with here has to do with instrumental observations and in particular with the parallactic photographic data of Størmer and his colleagues (Egeland and Omholt, 1966). The 12,000 individual data points in this survey were taken at various intervals between 1910 and 1943. The data are not grouped according to the position of the measurement on the auroral form (i.e., lower edge, top of ray, etc.) but represent the collection of all portions of the auroral arc distinct enough on the photograph to enable a height measurement to be made. One may assume, however, that lower borders are most distinct and that upper level observations are probably balanced out with the lower ones. In any case, the higher observations indicate the existence of relatively large fluxes of low energy particles. These data have been grouped according to sunspot activity and show little or no correlation with sunspot number.

Auroral height can be shown to be related to solar activity, however, by considering a different measure of solar activity, called the "recurrence index". This index is a measure of the recurrence probability of geomagnetic activity associated with solar rotation developed by Sargent (1978). Figure 2 shows the altitude of the aurora as a function of the concurrent recurrence

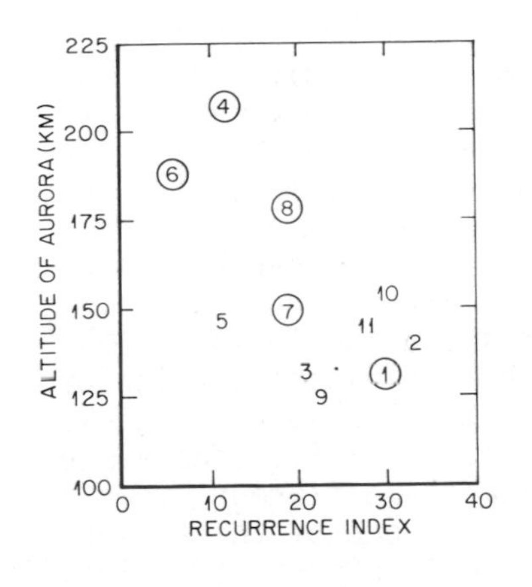

Figure 2. Showing the relationship between the magnetic recurrence index (Sargent, 1978) and the altitude of the aurora determined by parallactic photography by Størmer (Omholt and Egeland, 1966). The numbers refer to the year in the sunspot cycle starting from minimum. The circled numbers are those years for which there are more than 400 data points.

index and grouped according to the year of the sunspot cycle from minimum to minimum for the three cycles included in the Størmer data. Years for which there are more than 400 data points are circled. It is apparent that there is a distinct tendency for higher aurorae during sporadic storms while recurrent storms are associated with lower altitude aurorae.

In summary, the study of individual auroral forms for the purpose of determining the energy of the incoming particles is a valuable means of relating auroral activity to magnetospheric and solar activity. Of particular interest is the value of visual observations in determining solar activity and the relationship between the altitude of the aurora and the tendency for a geomagnetic storm to recur. Because the recurrent storms are associated with "coronal holes" on the sun and the sporadic storms with solar flares (Simon 1979), we infer that coronal holes on the sun are correlated with lower altitude aurorae and solar flares with higher aurorae. This is consistent with the visual observers' impression that there tends to be more active, red lower border aurorae nearing sunspot minimum and a sign of approaching maximum is the appearance of blood red, diffuse upper borders (Elvey 1958). It should also be noted that particle streams associated with coronal holes are characterized by a low flux of higher energy particles while solar flares give out much larger fluxes of lower energy particles. We therefore propose that the aurora reflects directly the source of particles on the sun and that the magnetospheric acceleration processes do not appreciably change the relative number flux or the relative energy distribution from that at the source on the sun.

MACROSCOPIC MORPHOLOGY

Macroscopic morphological observations may be divided roughly into two groups according to the means of study: 1) Observations from space vehicles providing instantaneous views with a time resolution of approximately two hours. Such data may be combined to give a picture of the locus of precipitation with a severe loss in spatial resolution due to the dynamic nature of the aurora at time scales less than two hours. 2) Ground-based observations providing high resolution in local time and space. The difficulty involved in accumulating enough simultaneous data from a large enough density of ground stations has precluded an extensive accumulation of data showing Universal Time variations over a significant portion of the auroral precipitation zone, although valuable constant-local-time observations have been made from jet aircraft. The aim of macroscopic morphology is to combine the different views to produce a realistic picture in time and space of the entire auroral process. The following is an attempt to present a simplified diagram of auroral occurrence and to isolate physically separable parameters. Readers are referred to a complete treatese on the substorm by Akasofu (1978) and recent reviews of the subject by Swift (1979) and McPherron (1979).

Figure 3 shows a composite of auroral occurrence as a function of local geomagnetic time and geomagnetic latitude. The occurrence and spatial distribution of various types of aurorae depend strongly on local and substorm time, so there are some effects shown on the diagram which would not be together in time or space in a real situation.

The "diffuse aurora" is called so because its very small, striated structure was not observable in the satellite data where

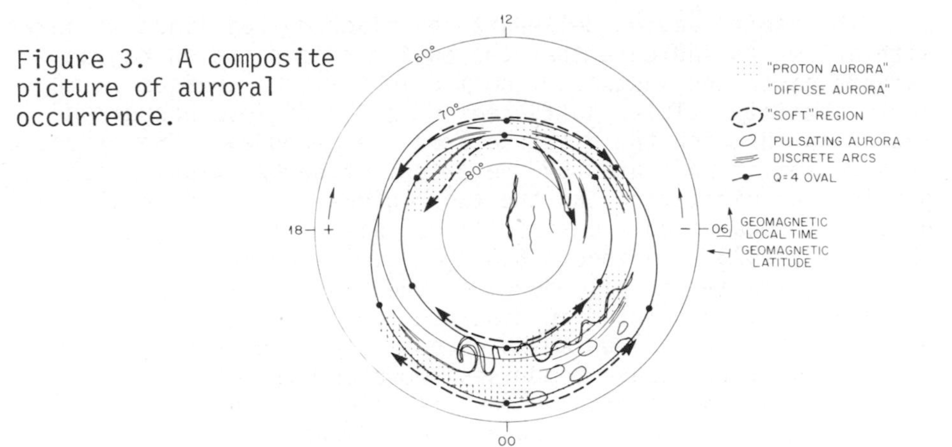

Figure 3. A composite picture of auroral occurrence.

it was first documented as a separate form. It is associated in the evening with the region of greatest proton energy flux and lies just inside the boundary of stable trapping which explains its more or less circular form about the pole, i.e. it is produced by higher energy particles bouncing from one hemisphere to another and drifting in longitude, electrons to the east and protons to the west, losing energy to the atmosphere by diffusing into "loss cone" pitch angles. It moves equatorward and grows in intensity with the buildup of ring current and is always observed to be conjugate. The region between this form and the discrete aurora on the dayside is filled in with drifting particles after the onset of a storm or the nightside, but this population quickly abates since the particles are outside the trapping boundary defined by the possibility to drift completely around the earth.

Also associated with closed field lines is the pulsating aurora (also conjugate) seen equatorward of the trapping boundary after it moves poleward with the passage of a substorm after magnetic midnight. The acceleration mechanisms for these and "discrete forms" which are the most visible manifestation of the substorm are probably on the field lines occupied by the particles and in some cases quite near the earth.

There is proton precipitation wherever there is electron precipitation (maintaining charge neutrality at least on a large scale) but the greatest energy flux produces the most hydrogen emission which is what is observed and plotted here. The "proton aurora" is broad and diffuse because the incoming protons charge-exchange with atmospheric constitutuents and spend perhaps half the time as hydrogen atoms unconstrained by magnetic field lines. The other half of the time is spent as protons ionizing atmospheric constituents in the same manner as electrons. The "proton aurora moves poleward of the trapping boundary sometime after the poleward expansion of the discrete arcs associated with the substorm.

The "soft" region delineated by black dashed lines is marked with arrows to indicate that the entire precipitation region extends east- and westward from midnight with increases in magnetic activity. This "soft" region is the diffuse background radiation in which the discrete arcs are embedded. The extent of this region is indicated by the extent of 6300A emission due to particle precipitation and the two regions of "soft" precipitation have been documented from satellites, aircraft and the ground (Shepherd, 1979). Although 6300A[OI] emission is created by auroral primaries of all energies, the higher energy particles are generally restricted to discrete forms or regions. We propose here that the 6300A emission not associated with the forms or regions listed above comprises two elongated regions of variable length and width centered on magnetic midnight and noon and is produced by particles of "magnetosheath" energy. The region on the day side has been associated with the magnetospheric cusp or cleft. Associated with this region is a series of discrete arcs appearing (when viewed from above) to emanate from a "radiant point" near magnetic noon. These arcs are produced by higher energy particles than those producing the 6300A region, but the average energy is far more variable than on the nightside. Altitudes range from 100 to 200 km and the intensity and occurrence of the arcs varies directly as the magnitude of the auroral substorm on the nightside. Although they seem to be separate from the nightside arc system (Akasofu, 1980). Discrete arcs in the polar cap are also short-lived arcs of the same spectral type as the dayside arcs. The polar cap arcs occur at very quiet times and disappear with increasing activity. These arcs may be found over the entire polar cap but they seem to be most prevalent in the early morning sector.

Ground-based and satellite observations of the dayside aurora have not clearly identified the relationship between the night- and dayside aurora mainly because all-sky cameras are not sensitive to the diffuse 6300A[OI] emission region and satellite observations do not have adequate time resolution. The establishme of an international "trans-polar" chain of stations during the International Magnetospheric Study allowed the observation of auroral activity in the day and night sectors, simultaneously. Photometer scans of the north-south geomagnetic meridian may be aligned to show the N-S motion of the aurora from a single station as a function of local time. The 6300A data for December 13, 1979, is given in Figure 4 for Longyearbyen, Svalbard (LYR) and Poker Flat, Alaska (PKR). The N-S scans are "stacked" in Universal Time with magnetic midnight and noon displaced by 2 hrs because the two stations are not 180° apart either geographically or geomagnetically. The most striking feature of the simultaneous observations is the simultaneity and similarity of the auroral storm. The slow curve of the maximum through noon is the passage of the station under the oval, but the poleward expansion on the nightside coincides with equatorward expansion on the dayside. Other examples show that the quiet period can occur ± 3 hrs on either side of magnetic noon. Analysis of the geomagnetic disturbance field across the transpolar chain has confirmed these observations and shown evidence for a Harang discontinuity on the dayside (Sandholt, et al. 1980; Scourfield and Nielsen, 1980).

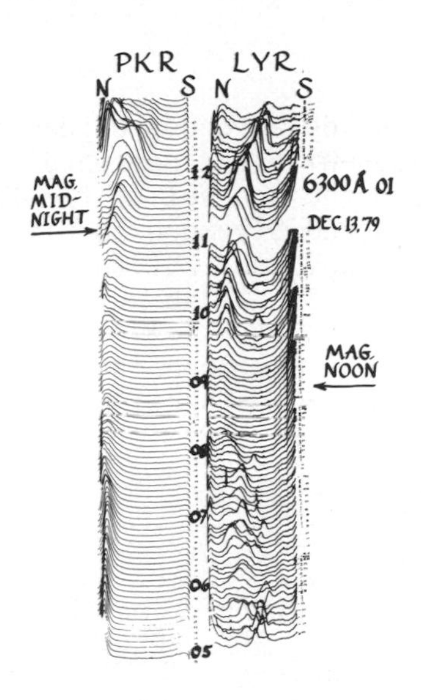

Figure 4. A comparison of simultaneous meridian photometer scans from dayside (LYR) and nightside (PKR) stations. The vertical scale is hours in UT. Each scan represents the slant intensity of 6300A[OI] emission in the N-S meridian of the station.

These results indicate a more intimate connection between the night- and dayside aurora than one would anticipate assuming the separate and direct access of magnetosheath particles to the atmosphere provided by the magnetospheric cleft. There is at least some evidence in satellite data (Burrows, et al. 1976; Formisano and Domingo 1979) and field calculations (Shepherd and Shepherd; 1979) for the existence of another source region for the particles responsible for most of the dayside aurora and for an associated Harang-type discontinuity. Perhaps the actual cleft precipitation is seen only for the short period characterized by low activity near magnetic noon in Figure 4. It is tempting, therefore, to propose a source for the major portion of the dayside aurora in the plasma mantle of the geomagnetic tail. A current system consistent with the observations is shown in Figure 5. This is patterned after a model of the magnetospheric substorm by Yasuhara, et al. (1975). Interruption of the cross-tail current sheet connects the aurora with the MHD generator which is the power source for our television picture of the magnetosphere and the sun. Because the cross-tail current is completed over top of the geomagnetic tail, interruption there

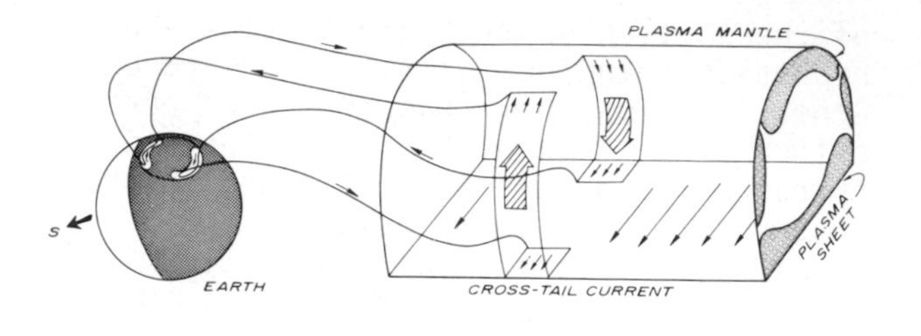

Figure 5. Showing a proposed substorm current system based on a model by Yasuhara, et al, (1975).

for the dayside aurora provides equatorward motion on the dayside simultaneous with the poleward expansion on the nightside. Thus, even though the concept of the TV picture is still valid, it is probably incredibly more complicated and interesting than we at first had imagined. But that is what science and in particular space science is all about.

This work was supported by the National Science Foundation, Atmospheric Science Section, Grants ATM77-24837, ATM77-24838 and ATM78-25249 .

REFERENCES

Akasofu, S.-I., 1980. Private communication.
Akasofu, S.-I., 1979. The Physics Teacher, p. 228.
Akasofu, S.-I., 1977. Physics of Magnetospheric Substorms, Reidel, Dordrecht.
Burrows, J. R., Wilson, M. D. and McDiarmid, I. B., 1976. in Magnetospheric Particles and Fields, McCormac, Ed. Reidel.
Egeland, A. and Omholt, A., 1966. Geophys. Publ., 26, 1.
Elvey, C. T., 1958. Private communication.
Formisano V. and Domingo, V., 1979. Planet. Space Sci., 27, 1979.
McPherron, R. L., 1979. Revs. Geophys. Space Phys., 17, 657.
Sandholt, P. E., Henriksen, K., Deehr, C. S., Sivjee, G. G. and Romick, G. J., 1980. J. Geophys. Res., in press.
Sargent, H. H., 1978. IEEE Vehicular Technology Conference, 490.
Scourfield, M. W. J. and Nielsen, E., 1980. EOS, 61, no. 17, 345. Shepherd, G. G., 1979. Revs. Geophys. Space Phys., 17, 2017.
Shepherd, M. M. and Shepherd, G. G., 1979. in Proc. Magnetospheric Boundary Layers Conf., Alpbach. ESA SP-148.
Simon, P. A., 1979. Solar Physics, 63, 399.
Swift, D. W., 1979. Revs. Geophys. Space Phys., 17, 681.
Yasuhara, F., Kamide, Y. and Akasofu, S.-I., 1975. Planet. Space Sci., 23, 575.

ELECTRIC FIELDS IN THE DAYSIDE AURORAL REGION

T. Stockflet Jørgensen

Danish Meteorological Institute
Lyngbyvej 100, 2100 Copenhagen
Denmark

INTRODUCTION

The dayside auroral oval is located at or near the feet of magnetic field lines in the magnetospheric boundary region separating closed field lines connecting conjugate points on the surface of the earth, and open field lines extending from the polar regions into the magnetospheric tail.

This boundary region, called the cusp, or the cleft, is of particular interest, because it plays a major role for the coupling between the solar wind and the earth's upper atmosphere. The reason is of course, that the magnetic field lines, which here directly connect the magnetosheath with the ionosphere, offer exceptional possibilities for energy transfer from the solar wind to the earth's atmosphere. Such a coupling has been experimentally verified. Thus Svalgaard (1968) and Mansurov (1969) discovered, that the solar wind magnetic field sector structure can be inferred from ground-based magnetic observations at high latitudes, and Heikkila and Winningham (1971) and Frank and Ackerson (1971) found evidence of direct entry of magnetosheath plasma into the dayside polar ionosphere.

Whatever the nature is of the mechanisms coupling the inner and outer plasma regions in the cusp together, the electric fields there are of importance. This is realized in models of auroral electrodynamics, which emphasize the close relationship between electric fields and currents and energization of charged particles.

It is the purpose of this paper to review present knowledge of electric fields perpendicular to the earth's magnetic field

C. S. Deehr and J. A. Holtet (eds.), Exploration of the Polar Upper Atmosphere, 267–280.

in the dayside auroral region at relatively low altitudes, i.e. below about 1000 km.

OBSERVATIONS

Survey of Experiments

Electric fields in the dayside auroral region have been investigated experimentally by means of satellites, rockets, balloons and recently also by incoherent scatter radars.

Symmetrical double electric field probes (Fahleson 1967) were flown on the Injun 5 and Hawkeye 1 satellites, and results have been reported by Cauffman and Gurnett (1971; 1972), Gurnett (1972), Gurnett and Frank (1972; 1973), Kintner et al. (1978), and Saflekos et al. (1979). Observations of ion drift caused by convection electric fields in the dayside auroral oval have been made from the Cosmos 184 and the Atmosphere Explorer C Satellites and reported by Galperin et al. (1974; 1978), Heelis et al. (1976), and Heelis (1980).

Rockets possess the attractive feature, that they can be launched in a desired specific situation, and they can have sufficient range to traverse the latitudinal extent of the auroral oval. Their horizontal velocity can be nearly an order of magnitude lower than satellite velocities, and hence they have the capability of much more detailed measurements of this narrowly confined and very active region.

Releases of barium vapor from rockets in the ionosphere (Haerendel et al. 1967; Föppl et al. 1967) are well suited for observations of plasma drift and equivalent electric fields. In this technique barium is vaporized in the ionosphere usually by means of a chemical reaction involving CuO. The experiment is carried out, when the sun is below the horizon of a ground observer but still illuminates the barium cloud, making it visible and partly ionizing it. A greenish cloud of neutral barium vapor is easily distinguished from the red-violet cloud of Ba^+.

To the knowledge of the author a total of 15 rocket-borne electric field experiments have been made in the dayside auroral region. They are listed in Table 1.

Table 1. Rocket experiments for electric field observations in the dayside auroral region.

Exp.	Date (d/m/y)	Site*	Inv. Lat.	MLT	Method	References
1	15/03/71	F.M.	∿ 80	∿ 11	Double probe	Maynard and Johnstone 1974
2	18/03/71	F.M.	∿ 80	∿ 11	Double probe	Maynard and Johnstone 1974
3	10/12/72	S.S.	76-78	10-11	Barium cloud	Mikkelsen and Jørgensen 1974
4	02/07/74	S.S.	75-77	∿ 10	Double probe	Ungstrup et al. 1975 Petelski et al. 1978
5	08/07/74	S.S.	76-78	∿ 13	Double probe	Olesen et al. 1976 Primdahl et al. 1979
6	17/12/74	S.S.	76-77	∿ 09	Barium jet	Jørgensen et al. 1980
7	18/12/74	S.S.	75-77	∿ 09	Double probe	Jørgensen et al. 1980
8	06/01/75	C.P.	∿ 76	∿ 13	Barium jet	Jeffries et al. 1975 Wescott et al. 1978
9	11/01/75	C.P.	∿ 78	∿ 14	Barium jet	Jeffries et al. 1975
10	11/01/75	S.S.	74-75	∿ 09	Ba.jet D.probe	Jørgensen et al. 1980
11	25/11/75	C.P.	76-81	∿ 13	Ba.jet Ion det.	Walket et al. 1978 Daly and Whalen 1978
12	28/11/75	C.P.	76-78	12-13	Ba.jet Ion det.	Walker et al. 1978 Daly and Whalen 1978
13	22/08/76	S.S.	75-76	∿ 16	Double probe	Fahleson 1977
14	27/08/76	S.S.	75-76	∿ 09	Double probe	Fahleson 1977
15	04/12/77	C.P.	Cusp	∿ 13	Ion det.	Yau and Whalen 1980

* F.M. = Fox Main, S.S. = Søndre Strømfjord, C.P. = Cape Perry

Whereas rockets permit studies of small-scale structures of the electric field, such as the relationship of the field to auroral arcs, balloon-borne electric field probes on the other hand, allow one to sample the field below the ionosphere only and as such, they are insensitive to structures less than about 100 km in extent. However, observations are possible for several days and by launching balloons simultaneous from different locations, it is possible to determine something about the spatial variation. Observations by means of balloon-borne sensors in the dayside auroral region have been made by Mozer et al. (1974), Iversen et al. (1975), and Møhl Madsen et al. (1976).

The ground-based incoherent scatter technique (Evans 1969) depends upon observing by radar the mean motion of the ambient plasma at F-region height. It has the advantage, that a wide range of latitudes and all local times can be covered during the cause of one day. Electric fields at dayside auroral latitudes observed with the Chatanika and Millstone Hill incoherent scatter radars have been reported recently by Evans et al. (1979), Doupnik (1980), Evans et al. (1980), and Foster (1980).

General Results

The observations referred to above have shown, that the plasma convection at dayside auroral latitudes is primarily E-W and comprises an equatorward zone with convection towards noon and a poleward zone with convection away from noon. In a few hours wide zone near noon the plasma flow is generally poleward. The zones of E-W convection are a few degrees of latitude wide near noon, but they increase in width toward evening and morning. This convection pattern is illustrated in Figure 8 of Heelis et al. (1976) and it is consistent with the features of a high-latitude two-cell convection pattern corresponding to the original idea of magnetospheric convection by Axford and Hines (1961).

The results of all the rocket experiments listed in Table 1 are illustrated in the upper part of Figure 1, where average convection velocities measured in the various experiments are shown relative to the location of the cusp defined as a region of low energy ($E > 200$ eV) electron precipitation at an altitude of about 550 km (Craven and Frank 1978). The rocket observations are quite consistent with Gurnett's (1972) statement, that on the dayside of the magnetosphere the electric field reversal is observed to coincide with the equatorward boundary of the polar cusp.

The question of the location and extend of the region in which the noonward convection turns antisunward or poleward is an interesting one for which no simple answer seems available. In Heppner's (1977) models the noonward flow may enter the polar cap region and turn antisunward at all local times between dawn and dusk, whereas in the model by Heelis et al. (1976) antisunward polar cap flow in particular seems to enter through a region only a couple of hours wide and located near local magnetic noon. Reiff et al. (1978) have named this region the "throat".

One difficulty by accepting a narrow throat through which all plasma convect is, that since the plasma is essentially incompressible at ionospheric heights, it is only possible for all of the

ROCKET OBSERVATIONS OF PLASMA FLOW IN THE CUSP REGION

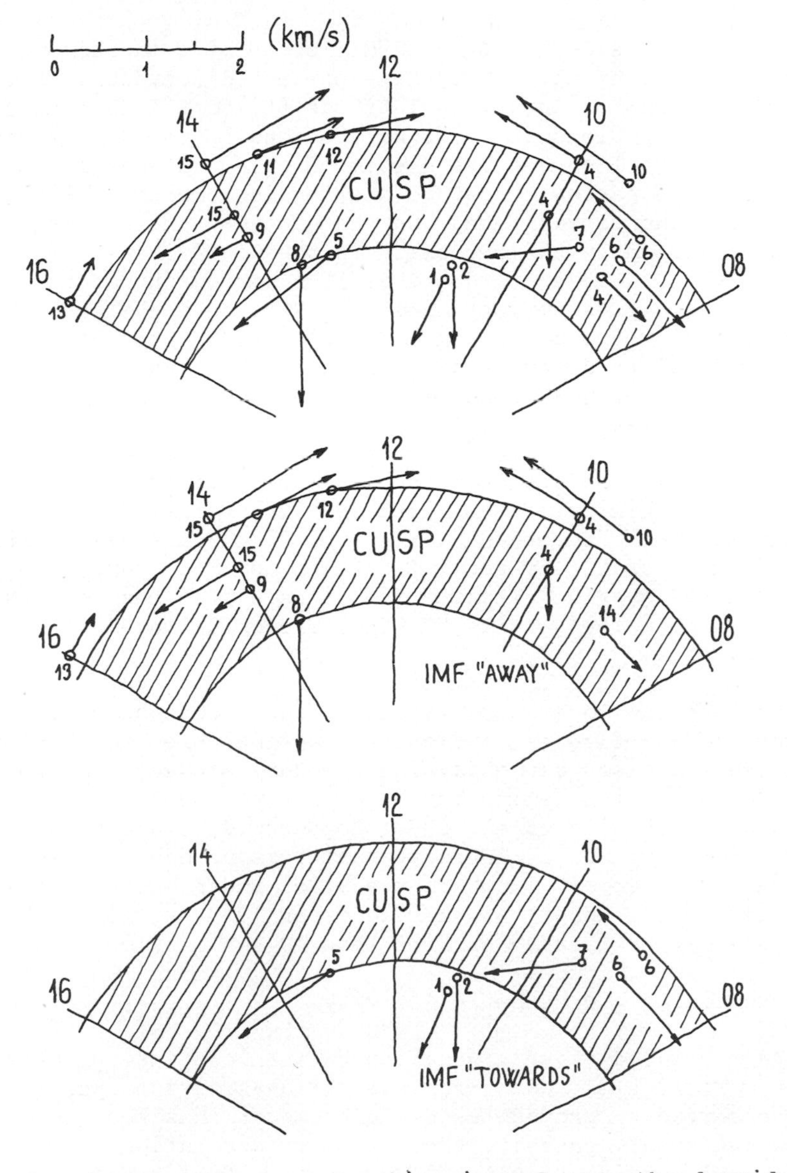

Figure 1. Average plasma convection in and near the dayside cusp observed in the rocket experiments listed in Table 1. Upper part shows results of all experiments. Middle and lower part shows observations when the interplanetary magnetic field was away and towards the sun respectively.

plasma to convect through a narrow throat region if the velocity there is extremely high. Observations indicating that this is the case are hard to find.

Using the recently upgraded Millstone Hill incoherent scatter radar Evans et al. (1980) have presented convection patterns between 60° and 75° invariant latitude (Λ) for 11 moderately quiet 24 hour periods in 1978. All but one of the days exhibited a simple two-cell convective system. Little evidence for the existence of a narrow throat on the dayside through which all plasma must funnel was found. On the contrary the electric fields seemed consistently their weakest in the 08-12 local time sector. Evans et al. (1980) conclude, that if a narrow throat region exists at all times, it may lie at $\Lambda \geq 75^o$, or the electric fields in its vicinity are so random and disordered, that the radar measurements fail to describe the flow properly. Evans et al. (1980) believe both hypothesis can be discounted on the grounds that the observed plasma flow on the dayside usually fits Heppner's (1977) models quite well. The reason for the weakness of the field in the 08-12 local time sector may, however, result in part from the variability of the field in this region. Heppner (1977) notes that it is "turbulent", i.e. the fields are highly irregular with many spatially small fluctuations. Such very irregular high-latitude dayside electric fields have also been observed by Mikkelsen and Jørgensen (1974). They are illustrated in Figure 3 of Kelley et al. (1977).

Whereas the local time extent of the throat or region of poleward plasma flow is difficult to define, the center of this region, which also is the dayside boundary between the two high latitude convection cells, may be easier to locate. Satellite observations (Heelis et al. 1976; Heppner 1977) and rocket observations (Figure 1) indicate that this boundary is in the 10-12 local time sector. Incoherent scatter radar observations (Doupnik 1980; Evans et al. 1980; Foster 1980) suggest, that it sometimes may be located earlier than 10 local time.

Electric Fields and Magnetic Activity

As mentioned above a relation between the interplanetary magnetic field (IMF) and high-latitude dayside ionospheric currents observed by ground-based magnetometers has been shown to exist, and so of necessity there must be a relation also between the IMF and dayside high-latitude electric fields and equivalent plasma convection. The question is how the convection depends on the IMF.

Heppner's (1977) models of high latitude plasma convection are based mainly on satellite electric field probe observations in the dawn, night, and dusk auroral regions as well as in the central polar cap. Not many observations have been available for his modelling of the dayside high-latitude region, so this must be considered as the least reliable part of the models. However, Heppner's (1977) models indicate that the center of the throat is located at noon when IMF-By<0, and that it occurs one or two hours before noon for IMF-By>0.

Heelis et al. (1976) found from Atmosphere Explorer C ion drift data, that gives a uniform sampling around local noon of the dayside high-latitude region, a tendency for the plasma flow within the dayside polar cap to be preferentially directed to either the morning side or the evening side consistent with asymmetries in the polar cap field observed by Heppner (1972; 1977) near the dawn-dusk meridian. Heelis et al. (1976) conclude that this polar cap flow may result from a larger region of rotational flow that is not symmetric about local noon or a displacement of a smaller region to either the morning or the afternoon side, and they state that the latter is most likely.

In a conceptual diagram presented by Fairfield (1977) - see his Figure 13 - the results of Heelis et al. (1976) are used to suggest a possible convection pattern for the northern polar region for conditions of a positive IMF-By component. The fact that the throat - or more precisely the dayside boundary between the two high-latitude convection cells - in Fairfield's (1977) model is located after noon and in Heppner's (1977) model before noon indicates that the problem about how the throat location depends on the IMF-By component is not quite settled.

As illustrated in Figure 1 the rocket experiments have been used in an attempt to look after a throat location dependence on the sign of IMF-By, but no one is found. It may be remarked that the direction of convection observed at the cusp equatorward boundary near 13 local time for IMF "away" conditions, which corresponds to IMF-By>0, does not resemble the flow direction proposed by Fairfield (1977), but quite well the flow direction in Heppner's (1977) model.

That the throat in Fairfield's (1977) and Heppner's (1977) model moves in opposite directions in local time for a given change of IMF-By does not mean, that one of these models is right and the other wrong. They may both be wrong as far as the flow in the throat region is concerned.

A problem with both models is, that they show that the direction of flow near the noon meridian equatorward of the cusp, i.e. in the latitude range $\Lambda \simeq 70^{o}$-75° is IMF-By dependent. It seems difficult to find observational evidence for this.

A model of the dayside high-latitude convection pattern

which in the author's opinion would fit available observations can be described as follows: The center of the throat, which is also the boundary between the two convection cells, is located near the noon meridian, and it does not move with changes in IMF-By. Neither is there any change of the flow pattern equatorward of the cusp as function of the IMF-By. Noonward flowing plasma equatorward of the cusp, which moves into the cusp about an hour and more away from the center of the throat, rotates almost 180° and flows back towards the morning and the evening parts of the polar cap independent of IMF-By, whereas noonward flowing plasma which reaches the central region of the throat, flows into the polar cap and towards the morning or the evening side depending on IMF-By conditions, namely towards morning for IMF-By>0 and towards evening for IMF-By<0 in the northern polar region and oppositely in the southern polar region.

It is known that at least a part of the dayside aurora occurs in the region of noonward plasma flow (Jørgensen et al. 1980). If the so-called auroral midday gap is spatially related to noonward convection, the conclusion by Dandekar (1979) that the role of the IMF-By component on the morphology of the midday auroral gap could not be detected, is not inconsistent with the dayside convection suggested here.

Ground based magnetometer observations (Maezawa 1976) have indicated that at $\Lambda > 80^{\circ}$ the generally occurring antisunward convection may become sunward for IMF-Bz>1 nT, and balloon (Mozer and Gonzales 1973) and satellite (Burke et al. 1979) electric field probe measurements have confirmed that sunward convection occurs for such IMF conditions. These results do not imply a corresponding reversal of convection in the dayside auroral region, and no observations indicating an IMF-Bz effect on convection in this region seem to exist.

Magnetospheric convection has often been inferred from magnetic disturbances observed on ground stations (e.g. Axford and Hines 1961; Maezawa 1976). The idea is, that the direction of the ionospheric electric field $\bar{E}_{\perp}$, and then the direction of convection is known, if the horizontal magnetic disturbance vector on the ground $\Delta\bar{H}$ is due to an overhead ionospheric Hall current.

In the evening to early morning part of the auroral oval comparisons of results of rocket-borne electric field experiments with ground magnetic observations have shown, that the assumption that $\Delta\bar{H}$ is caused by Hall currents is quite good (e.g. Heppner 1972).

Since plasma density height profiles are different in the nightside and the dayside auroral oval, the conductivities will also be different, and so it cannot be taken for granted that the above mentioned relationship between $\bar{E}_{\perp}$ and $\Delta\bar{H}$ in the nightside auroral region also will exist in the dayside auroral region.

In order to investigate this problem, all dayside rocket electric field experiments for which a comparison of $\bar{E}_{\perp}$ and $\Delta\bar{H}$ has been possible have been studied. The results, which are presented in Table 2, show, that $\Delta\bar{H}$ in the dayside auroral region in less than half of the experiments are caused by Hall currents exclusively, and so it seems not advisable to infer direction of convection from ground magnetic observations in this region.

Table 2. Source of ground horizontal magnetic disturbance $\Delta\bar{H}$ during rocket experiments in the dayside auroral region.

Date of exp. (d/m/y)	
15/03/71	Hall current
18/03/71	Neither Hall nor Pedersen current
10/12/72	Hall current
02/07/74	Combination of Hall and Pedersen currents
08/07/74	Hall current
17/12/74	Part time Hall current, part time neither Hall nor Pedersen current
18/12/74	Uncertain due to large and fast variations of $\bar{E}$ and $\Delta\bar{H}$
06/01/75	Neither Hall nor Pedersen current
11/01/75	(Exp. 10) Hall current generally
22/08/76	Hall current
27/08/76	Neither Hall nor Pedersen current

Electric Fields and Aurora

The aurora and electric fields in the ionosphere and magnetosphere seem to be closely related. Very likely a certain electric field structure in space is required in order to have aurora produced.

To the author's knowledge simultaneous observations of electric fields and aurora at dayside high latitude have been made in connection with rocket experiments only. Referring to Table 1 and Figure 1 such simultaneous observations have been reported for experiments 6, 7, 10-12, and 15.

In all these experiments noonward convection was observed, and the general direction of the convection outside auroral arcs was parallel to the arcs. In experiments 6, 7, 10, and 12 auroral

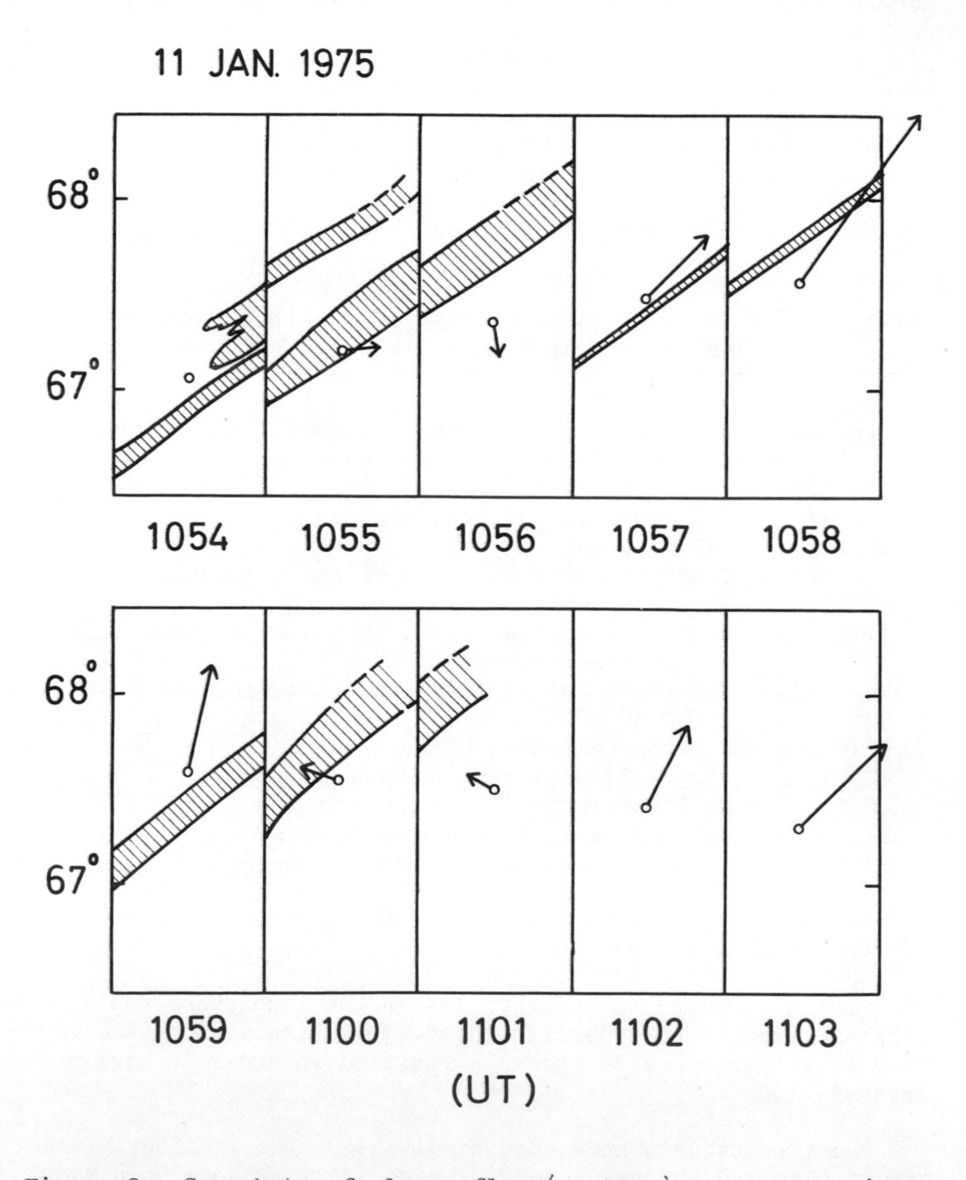

Figure 2. Snapshots of plasma flow (vectors) observed during a rocket experiment in which 3 dayside auroral arcs (hatched) were overflown (Jørgensen et al. 1980). The circles indicate positions of the payload. Speed of flow at 1058 UT was about 0.7 km/s. No measurements were made at 1054 UT. Latitudes indicated are geographic.

arcs occurred equatorward of noonward plasma flow showing, that at least partly the dayside auroral oval is located in the region of noonward convection.

Convection away from noon was observed in experiments 6 and 15, and in the first of these aurora was observed poleward of this kind of flow indicating, that the dayside auroral oval may extend polewards into the region of poleward and antisunward flow.

In experiments 10, 11 and 15 only electric fields were observed inside auroral structures. These experiments show, that the electric field is of the order of 10 mV/m inside the auroral arcs, and that the electric field magnitude there is smaller than outside the arcs. The observations also show, as it appears in Figure 1 of Daly and Whalen (1978) and in Figure 11 of Jørgensen et al. (1980), which is reproduced as Figure 2 here, that there is plasma flow across the auroral arcs, indicating that the arcs are not equipotentials. The observations further indicate, that the plasma flow changes direction during the passage of an auroral arc.

These results are in disagreement with models in which auroral arcs are located in tangential discontinuities or shears in the plasma flow (e.g. Reiff et al. 1978), whereas they indicate, that auroral arcs occur in rotational discontinuities in the plasma flow. Implications of this are, that auroral arcs are not flow-aligned, and that horizontal currents then are allowed along auroral arcs, and so field-aligned currents do not necessarily have to occur everywhere over an arc.

CONCLUSIONS

1. Plasma convection at dayside high latitudes is largely towards noon equatorward of the cusp defined as a region with low energy electron precipitation. At all local times between dawn and dusk, the noonward flow may cross the equatorward cusp boundary and move in a more or less antisunward direction. While the convection equatorward of the cusp is quite symmetric around the noon meridian, and apparently independent on the direction of the IMF, the convection in and poleward of the cusp generally is asymmetric with respect to the noon meridian and dependent on the IMF. This is in particular true for the flow near noon, which in the northern hemisphere is directed towards the dawn side of the polar cap when IMF-By > 0, and towards the dusk side for IMF-By<0. In the southern polar cap the opposite is the case. At $\Lambda \lesssim 80^{\circ}$ the convection seems independent on IMF-Bz.

2. Horizontal magnetic disturbance vectors observed on the ground in the dayside auroral region are sometimes due to iono-

spheric Hall currents only but often they are not. Therefore, one should be cautious about deriving the direction of the perpendicular electric field and the direction of plasma convection from ground based magnetic observations.

3. The dayside auroral oval is at least partly located in the region of convection towards noon, and it may extend polewards into the region of antisunward flow.

4. The electric field magnitude inside dayside auroral arcs has been measured to be about 10 mV/m and to be smaller than the field outside the arcs. Observations have shown plasma flow through dayside auroral arcs, and that the flow changes direction during the passage of the arcs. So, dayside auroral arcs may delineate contours of rotational discontinuities in the convective plasma flow.

REFERENCES

Axford, W.I., and Hines, C.O. 1961, Canad. J. Phys., 39, 1433.
Burke, W.J., Kelley, M.C., Sagalyn, R.C., Smiddy, M., and Lai, S.T. 1979, Geophys. Res. Lett., 6, 21.
Cauffman, D.P., and Gurnett, D.A. 1971, J. Geophys. Res., 76, 6014.
Cauffman, D.P., and Gurnett, D.A. 1972, Space Sci. Rev., 13, 369.
Craven, J.D., and Frank, L.A. 1978, J. Geophys. Res., 83, 2127.
Daly, P.W., and Whalen, B.A. 1978, J. Geophys. Res., 83, 2195.
Dandekar, B.S. 1979, J. Geophys. Res., 84, 4413.
Doupnik, J. 1980, p. 12 in Chapman Conference, High Latitude Electric Fields (Ed. J.C. Foster). Center for Atmospheric and Space Sciences, Utah State University.
Evans, J.V. 1969, Proc. IEEE, 57, 496.
Evans, J.V., Holt, J.M., and Wand, R.H. 1979, J. Geophys. Res. 84, 7059.
Evans, J.V., Holt, J.M., Oliver, W.L., and Wand, R.H. 1980, J. Geophys. Res., 85, 41.
Fahleson, U. 1967, Space Sci. Rev., 7, 238.
Fahleson, U. 1977, p. 73 in SEC-ESA/CUSP 1976 Rocket Programme Progress Report, (Ed. F. Spangslev) Geophysical Papers B-14, Danish Meteorological Institute.
Fairfield, D.H. 1977, Rev. Geophys. Space Phys., 15, 285.
Foster, J.C. 1980, p. 26 in Chapman Conference, High Latitude Electric Fields (Ed. J.C. Foster). Center for Atmospheric and Space Sciences, Utah State University.
Frank, L.A., and Ackerson, K.L. 1971, J. Geophys. Res., 76, 3612.
Föppl, H., Haerendel, G., Haser, L., Loidl, J., Lütjens, P., Lüst, R., Melzner, F., Meyer, B., Neuss, H., and Rieger, E. 1967, Planet. Space Sci., 15, 357.
Galperin, Yu. I., Ponomarev, V.N., and Zosimova, A.G. 1974, Ann. Geophys., 30, 1.

Galperin, Yu.I., Ponomarev, V.N., and Zosimova, A.G., 1978, J. Geophys. Res., 83, 4265.

Gurnett, D.A. 1972, p. 123 in Critical Problems of Magnetospheric Physics (Ed. E.R. Dyer). IUCSTP Secretariat, National Academy of Science, Washington D.C.

Gurnett, D.A., and Frank, L.A. 1972, J. Geophys. Res., 77, 172.

Gurnett, D.A., and Frank, L.A. 1973, J. Geophys. Res., 78, 145.

Haerendel, G., Lüst, R., and Rieger, E. 1967, Planet. Space Sci., 15, 1.

Heelis, R.A., Hanson, W.B., and Burch, J.L. 1967, J. Geophys. Res., 81, 3803.

Heelis, R.A. 1980, p. 27 in Chapman Conference, High Latitude Electric Fields (Ed. J.C. Foster). Center for Atmospheric and Space Sciences, Utah State University.

Heikkila, W.J., and Winningham, J.D. 1971, J. Geophys. Res., 76, 883.

Heppner, J.P. 1972, J. Geophys. Res., 77, 4877.

Heppner, J.P. 1977, J. Geophys. Res., 82, 1115.

Iversen, I.B., D'Angelo, N., and Olesen, J.K. 1975, J. Geophys. Res., 80, 3713.

Jeffries, R.A., Roach, W.H., Hones, E.W., Jr., Wescott, E.M., Stenbaek-Nielsen, H.C., Davis, T.N., and Winningham, J.D. 1975, Geophys. Res. Lett., 2, 285.

Jørgensen, T.S., Mikkelsen, I.S., Lassen, K., Haerendel, G., Rieger, E., Valenzuela, A., Mozer, F.S., Temerin, M., Holbach, B., and Björn, L. 1980, J. Geophys. Res., in press.

Kelley, M.C., Jørgensen, T.S., and Mikkelsen, I.S. 1977, J. Atmos. Terr. Phys., 39, 211

Kintner, P.M., Ackerson, K.L., Gurnett, D.A., and Frank, L.A. 1978, J. Geophys. Res., 83, 163.

Maezawa, K. 1976, J. Geophys. Res., 81, 2289.

Mansurov, S.M. 1969, Geomagn. Aeron., 9, 622.

Maynard, N.C., and Johnstone, A.D. 1974, J. Geophys. Res., 79, 3111.

Mikkelsen, I.S., and Jørgensen, T.S. 1974, EOS Trans. AGU, 55, 70.

Mozer, F.S., Gonzales, W.D., Bogott, F., Kelley, M.C., and Schultz, F. 1974, J. Geophys. Res., 79, 56.

Mozer, F.S., and Gonzales, W.D. 1973, J. Geophys. Res., 78, 6784.

Møhl Madsen, M., Iversen, I.B., and D'Angelo, N. 1976, J. Geophys. Res., 81, 3821.

Olesen, J.K., Primdahl, F., Spangslev, F., Ungstrup, E., Bahnsen, A., Fahleson, U., Fälthammar, C.-G., and Pedersen, A. 1976, Geophys. Res. Lett., 3, 711.

Petelski, E.F., Fahleson, U., and Shawhan, S.D. 1978, J. Geophys. Res., 83, 2489.

Primdahl, F., Walker, J.K., Spangslev, F., Olesen, J.K., Fahleson, U., and Ungstrup, E. 1979, J. Geophys. Res., 84, 6458.

Reiff, P.H., Burch, J.L., and Heelis, R.A. 1978, Geophys. Res. Lett., 5, 391.

Saflekos, N.A., Potemra, T.A., Kintner, P.M., Jr., and Lauer, Green, J. 1979, J. Geophys. Res., 84, 1391.

Svalgaard, L. 1968, Geophysical Papers, R-6, Danish Meteorological Inst.

Ungstrup, E., Bahnsen, A., Olesen, J.K., Primdahl, F., Spangslev, F., Heikkila, W.J., Klumpar, D.M., Winningham, J.D., Fahleson, U., Fälthammar, C.-G., and Pedersen, A. 1975, Geophys. Res. Lett., 2., 345.

Walker, J.K., Daly, P.W., Pongratz, M.B., Stenbaek-Nielsen, H.C., and Whitteker, J.H. 1978, J. Geophys. Res., 83, 5604.

Wescott, E.M., Stenbaek-Nielsen, H.C., Davis, T.N., Jeffries, R.A., and Roach, W.H. 1978, J. Geophys. Res., 83, 1565.

Yau, A.W., and Whalen, B.A. 1980, p. 32 in Chapman Conference, High Latitude Electric Fields, (Ed. J.C. Foster). Center for Atmospheric and Space Sciences, Utah State University.

ELECTRIC FIELDS AND ELECTROSTATIC POTENTIALS IN THE HIGH LATITUDE IONOSPHERE

P.M. Banks*, J.-P. St. Maurice*
R.A. Heelis+, W.B. Hanson+
*Center for Atmospheric and Space Sciences
Utah State University, Logan, Utah 84322
+Department of Space Science
University of Texas at Dallas
Richardson, Texas

ABSTRACT

Ion drift velocity measurements from the Atmosphere Explorer-C satellite have been used to derive curves of electrostatic potential for high latitude segments of the satellite's orbit. The potential curves are shown to be useful in interpreting the character of the global electrostatic potential pattern, particularly with respect to the separation of different convective cells. Results for six orbits are presented with emphasis upon the mid-day auroral region.

1. INTRODUCTION

In this report we describe recent interpretive studies of electric field-driven ionospheric plasma convection data obtained with the Atmosphere Explorer-C (AE-C) satellite. The instruments pertinent to this study include an ion drift meter and an ion retarding potential analyzer, both described in detail elsewhere [1] and which have been used frequently in geophysical studies [2,3]. With these instruments it is possibile to make frequent measurements of the velocity of ionospheric plasma along the orbital path of the satellite. In the case of the AE-C, the orbital inclination of 68.4° insures frequent penetration of the auroral oval and there are, at times, extensive paths which lie within the polar cap. Since information is also gathered on electron and ion temperatures and densities, neutral gas composition and the fluxes of energetic particles, the AE-C data set can be used to reconstruct with reasonable accuracy the physical state of the

C. S. Deehr and J. A. Holtet (eds.), Exploration of the Polar Upper Atmosphere, 281–291.

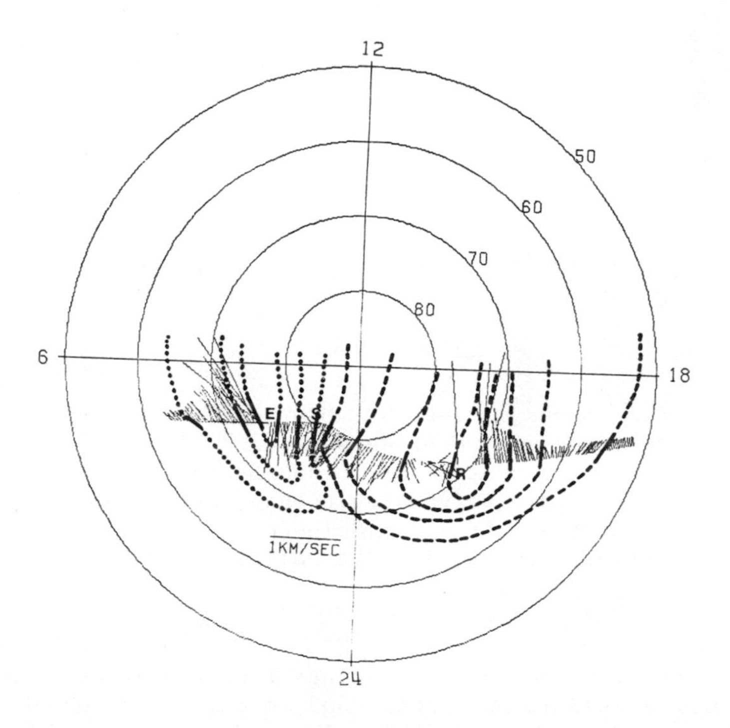

Figure 1. Ion convection velocities obtained from the AE-C satellite for 2 June 1976 at 1747 UT. The coordinates are Invariant Latitude and Magnetic Local Time. The superimposed lines represent contours of electrostatic potential and/or the stream lines of convective plasma drift. The point S is the separatrix between the afternoon and morning convective cells.

auroral and polar ionosphere during any one particular pass.

Interpretation of the measured plasma velocities to determine the general pattern of high latitude plasma convection is often difficult owing to the various swirls and confusing changes in direction frequently seen in the data. We have found, however, that this difficulty can often be overcome by using the plasma velocity data to compute electrostatic potentials which give considerable insight to the nature of the high latitude convection pattern intercepted by the satellite trajectory. With this electrostatic potential method, much of the apparent complexity seen in the drift velocity data can be readily explained in terms of simple plasma convection patterns [4].

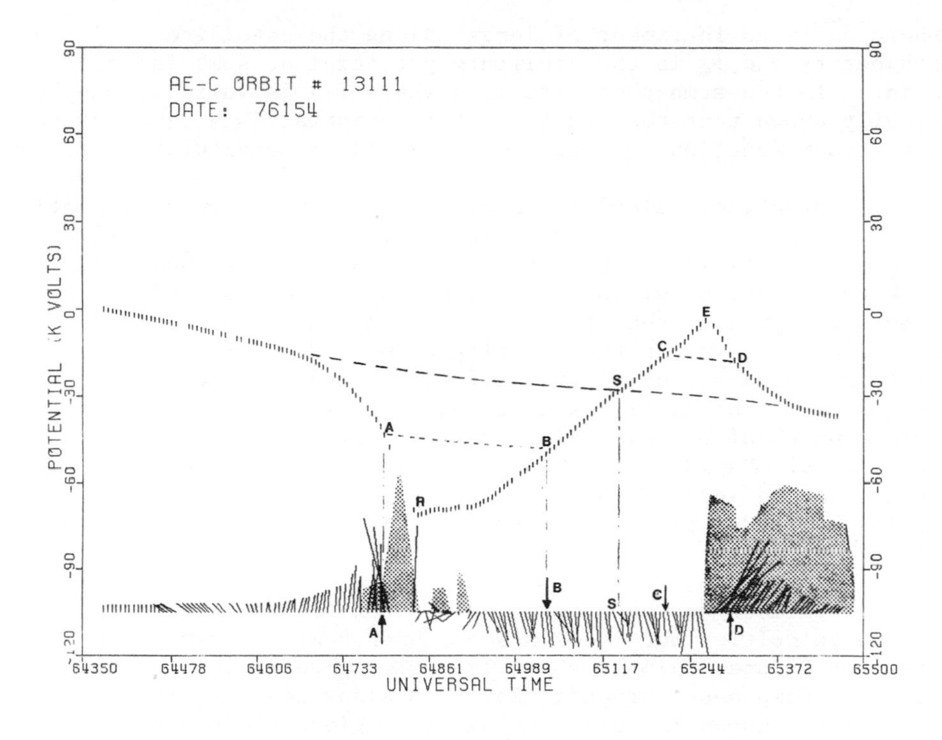

Figure 2. Electrostatic potentials obtained from the data of Figure 1. The symbols are discussed in the text. The dotted area gives the regions where significant auroral particle fluxes were observed in relation to the ion convection velocities.

2. METHOD

Above 160 km altitude, ion-neutral collisions are sufficiently infrequent that the convection electric field, $\bar{E}$, which drives the observed plasma motions can be deduced from the relation

$$\bar{E} = -\bar{V} x \bar{B} \quad (1)$$

where $\bar{B}$ is the magnetic field and $\bar{V}$ is the observed ion drift velocity. In the electrostatic approximation $\bar{E}$ is related to an electrostatic potential, ϕ, through the relation $\bar{E} = -\nabla\phi$. Thus, the potential can be deduced from satellite measurements through the expression

$$\phi(s) = \phi_o + \int \bar{V} x \bar{B} \cdot d\bar{s} \quad (2)$$

where $d\bar{s}$ is an increment of length along the satellite trajectory and ϕ_o is the arbitrary potential at some initial point. In the atmospheric regions where (1) is valid it can be readily shown that the electrostatic potential is equivalent to the stream function for motions of plasma perpendicular to $\bar{B}$.

In practice, calculations of ϕ are begun at low geomagnetic latitudes where ϕ_o is taken as zero. The integration of (1) then gives $\phi(s)$ along the trajectory to higher latitudes and back to midlatitudes, where one expects on the basis of theoretical models that the potential should be approximately the same as it was at the starting point. However, deviations of this ending potential from zero are found to occur frequently, most likely as a consequence of vehicle attitude errors which affect the values deduced for the ion drift velocities. Rapid changes in the global convection pattern could also be influential, but the systematic variation of $\phi(s)$ at low latitudes is too pervasive to allow this to be a common effect.

To illustrate the present method, of Figure 1 shows the plasma velocities obtained in a southern hemisphere nightside pass for 2 June, 1976 at about 1747 UT. Integration of the velocity component perpendicular to $\bar{B}$ along the satellite trajectory (shown in MLT/Λ coordinates) gives the potential curve shown in Figure 2. The dashed line in this figure gives a baseline for excursions of the potential. The systematic variation in potential along s even in the absence of convection is readily apparent and cannot be explained in any simple way other than by noting that small errors in satellite attitude can yield such an effect. To avoid this problem, in the following examples we have attempted to measure the potential variations referenced to a modified baseline which assumes a smooth variation of the background potential with superposed auroral fluctuations.

To see the importance of the curve of electrostatic potential given in Figure 2, we note that within the electrostatic approximation, points of equal electric potential along the satellite trajectory must be connected by one potential contour. Thus, points A and B on the lower panel of Figure 2 are so connected. Similarly, pairs of points interior to A and B are also connected by contours having different potentials, and each of these equipotential contours does not cross any other. On this basis, we see that the negative potential extremum at point R must represent a reversal or pivot point in the plasma flow in the sense that at this point the component of flow perpendicular to the satellite trajectory undergoes a reversal. The extension of the potential contours to the regions away from the satellite path can be accurately

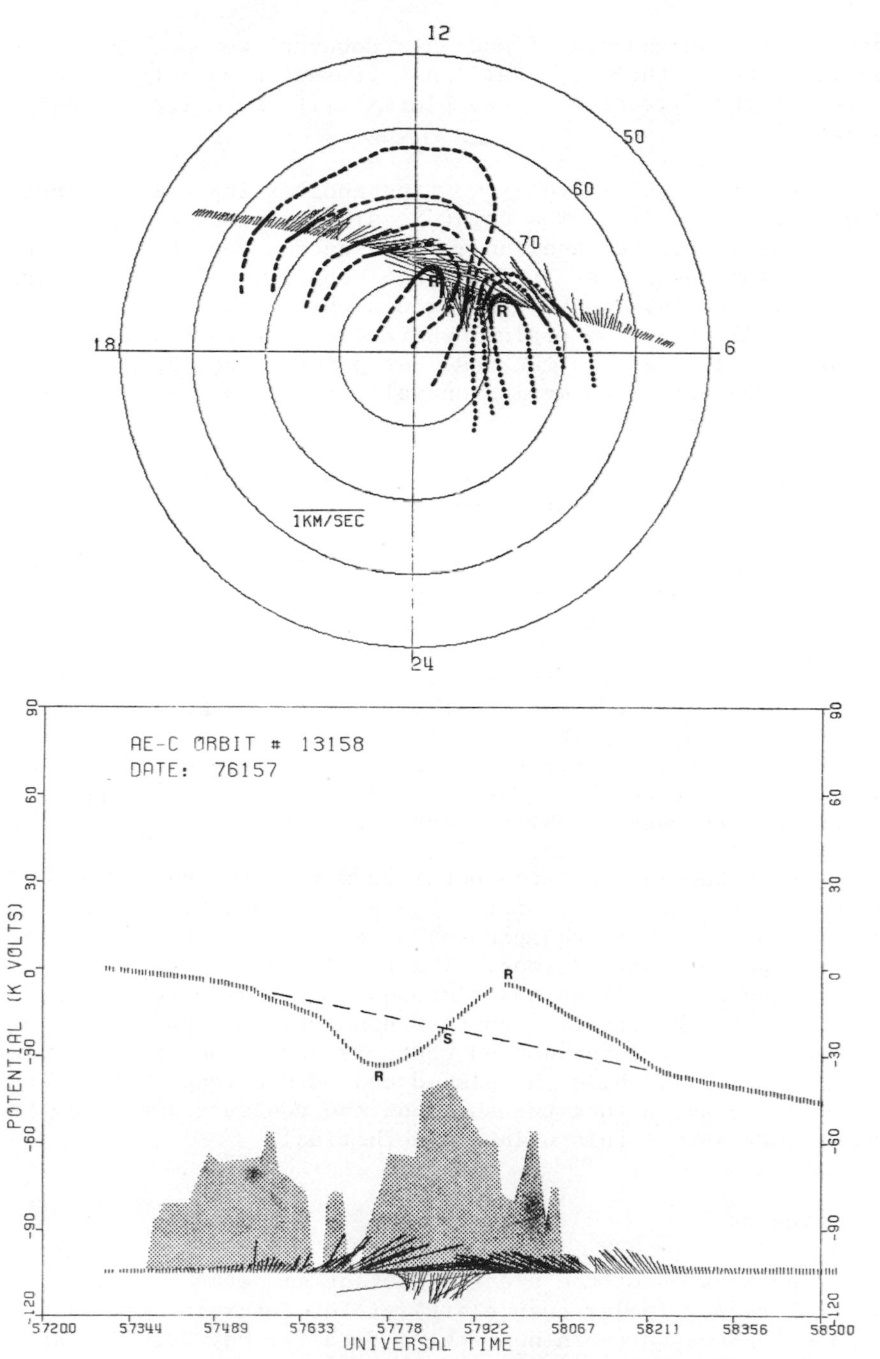

Figure 3. Ion velocity, electrostatic potential and auroral precipitation data for 29 May 1976 at 1555 UT, northern hemisphere.

done only with the aid of models. However, we note that the angle at which the potential lines cross the satellite path is given by the direction of the plasma drift velocity at that point.

Returning to Figure 2, we find another important feature of the potential curve. The point S, since it lies at zero potential (modified baseline), represents a separatrix for the flow in the sense that it marks the boundary point of different convective cells. Thus, the regions where $\phi(s) < 0$ form one closed cell while the regions where $\phi(s) > 0$ form another separate cell. As was the case for points A and B, points C and D are connected by an equipotential line and at point E a flow reversal occurs.

From the foregoing, we see that an important, immediate, result of this method of analysis is that it permits one to identify the separatrix between distinct flow cells; e.g., one sees in Figure 1 that the line of separation between the evening and morning convection cells occurs in an otherwise undistinguished zone of antisunward midnight sector flow. To give a more complete geophysical context for this example the intensity of energetic particle precipitation is indicated by the dotted areas immediately adjacent to the satellite path. It is clear that a substantial portion of the trajectory was within the polar cap. Furthermore, the flow reversals given by points R and E fall at the high latitude edge of the precipitation zones, a circumstance which occurs frequently in the AE-C data.

Using the information contained within the potential curve, we have attempted to reconstruct a global potential pattern consistent both with the present data and current conception of how the pattern should look. The results are shown as the dotted and dashed lines superimposed on the velocity data of Figure 1. Each potential contour connects to proper point on the trajectory and the dotted contours correspond to the evening convective cell, while the dashed contours belong to the morning cell. Note again that the shape of the contours away from the trajectory is, at this point, hypothetical.

3. RESULTS

Application of the electric potential method to the mid-day auroral region (polar cusp/cleft) yields interesting results. Figure 3 gives northern hemisphere data for May 29, 1976 at about 1555 UT. The potential curve (lower panel) shows the presence of a two cell convection structure with a separatrix at S and two reversal points. The soft particle precipitation zone is indicated by the dotted area. The two cell flow shown here is

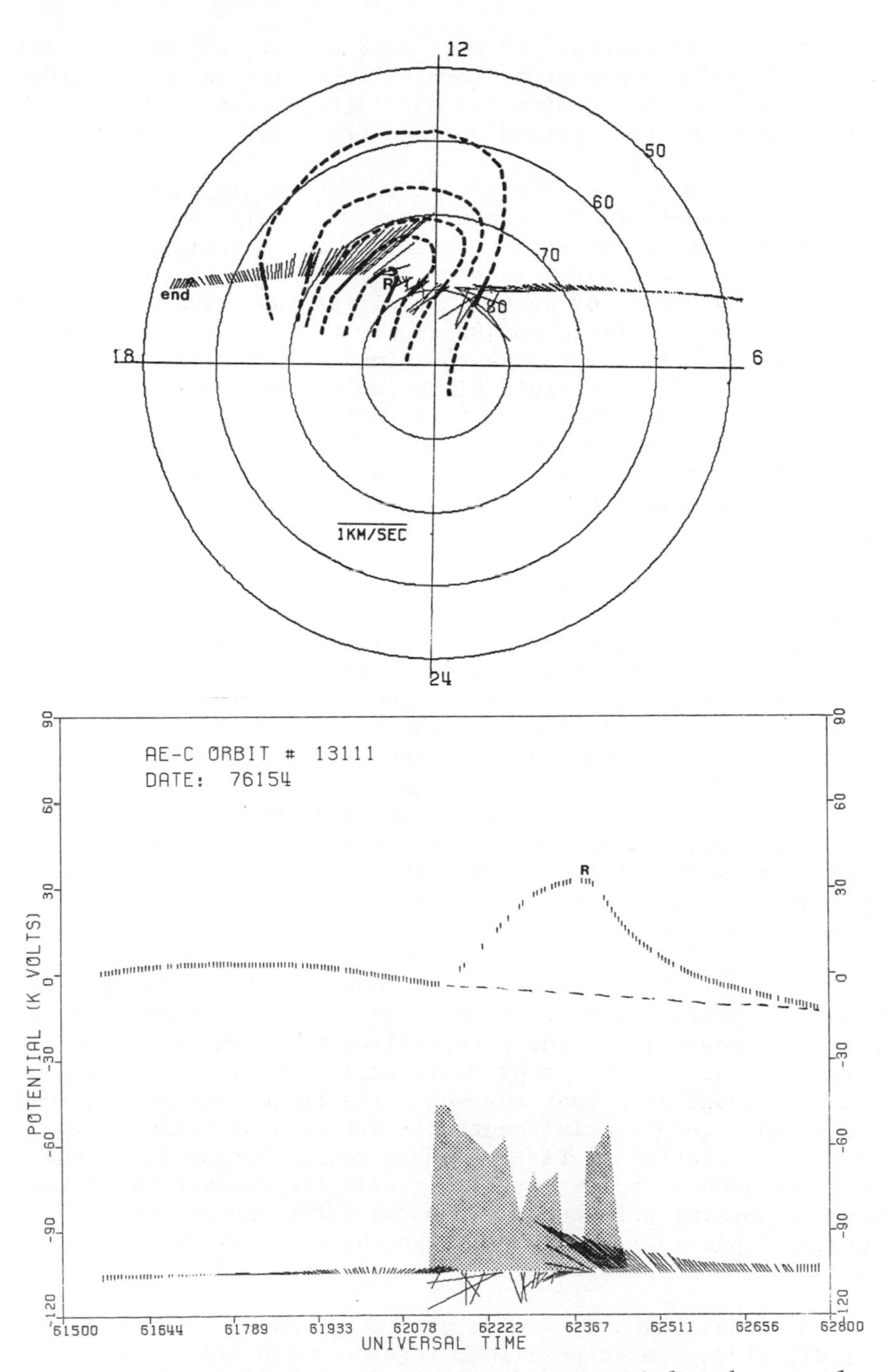

Figure 4. Ion velocity, electrostatic potential and auroral precipitation data for 3 June 1976 at 1714 UT, northern hemisphere

similar to that described by Heelis et al. [2] and the present case is a good example of a "throat" condition in the dayside cleft with both the evening and morning convective cells contributing to the observed antisunward flow.

A more commonly seen plasma flow pattern is shown in Figure 4 where afternoon sector flow completely dominates such that the potential curve is virtually that of a single cell. It is seen that the large zone of afternoon flow occurs somewhat equatorward of auroral precipitation. The antisunward flow, however, occurs in conjunction with precipitation. The potential contours illustrate the single cell character of the observations. In fact, this situation is very common in the AE-C data set and suggests that the morningside convective cell frequently must be considerably displaced towards 06 to 10 MLT and that continuity of the plasma flow must occur across the auroral precipitation zone in this local time sector.

The peculiarity of this situation is emphasized by noting that the data shown in Figure 4 represent mid-day northern hemisphere results obtained 40 minutes prior to the midnight sector southern hemisphere data of Figures 1 and 2. Comparison of the results in the two hemisphere indicates that while there is nothing out of the ordinary in the night-time convective cells, the dayside morning cell was completely missed, implying that it occupied a small area which did not extent to the AE-C orbit path (78° Λ). Of course, one can always admit the possibility of grossly non-symmetric convection patterns between the polar caps, but it seems more likely that these observations simply demonstrate the geographical changes which are associated with the morning sector convective cell for different inter-planetary conditions.

The power of the present technique is well demonstrated through interpretation of the complex flow velocities seen in Figure 5. Analysis of the potential curve shown in Figure 5, indicates that a single convective cell is present with a bean-like shape such that the satellite trajectory cuts across some of the equipotential contours four separate times. The cleft precipitation is first seen at noon, then again at the late afternoon velocity reversal. From the present analysis, there is nothing particularly unusual about the afternoon pattern of plasma flow, even though the velocity data themselves give a tortuous appearance.

The final example, a noon pass on 30 March 1977 at about 0211 UT, gives a most puzzling, regular potential variation. The velocity data of Figure 6 show a strong, sunward flow at noon (MLT flanked first by antisunward flow, then sunward flow. The potential curve indicates there is a superposition of a

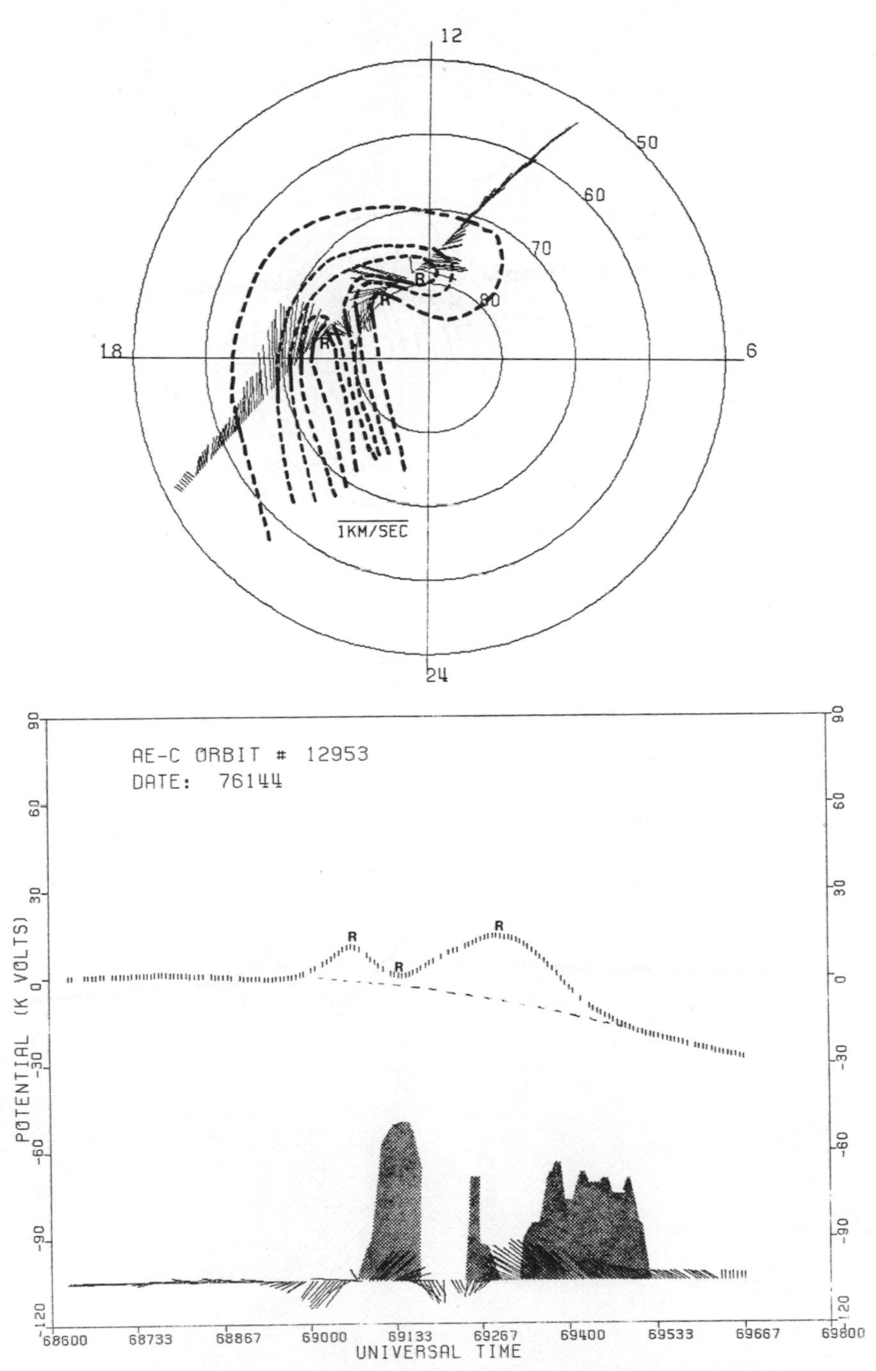

Figure 5. Ion velocity and potential data for 24 May 1976 at 1903 UT, northern hemisphere.

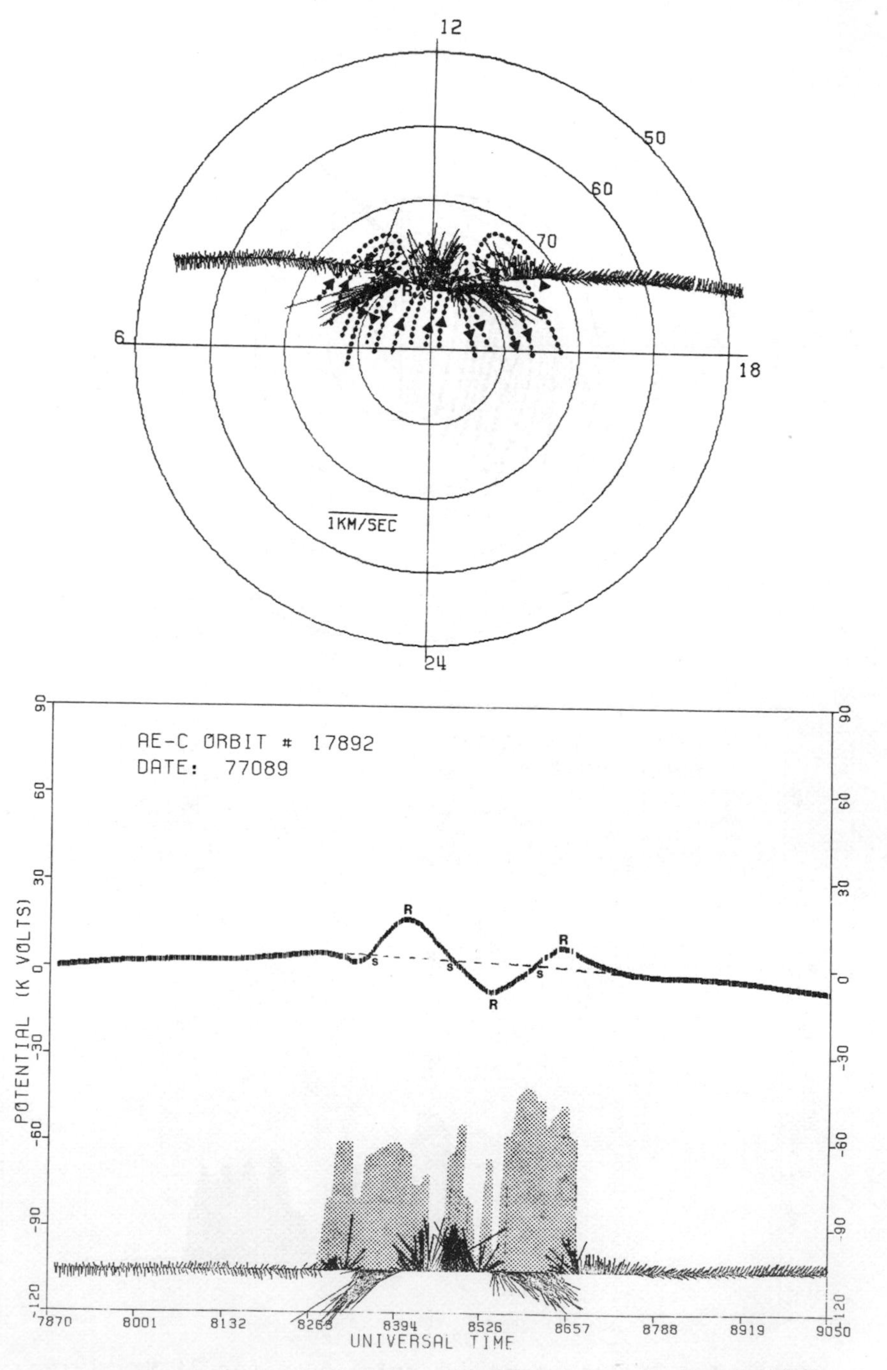

Figure 6. Ion velocity and potential data for 30 March 1977 at 0211 UT, southern hemisphere. Satellite nutation is seen in the velocity data.

normal throat-like flow plus the novel feature of two vortex-like flows centered at about 80° which yield the distinctive sunward flow between them. This phenomenon has been reported previously [5] in connection with a strong northerly component of the IMF. Two subsequent AE-C satellite passes demonstrated that this feature weakened with time.

4. CONCLUSIONS

Electrostatic potentials computed from satellite ion drift velocities provide an important new means for studying high latitude plasma convection. When applied to studies of the cleft region, the results indicate the frequent dominance of the afternoon convective cell. In one case, a peculiar pattern of sunward polar cap flow has been seen. Applications of this method to other local time sectors are underway.

ACKNOWLEDGEMENTS

This work was supported, in part, through NASA grant NSG-7289 and NSG-5215.

REFERENCES

(1) Hanson, W.B., D.R. Zuccaro, C.R. Lippincott and S. Sanatani, Radio Sci., 8, 333, 1973.

(2) Heelis, R.A., W.B. Hanson, and J.L. Burch, J. Geophys. Res., 81, 3803.

(3) Spiro, R.W., R.A. Heelis, and W.B. Hanson, J. Geophys. Res., 83, 4255.

(4) Banks, P.M., J.-P. St-Maurice, R.A. Hellis, W.B. Hanson, Conference Proceedings, The High Latitude Electric Fields Chapman Conference, Yosemite, California, January 30 - February 2, 1980, p. 28.

(5) Burke, W.J., M.C. Kelley, R.C. Sagalyn, M. Smiddy and S.T. Lai, Geophys. Res. Letts., 6, 21, 1979.

SPATIAL VARIATIONS OF IONOSPHERIC ELECTRIC FIELDS AT HIGH LATITUDES ON MAGNETIC QUIET DAYS

Zi Minyun* and E. Nielsen

Max-Planck-Institut für Aeronomie
3411 Katlenburg-Lindau 3, F.R.G.

* Perm. Address: University of Beijing, Beijing, P.R.C.

ABSTRACT The Scandinavian Twin Auroral Radar Experiment, STARE, has been used to determine the steady state spatial distribution of ionospheric electric fields as a function of solar local time in the invariant latitude interval from $\sim 65°$ to $70°$ north. Several magnetically quiet days ($kp < 3$) have been averaged together in order to minimize the effects of transient phenomena. We have compared the result of this analysis with several model predictions published in the literature, and found the gratifying result, that the more realistic the models became the more features of the observed fields could be reproduced. We suggest that the model calculations should be redone, and the comparison with STARE data used to determine the optimum values of the parameters that govern the electric potential at high latitudes.

1. INTRODUCTION

It is well known that the plasma flow in the ionosphere of the auroral zone is associated with magnetospheric convection (Axford, 1961; Dungey, 1961). Results of model calculations predicting the plasma flow and electric potential distribution at high latitudes (Fejer, 1964; Vasyliunas, 1970; Wolf, 1975; Kamide and Matsushita, 1979) have in the past been compared with all sky camera, magnetometer, rocket, satellite and incoherent radar measurements (Heppner, 1977; Evans, et al., 1980). Such comparisons between theory and observations are important, as they represent a test of how good the assumptions about the height integrated conductivities and the field-aligned currents are that go into the model calculation.

C. S. Deehr and J. A. Holtet (eds.), Exploration of the Polar Upper Atmosphere, 293–304.

A few years ago, a new radar system named STARE (Scandinavian Twin Auroral Radar Experiment) was put into operation. For a detailed description of this system see (Greenwald, et al., (1978)). The STARE measurements are based on observations of electron drift in the E-layer through detection of the phase velocity of plasma instabilities associated with the drifts (Sudan, et al., 1973). Because the electrons to a good approximation are $\underline{E} \times \underline{B}$ drifting in the E-layer the E-field can be calculated when the drift velocity is known. Comparison between electric fields derived from the STARE data and in-situ measurements indicated that STARE measurements yield a reasonable approximation of the ionospheric E-field (Cahill, et al., 1978).

STARE allowed for the first time the plasma flows and the associated electric fields at high latitudes to be observed on a continuous basis over a large area (400 x 400 km^2), with good temporal (20s) and spatial (20 x 20 km^2) resolution. The STARE field of view extends about 10° in longitude and 5° in latitude. As the earth rotates during a 24 hour interval, will the field of view have traced out a circular area, covering an angular interval of 360° in longitude ($\sim$ 24 hour) and about 5° in latitude.

In this paper we will determined the observed electric field pattern within this area and compare it with auroral model calculations.

2. STATISTICAL ANALYSIS

Since we are primarily interested in the gross features of the high latitude electric potential distribution, on magnetic quiet days we will therefore use data which have been averaged both in time and space for auroral days to minimize the effects of small scale spatial and temporal effects. One must however be cautious before analyzing average data values to insure that the averages are meaningful. Whether or not this is the case depends on the statistical distribution of the data. Thus before calculating average values, we will first investigate the statistical distribution of the observed velocities.

A first investigation showed that on magnetic quiet days within the STARE field of view the gradient of the electric field in the east-west direction is much smaller than in the north-south direction. We therefore neglect the east-west gradient or rather assume it to be time stationary so that it would be measured as the STARE field of view rotated with the earth. We have used STARE data from the latitudinal interval from 67.6° N to 72.6° N averaged over the longitudinal interval from 18° E to 20° E. This latitudinal region was subdivided into 11 smaller partially overlapping intervals each 0.8° wide. Electric field values measured within each of these latitude intervals during half hour time interval (7.5° in longitude) for 5 - 10 days were used to determine the statistical distribution representing the conditions in that interval during that time period.

Some statistical distributions of drift velocities are shown in Figures 1a and 1b. Even without any statistical tests it is clear that both the N-S and E-W velocity component, which are associated with the E-W and N-S component of E-field respectively, are well represented by a Gaussian distribution.

We have investigated many samples. The results are generally very similar to the ones shown. The standard deviation of the east-west velocity is typically, σ = 15 - 140 m/s, and the north-south velocity, σ = 25 - 90 m/s. The standard deviation of the mean is in most cases between 2 and 10 m/s. According to this investigation, one can use the average values within the same time and latitude interval mentioned above to describe the average electron drift and E-field structures.

3. OBSERVATIONS

The selected time intervals were characterized by low to moderate magnetic activity (k_p < 3). Radar aurora in the electrojet regions occur only if the electric fields exceed a threshold of 15-20 mV/m. This seems hardly to happen when K_p < 1. So in fact, our results are typical for the condition 1 < k_p < 3. The results are plotted in Figure 2 in geomagnetic coordinate. The directions

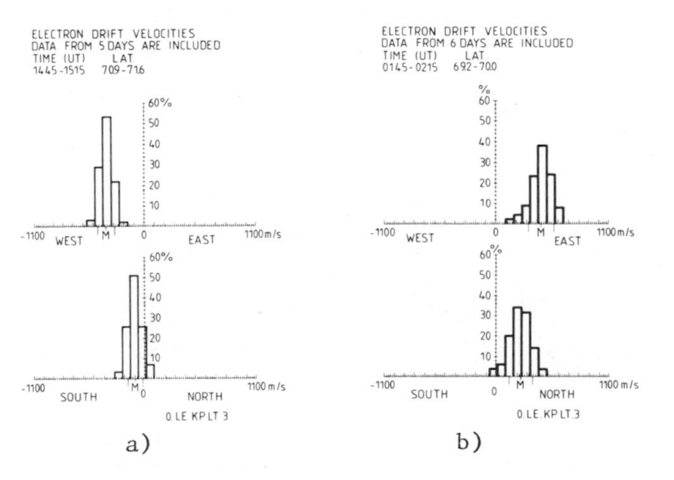

Fig. 1. Distribution of drift velocities observed with STARE, (a) 1445-1515 (UT), 70.8°-71.6° (Geogra. lat) including data from 5 days. (b) 0145-0215 (UT), 69.2°-70.0° (Geogra. lat) including data from 6 days.

of drift velocities (corresponding approximately to the direction of the electric equipotential lines) are shown in Figure 2a. Figure 2b displays the observed electric fields.

Regions displaying no data indicate that the electric field was below the threshold value. The electric fields in the STARE field of view typically exceed the threshold between 1600 and 0430 MLT. Sometimes, usually in the high latitude part of the STARE field of view we also measure fields between 1400 and 1600 MLT or between 0430 and 0600 MLT. But we never observed radar aurora between 0600 and 1400 MLT. When backscatter between 0430 and 0600 MLT, was observed the local k-index in Tromsö (near the center of the field of view) was > 4, although $k_p < 3$. The flow pattern at such times is probably not representative of a steady state and observations following 0430 MLT have not been used.

From 1600 to 2100 are the E-fields nearly northward pointing with a small eastward component early on and a small westward component later. After 0100 MLT are the E-fields predominantly southward, with corresponding eastward electron drifts (see Figure 2b). During times when the Harang discontinuity is in the field of view the E-field structure is very changeable. The standard deviation of the velocities can be quite large during those times. However, one can still recognize the general structure of the Harang discontinuity in Figure 2b.

Both during the afternoon to evening and the morning to dawn sector are the magnitudes of the E-field and its north-south component are always larger at higher than at lower latitudes. Before

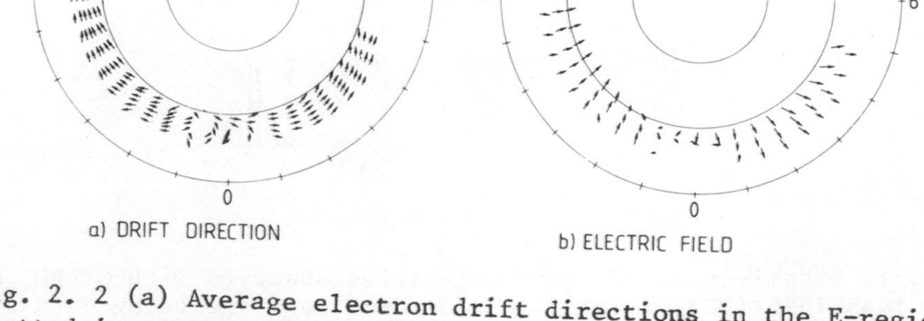

Fig. 2. 2 (a) Average electron drift directions in the E-region plotted in geomagnetic latitude versus MLT, $1 < K_p < 3$. 2 (b) The same data as in (a) transformed into electric fields.

dusk is the magnitude of the E-field increasing gradually with time and reach its maximum at 1800 MLT. Then it starts to decrease slowly at lower latitude, while it stays nearly constant at the higher latitudes. After midnight, is the magnitude of the E-field increasing from 0100 to 0230 MLT. The field is fairly constant between 0230 - 0400 MLT except at lower latitudes where it is decreasing with time. Between 0200 and 0400 MLT can the magnitudes of the E-fields at the higher latitudes be even larger than during dusk. However, the fields decrease steeply with time and can not be traced after 0430 MLT.

4. THEORETICAL MODELS

As summarized by many authors (e.g. Vasyliunas, 1970; Kamide and Matsushita, 1979) are the E-fields in the high latitude ionosphere part of the coupling between the magnetosphere and the ionosphere. This coupling can be expressed by a closed self-consistent chain of equations. In general, are the E-fields (or the associated equipotential contours) in the ionosphere presented as a result of the model calculations under various assumptions of the driving E-field and of the spatial variations of field-aligned currents and ionospheric conductivities. Since we have measured the average E-field structure directly, we will take it as a start point and modify the assumptions made in the models until a reasonable fit is achieved.

The equipotential contours in several of the following figures are plotted in invariant latitude. The latitude range covered by STARE in those coordinates is $\sim 65° - 70°$ N.

4.1 Comparisons with Vasyliunas models

We begin the comparison with Vasyliunas models (Vasyliunas, 1970, 1972) because they are relatively simple and easy to understand on the one hand, and consider on the other hand also the main causes involved. Three groups of models were considered: (1) the field-aligned currents (J_{II}) are flowing only on the polar cap boundary ($\lambda = 72°$). The height integrated Hall (Σ_p) and Pedersen (Σ_p) conductivities are constant with a ratio of 3.5. (2) J_{II} are flowing on the polar cap boundary. Σ_H and Σ_p are increased by some factor inside the auroral zone. Σ_H/Σ_p is kept constant at 3.5. (3) The effects of the ring current are taken into account, so that J_{II} now flow between the polar cap boundary and λ_{RC}, (Polar cap bounday = outer edge of the ring current, and λ_{RC} = the invariant latitude of the inner edge of the ring current). The intensity of J_{II} varies with Σ^*, a ring current parameter which is proportional to the proton number density along the path of the field-aligned current. Σ_H and Σ_p are constant with a ratio of = 3.5.

In all these models are the driving electric field, which represent the effect of the interaction of the solar wind and the geomagnetic field, introduced as a boundary condition for the electric field potential, ϕ, at the polar cap boundary:

$$\phi\ (\lambda = 72°, \varphi) = \phi_o \sin \varphi \tag{1}$$

where λ = invariant latitude, φ = solar local time. ϕ_o is an undetermined constant, essentially a scaling factor. This boundary condition is equivalent to postulating (as done above) field-aligned currents flowing only at the polar cap boundary directed into the dawn hemisphere and out of the dusk hemisphere.

Model 1 shown in Figure 3a represents a basic two cell convection pattern. It is clearly too simple to reproduce our measurements. The main disagreement is that while the observed east-west component of the E-field is generally small compared to the north-south component, is that only the case in this model around 1800 and 0600 MLT.

Model 2 is illustrated in Figure 3b. Σ_H and Σ_p are assumed to be enhanced by a factor of $\sim$ 10 in the auroral zone. This model may be more realistic than model 1, since the height integrated conductivities are known to be enhanced in the auroral zone. But compared with STARE measurements is the model predicting a too large east-west component of the electric field prior to 1800 MLT.

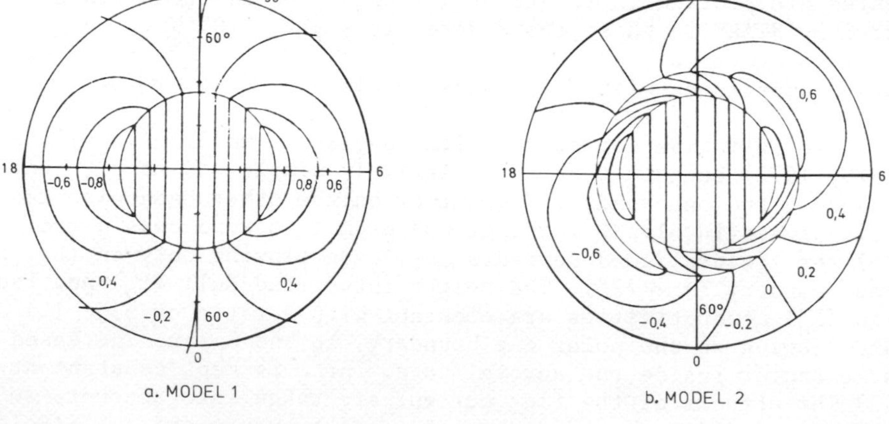

Fig. 3. (a): Calculated equipotential contour in the ionosphere for a simple model with homogenous conductivity. Replotted from (6). By values in unit of Ø. (b): Same as (a) but including effects of enhanced conductitivity between λ = 65° and 72°. The three circle are λ = 5o, 67, 72°, respectively. (Replotted from Vasyliunas, (1970).

Later is the density of the equipotential contours decreasing, so that the electric field magnitude becomes too small.

The electric potential configuration predicted by Model 3, for parameter values $\lambda_{RC} = 64°$, $\Sigma_H/\Sigma_p = 3.5$ and $\Sigma^*/\Sigma_H = 2.9$, turns out to be quite similar to the STARE measurement. We will therefore compare this model quantitatively with the observations.

The theoretical electric potential ϕ has the form of a sinus function of the local time (φ) and with a latitude dependent amplitude, $A(\lambda)$, and phase, $\delta(\lambda)$,

$$\frac{\phi(\lambda, \varphi)}{\phi_o} = A(\lambda) \sin(\varphi - \delta(\lambda)) \tag{2}$$

Calculated values of $A(\lambda)$ and $\delta(\lambda)$ were given (Vasyliunas, private communication) in a grid of points covering the surface of the Earth. We have used linear interpolation among these values to calculate the electric field using the relation

$$\bar{E} = -\nabla\phi \tag{3}$$

The model values and the measurements are comparable if we assume that ϕ_o = 20 kV for the afternoon to evening hours and ϕ_o = 30 kV for the morning hours. These values correspond to potential drops across the polar cap of 40 kV and 60 kV, respectively.

Figures 4 and 5 show the comparison between the magnitude of the E-field measured with STARE (thick line) and the model calculated values (dot-dashed lines). Figure 4 show the comparison of the time between 1600 and 0400 MLT (except the discontinuity time) at different geomagnetic latitudes. Figure 5 display the latitude variations at different MLT.

The good agreement between the model and the STARE measurements is striking. The time variation has the same tendency in the two sets of data so are the latitude variations. There are also some disagreements. Before dusk the model values are always larger than the measurements while after dusk the measurements are the largest. Similarly, before 0300 MLT the model values are too small compared to the measurements while they later become too large.

We suggest that the reason for these discrepancies is that the location of the polar cap boundary is assumed in the model to be at a fixed latitude (72°), while it is well known that this boundary is located at higher latitudes between dawn and dusk, than between dusk and dawn (Vampola, 1973). But, as suggested by Vasyliunas (private discussions), one can use the model cal-

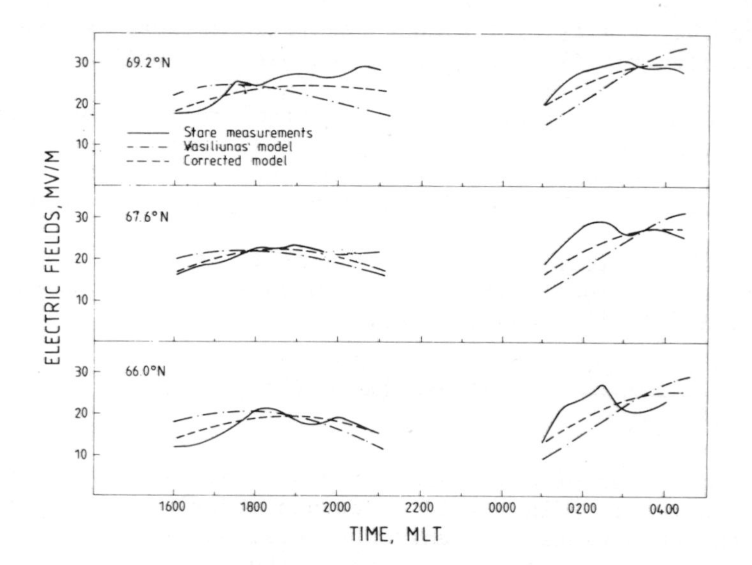

Fig. 4. Comparison between the time variations of the electric field magnitudes observed with STARE (thick lines, with that of the Vasyliunas model (dot-to-dash-line) and that of the corrected model (dashed line). The comparison is made at different geomagnetic latitudes.

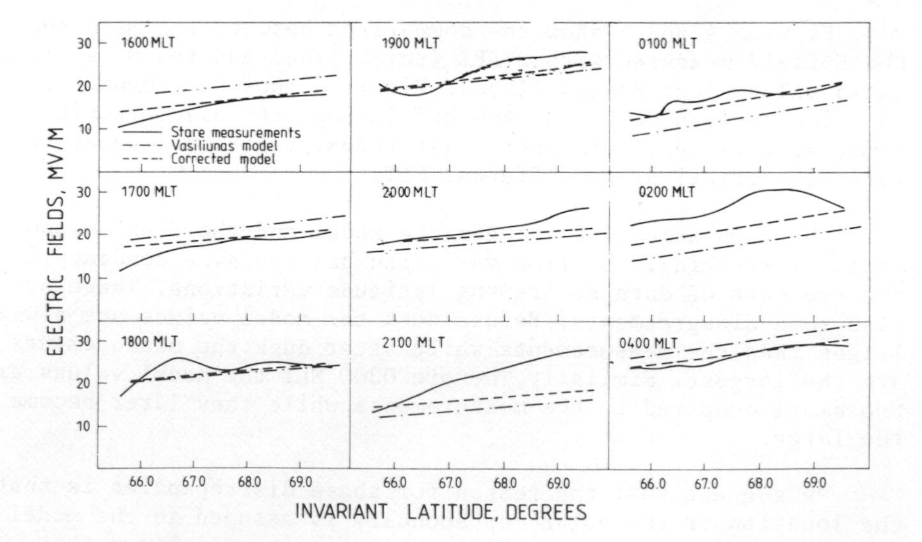

Fig. 5. Comparison of the latitude variation of STARE measured electric field magnitudes at different MLT.

culations the following way: Consider $\phi(\lambda)$ in equation 2, not as the potential value at the latitude λ, but as the potential value at the latitude which is $\Delta\lambda = 72° - \lambda$ degree lower than the actual location of the polar cap boundary. Then the potential distribution can be written as

$$\phi'(\lambda) = \phi(72° - (\lambda'_o(\varphi) - \lambda)) \qquad (4)$$

where ϕ' is the potential value at latitude λ, and $\lambda'_o(\varphi)$ is the latitude of the polar boundary at time, φ (when $\lambda'_o(\varphi)$ then $\phi'(\lambda) = \phi(\lambda)$). Determining $\lambda'_o(\varphi)$ using (22), we found the results which are drawn into Figures 4 and 5 (dashed lines for the corrected model).

Using this corrected model the agreement is much improved. Significant disagreements occur only at the highest latitude edge between 2000 and 2100 MLT (near the Harang discontinuity) and around 0200 MLT where the calculated values are too small. After 0400 MLT where model values are still increasing to reach another maximum near dawn is the measured field rapidly decreasing. After 0500 the fields are too weak to be observed with STARE.

In this model the effects of the ring current to the ionospheric E-field is mathematically equivalent to the effects of an enhanced Hall conductivity in that region of the ionosphere magnetically connected to the ring current (Vasyliunas, 1972). Thus the measured feature of the E-fields can also be explained by an enhanced Hall conductivity region. Whether the ring current or the enhanced Hall conductivity is the most important one can not be answered on the basis of our data alone. Possibly a combination of the two effects would have to be considered.

4.2 Comparisons with Kamide and Matsushita models

The STARE fields observed around dawn are not in agreement with the simple models so far considered. A resolution to this phenomena can be illustrated by considering one of the models recently published by Kamide and Matsushita (1979), which is realistic in considering day-night spatial changes in ionospheric conductivities in addition to including enhanced conductivities in the auroral zone, and pairs of field-aligned current sheets flowing both into and out of the ionosphere during evening and morning hours. The spatial distribution of field-aligned currents used in the model calculations is shown in Figure 6a. The poleward currents (region 1 in Figure 6a) are flowing into the morning half and out from the evening half of the auroral ionosphere. The equatorward currents (region 2 in Figure 6a) are flowing conversely. These are two current density maxima at 0600 and 1800 MLT. The peak intensity of the poleward field-aligned current is twice as large as that of the equatorward current, and is assumed to be $2.0 \cdot 10^{-7}$ A/m².

Before comparing this model with STARE measurement we will attempt to make corrections for the spatial distribution of field-aligned current used in the model, since the observed pattern of field-aligned currents (Iijima and Potemra, 1976), is Figure 6b. In doing so we will make use of the similar idea, which we already have used in comparing our observations with the Vasyliunas model by arguing that it is the relative position with respect to the double field-aligned current sheets and the conductivity distribution that is important. This procedure amount essentially to a coordinate transformation.

We determine the center position, lo, of the observed field-aligned currents in region 1 at different local times and use that as the reference position since the region 1 currents are always more intense than currents in region 2. Then we calculate the latitude difference between the lo and the high latitude boundary of the STARE field of view (Δl). At last we can deduce the relative position of the STARE measurement in Kamide and Matsushita's model by placing the high latitude boundary of the STARE measurement at 67.5+Δl. The latitude range covered by the STARE measurements are shown in Figures 6a and b with dotted lines.

The computed polar cap potential (Kamide and Matsushita, 1979) is shown in Figure 7. The transformed area swept out by STARE is located within the dashed lines in the figure. The model indicates that the E-fields are rather strong at dawn, between 0200 and 0400 MLT before they rapidly decrease and get very weak after 0500. This feature, which is clearly present in the STARE measurements, was not part of the other models discussed. Notice that the maxi-

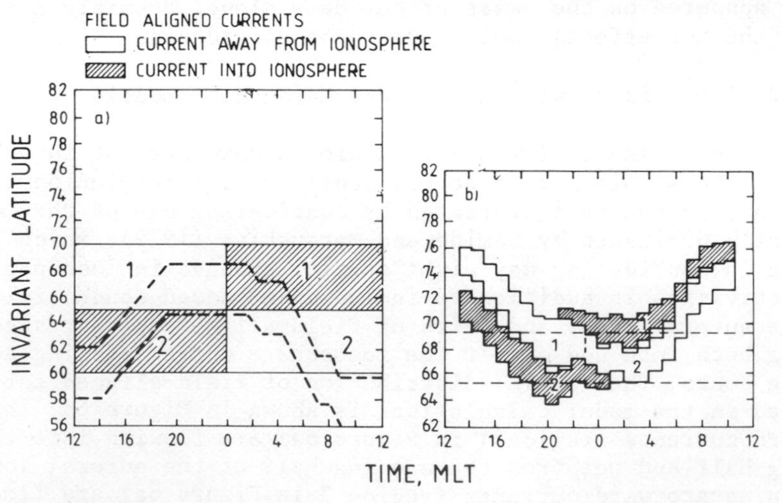

Fig. 6. (a). The location of the field-aligned currents in the model of Kamide and Matsushita (1979). (b) the location of the field-aligned currents observed with Triad 2 < k_p < 4, replotted from Iijima and Potemra (1976). The dotted lines show the area of the STARE observation.

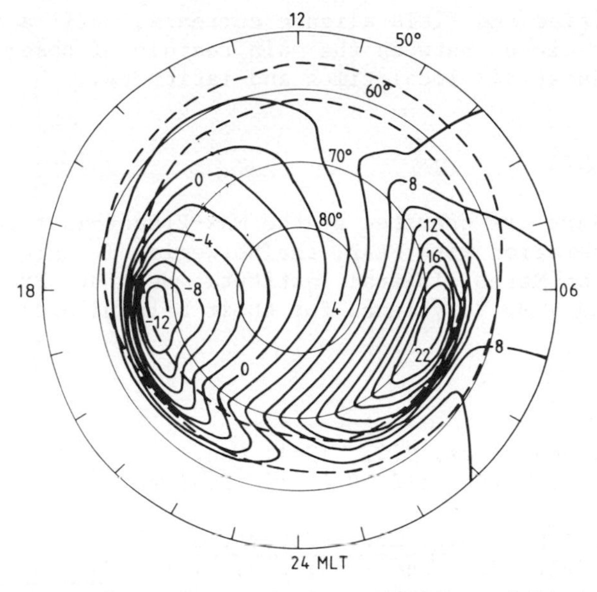

Fig. 7. Equipotential contours after Kamide and Matsushita (1979) Dotted lines enclose the area in which STARE observations were made.

mum potential has been displaced toward midnight, thereby causing the rapid decrease of the field magnitudes near dawn. This displacement occurred owing to the general low conductivity level at night.

There are some disagreements between the model and the STARE measurements in the time interval between 2000 - 0100 MLT, before and during the presence of the Harang discontinuity. The discontinuity occurs at 2000 - 2330 MLT in the model, i.e. earlier than in the STARE data. A possible reason for this is that the real field-aligned current structure is more complicated near the discontinuity than assumed in the model (Iijima and Potemra, 1976).

5. SUMMARY

Comparisons between these models and the STARE measurements have lead to the gratifying result and provide better insight into the processes controlling the main features of the ionospheric electric fields.

However, there are of course limits to what can be achieved by applying the method used in this paper. To continue this investigation is it necessary to calculate the high latitude electric fields for different realistic spatial distributions of ionospheric conductivities and field aligned currents, until a reasonable good fit is achieved between the main feature of observed and calculated fields at all local times and latitudes.

ACKNOWLEDGEMENTS

The STARE radars are operated by the Max-Planck-Institut für Aeronomie in cooperation with ELAB, the Norwegian Technical University and the Finnish Meteorological Institute. We thank Profs. J.A. Fejer and V.M. Vasyliunas for their helpful discussions.

REFERENCES

Axford, W.I., and Hines, C.O., 1961, Can. J. Phys., 39, 1464.
Cahill, L.T. jr., Greenwald, R.A. and Nielsen, E., 1978, Geophys. Res. Lett. 5
Dungey, J.W., 1961, Phys. Rev. Letters, 6, 47.
Evans, J.V., Holt, J.M. and Wand, R.H., 1980, J. Geophys. Res. 85.
Fejer, J.A., 1964, J. Geophys. Res., 69, 123.
Greenwald, R.A., Weiss, W., Nielsen, E. and Thomson, R.N., 1978, Radio Science 13.
Heppner, J.P., 1977, J. Geophys. Res., 82, 1115.
Iijima, T., and Potemra, T.A., 1976, J. Geophys. Res., 81, 2174.
Kamide, Y., and Matsushita, S., 1979, J. Geophys. Res. 84, 4083.
Sudan, R.N., Akinrimisi, J. and Farley, D.T., 1973, J. Geophys. Res., 78, 240.
Vampola, 1973, Aerospace Report No. TR-0074 (4260-201-5).
Vasyliunas, V.M., 1970, p. 60, in "Particle and Fields in Magnetospheric Processes". (Ed. B.M. McCormac), D. Reidel, Dordrecht, Holl.
Vasylinuas, V.M., 1972, p. 29, in "Earth's Magnetospheric Processes". (Ed. B.M. McCormac), D. Reidel, Hingham, Mass.
Wolf, R.A., 1975, Space Sci. Rev., 17, 537.

ELECTRIC FIELD MEASUREMENTS WITH BALLOONS

Iver B. Iversen

Danish Space Research Institute
Lundtoftevej 7
2800 Lyngby , Denmark

ABSTRACT

Stratospheric balloons have in the last decade been used for investigating the electric field of the magnetosphere. The idea behind this technique is outlined and problems involved are discussed. A number of examples of measurements related to various subjects are given.

1. INTRODUCTION

The electric field, which is a parameter of utmost importance in magnetospheric physics, was for a long time after the beginning of the space age poorly (or not at all) measured as compared with other significant parameters (energetic particles etc.).

Satellite measurements have encountered considerable difficulties and succeeded only recently. It is only within the last few years that extensive satellite measurements of a sufficient quality have been published. Urged by this situation, but also because satellite measurements are not well suited in all cases (and because they are expensive) other methods were developed. A general survey is given by Stern (1977) in which paper a comprehensive list of references can be found as well.

Balloon measurements of the electric field were first suggested by Kellogg and Weed (1968) and is the subject of this paper. We are dealing with a technique which is an excellent example of how the necessary equipment (in this case amplifiers, balloons etc.) can be available for years but is never put into use simply because the practicability is not recognized.

C. S. Deehr and J. A. Holtet (eds.), Exploration of the Polar Upper Atmosphere, 305–314.

2. THE MAPPING PROBLEM

A crucial question is how the high latitude magnetospheric electric field is related to the field observed by a balloon instrument. Inside the magnetosphere the perpendicular component of the electric field, $E_{\perp}$, is mapping along the magnetic field lines due to the high conductivity (the possible existence of parallel electric fields is not considered here). In the lower ionosphere the conditions change gradually and the question is then whether the field can be observed even by a balloon instrument. At this location the conductivity of the atmosphere has become very high and isotropic. The answer is, that due to the exponential variation of the conductivity versus height within the lower atmosphere the mapping is still so effective that balloon experiments are possible, Mozer and Serlin (1969).

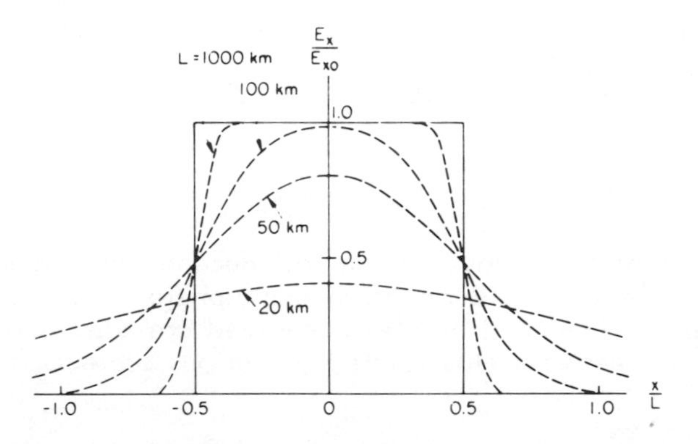

Figure 1. The mapping efficiency of a step function electric field distribution down to balloon altitude (from Park, 1976).

The mapping problem can be solved theoretically by assuming proper boundary conditions. Figure 1 is from Park (1976) and is one way of illustrating the result from such calculations. The plot shows how a square field configuration at 150 km's height is mapping down to a balloon at 30 km's height for different values of scale size. We see that for structures of 100 km or larger the mapping is quite good and 100 km is a reasonable lower limit to define for the resolutional power of this kind of experiment.

3. THE ATMOSPHERIC FIELD

A vertical electric field is present in the lower atmosphere due to a potential difference between the Earth and the ionosphere maintained by the worldwide occurrence of thunderstorms. It has been measured to about 100 volts/meter at the ground and is decreasing upwards according to the variation of the conductivity. At balloon height it is still 100 - 200 mv/meter. This atmospheric field can course trouble to the measurements in several ways as described in the following.

One problem is instrumental. Usually we measure the horizontal field with an electrostatic probe some meters long which is spinning in a horizontal plane in order to obtain the amplitude and the direction of the field as described later. If the spin plane is not exactly horizontal an error is introduced because a component of the vertical field is measured and cannot be distinguished from a horizontal field. Therefore it is very important that the spin plane is maintained horizontal during the flight.

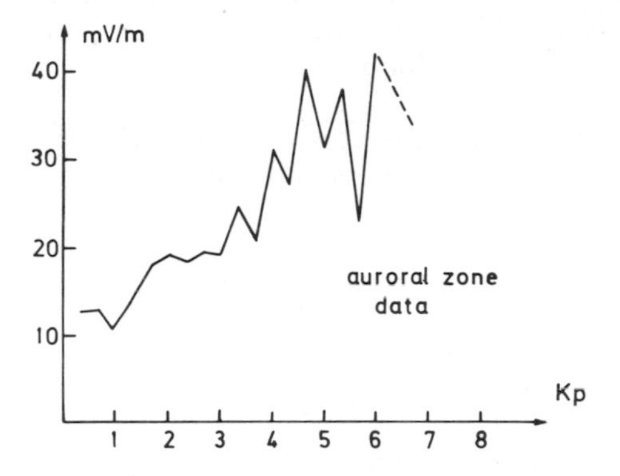

Figure 2. The average value of the electric field vs. K_p

The atmospheric field can be disturbed by inhomogeneities (conductivity irregularities) within the atmosphere causing the field to be no longer strictly vertical and thereby causing an erroneous measurements on the horizontal probes. These problems have been examined theoretically and calculations are published in the literature. Park (1976) finds that, except for extreme cases, the vertical electric field will not, due to conductivity irregularities, project on a horizontal plane at 30 km's height with more than a few millivolts per meter at most, which is an order of magnitude less than typical auroral and polar cap values. This result is confirmed experimentally by several thousand hours at balloon measurements showing that: 1) during geomagnetically quiet conditions we measure very low fields in general. Figure 2 shows how the average electric field strength in the auroral zone is dependent on the K_p value, 2) under disturbed conditions the electric field measurements correlate in most cases well with other geomagnetic measurements, when available. Examples of such cases are presented in a later section.

More severe is the occurrence of thunderclouds. They are surrounded by strong dipole or multipole fields and measurements are then not possible. Figure 3 shows how the vertical component can be disturbed for several hours due to the appearance of a thundercloud. Fortunately such clouds are not frequently encountered on high latitudes. During a series of balloon flights in 1975 four cases were observed during 1200 flight hours.

The presence of very high mountains near the balloon may cause

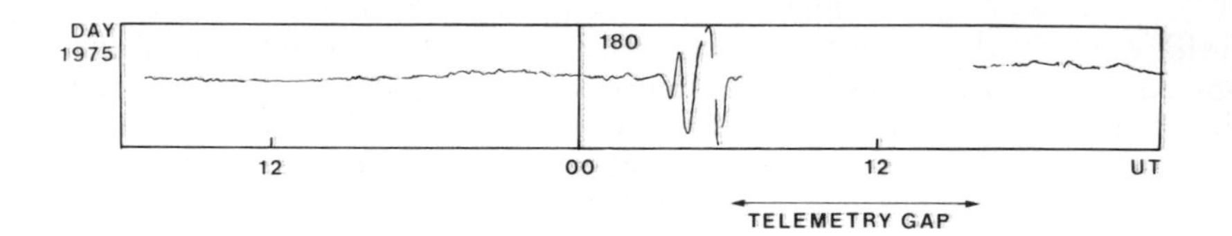

Figure 3. Recording of the atmospheric electric field in arbitrary units. A thundercloud is observed around 05 UT on day 180 over Greenland.

the atmospheric field to deviate from vertical,(Hoppel, 1971), but on high latitude experiments it has not been observed.

4. ADVANTAGES OF THE BALLOON METHOD

Detailed in-situ measurements (as with a rocket or with a satellite) are, of course (as described previously when dealing with the mapping problem), not possible, but attention should be drawn to the following characteristic features of this kind of technique.

1. With relatively few resources one can on a short notice set up the experiment almost anywhere. This facilitates use for special purposes such as to correlate with a rocket experiment etc. Also, simultaneous measurements from more than one location are reasonable and have been carried out on several occations (Ullaland, 1979).

2. It is possible to continue a measurement at a selected location for hours (limited by the drift of the balloon, which can be very slow). This contrasts with a fast moving satellite.

3. Difficulties characteristic for satellite and rocket experiments, such as the v×B field and local plasma disturbances created by the vehicle, do not constitute a problem when using a balloon instrument.

4. The instrument can be made simple and inexpensive and a balloon campaign (the activity in the field) can be performed with the help of very few people.

5. THE BALLOON INSTRUMENT

An electric field instrument to be flown on a balloon is essentially constructed like this: On the payload body are, at the end of insulating booms, mounted two metallic spheres of ~ 20 cm diameter separated by a distance of a few meters (typical ~ 3 meters) and connected to a high impedance amplifier. The payload is made to rotate around a vertical axis with a period of a few tens of seconds (typically 20 seconds). The output voltage is thus sinusoidally modulated at that period, and amplitude and phase give the electric field vector.

As discussed by Mozer and Serlin (1969), the input impedance of the preamplifiers must be $>10^{13}\Omega$ in order to avoid shortcircuiting of the electric field by the measuring device. Today this value is easily obtained with available components. The total weight of the payload is often less than 10 kg .

In order to avoid or minimize distortions of the electric field near the instrument by the balloon itself a rather long (at least one balloon diameter) load line between the balloon and the payload should be used. This line must of course, be made of insulating material (nylon).

6. DRIFT OF STRATOSPHERIC BALLOONS

When planning for an electric field balloon flight (as for any balloon flight) the wind conditions should, of course, be considered. Fortunately the stratospheric wind has a regular annual variation so that a balloon is not drifting at random but a reasonable estimate can be made before launch.

The stratospheric wind is essentially blowing west during the summer and east during the winter with two short so called turn-around periods in between.

If measurements are needed on a fixed location for a longer time (say one day) the balloon must be launched during one of the turn-around periods. Experience shows, however, that these periods are not easily predictable. They may come a few days sooner or later and they may also be very short so that the wind is shifting quickly to the opposite direction.

It is much easier not to be dependent on the balloon being more or less stationary. Then it can be launched around solstice and the drift path is from experience well known. As an example can be mentioned a series of balloon flights in the summer 1975 carried out by the Danish Space Research Institute together with the University of Bergen. Nine balloons were launched from North Scandinavia and drifted westwards to Arctic Canada which is a distance of about 4500 km . The average drift speed varied between 0.88 and 1.08 longitudinal degrees per hour and no balloon diverged more than two degrees from the nominal latitude given by the launch site (69.3 N). Because of the midnight sun all balloons stayed at high altitude for five days without any ballast drop and the electric field could be measured continuously for three weeks Figure 4 shows the detailed trajectories for four similar flights carried out in 1974.

Experience has also been obtained in flying balloons during the

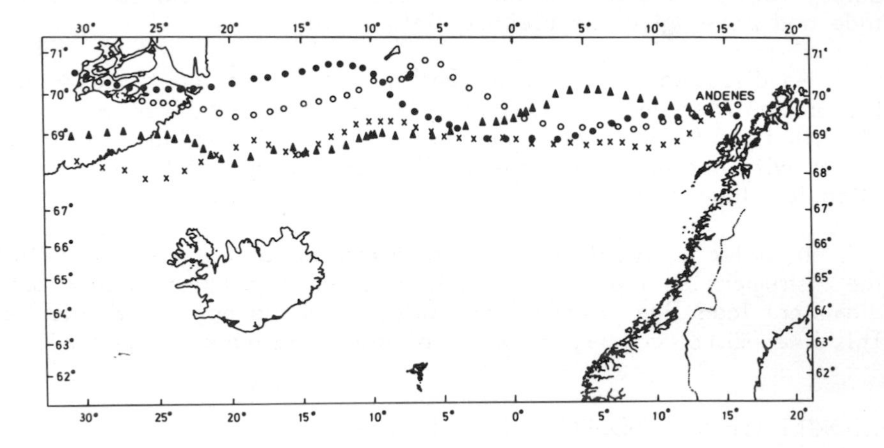

Figure 4. The drift paths for four balloons at midsummer 1974. One hour is separating each symbol.

winter (Zhulin, 1979). The stratospheric wind is then blowing eastwards and it should be noted that the wind speed is considerably higher than during the summer.

7. EXAMPLES OF MEASUREMENTS

Valuable information has during the last decade been obtained on a variety of the most interesting problems of magnetospheric and ionospheric physics through balloon measurements of the electric field. This section is dealing with a few examples including subjects as magnetospheric convection, the question of reconnection, ionospheric currents, pulsations and mapping along the geomagnetic field lines.

7.1 Magnetospheric plasma convection

Balloon measurements are well suited for studying the large scale electric field over the polar cap. This field is closely associated with the magnetospheric convection, a phenomenon first observationally verified by Axford and Hines (1961). The subject has attracted much interest e.g. in the discussion of the possible reconnection by the interplanetary magnetic field.

The general structure of the convection is antisunward flow over the polar cap combined with a return flow equatorward of the cap in which way a two-cell configuration appears (Heppner 1972 ; Mozer et al 1974 ; Madsen et al 1976).

Electric field measurements carried out at high latitude by the Danish Space Research Institute in the period June 20. to July 10., 1975 are collected in figure 5. A corrected geomagnetic latitude and magnetic local time frame has been used, and in order to demonstrate the displacement of the convection pattern as a function of the solar wind magnetic field polarity, only data collected during "away" conditions (B 0) has been used for the plot. The convection pattern is clearly unsymmetrical with the strongest electric fields on the dawn side of the cap as found also by other investigators (Iversen and Madsen 1978).

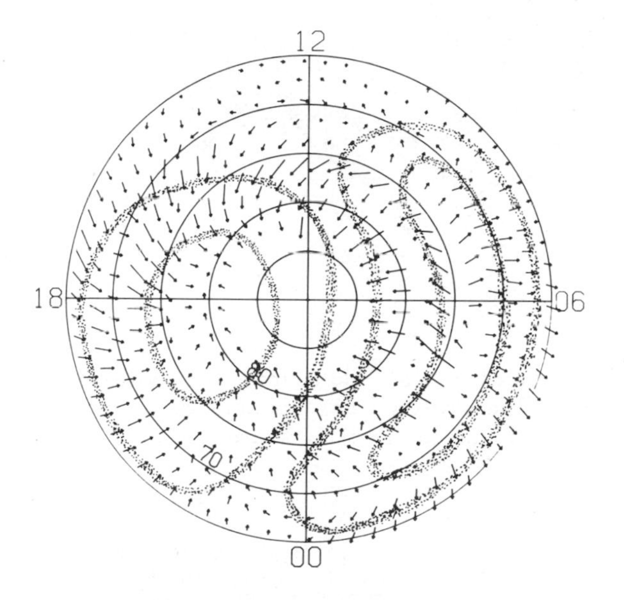

Figure 5. Electric field data obtained when the interplanetary magnetic field is directed away from the sun. The dotted lines indicate the convection pattern.

The convection pattern is not at all times a simple two-cell system. Strong deviations occur (see e.g. Banks, this volume), but it is not easy to obtain a snapshot of the whole system. A balloon instrument can record the time variations, but on one location only, whereas a satellite instrument can give a quick cross section of the system but will miss the time variations. A combination of the two methods would be useful but is not easy to arrange for. Simultaneous balloon flights are more realistic and have been carried out on a number of occations, one example can be seen on figure 6. Measurements from two balloons flown simultaneously, one inside the polar cap and the other outside, are shown and again we see that the polarity of the dawn-dusk component reverses over the auroral oval. The dotted lines are connecting field vectors of the same universal time.

Most electric field balloon flights have been carried out in the northern hemisphere, but also flights in the antarctic region have been described (Tanaka et al 1977; Berthelier and Mozer 1975).

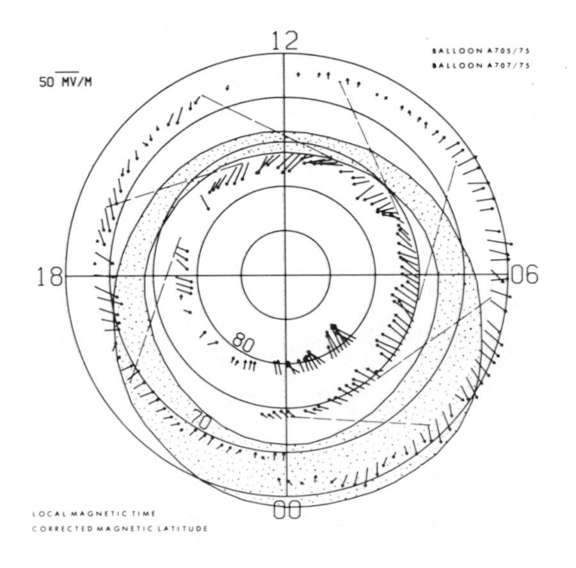

Figure 6. Electric field data from two balloons flown simultaneously. The dotted lines connect points of the same UT, and the Q=3 auroral oval is shown.

7.2 Auroral currents

The data on the morning side in figure 6 has been collected during a magnetic storm. This situation is better illustrated in figure 7, where the electric field amplitude is plotted vs. UT together with a ground based magnetometer recording. During the night of July 8. 1975 the magnetometer x-component shows a substantial deviation indicating the flow of a strong auroral electrojet. Some hours later, in the early afternoon, the return current (the eastward electrojet) is observed.

We note a pronounced correlation with the electric field (figure 7 bottom). A more detailed investigation reveals that a Hall current is flowing because the equivalent current as defined by the magnetometer is perpendicular to the electric field vector.

7.3 Geomagnetic micropulsations

It has been observed by balloon measurements that electric field pulsations can be associated with Pc-5 micropulsations. In D'Angelo et al (1975) is suggested a model that can explain the connection between a geomagnetic pulsation of the order of 100 γ and an electric field pulsation of around 20 mV/m., which is the observed value.

In Fahleson et al (1979) an electric field pulsation with a period of $\sim$6 min correlates with several ground based magnetic field measurements and also with particle and field measurements on the geostationa-

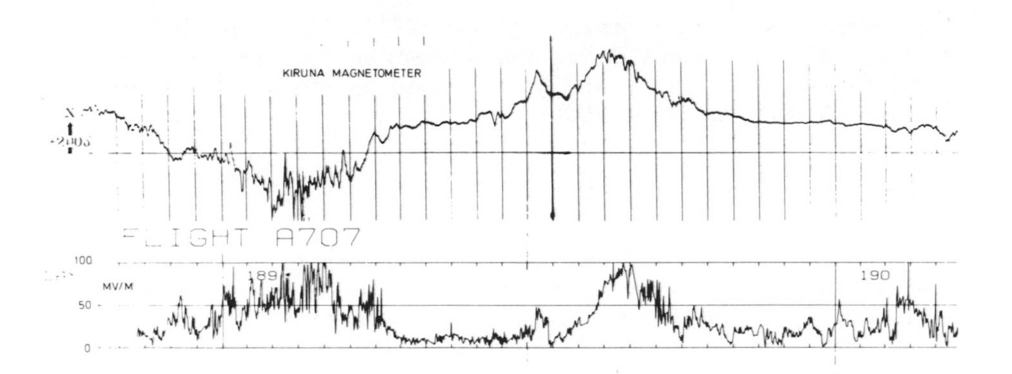

Figure 7. Upper : Kiruna magnetometer x-component. Lower : Electric field amplitude as observed by the balloon instrument.

ry satellite GEOS-2 positioned on a field line, which also passes close to the balloon.

7.4 The electric field and plasma instabilities

Electric field measurements on balloons have been used for explaining the occurrence of certain kinds of instabilities in the ionosphere. An association between the 'slant E condition' (SEC) in polar cap ionograms and the occurrence of the Farley instability in the E-region was suggested by Olesen (1972). On some occations a balloon with an electric field experiment has passed over a back-scatter facility, and correlation studies have then been possible. Figure 8 is from Iversen et al (1975) and illustrates the correlation between SEC and the existence of electric fields. The Farley instability has a threshold of about 25 mV/m in the polar ionosphere, but a perfect correlation can of course not be expected because the balloon is mooving and cannot for a longer period stay exactly above the ground based back-scatter installation.

7.5 Magnetospheric substormes

The electric field has on several occasions been measured by bal-

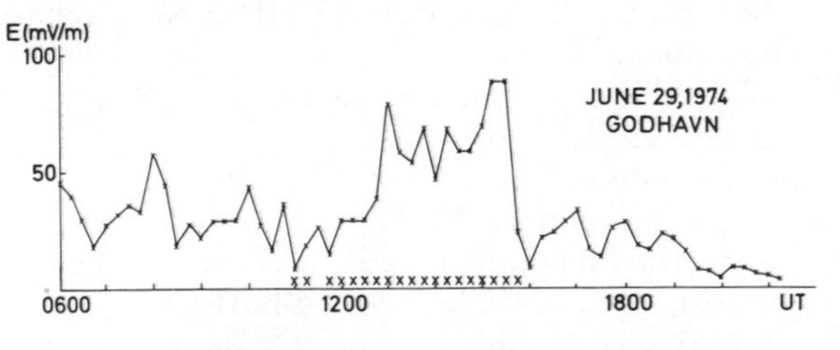

Figure 8. The electric field amplitude near Godhavn, Greenland. Crosses along the abscissa indicate, when the auroral back-scatter is being observed.

looninstruments during substorms (Mozer 1971). As one result from extensive balloonmeasurements during the International Magnetospheric Study (IMS) further observations will be published on this subject in the near future (e.g. Kremser et al 1980).

8. CONCLUDING REMARKS

It has not been possible in this paper to include all subjects related to balloon measurements of the electric field. Extensive measurements have been carried out recently, and the reader is recommended to pay attention to results which are expected to be published in the early 80's.

REFERENCES

Axford, W.I., and Hines, C.O. 1961, Can.J.Phys., 39, 1433.
Berthelier, J.J., and Mozer, F.S. 1975, paper presented at UGGI XVI General Assembly, Grenoble, France.
D'Angelo, N., Iversen, I.B., and Madsen, M.M. 1975, J. Geophys. Res., 80, 1352.
Fahleson, U., Grard, R., Madsen, M.M., Iversen, I.B., Tanskanen, P., Niskanen, J., Korth, A., Kremser, G., Torkar, K., Riedler, W., Ullaland, S., Brønstad, K. 1979, paper presented at First International Symposium on IMS Results, Melbourne, Australia.
Heppner, J.P., 1972, J. Geophys. Res., 77, 4877.
Hoppel, W.A., 1971, Pure Appl. Geophys., 84, 57.
Iversen, I.B., D'Angelo, N., and Olesen, J.K. 1975, J. Geophys. Res., 80, 3713.
Iversen, I.B., and Madsen, M.M. 1978, p. 293 in Space Research, vol. XVIII, Pergamon Press, Oxford and New York.
Kellogg, P.J., and Weed, M. 1968, paper presented at 4th International Conf. on the Univers. Aspects of Atm. Electricity, Tokyo, Japan.
Kremser, G., Bjordal, J., Block, L.P., Brønstad, K., Iversen, I.B., Kangas, J., Korth, A., Madsen, M., Moe, T., Riedler, W., Stadsnes, J., Tanskanen, P., Ullaland, S. 1980, Space Research XXI, to be published.
Madsen, M.M., Iversen, I.B., and D'Angelo, N. 1976, J. Geophys. Res., 81, 3821.
Mozer, F.S., and Serlin, R. 1969, J. Geophys. Res., 74, 4739.
Mozer, F.S. 1971, J. Geophys. Res., 76, 7595.
Mozer, F.S., Gonzalez, W.D., Bogott, F.H., Kelley, M.C., and Schutz, S. 1974, J. Geophys. Res., 79, 56.
Olesen, J.K. 1972, AGARD Conference Proceedings p. 97.
Park, C.G. 1976, J. Geophys. Res., 81, 168.
Stern, D.P. 1977, Rev. Geophys. Space Phys., 15, 156.
Tanaka, Y., Ogawa, T., and Kodama, M. 1977, J. Atmos. Terr. Phys., 39, 921.
Ullaland, S. 1979, Scientific Ballooning, (Edt. W. Riedler), Pergamon Press p. 83.
Zhulin, I.A., and Lazutin, L.L. 1979, Scientific Ballooning (Edt. W. Riedler) Pergamon Press, p. 89.

HIGH LATITUDE IONOSPHERIC CURRENTS

E. Friis-Christensen

Geophysical Department
Danish Meteorological Institute
Copenhagen, Denmark

ABSTRACT

The high latitude region inside the auroral zone has not been studied as extensively as the auroral zone with its spectacular phenomena like magnetic and auroral substorms. Contrary to what is often thought, the polar cap and in particular the projection of the magnetospheric cusp into the dayside auroral oval, is a region of very structured magnetic variations, even in case of a magnetosphere in a state of minimum activity.

In this paper a summary is given of the very high latitude magnetic variations and their relation to ionospheric and field-aligned currents.

INTRODUCTION

Much of our knowledge about ionospheric currents has traditionally been obtained by ground based observations of magnetic perturbations and interpretation of these perturbations in terms of equivalent currents. Equivalent, because from ground based magnetic observations alone there is no unique solution by which we can derive the actual electric current system.

In recent years observations from satellite-borne magnetometers have confirmed that the actual current systems are in fact three-dimensional with a considerable contribution from currents along the magnetic field-lines, the Birkeland currents.

Also in recent years it has been possible to derive the ionospheric current densities on continuous basis using the incoherent scatter radar technique to observe the height-integrated conductivities and electric fields. Comparisons between ionosphe-

C. S. Deehr and J. A. Holtet (eds.), Exploration of the Polar Upper Atmosphere, 315–328.

ric current densities derived in this way and the observed magnetic variations have shown that in the auroral zone region there is a pronounced difference between the radar observations of the north-south ionospheric current and the ground D-perturbations (Kamide et al. 1976) which indicates that part of the ground D-perturbation must be due to Birkeland currents.

Inside the auroral zone, however, up till now ground based magnetic observations provide the only continuous observations of currents. This region comprises the polar cap and the projection into the ionosphere of the polar cusp and will be the main topic of this review. While the auroral zone has been intensively studied during the past years the polar cap is still not sufficiently described and it has for a long time been regarded as a relatively uniform area. However, even when the magnetosphere seems to be in a state of minimum activity, defined by geomagnetic indices like Kp or AE, or defined by prolonged periods of a northward directed interplanetary magnetic field (IMF), the polar cap, and in particular the dayside part of the auroral oval, will still show magnetic perturbations suggesting currents of considerable amplitude.

These variations are closely connected to the varying interplanetary parameters, in particular the magnetic field and the solar wind velocity.

It is evident from the observations of magnetic perturbations in the polar region that the responsible currents must display highly localized features as well in space as in time. The available models for processes in the polar region are at present not capable of explaining these detailed features - and do not intend to do so - so it would be rather meaningless to try to explain the individual observations by means of general models. For this reason it is necessary to filter the observations in some way so we may derive the general and repeatable features which then may be compared with theoretical models.

PRESENTATION AND INTERPRETATION OF MAGNETIC PERTURBATION VECTORS.

The fundamental quantity is the observed magnetic perturbation vector, reckoned from the undisturbed level. How the undisturbed level is defined depends on the quality of the observations, the amount of data available and the required resolution.

The magnetic perturbation vector is a superposition of contributions from different sources among which we can mention local ionospheric currents, Birkeland currents, distant currents and effects from subsurface induction currents. It is in principle impossible to derive the actual current system in a unique way from magnetic ground observations. For this reason it has been widely used to present observations in terms of ionospheric equivalent current vectors or current systems.

The disadvantage of equivalent current systems is, that they include the effects from non-horizontal currents. Therefore there has been many attempts to use ground based measurements together with certain assumptions to separate horizontal and field-aligned currents (Kisabeth and Rostoker 1977; Hughes and Rostoker 1979; Akasofu et al. 1980a). Although the resultant three-dimensional current systems may be more realistic they are still not real current system but just three-dimensional equivalent current systems because of their inherent dependence on the various assumptions made.

Since so far no unique separation of ionospheric and Birkeland currents has been presented I will in this paper present magnetic observations in the least prepared form, namely as equivalent current vectors, which simply represent the observed magnetic perturbation vectors rotated 90 degrees clockwise. However, I want to stress that these equivalent current vectors should not be mistaken for the real ionospheric currents.

In the interpretation we may, however, under certain assumptions, make statements about the ionospheric part of the equivalent current. In case of a homogeneous conductivity in the ionosphere and vertical magnetic field lines, assumptions which may be realistic in the dayside polar cap where photo ionization is the dominant source of ionospheric conductivity, the theorem of Fukushima (1976) states that the magnetic effect on the ground of the field-aligned currents will cancel the magnetic effects of the ionospheric Pedersen currents and so the magnetic effect on the ground will be caused solely by Hall currents in the ionosphere. Rocket observations in Greenland (Primdahl and Spangslev 1977) have indicated that this assumption is realistic in the dayside polar cap while Barium-release experiments, which, due to experimental limitations, have to be carried out in the zone between daylight and night, have shown that ground magnetic observations may not always be due to ionospheric Hall currents. In particular in case of large activity accompanied by particle precipitation in the auroral oval, homogenious conductivity is not likely to be present.

Troschichev et al. (1979) modelled the magnetic effects from Birkeland and ionospheric currents and concluded that for low activity the magnetic effects produced by Pedersen currents are mainly cancelled by the distant effects from the Birkeland currents.

As it will appear from the following sections, we do not yet have sufficient knowledge to explain the various phenomena which are characteristic for the polar cap and the dayside auroral oval magnetic perturbations. Our first objective must be to describe the magnetic variations and their dependence on external parameters like parameters of the solar wind. When this has been achieved and when the general and repeatable features have been separated out, we may go further in a detailed modelling using three-dimensional current systems. But this will be beyond the scope of the present paper.

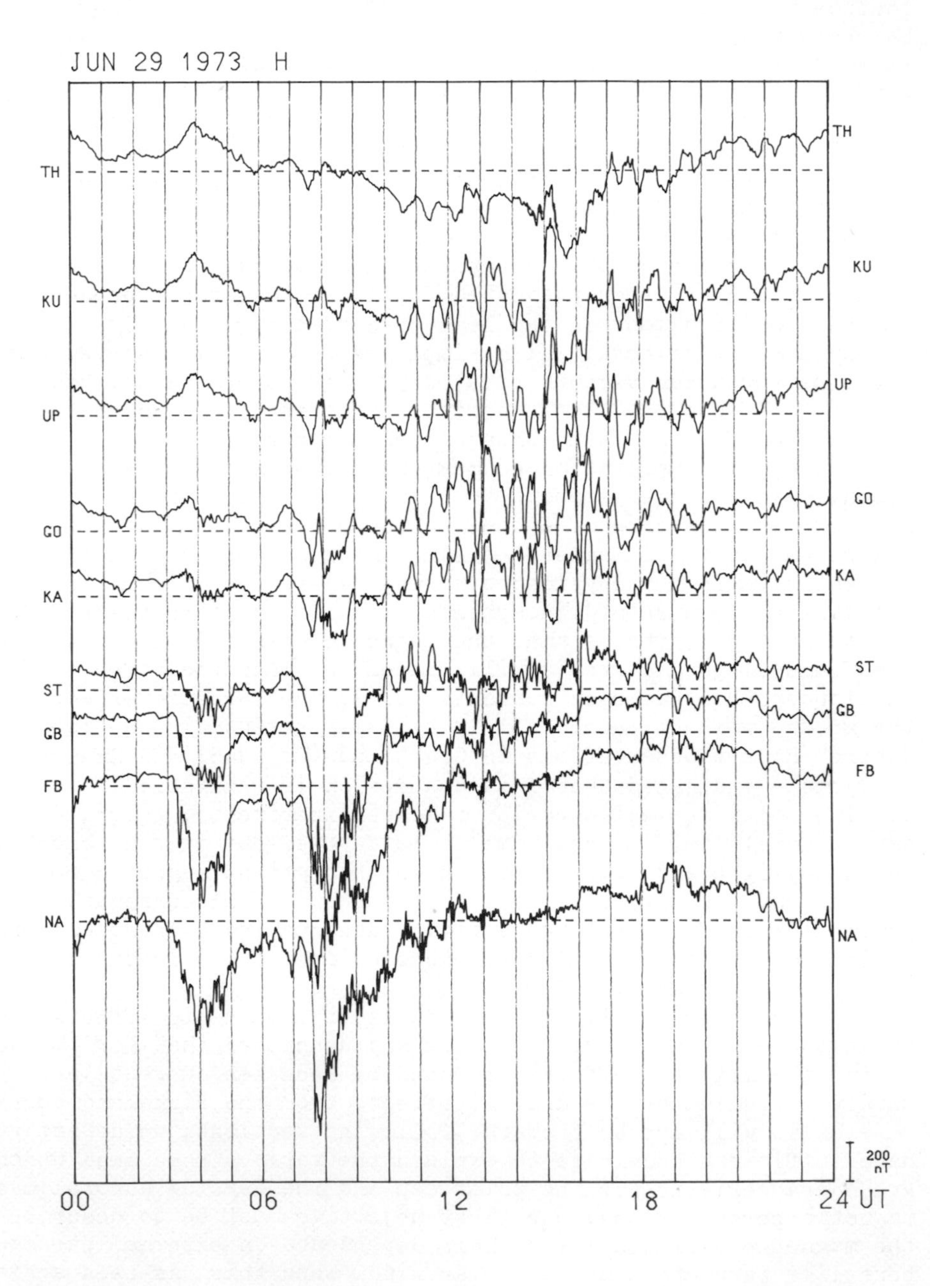

Fig. 1 Stacked plot of H-component magnetometer recordings from the westcoast of Greenland covering invariant latitudes from 68.5° to 86.5°. Local magnetic noon is at 1400 UT.

CHARACTERISTICS OF THE MORPHOLOGY OF MAGNETIC PERTURBATIONS IN THE POLAR CAP AND CUSP

Meridian chain of magnetometers inside the auroral oval, The Greenland Array 1972-1973 (Wilhjelm and Friis-Christensen 1976) and the IMS meridian chains in Greenland, Canada and Alaska (Akasofu et al. 1980b) have shown that the ionospheric currents especially in the dayside near local magnetic noon are highly structured in space as well as in time. From the stacked plot of magnetograms from Greenland shown in Figure 1 it is seen, that the region covered by the magnetic chain may be divided into three fundamentally different regions each with specific characteristics. The three regions, counted from the equatorward side, and from the bottom of the plot, are:

I: The closed field line region, characterized by the eastward and westward electrojets, a minimum of activity near local magnetic noon, and a considerable contribution of short period (less than 5 minutes) fluctuations.

II: The intermediate region, supposed to be connected to the solar wind magnetic field through the boundary layer magnetic field lines. This region displays a maximum of activity near local magnetic noon and the activity is dominated by periods from 15-60 minutes, closely correlated with variations in the interplanetary field.

III: Tail lobe field line region, the most poleward region showing a more smooth sinusoidal daily variation of the horizontal perturbation field. This region is characterized by complete absence of the short period fluctuations seen in the closed field line region (I).

Of course the boundaries between these three regions are not fixed and may vary even during the day with the z-component of interplanetary magnetic field, Bz.

INFLUENCE OF THE INTERPLANETARY FIELD

The observations by Nishida (1968) of a polar cap disturbance field which was correlated with the southward component of the interplanetary magnetic field, IMF, and the findings of Svalgaard (1969) and Mansurov (1969) that the IMF sector polarity gave rise to different daily variations of the geomagnetic field in the polar cap, indicated that the "open" model of the magnetosphere (Dungey 1961) possibly was an important key to understand the morphologi of the ionospheric currents in the polar cap.

Friis-Christensen et al. (1972) demonstrated that the critical component of the IMF producing the effect discovered by Svalgaard and Mansurov was the east-west component or the By component, while there was no significant effect due to the sunward component Bx.

Using standard polar cap observatory data Friis-Christensen and Wilhjelm (1975) separated the By and Bz effects on the steady state patterns of equivalent current vectors in the polar cap. The effect due to By was called the DPY-disturbance and was suggested to be caused by a Hall current flowing in the day sector around the magnetic pole in a direction which changes with the sign of the By component.

The two-cell equivalent current system is related to the Bz component and decreases when the IMF becomes more northward. The DPY current keeps its amplitude and orientation, but moves to higher latitudes with an increasing Bz component.

For a very northward directed IMF the two-cell current system vanishes and even seems to reverse (Maezawa 1976).

Although the observations in the polar cap are sparse and widely scattered it has been possible by different statistical methods to derive global features of the steady state condition of the equivalent currents in the polar cap and their dependence on the interplanetary magnetic field. More detailed reviews of these features have been given by Nishida (1975), Feldstein (1976), Mishin (1977) and Fairfield (1977).

Apart from their different dependence on the IMF components the two-cell equivalent current and the DPY current exhibit different seasonal variations which indicates that the DPY current probably is an ionospheric current since it decreases by a factor of 5-10 from the summer season to the winter season following the decrease in the ionospheric electric conductivity. Using satellite observations above the ionosphere Langel (1974) also showed, that the DPY is caused by an ionospheric current, whereas a considerable part of the two-cell current system probably is caused by Birkeland currents.

Using data from the magnetometer chain in Greenland during the summer season of 1972 and 1973 and simultaneous 20 minute average IMF data from the HEOS-2 satellite, Friis-Christensen (1979) derived patterns of the magnetic perturbations in the polar cap and cusp for different directions of the IMF. Due to the limited data available, it was necessary to use a regression technique in order to make efficient use of the data. The technique requires the assumption of a linear relationship between a perturbation vector and the IMF-components for a given latitude (station) and magnetic local time and season. This technique has been used on standard observatory hourly average data by Maezawa (1976), Mishin et al. (1978) and Belov et al. (1978). While a linear relationship as indicated above is a good approximation for the By component it seems not to be the best approximation for the Bz component because there is a definite asymmetry between the data for Bz>0 and Bz<0. For this reason Friis-Christensen (1979) divided the dataset into two independent datasets according to the sign of Bz.

The results, which in general are similar to the results of

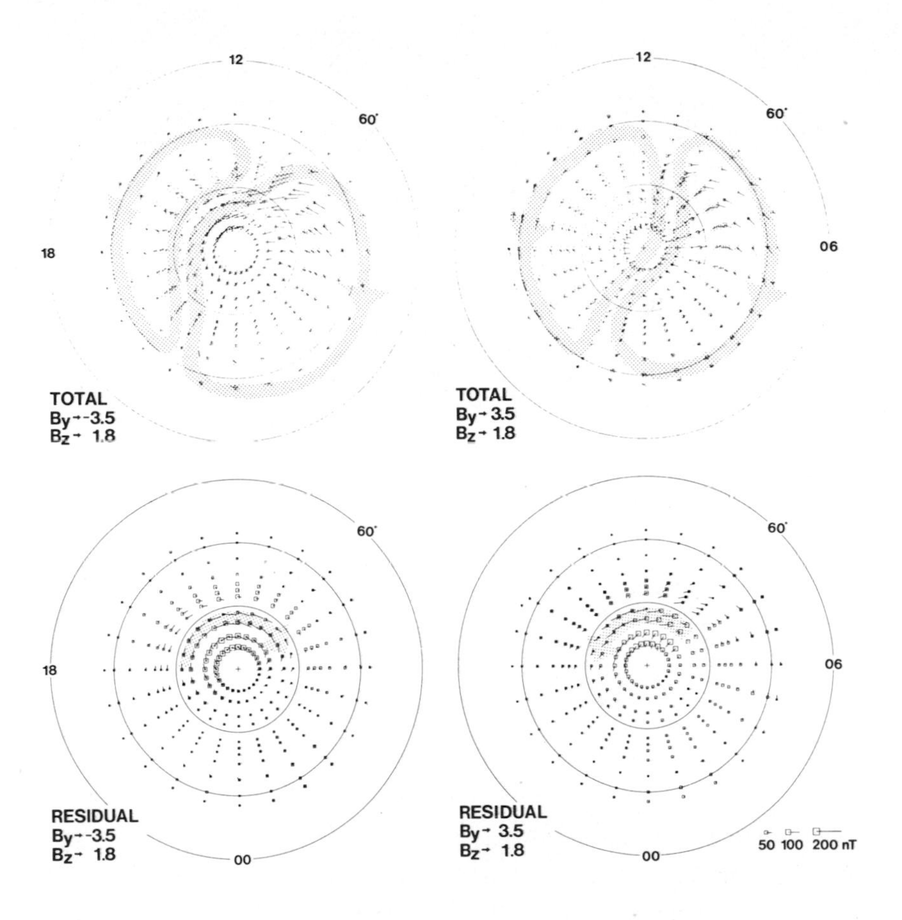

Fig. 2 Equivalent current vectors for Bz = 1.8 and By = ±3.5 nT. The upper panel shows the total vectors. On the lower panel the equivalent current vectors for (By, By) = (0.0) have been subtracted.

the previous studies using hourly average standard observatory data, indicate the presence of a permanent two-cell equivalent current system, independent on the IMF, which is similar to the well-known S_q^p-system.

Superposed upon this are patterns of magnetic perturbations proposed to be created by the convection electric field caused by merging of the interplanetary and geomagnetic field lines.

A S_q^p-like current system existing independently of the IMF indicates that besides convection related to magnetic merging also viscous interaction between the solar wind and the magnetosphere plays an important role as a cause of polar cap magnetic

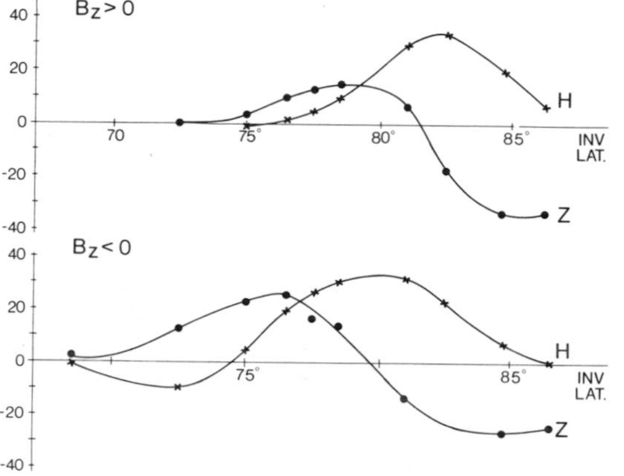
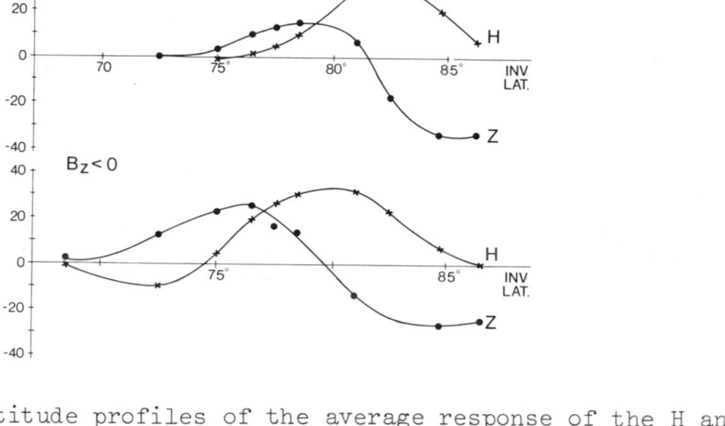

Fig. 3 Latitude profiles of the average response of the H and Z components to the IMF By-component. The average change of the respective component due to a change in By of 1 nT is given. The time is local magnetic noon.

activity.

The regression technique makes it possible to estimate the equivalent current vectors for given values of By and Bz. In Figure 2 is shown the equivalent current vectors derived in this way for By = $\pm$3.5 and Bz = 1.8 nT. On the bottom panel is shown the residual equivalent current vectors, which have been defined by subtracting the IMF independent part (the S_q^p-type). In this way we can separate the typical characteristics of the DPY-current, a zonal current on the dayside poleward of the 80° invariant latitude-circle.

The latitude of the DPY-current moves according to the Bz-component. This is illustrated in Figure 3 showing latitude profiles of the H- and Z-components of the magnetic perturbations. The H- and Z-components have been normalized to correspond to a change of +1 nT of the IMF By-component and have been given separately for two ranges of Bz, positive and negative values respectively. The DPY current moves to higher latitudes for increasing Bz. Also a typical feature is the asymmetry of the Z-component relative to the maximum of the H-component. This is due to the fact that the DPY is not a line current, but is located along magnetic latitudes. The curvature of the DPY increases the Z-component response poleward of the current.

Results obtained using statistical methods tend to simplify the model used to explain the data. It is therefore important to test the model on individual data also. For this purpose it is

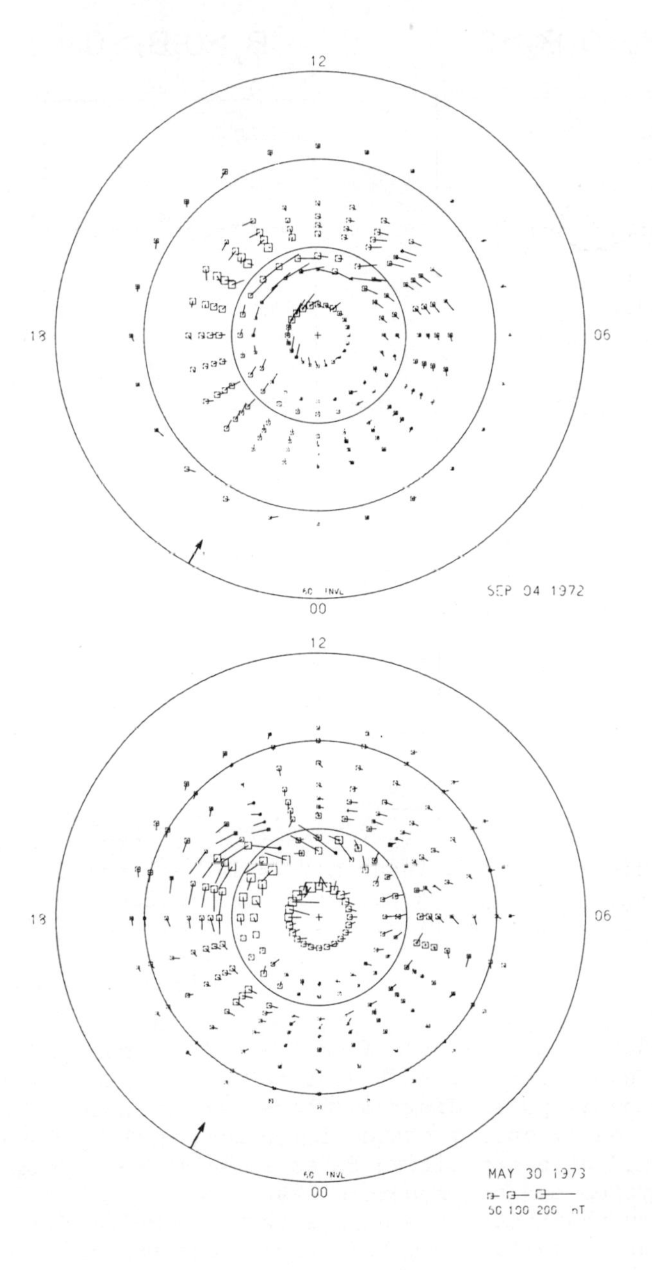

Fig. 4 Equivalent current vectors for 24 hours during two selected days.
On Sept. 04, 1972 By<0, and on May 30, 1973 By>0.
The unit of the equivalent current and the vertical perturbation component (boxes) are indicated.

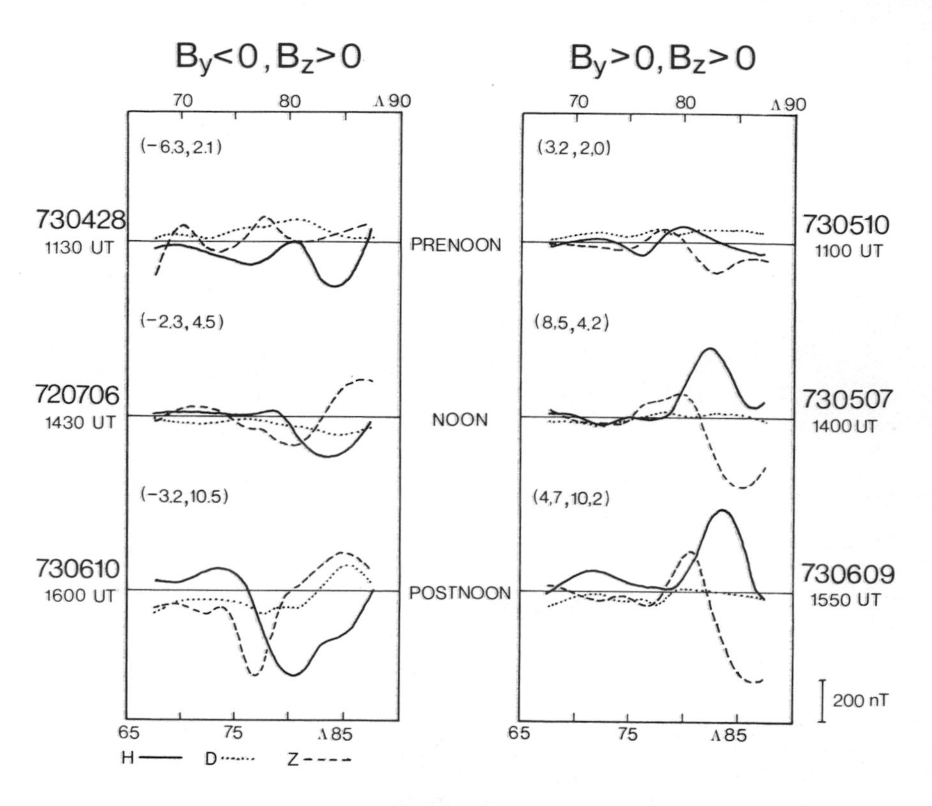

Fig. 5 Individual latitude profiles, H, D and Z components, where the H-component is the horizontal component perpendicular to the invariant latitudes towards the invariant pole. By and Bz values are indicated in brackets on each profile.

particularly useful to have data from a meridian chain of stations. Figure 4 shows equivalent current vectors on two selected days plotted in the usual polar diagram. These days were selected among days showing a nearly steady state of the IMF, with By<0 and By>0 respectively, and with a positive Bz to avoid effects from the DP2 two-cell system and the substorm system.

It is noted that these two days show the simultaneous presence of the very high-latitude DPY current and the westward electrojet in the morning and the eastward electrojet in the afternoon.

This may be further elucidated by individual latitude profiles selected for given signs of By and Bz. In Figure 5 latitude profiles in the prenoon, noon and postnoon sectors are given for all three components of the magnetic perturbation vectors.

Although the amplitudes of the actual IMF components vary

from sample to sample creating differences in the corresponding profile amplitudes, it is noted, that the individual profiles at noon are similar to the average profiles of Figure 3, although they are more pronounced.

In the prenoon sectors we are able to see the westward electrojet, independent of the sign of By; similarly in the postnoon sector the eastward electrojet is present, independent of the existence and direction of the DPY current at higher latitudes.

RELATION TO BIRKELAND CURRENTS

While there is a definite response in the magnetic variations to the IMF By component, the mechanism for this is still not fully explained.

It is known that the electric field at ionospheric as well as at satellite height shows an asymmetry according to the sign of the By component. Contrary to magnetic observations, simultaneous and continuous electric field measurements at several locations inside the polar cap have not been performed so the electric field distribution is not known in details.

Measurements of the magnetic perturbations on board satellites like TRIAD and ISIS have indicated the presence of field-aligned or Birkeland currents also in the polar cusp.

Wilhjelm et al. (1978) compared the ground based observations of the DPY current with the existence and direction of simultaneous overhead measurements of Birkeland currents from the TRIAD satellite.

Several authors had from theoretical reasons predicted a correspondence between a suggested ionospheric Hall current and closing Birkeland currents. See for example Wilhjelm and Friis-Christensen (1972), Leontyev and Lyatsky (1974) and Volland (1975).

Figure 6, taken from Wilhjelm et al. (1978), shows that there

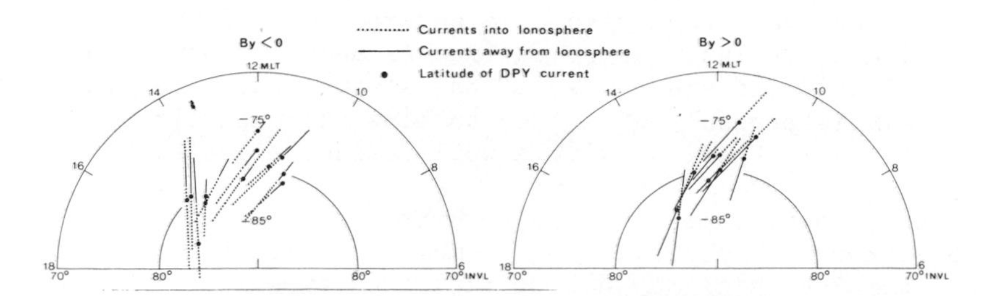

Fig. 6 Polar plot of the location and direction of field-aligned currents and the simultaneous location of the center of the DPY current. (After Wilhjelm et al. 1978).

is a close connection between the latitude of the DPY current and the latitude of Birkeland currents in the cusp. Furthermore when By changes sign, the DPY current as well as the Birkeland currents change sign accordingly. The direction of the DPY-current, the electric field directions and the direction of the Birkeland currents are systematically in agreement with the assumption that the DPY-current is a Hall current, sandwiched between two antiparallel field-aligned current sheets.

The general system of field-aligned currents, the Region 1 and Region 2 currents (Iijima and Potemra 1976) which border the auroral oval at all longitudes except near local magnetic noon, is a permanent system which is independent or very little dependent on By. It is therefore still an unsolved problem whether the cusp Birkeland current sheets indicate the existence of a separate system or if they are part of the general system.

Undoubtedly the answer to this is closely connected to the problem of how the DPY ionospheric current is related to the east- and westward electrojet currents in the post- and prenoon sectors.

DISCUSSION

Hughes and Rostoker (1979) have presented a very comprehensive model of the high latitude three-dimensional current configuration.

This model based upon data from the University of Alberta meridian line of magnetometers presents very detailed features of the ionospheric currents in the auroral oval and the field-aligned currents which flow into and out of the boundaries of the auroral oval.

It is characteristic of the model that poleward of 75° invariant latitude no ionospheric currents are present. Probably this is caused by three facts.

1. Only winter data have been used which means that the polar cap is in darkness.
2. The line of magnetometers has no good coverage poleward of 75° (only two stations: Cambridge Bay and Resolute Bay).
3. The data have been used in a superposed epoch analysis using a key latitude of maximum ΔX magnetic perturbation, indicative of the center of the electrojet. This kind of analysis probably will tend to bias the analysis towards more active periods with a well defined electrojet.

The present paper intends to supplement the model of Hughes and Rostoker with observations of magnetic perturbations at very high latitudes which may not be explained by effects from non-local currents. To present magnetic effects specific for the polar cap, special attention has been paid to periods of a quiet magnetosphere due to prolonged periods of a northward IMF.

For such periods the magnetic activity is confined to the

dayside as is seen from the Figures 2 and 4. The equivalent current system consists of a strong westward or eastward cusp current (DPY) at latitudes poleward of 80° invariant latitudes. The direction of this current is uniquely fixed by the sign of the By component of the IMF. In addition to this current the westward and eastward electrojets at latitudes from 75°-80° are present, but they are generally weaker and are confined to the dayside prenoon and postnoon sector respectively.

Hughes and Rostoker (1977) have inferred persistent net downward field-aligned currents in the noon sector. The basis for this was that latitude profiles near noon showed a consistent level shift of the D-component of the magnetic perturbation around 75° latitude. These net downward field-aligned currents have been proposed to feed the westward as well as the eastward electrojets.

The results from the Greenland magnetometers do not support a strong net downward field-aligned current near noon since the latitude profiles for the D-component are always of minor amplitude relative to the H- and Z-profiles. Furthermore the equivalent current vectors show no simple continuation of the westward or eastward electrojet into the DPY-current as has been proposed by Rostoker (1980)

On the contrary the difference in the response to the IMF of the DPY-current and of the west- and eastward electrojets indicates that the DPY-current is of a different origin. Of course, in a steady state the DPY current will blend into the other ionospheric currents and it may be difficult to separate the currents when using average data like hourly mean values.

The latitude profiles presented in this paper represent 10-minut average values for the summer season while the latitude profiles of Hughes and Rostoker (1977) represent hourly averages from the winter season and this might explain some of the discrepancies in the interpretations of the Canadian and Greenland data.

ACKNOWLEDGEMENT

The establishment of the magnetometer chain in Greenland and the reduction of the data was done partly under a grant from the Danish Space Research Board.

Dr. P.C. Hedgecock kindly supplied the HEOS-2 IMF-data used in this study. The HEOS magnetometer experiment was supported by the British Science Research Council.

REFERENCES

Akasofu, S.-I., Kisabeth, J., and Byung-Ho Ahn 1980a, J. Geophys. Res. (in press).

Akasofu, S.-I., Romick, G.J., Kroehl, H.W. 1980b, J. Geophys. Res. (in press).

Belov, B.A., Afonina, R.G., Levitin, A.E., and Feldstein, Y.I. 1978, Geomagn. Aeron., 18, 471.

Dungey, J.W. 1961, Phys. Rev. Lett., 6, 47.

Fairfield, D.H. 1977, Rev. Geophys. Space Phys., 15, 285.

Feldstein, Y.I. 1976, Space Sci. Rev., 18, 777.

Friis-Christensen, E., Lassen, K., Wilhjelm, J., Wilcox, J.M., Gonzalez, W., and Colburn, D.S. 1972, J. Geophys. Res., 77, 1972.

Friis-Christensen, E., and Wilhjelm, J. 1975, J. Geophys. Res., 80, 1248.

Friis-Christensen, E. 1979, Proc. International Workshop on Selected Topics of Magnetospheric Physics, Tokyo, 290.

Fukushima, N. 1976, Rep. Ionos Space Res., Jpn., 30, 35.

Hughes, T.J. and Rostoker, G. 1977, J. Geophys. Res., 82, 2271.

Hughes, T.J. and Rostoker, G. 1979, Geophys. J.R. Astr. Soc., 58, 525.

Iijima, T., and Potemra, T.A. 1976, J. Geophys. Res., 81, 2165.

Kamide, Y., Akasofu, S.-I., and Brekke, A. 1976, Planet. Space Sci., 24, 193.

Kisabeth, J.L. and Rostoker, G. 1977, Geophys. J.R. Astr. Soc., 49, 655.

Langel, R.A. 1974, Planet. Space Sci., 22, 1413.

Leontyev, S.V., and Lyatsky, W.B. 1974, Planet. Space Sci., 22, 811.

Mansurov, S.M. 1969, Geomagn. Aeron., 4, 622.

Maezawa, K. 1976, J. Geophys. Res., 81, 2289.

Mishin, V.M. 1977, Space Sci. Rev., 19, 621.

Mishin, V.M., Bazarzhanov, A.D., Anistratenko, A.A., and Aksenova, L.V. 1978, Geomagn. Aeron., 18, 516.

Nishida, A. 1968, J. Geophys. Res., 73, 1795.

Nishida, A. 1975, Space Sci. Rev., 17, 353.

Primdahl, F. and Spangslev, F. 1977, J. Geophys. Res., 82, 1137.

Rostoker, G. 1980, J. Geophys. Res. (in press).

Svalgaard, L. 1969, Dan. Meteorol. Inst. Geophys. Pap. R-6.

Troshichev, O.A., Gizler, V.A., Ivanova, I.A., and Merkuryeva, A.Yu. 1979, Planet. Space Sci., 27, 1451.

Volland, H. 1975, J. Geophys. Res., 80, 2311.

Wilhjelm, J. and Friis-Christensen, E. 1972, Dan. Meteorol. Inst. Geophys. Pap. R-31.

Wilhjelm, J. and Friis-Christensen, E. 1976, Dan. Meteorol. Inst. Geophys. Pap. R-48.

Wilhjelm, J., Friis-Christensen, E., and Potemra, T.A. 1978, J. Geophys. Res., 83, 5586.

ISIS OBSERVATIONS OF AURORAL PARTICLES AND LARGE-SCALE BIRKELAND CURRENTS

D. M. KLUMPAR

Center for Space Sciences, University of Texas
at Dallas, Box 688, Richardson, Texas 75080

ABSTRACT

The relationships between primary and secondary auroral particle distributions (5eV to 15keV) and the large-scale Birkeland currents are investigated. Observations of electrons and positive ions taken simultaneously with single component magnetic perturbations on the ISIS-2 satellite are used to compare and contrast these relationships in the pre- and post-midnight local time sectors. The various regions of the large-scale Birkeland current system do not appear to be uniquely related to particular regions of auroral particle distributions although some repeatable systematics are observed in regions of upward directed current. Little evidence exists in either local time sector for direct detection of the current carriers associated with the downward current.

I. INTRODUCTION

Based upon statistical analyses of the TRIAD satellite magnetic observations it has become customary to separate the large-scale Birkeland current system into two general regions containing oppositely directed currents (Zmuda and Armstrong 1974). This generalization includes a high latitude ring of field-aligned current directed into the ionosphere over the dawn hemisphere and outward over the dusk hemisphere referred to as the region 1 currents (Iijima and Potemra 1976a). At somewhat lower latitude an oppositely directed current, referred to as region 2, exists in each local time hemisphere. Near noon

C. S. Deehr and J. A. Holtet (eds.), Exploration of the Polar Upper Atmosphere, 329–336.

and near midnight these current sheets may overlap or additional sheet currents of more limited longitudinal extent may appear (Iijima and Potemra 1976b; McDiarmid et al. 1979).

Similarly, it has become accepted practice to discuss the optical features of the auroral oval in terms of two regions usually referred to as the discrete and diffuse aurora (Lui and Anger 1973). Typically, the diffuse aurora form a nearly circular belt of relatively uniform luminosity within which and poleward of which lie the discrete aurora (Lui et al. 1975). It has furthermore been found that these diffuse and discrete auroral regions correspond to two distinct regions in the pattern of precipitation of auroral particles referred to by Winningham et al. (1975) as the CPS (central plasma sheet) and BPS (boundary plasma sheet) respectively (Deehr et al. 1976).

With these two general pictures of the polar ionosphere in mind there is a natural tendency to relate one to the other. It is indeed inviting to associate the high latitude region 1 currents with the discrete auroral belt and the lower latitude region 2 currents with the diffuse aurora. Such an association is, in general, an erroneous one.

This study summarizes and extends previously published results (Klumpar et al. 1976; Klumpar 1979) utilizing simultaneous measurements of the particle distributions associated with the aurora and magnetic perturbations containing the signatures of field-aligned currents to investigate the spatial relationships between these phenomena. The results are restricted to magnetic local times nightward of the dawn-dusk meridian plane well away from the noon and midnight zones of confusion. The region 1 and region 2 currents are found not to be uniquely related to the discrete (BPS) and diffuse (CPS) auroral particle distributions respectively. The organization of this paper will be to discuss first the premidnight sector, proceeding poleward across the region of interest, and then turn to the postmidnight sector again proceeding from low latitude to high.

II. OBSERVATIONS

The main features of the particle-current relationship in these two local time sectors will be illustrated by means of the single dusk to dawn satellite pass shown in figure 1. Although only one example is shown the results quoted here have been garnered from a large number of satellite passes near the dawn-dusk meridian, mostly during the northern hemisphere summer of 1972. Figure 1 displays a composite of data from the ISIS-2 soft particle spectrometers (SPS) and the on-board magnetometer. The SPS data appears in the form of an electron energy versus

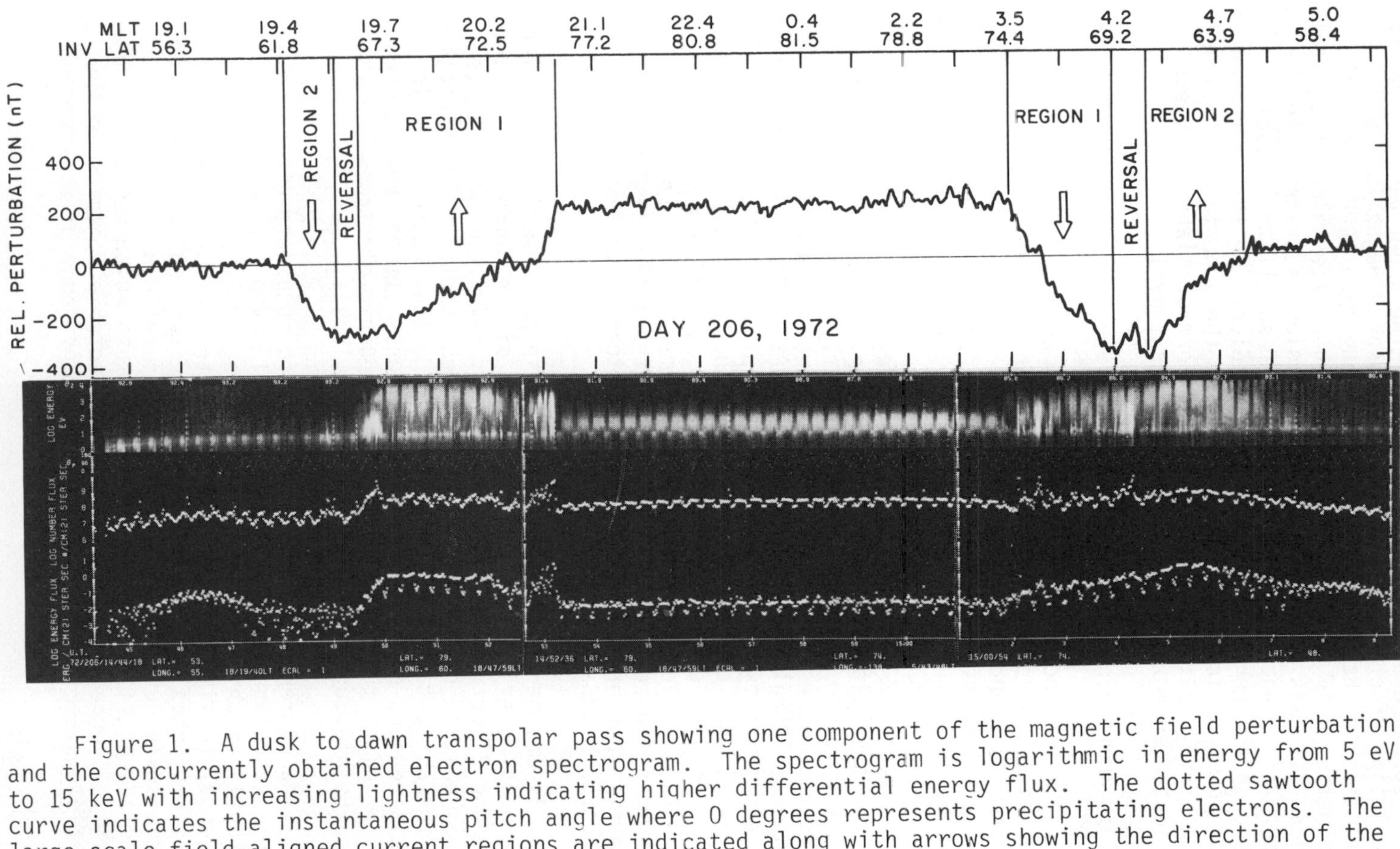

Figure 1. A dusk to dawn transpolar pass showing one component of the magnetic field perturbation and the concurrently obtained electron spectrogram. The spectrogram is logarithmic in energy from 5 eV to 15 keV with increasing lightness indicating higher differential energy flux. The dotted sawtooth curve indicates the instantaneous pitch angle where 0 degrees represents precipitating electrons. The large-scale field-aligned current regions are indicated along with arrows showing the direction of the current in each region.

time spectrogram in the lower panel. The upper panel shows the deviation of one component of the measured magnetic field relative to the value expected from the 1965 IGRF (updated) magnetic field model. The component used lies along the satellite spin axis and for the passes studied is oriented in the approximate east-west direction as the satellite crosses through the perturbation region. Gradients in this component of the magnetic field are taken to be the signatures of large-scale field-aligned currents in the form of slabs or sheets oriented approximately along the auroral oval.

II.1 Premidnight

In the local time sector between about 1800 and 2100 MLT the equatorward boundary of the region 2 large-scale current system is found to be persistently located equatorward of the low altitude termination of electrons precipitating from the plasma sheet. Electrons at an energy of 1keV are used here to define this boundary. The poleward boundary of this large-scale region 2 current, which may have a width on the order of a thousand kilometers along a meridian, is most often found to reach poleward into the region of nearly isotropic (> 1keV) electron fluxes that are normally taken to indicate the low altitude signature of plasma sheet (CPS). Thus, in spite of the upward current carried by these precipitating electrons, the net field-aligned current inferred from the magnetometer is downward. This apparent discrepancy is resolved by noting that only particles above 5eV are measured. Some cases (such as that shown in figure 1) do exist in which the entire region 2 magnetic perturbation is located equatorward of the 1keV electron boundary.

It may be generally stated that in this local time sector there is no evidence for the carriers of the downward directed (region 2) current in the 5eV to 15keV electrons or positive ions measured by the soft particle spectrometer. The direction of the current requires a net flux of either downward moving positive ions or upward moving electrons. It has been argued in the past, that upward moving ionospheric electrons are the responsible current carriers (Klumpar 1979).

The current reversal from downward to upward directed current separating region 2 from region 1 may occur quite abruptly in which case it appears effectively as a shear reversal to within the resolution of the measurements ($\sim$30km). On other occasions the oppositely directed region 2 and region 1 large-scale current systems may be separated by several tens to hundreds of kilometers by a region within which no significant parallel current is detected. No detectable features in the particle distributions within the energy range of measurement

(5eV to 15keV) have been found to correlate with either types of current reversal in this premidnight sector.

The region 1 upward directed current in this local time sector is typically bounded on its poleward edge by an abrupt change in the character of the auroral electrons. This characteristic in the electron distributions has been noted before as the poleward boundary of the boundary plasma sheet precipitation (BPS) (Winningham et al. 1975) and the equatorward boundary of the polar rain (Winningham and Heikkila 1974). Since the magnetic perturbation terminates at this boundary, with the magnetic gradient lying equatorward of it, one may conclude that the region 1 field-aligned currents flow along flux tubes containing at least the poleward portion of the BPS electron precipitation. The direction of this current and generally speaking its magnitude is consistent with the identification of these precipitating boundary plasma sheet electrons as the major carriers of this upward directed current.

Returning momentarily to the lack of an identifiable feature in the particle distributions associated with the region 2 - region 1 reversal it is now possible to see how this may arise. The net field-aligned current at any location is the sum of all parallel currents of either sign on that flux tube. One may conceive of two regions of space that are not prohibited from overlapping onto one another with one region dominated by downward currents carried for example by upward flowing electrons (ionospheric and or photoelectrons) and another containing precipitating plasma sheet electrons carrying an upward current. The final demarcation between the two regions and its detailed shape will depend entirely upon the relative magnitudes of the current densities carried by the particles in each plasma regime and by the extent of their overlap. In particular the reversal location will be highly sensitive to these factors, and will not explicitly relate to variations in the auroral particle distributions, as observed.

II.2 Postmidnight

In the postmidnight sector the sense of direction of the field-aligned currents is reversed with respect to the premidnight currents. The lower latitude region 2 currents are directed upward consistent with downcoming electrons as the carriers of the current. Comparisons of the spatial relationships between this current region and the plasma sheet electron precipitation indicate that these region 2 (postmidnight) currents have their equatorward boundary coincident with the equatorward termination of the precipitating plasma sheet electrons. This occurs in figure 1 just after 4.7 MLT and 63.9 invariant latitude and is most apparent as a change in the character of the spin modulation of the electron energy flux. From the data available it appears that electrons between 1keV and 15keV are the dominant carriers of this large-scale current.

In contrast, the large-scale region 1 current, directed downward in this local time sector, shows little responsiveness to the 5eV to 15keV particle distributions measured by the soft particle spectrometer. In general the carriers of this current system are not found in the energy range measured. Notable exceptions are at times observed where large fluxes of upward streaming electrons are measured and where the net current determined by direct detection of 5eV to 15keV charged particles, when properly integrated over all pitch angles, is downward. In such cases large fluxes of electrons below a few hundred eV are the dominant current carriers. Nevertheless, such instances occur infrequently enough to conclude that the dominant carriers of the large-scale postmidnight region 1 currents are not generally within the range of energies quoted above.

III. DISCUSSION

The large-scale field-aligned currents observed above the polar ionosphere must map into the magnetosphere to locations where magnetospheric processes are acting to drive the current. Their association with low altitude particle profiles described in the previous sections allows us to speculate how that mapping may occur.

The relationship of the region 2 currents to the low altitude signature of the electron plasma sheet indicate that in the premidnight sector the region 2 currents flow along field lines that cross the equatorial plane radially inward of the electron plasma sheet boundary. Models of the magnetospheric convection in the equatorial plane (Schield et al. 1979) show that the drift paths of ions reach closer to the earth than electrons at these local times setting up a positive space charge region in this part of the magnetosphere. Neutralization of this space charge by upflowing electrons from the ionosphere is, qualitatively, an appealing explanation for the source of the premidnight sector region 2 currents and is consistent with the observed equatorward displacement relative to the plasma sheet electrons.

Similarly in the postmidnight magnetosphere, electron drift paths reach inward of the ions and a negative space charge region will tend to develop. In this case the energetic electrons themselves provide the current carriers for the region 2 current as their mirror points are lowered by the space charge electric field causing them to precipitate into the atmosphere.

The mapping of the higher latitude, region 1, currents is somewhat more speculative but it seems reasonable to suggest that in the local time sectors under consideration these currents map into the low latitude boundary layer described by Eastman et al. (1976).

IV. SUMMARY

Table 1 summarizes the characteristics of the large-scale currents described in this paper. The region 1 and region 2 currents do not in general correspond respectively to the discrete and diffuse auroral particle domains, particularly with respect to the regions where the predominant current flow is downward. Precipitating electrons in the poleward portion of the premidnight BPS particle regime, and the diffuse auroral electron precipitation in the postmidnight sector do seem to indicate a systematic relationship with the large-scale upward field-aligned currents observed in these regions. In these limited regions the precipitating auroral electrons are consistent with the inferred currents.

TABLE I. CHARACTERISTICS OF LARGE SCALE FIELD-ALIGNED CURRENTS

	PREMIDNIGHT (1800-2100 MLT)		POSTMIDNIGHT (0300-0600 MLT)	
	REGION 2	REGION 1	REGION 1	REGION 2
CURRENT DIRECTION	DOWNWARD	UPWARD	DOWNWARD	UPWARD
POSSIBLE CURRENT CARRIERS	IONOSPHERIC ELECTRONS	AURORAL ELECTRONS (5eV to 15 keV)	IONOSPHERIC ELECTRONS; COLD AND/OR ACCELERATED	STEADY PLASMASHEET ELECTRON PRECIPITATION (>1 keV)
TOPOLOGICAL CONNECTION TO MAGNETOSPHERE	INNER PORTION OF ALFVÉN LAYER; ION DOMINATED DRIFT SHELLS	LOW LATITUDE BOUNDARY LAYER	LOW LATITUDE BOUNDARY LAYER	INNER PORTION OF ELECTRON PLASMASHEET; ION DRIFT EXCLUSION REGION

ACKNOWLEDGEMENTS

The author is grateful to Dr. J. R. Burrows for the magnetic deviations used in this analysis. The author's participation in the institute was supported by the National Aeronautics and Space Administration under grant number NSG 5085 and by the NATO Scientific Affairs Division. This work has been supported by the above named NASA grant and by the National Science Foundation under grant number ATM 75-03985.

REFERENCES

Deehr, C.S., Winningham, J.D., Yasuhara, F., and Akasofu, S.-I. 1976, J. Geophys. Res., 81, 5527

Eastman, T.E., Hones, E.W., Bame, S.J., and Asbridge, J.R. 1976, Geophys. Res. Lett., 3, 685

Iijima, T., and Potemra, T.A. 1976a, J. Geophys. Res., 81, 2165

Iijima, T., and Potemra, T.A. 1976b, J. Geophys. Res., 81, 5971

Klumpar, D.M. 1979, J. Geophys. Res., 84, 6524

Klumpar, D.M., Burrows, J.R., and Wilson, M.D. 1976, Geophys. Res. Lett., 3, 395

Lui, A.T.Y., and Anger, C.D. 1973, Planet. Space Sci., 21, 799

Lui, A.T.Y., Anger, C.D., Venkatesan, D., Sawchuk, W., and Akasofu, S.-I. 1975, J. Geophys. Res., 80, 1795

McDiarmid, I.B., Burrows, J.R., and Wilson, M.D. 1979, J. Geophys. Res., 84, 1431

Schield, M.A., Freeman, J.W., and Dessler, A.J. 1969, J. Geophys. Res., 74, 247

Winningham, J.D., and Heikkila, W.J. 1974, J. Geophys. Res., 79, 1393

Winningham, J.D., Yasuhara, F., Akasofu, S.-I., and Heikkila,W.J. 1975 J. Geophys. Res., 80, 3148

Zmuda, A.J., and Armstrong, J.C. 1974, J. Geophys. Res., 79, 4611

SOME ASPECTS OF ULF WAVES OBSERVED ONBOARD GEOS RELATED TO CONVECTION, HEATING AND PRECIPITATION PROCESSES

Roger GENDRIN

Centre de Recherches en Physique de l'Environnement
CNET - 92131 Issy-les-Moulineaux (France)

ABSTRACT

Because of their very sensitive wave and particle detectors the GEOS-1 and -2 spacecraft were well suited for a comprehensive study of wave-particle interactions leading to fast particle precipitation or heating on field lines mainly connected with the auroral zone. Data obtained in the ultra low frequency range (ULF) are presented here. Five types of phenomena are described, involving either quasi-monochromatic emissions or wide band noise. These phenomena involve high energy protons ($\sim 5 < E < 100$ keV), thermal Helium ions of ionospheric origin, or supra-thermal electrons ($E < 200$ eV).

1. INTRODUCTION

In the global process of magnetospheric convection and magnetosphere-ionosphere coupling, the precipitation of particles, both electrons and protons, in (or in the vicinity of) the auroral zone plays an important role. Wave-particle interactions in the equatorial region connected to auroral field lines are the basic mechanisms by which these precipitations can occur, because of the fast pitch-angle diffusion which they generally induce. This paper deals with recent in situ observations of these mechanisms in the ULF range ($\sim 0.1 < f < 10$ Hz), obtained via the ESA spacecraft GEOS-1 and -2.

ULF waves have for long been recognized as a typical characteristic of magnetospheric substorms (Gendrin 1970). They are also supposed to play an important role in the loss of ring

C. S. Deehr and J. A. Holtet (eds.), Exploration of the Polar Upper Atmosphere, 337–354.

current protons (Cornwall et al.1970; Gendrin 1975) in the generation of the SAR arcs (Cornwall et al. 1971), or of the Quasi-Periodic VLF emissions, with their associated electron precipitations (Kimura 1974). However, and apart from the long period pulsations ($T > 10s$) which can be detected with fluxgate magnetometers, not much experimental data were obtained in space concerning the shorter period pulsations ($0.1 < T < 10s$). Because of the high sensitivity of its search coils and of the use of a despin system, the GEOS ULF experiment has brought up a lot of new results concerning wave phenomena occurring in the deep magnetosphere ($L \simeq 5-8$) in this frequency range.

We will present here a selection of results showing the wide variety of ULF emissions that have been detected at (or in the vicinity of) the geosynchronous orbit with the two spacecraft GEOS-1 and GEOS-2. All these emissions are linked with phenomena of geophysical importance : convection, heating and precipitation.

Section 2 of this paper will be devoted to a general overview of the GEOS capabilities and to the characteristics of the ULF equipment and processing. Sections 3 and 4 will be devoted to the study of five ULF phenomena, namely
- short irregular pulsations (SIP's) associated with the break-up phase of the substorm,
- ULF-associated ELF electrostatic waves, which occur in conjunction with low energy ($E < 20$ eV) field-aligned electron fluxes,
- harmonically-related magnetoacoustic waves, which are linked to the existence of large gradients and strong anisotropy of high energy protons,
- Helium-associated ULF events, which may be at the origin of part of the energetic He^+ ions in the equatorial magnetosphere,
- magnetosheath ULF turbulence, which may partly solve the 'energy crisis' of present magnetopause reconnection theories.

Most of these phenomena bring out more questions than answers, but they are a vivid proof of the geophysical importance of magnetospheric ULF waves.

2. OVERVIEW OF THE GEOS CAPABILITIES

GEOS-1 was launched April 1, 1977, but it did not reach its nominal geostationary orbit. Instead, it had an elliptical orbit (perigee : 2050 km, apogee : 38.000 km, inclination : 26°) with a 12 hours period. It operated until June 23, 1978, with some brief reactivations. GEOS-2 was a truly geostationary satellite which was launched July 14, 1978; it will operate until the end of July 1980 and be reactivated for 6 months operation at the beginning of 1981, in conjunction with the European Incoherent Scatter Facility (EISCAT). The apogee of both satellites were shifted in longitude many times so as to allow conjugate point studies with different regions of the Earth (Northern Scandinavia, Iceland, Antarctica), cooperative studies with other geostationary satellites (SCATHA for instance) or special studies at different geomagnetic latitudes. The spin axes of the spacecraft were aligned with the Earth's axis with the possibility of inverting the spin axis direction (to avoid interferences caused by shadowing of the solar celles by the radial booms or vice-versa) or even staying for a while in any other orientation (in order to have a different pitch angle coverage for some detectors).

2.1. The measured parameters

The payloads of both spacecraft were identical (except for some minor modifications). They consist of a set of wave and particle detectors which allow for a very comprehensive study of magnetospheric processes (Knott 1975, 1979).

In the field of wave-particle interactions, we need the simultaneous measurements of different parameters which are measured by the different experiments with a reasonable time resolution. These are :
- the magnetic field which is provided by a three components fluxgate magnetometer,
- the AC electromagnetic field of the wave, which in the ULF range is provided by two experiments (see Section 3),
- the cold plasma density, which can be obtained from 6 different techniques , two of them (the relaxation sounder and the mutual impedance experiments) being original (Etcheto and Bloch 1978; Higel 1978; Décréau et al. 1978a). When the plasma density is large ($N_e \gtrsim 2\ cm^{-3}$) there is a reasonable agreement between all the measurements of N_e but, because of satellite potential or photoelectrons or wake effects, the values of N_e which are given by the two active experiments are the more reliable when $N_e \lesssim 2\ cm^{-3}$ (Décréau et al. 1978b),
- the plasma composition which, as will be shown, plays an important role in the generation of some ULF waves, is measured with a good accuracy by a mass and energy spectrometer (Geiss et al. 1978), at least as far as the suprathermal ($E \gtrsim 5$ eV) component is concerned,
- energetic particles (both protons and electrons) are measured in different energy ranges by 3 experiments : from ≈ 0 to 0.5 keV (Johnson et al. 1978), from ~ 0.2 to ~ 20 keV (Borg et al. 1978) and from ~ 17 keV to ~ 200 keV (Wilken et al. 1978). Up to now intercomparison between particle distribution functions and wave spectra have been done mainly on the basis of the third experiment.

2.2. The ULF experiment

The passive wave experiment onboard GEOS is part of the S-300 experiment which is a cooperative adventure to measure the DC and AC electric fields and the AC magnetic fields in the widest frequency range, by sharing elaborated onboard analysis techniques and telemetry systems. The technical capability of the S-300 experiment is described elsewhere (Jones 1978; S-300 Experimenters 1979) as well as the preliminary results obtained in the VLF range (Christiansen et al. 1978a; Cornilleau-Wehrlin et al. 1978a, b, c; Ungstrup et al. 1978). In this paper we will restrict ourselves to ULF observations.

Magnetic sensor characteristics. The magnetic sensors are search coils with a flux feed-back system to flatten the frequency response (Stéfant 1963). GEOS has 6 such antennas, orthogonal two by two. Three of these antennas are used to detect ULF waves in the frequency range ~ 0.1 - 11.5 Hz (the telemetry sampling rate is $23 s^{-1}$).

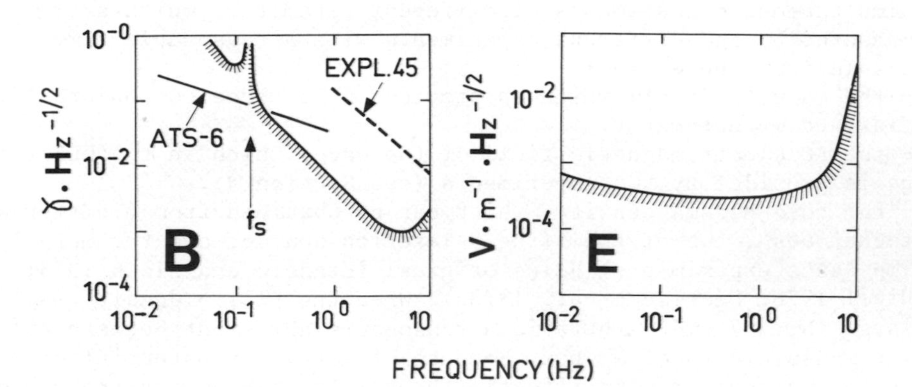

Figure 1. Sensitivity of the GEOS magnetic and electric antennas in the ULF range (S-300 Experimenters 1979).

The sensitivity of the magnetic ULF experiment is one of the best which has ever been achieved in space (Figure 1). Due to the spinning of the spacecraft (spin frequency ~ 0.17 Hz) it was necessary to include a despin system in order to reject the signal induced on the X and Y antennas by the component of the DC magnetic field perpendicular to the spin axis. This component may be very large (up to 500γ in the internal regions of the magnetosphere on GEOS-1). The despin system is a notch filter whose central frequency is continuously locked on the spin frequency via pulses sent by the solar aspect sensor. This system works in a wide frequency range (the spin frequency of the satellite being adjustable) and it has a reasonable efficiency (between 20 and 35 dB, depending on the component). The output

of the despin system is also sent to the ground via the low speed telemetry channel, in order to enable us to have a direct information on the amplitude and azimuth of the perpendicular component of the magnetic field.
Magnetic data processing. The data are processed off-line at the operational center of Darmstadt (West Germany) and experimenters' summaries are produced, which contain frequency-time grey scaling plots (up to 1.5 Hz) and amplitude-versus time plots of the noise integrated in different frequency bands. The data are also processed later on at the French computing center of CNES in Toulouse where more refined algorithms are applied. The details of the technique which is used are described elsewhere (Robert et al. 1979) and we will just summarize here the principle of the polarization analysis.

Because of the satellite rotation the X and Y antennas are not looking in a fixed direction. Consequently circularly polarized waves are Doppler-shifted, upward in frequency for those waves which rotate in the opposite sense than the spacecraft, downward in frequency for those waves which rotate in the same sense. It may even happen that a wave whose frequency is smaller than the spin frequency and which is polarized in the same sense as the satellite rotation is seen by the two orthogonal antennas as a wave polarized in the other direction. The signal induced at the spin frequency by the DC magnetic component in the plane X,Y, is itself seen as a polarized wave, at the spin frequency, rotating in a sense opposite to the spin.

An easy way to overcome all these effects is to make a complex polarization analysis (Kodera et al. 1977) in which one makes the Fourier transform of the complex signal X + jY and, afterwards, a positive or negative translation in frequency to obtain the spectra of the waves polarized in both directions, right or left (right-handed waves are waves rotating in the same direction around the DC magnetic field as the electrons). An example of the efficiency of this processing may be found in Perraut et al.'s Figure 1 (1978). Because the two despin systems do not have, for harware reasons, the same attenuation on both components the signal at the spin frequency cannot be completely suppressed and a spurious line at twice the spin frequency appears on the circular component which is polarized in the same direction as the satellite rotates. More evolved techniques can be used to suppress these spurious signals (Robert et al. 1979).
The electric ULF component. The signal coming from the DC electric spheres placed at the end of the long radial booms (experiment S-328) is transmitted to the ground via the low-speed telemetry system. It is mainly used to measure the amplitude and azimuth of the DC electric field. But, once the signal at the spin frequency has been removed, it may be also Fourier-analysed in the same frequency range as the magnetic AC signals. The respective

sensitivities of electric and magnetic sensors can be deduced from the examination of Figure 1. If one assumes that the Alfven velocity is of the order of $10^6 m.s^{-1}$, corresponding to an Alfven energy $1/2\ mV_a{}^2$ of the order of 5 keV (which is easily obtained for reasonable day values at the geostationary orbit, e.g. n = 10 cm^{-3} and $B_o = 140\gamma$), one sees that the electric and magnetic sensors have an equivalent sensitivity in the very low frequency range ($f \lesssim 0.1$ Hz) for electromagnetic waves ($E/V_aB \lesssim 2$). On the contrary the magnetic antenna is 100 times more sensitive than the electric antenna at 1 Hz and 1000 times at 10 Hz. This explains why the ULF electromagnetic phenomena have been observed much more frequently on the magnetic sensors than on the electric ones. Obviously, as soon as the waves have a significant electrostatic behaviour, the electric antenna is more suited to the observation.

Besides, the electric antenna system suffers from another drawback. Because of the existence of only one antenna, it is not possible to make a polarization study of the observed waves. An elliptically polarized wave, for example, will give two peaks, at $f + f_s$ and $f - f_s$, and it is difficult to interpret individual electric spectra containing a large number of lines, as well as it is to compare them with the corresponding magnetic spectra. Consequently most of the results which will be presented below refer only to the magnetic component of the wave.

3. TYPICAL ULF PHENOMENA : QUASI-MONOCHROMATIC EMISSIONS

Because of its high sensitivity and its efficient processing, the GEOS ULF experiment has already provided a lot of interesting results. New types of waves have been discovered, their relationships with the magnetospheric parameters have been studied, as well as their conjugacy properties (Perraut et al. 1978; Gendrin et al. 1978). However, whereas the general properties of these ULF phenomena were easy to establish, their quantitative intercomparison with the associated variation of all the relevant parameters is more difficult and is still in progress. What is reported in this Section and in the following one is intended to give the reader an indication of the variety of ULF phenomena which can be detected at the geostationary orbit and a feeling of how important these phenomena are in the context of magnetospheric dynamics. By no way the tentative interpretations which are given are to be considered as definitive : there is still a lot of quantitative work to be done before establishing all the experimental facts.

The five categories of phenomena which we have selected to illustrate these ideas correspond to different magnetospheric situations. We have divided them into two classes : the quasi-monochromatic emissions which will be presented below and the wide band phenomena which will be dealt with in the next Section. The quasi-monochromatic emissions themselves can be divided into two sub-classes : emissions below the proton gyrofrequency which are clearly associated with the presence of Helium ions and emissions above the proton gyrofrequency (multi-harmonic emissions). Phenomena of the first type, when sufficiently strong, are also at the origin of modulated ELF electrostatic emissions, which are interesting to study (Section 3.3).

3.1. Low frequency ULF events

Before the GEOS launch, it was generally admitted that most of the ULF waves detected on the ground in the Pc1 frequency range (~ 0.1 - 3 Hz) were generated in the equatorial region of the magnetosphere via a gyroresonant interaction with an anisotropic and energetic proton population. The characteristic energy and the anisotropy of this population on the one hand, and the cold plasma density and magnetic field intensity on the other hand were the only parameters intervening in the emission frequency. Consequently all values of the frequency (below the proton gyrofrequency F_{H^+}) were equally possible. Similarly, it was admitted that the waves, generated in the left-hand mode, should not have any difficulty to propagate down to the ionosphere and to be dete-detected on the ground. The fact that sometimes waves with an opposite polarization were observed was attributed to ionospheric effects, although it was never satisfactorily explained (Gendrin et al. 1966; Gendrin 1970; Perraut 1974; Kodera et al. 1977).

One of the first surprises we had, by comparing the ULF emissions observed onboard GEOS-1 with the ones detected on the ground at the conjugate point of Husafell, was that there were ULF waves which were observed onboard GEOS and which were not seen on the ground (Gendrin et al. 1978). Besides, the wave polarization as seen onboard GEOS was often linear and in some occasions right-handed.

The second surprise was that the observed wave frequencies were often near the He^+ gyrofrequency F_{He^+} (as deduced from the magnetometer measurements) and that in almost all cases F_{He^+} was organizing the data : when two emissions were simultaneously present, there was a clear gap in frequency just around F_{He^+} (for examples, see Figure 2). It was also shown that these emissions were very often associated with an increase of the thermal ($E \lesssim 25$ eV)He^+ density as measured by the ion composition experiment (Young et al. 1979).

The experimental study and the theoretical interpretation of these Helium-associated emissions are still in progress (Young et al. 1980). It is worth mentioning here the basic properties

upon which such an interpretation could be based :
- In a multicomponent plasma (H^+ + He^+), the dispersion relation of ULF waves is drastically changed to what it is when there is

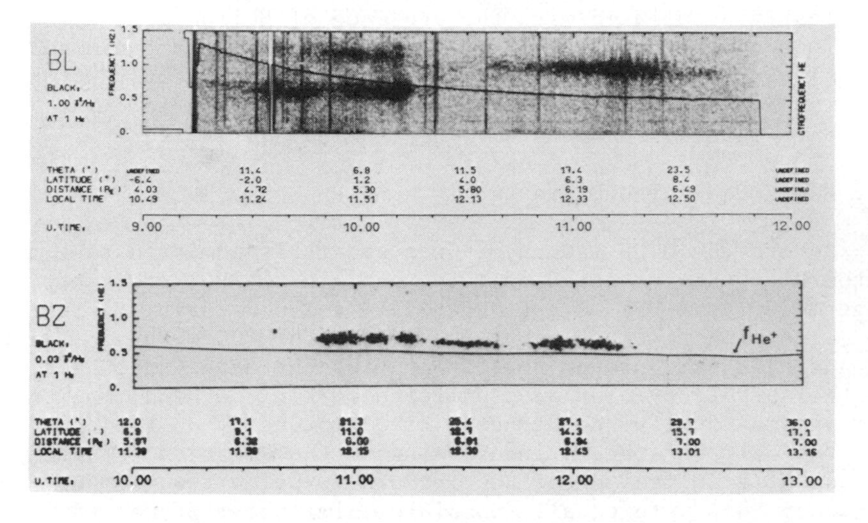

Figure 2. Examples of 'low-frequency' ULF events detected onboard GEOS-1 the 2nd and the 14th of aug.1977. Only one among the three available polarized components (BR, BL, BZ) has been represented. The upper case shows a 'two-frequency' emission, whose lower part cannot propagate to the satellite when it reaches the local value of F(He^+). In the second example the emission frequency follows the variation of F(He^+) (courtesy of S. Perraut).

only one ion (H^+) present. New resonances and cut-offs appear; there exists a cross-over frequency at which the polarization of a mode reverses (Smith and Brice 1964; Gurnett et al. 1965). Because of the variability of the plasma parameters along and across the field lines, these effects play an important role in the propagation of waves between the equator and the satellite (for observations off the equator) or between the equator and the ground. They may explain the great variety of polarization states observed at both places.

- In such a plasma the instability conditions for cyclotron waves generated by a hot anisotropic proton population are also modified : the growth rates are increased at some frequencies and they are decreased at some others (Märk 1974; Cuperman et al. 1975 a,b).The preferential appearance of emissions in the vicinity of the He^+ gyrofrequency is related to this effect.

- Finally the generated waves can resonate with cold He^+ ions and bring them, by quasilinear diffusion, to suprathermal ($E \gtrsim 100$ eV) energies. Their pitch-angle is also increased so that He^+ ions of ionospheric origin may become trapped (Brice and Lucas 1975; Gendrin and Roux 1980).

These Helium-associated ULF events being rather common, such a process may be one of the possible mechanisms by which He^+ ions of ionospheric origin are energized and trapped in the equatorial magnetosphere. A similar process, involving lower frequency waves, may also be at the origin of the capture and of the energization of O^+ ions. These processes have to be compared with the other mechanisms presently proposed to explain the origin of energetic ionospheric ions in the magnetosphere (see Prangé 1978 for a review).

3.2. Harmonically related emissions

Above the proton gyrofrequency, a new category of ULF waves was observed, mainly in the afternoon and dusk sectors. They consist of a series of monochromatic emissions, harmonically related,

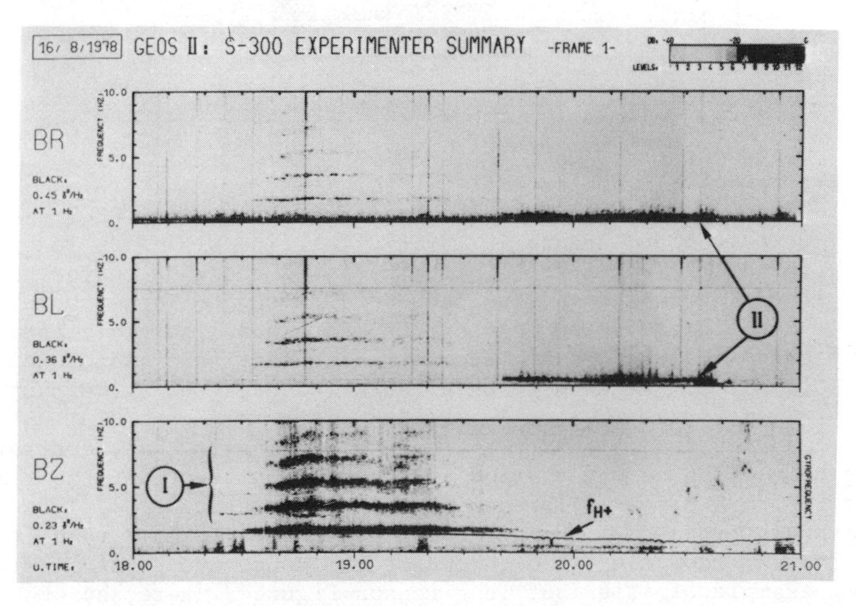

Figure 3. Harmonically-related 'high-frequency' events followed by a 'low-frequency' emission. The first event is mainly polarized along the Z axis (e.g. along B_o) whereas the second is mainly polarized in the X-Y plane, in the left-hand mode.

whose fundamental frequency lies in the vicinity of F_{H^+}. The emission often starts exactly at F_{H^+} but sometimes the initial frequency differs from the local gyrofrequency. The number of harmonics is very large : the emission are almost always observed up to the Nyquist frequency of the antenna (11.5 Hz) and even sometimes a strong aliasing occurs so that emissions above 11.5 Hz are detected in the frequency band of the system (see Perraut et al.'s (1978) Figure 8). The fundamental and the first harmonics are sometimes missing. Similar waves have already been reported to exist at lower L-values (Gurnett 1976) but their identification with respect to the characteristic frequencies of the plasma was not convincing and, to our knowledge at least, no more examples were given.

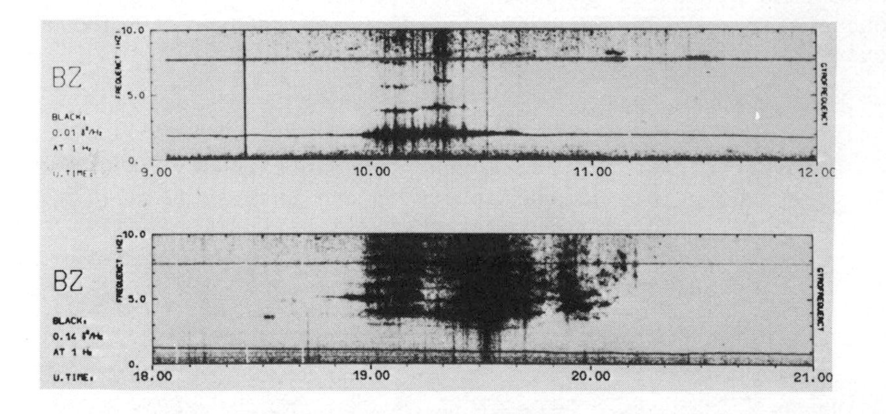

Figure 4. Examples of 'high-frequency' events (GEOS-2 : October 8 and August 10, 1978). The line at 7.7 Hz is due to interference. The line between ~ 1 and 2 Hz represents the H^+ gyrofrequency as deduced from the DC magnetometer. In the second example the fundamental and the two first harmonics are absent, demonstrating that this phenomenon is not due to saturation effects (courtesy of S. Perraut).

An important property of these emissions is that, contrary to the ones which were described in the previous Section, their magnetic field is almost aligned with the DC magnetic field. This means that they propagate in a direction perpendicular to B_o and it explains why they are not detected at the conjugate point. A striking example of this fact is given on Figure 3 where, by chance, such an emission was followed, within the same 3 hour period of the summary, by a low frequency event, with $k \parallel B_o$. Other examples of this kind of emissions are given in Figure 4. Again, in contrast with the low frequency waves, these emissions are observed in a restricted range of geomagnetic latitude (± 10°) which explains why they are so often seen onboard GEOS-2.

This phenomenon occurs in conjunction with sudden proton injection events with a larger increase of 90° pitch angle particle fuxes (see Perraut et al.'s (1978) Figure 10). A similar association was found with EXPLORER-45 by Taylor and Lyons (1976), although the frequency resolution of the ULF detector in the frequency range 1-30 Hz (3 bands only) did not allow a precise evaluation of the wave spectrum. One theory of these emissions involves the coupling of a ring distribution and the magnetosonic mode (Gul'elmi et al. 1975). Their role with respect to protons could be similar to that of electrostatic waves near $(n+1/2)f_{ce}$ with respect to electrons (Ashour-Abdalla and Kennel 1976).

3.3. ULF-modulated ELF waves

The modulation of VLF electromagnetic waves by strong ULF oscillations is a phenomenon which has been observed since long at the ground. These VLF emissions are known as 'quasi-periodic (QP) emissions' (see Kimura 1974, for a review) but they mainly involve long period ULF waves in the Pc-3, Pc-4 frequency range ($T \sim 20 - 100$ s). One of the presently available theories for explaining such a modulation effect is based upon the change of the loss cone angle induced by the change in the DC magnetic field, the Pc-4 wave being assumed to be compressional. A variation of the loss cone angle means a variation in the precipitation rate of energetic electrons and therefore a variation of the amplitude of the VLF wave generated by these electrons (Coroniti and Kennel 1970a,b). Such QP emissions have been observed both onboard GEOS and at the ground but the frequency of the associated ULF waves lies well beyond the frequency range of the S-300 ULF experiment.

Recently a new kind of modulated VLF emissions has been observed onboard GEOS, which involves ULF waves in the frequency range of our experiment. Besides the difference in the modulation frequency, there are other differences between this phenomenon and the QP emissions :
1. The VLF emission lies in the extremely low frequency range (~ 60-400 Hz), i.e. between the lower hybrid frequency and the ion plasma frequency, whereas the QP emissions occur at higher frequencies (~ 0.8 - 2 kHz);
2. The VLF emission is electrostatic and not electromagnetic;
3. The ULF-associated wave has almost no magnetic field component along the DC magnetic field.

Examples of these modulated ELF events can be found in Cornilleau-Wehrlin (1980). The ULF waves which initiate them are of the type described in Section 3.1. Because these waves are generated in the vicinity of a resonance, their electric component parallel to the DC magnetic field is sufficiently strong to induce a parallel drift of low energy electrons above the thermal velocity, therefore producing an instability in the magnetoacoustic mode. Such phenomena could be at the origin of field-aligned, suprathermal electron beams.

The frequency of the modulation (~ 0.2 - 0.6 Hz) lies well in the frequency range of the pulsating auroras and some indication exists that when pulsating auroras are detected on the ground, similar VLF modulated events are detected onboard GEOS (M.P. Gough and P. Rothwell, private communication). Besides it has been recently shown that fluctuations of the 4278 N^+_2 intensity recorded at Siple Station, Antarctica, are well correlated with Pc1 pulsations recorded at this same station and at its conjugate point in Canada (Mende et al. 1980). These are other proofs that ULF waves may be the driving mechanism of important magnetospheric phenomena.

4. TYPICAL ULF PHENOMENA : WIDE-BAND EMISSIONS

In this Section we will report on wide-band noise observed in the equatorial region during substorms or during magnetopause crossings.

4.1. Substorm events

During disturbed conditions the magnetic field variations cover the whole frequency spectrum from DC to/and above the proton gyrofrequency. In the ULF range the detailed frequency analysis of signals received on the ground has shown that Pi-1's mainly consist of auroral agitation, AA, and of short irregular pulsations, SIP's. The precise mechanisms by which these waves are generated (and even the precise region where they are generated) were not clearly identified (Gendrin 1970).

Measurements onboard GEOS indicate that these emissions are also present in the equatorial plane, with an intensity which is not much different from the ground intensity. AA is sometimes present, with a spectrum extending up to the Helium gyrofrequency (see Figure 5).

SIP's have the interesting property of being associated, within tens of seconds, with the break-up phase of substorms. This was evidenced during a ground campaign in Northern Scandinavia during which sensitive photometric and photographic equipments were operated. The photometer was able to detect fast enhancements and displacements of auroral arcs during the substorm break-up. Figure 5 gives an example of such an event. Before 22.35 UT, December 29, 1978, the pre-substorm period is clearly indicated by the decrease of high energy ($E > 22$ keV) electron flux and the increase of the component of the DC magnetic field perpendicular to the Earth's axis, as detected by the ULF despin signal (see Section 2.3.). The magnetic field takes a more and more tailward configuration. At the break-up time, the DC magnetic field quickly recovers its dipolar configuration (DX decreases), indicative

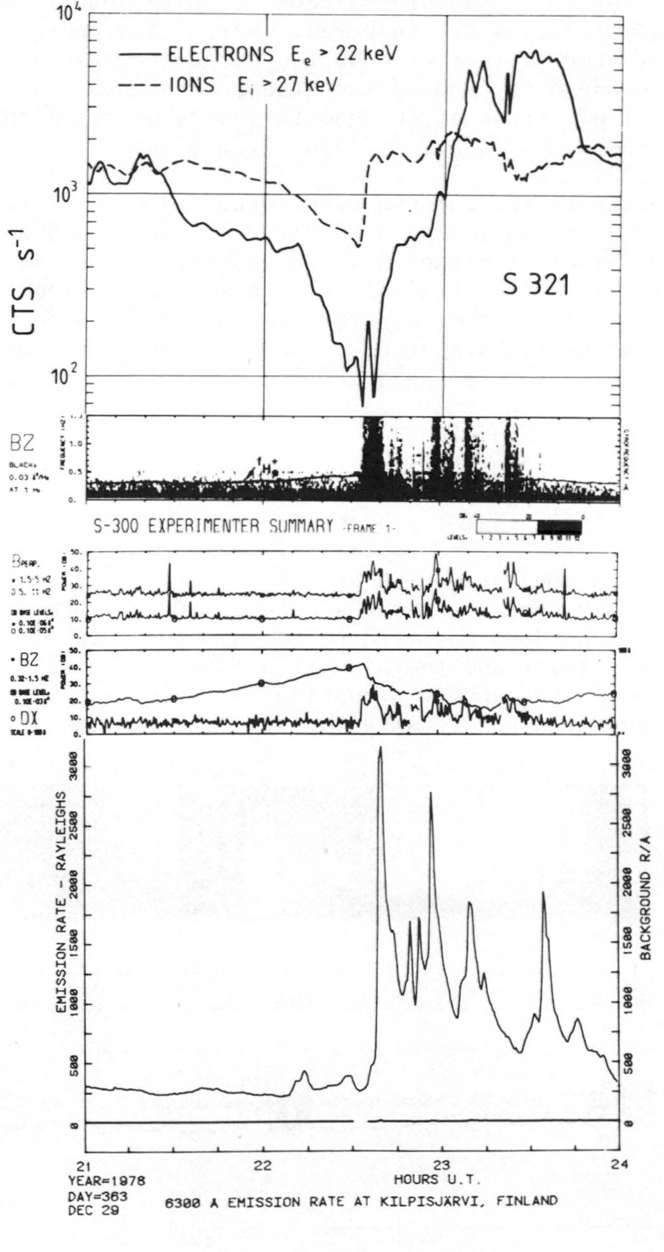

Figure 5. Ground photometric observation of substorm break-ups (bottom), in conjunction with changes in the GEOS high energy particle fluxes (top) and with occurrences of SIP's (middle). Note the building of a large tailward component of B, as measured by the despin system of the ULF experiment (DX signal) just before the break-up (from Shepherd et al. 1980).

of the ebb motion of the plasmasheet. Simultaneously, high energy protons and electrons are injected, whereas low energy electrons are precipitated as indicated by the large increase of the 630 nm line on the one hand, and by the disappearance of low energy (E < 500 eV) electrons at the geostationary orbit on the other hand (see Shepherds et al.'s (1980) Figure 1a).

During that time the ULF wave intensity becomes rather high ($\sim 0.1\gamma.Hz^{-1/2}$ around 1 Hz and $\sim 0.01\gamma.Hz^{-1/2}$ between 5 and 10 Hz). This is the proof of either a rapid displacement of currents crossing by the satellite trajectory (the magnetic antenna being sensitive to $\partial B/\partial t$) or of a strong turbulence induced at the edge of the plasmasheet. More precise spectral and polarization analyzes are needed before being able to conclude on this point.

4.3. Magnetopause crossings

From the GEOS-1 and GEOS-2 crossings of the magnetopause, it has been concluded that a region of strong ULF turbulence does exist within a thin boundary layer (Figure 6). The layer dimensions cannot be measured with one single spacecraft (see Russell and Elphic 1979) but it cannot exceed a ~500 - 1000 km thickness. Outside this boundary and within the magnetosheath, the noise level is much lower and the situation looks more or less similar to the inner magnetosphere situation, with the exception that He^{++} ions seem now to organize the data (Perraut et al. 1979).

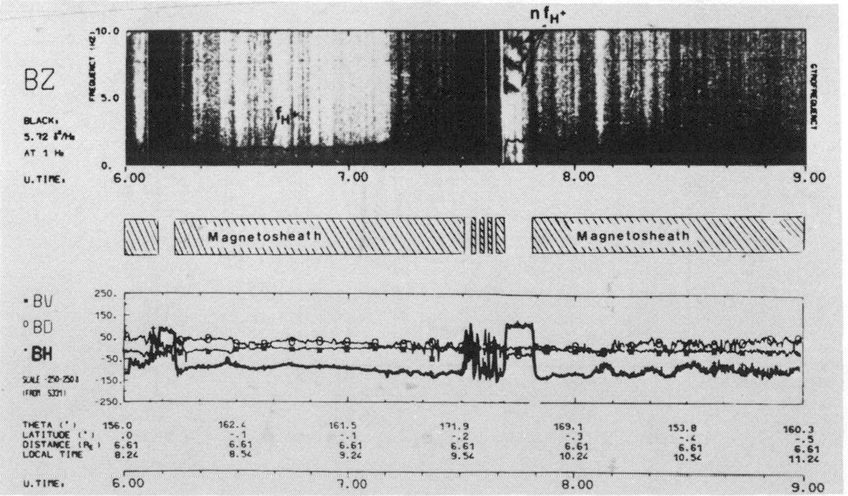

Figure 6. ULF turbulence as observed during consecutive magnetopause crossings by GEOS-2 the 28th of Aug. 1978. Crossings are identified by changes of the horizontal component BH of the DC magnetic field (Perraut et al. 1979).

Within the boundary layer, the level of turbulence is very high (up to 10γ within the 11 Hz bandwidth). It is anticorrelated with the intensity of the DC magnetic field, indicating that, when we are at the center of the magnetopause current sheet, the turbulence is stronger. Obviously the (quasi-periodic) displacement of this current sheet is itself able to induce in a dB/dt system some signal, but preliminary computations show that this effect should not be important above 2-3 Hz, whereas the turbulence is still strong above this level ($\sim 0.3\gamma.Hz^{-1/2}$ up to 11 Hz).

In their paper Perraut et al. (1979) have made estimations of the power flux which is transported by such intense ULF waves, and they have found that this power flux ($\sim 10^{-5}W.m^{-2}$) is not much different from the one which is dissipated by the magnetopause current in the presence of a reconnection electric field ($j \sim 8.10^{-2}$ $A.m^{-1}$, $E \sim 0.2$ $mV.m^{-1}$). The energy density ($\sim 4.10^{11}J.m^{-3}$) is also not much different from the magnetosheath particle enrgy density ($n \sim 20cm^{-3}$, $T \sim 10$-100 eV), so that in any case one is near the strong turbulence limit.

Other works have recently been devoted to the study of plasma flows in the vicinity of the boundary layer (Paschmann 1979; Paschmann et al. 1979; Richter et al. 1979). It seems that, with a precise information on the electromagnetic and electrostatic wave field, one would be able in the near future to solve the problem of the solar wind -magnetosphere interaction at the front of the magnetopause.

5. CONCLUSION

Because of their high sensitivity and their good frequency and time resolution, the wave detectors which were operating onboard GEOS have permitted to demonstrate the existence of a large variety of waves occurring at different local times or in various magnetospheric configurations and to establish their consequences as regards the dynamics of magnetospheric particles. It has been shown that ULF emissions not only involve interaction with protons but are also associated with changes in the cold plasma composition, with heavy ions heating, electron precipitation, etc..

In the global scheme of particle convection and precipitation, which is the result of the magnetosphere-ionosphere coupling, Wave-Particle Interactions (WPI) are the microscopie mechanisms which control the changes of particle distribution functions. The time constants of these changes (~2-20 mn) are generally faster than the time constants (~1-6 hrs) involved in the large scale exchange processes. Consequently WPI's play a sort of catalytic role similar to the one which, for instance, is played by chemical reactions of minor constituents in the dynamics of the neutral atmosphere.

The detailed measurements of ULF waves in space (as well as of VLF electrostatic or electromagnetic waves), combined with the measurements of all the plasma parameters make it now possible to understand the real importance of WPI's in the Earth-Magnetosphere system.

Acknowledgements. The data presented in this paper are the result of a very long and fruitful cooperation among all the GEOS experimenters. I would like to acknowledge especially S. Perraut, P. Robert and A. ROUX (C.R.P.E., France) for the ULF data, D.T. Young (University of Bern, Switzerland) for the ion composition data, and A. Korth and G. Kremser (Max-Planck Institut für Aeronomie, Germany) for their high energy proton measurements. The European Space Operation Center (Darmstadt) and the Centre National d'Etudes Spatiales (Toulouse) must also be thanked for the help they provided in operating the spacecraft and processing the huge amount of data in a very cooperative and efficient way.

REFERENCES

Nota : All references to Volume 22 of Space Sci. Rev., 1978 can be found in Advances in Magnetospheric Physics with GEOS-1 and ISEE (Eds. K. Knott, A. Durney and K. Ogilvie), D. Reidel Pub., Dordrecht, 1979.

Ashour-Abdalla, M., and Kennel, C.F. 1975, p. 201 in Physics of the Hot Plasma in the Magnetosphere, Eds. B. Hultqvist and L. Stenflo), Plenum Press, New-York.
Borg, H., Holmgren, L.A., Hultqvist, B., Cambou, F., Rème, H., Bahnsen, A., and Kremser, G.,1978, Space Sci. Rev., 22, 501.
Brice, N., and Lucas, C. 1975, J. Geophys. Res., 80, 936.
Christiansen, P.J., Gough, M.P., Martelli, G., Bloch, J.J., Cornilleau-Wehrlin, N., Etcheto, J., Gendrin, R., Béghin, C., and Décréau, P. 1978a, Nature, 272, 682.
Christiansen, P.J., Gough, M.P., Martelli, G., Bloch, J.J., Cornilleau-Wehrlin, N., Etcheto, J., Gendrin, R., Béghin, C., Décréau, P., and Jones, D. 1978b, Space Sci. Rev., 22, 383.
Cornilleau-Wehrlin, N. 1980, J. Geophys. Res., to be published.
Cornilleau-Wehrlin, N., Gendrin, R., Lefeuvre, F., Parrot, M., Grard, R., Jones, D., Bahnsen, A., Ungstrup, E., and Gibbons, W. 1978a, Space Sci. Rev., 22, 371.
Cornilleau-Wehrlin, N., Gendrin, R., and Tixier, M. 1978b, Space Sci. Rev., 22, 419.

Cornilleau-Wehrlin, N., Gendrin, R., and Perez, R., 1978c, Space Sci. Rev., 22, 443.
Cornwall, J.M., Coroniti, F.V., and Thorne, R.M. 1970, J. Geophys. Res., 75, 4699.
Cornwall, J.M., Coroniti, F.V., and Thorne, R.M. 1971, J. Geophys. Res., 76, 4428.
Coroniti, F.V., and Kennel, C.F. 1970a, J. Geophys. Res., 75,1279.
Coroniti, F.V., and Kennel, C.F. 1970b, J. Geophys. Res., 75,1863.
Cuperman, S., Gomberoff, L., and Sternlieb, A. 1975a, J. Plasma Phys., 14, 195.
Cuperman, S., Gomberoff, L., and Sternlieb, A. 1975b, J. Geophys. Res., 80, 4643.
Décréau, P.M.E., Béghin, C., and Parrot, M. 1978a, Space Sci.Rev., 22, 581.
Décréau, P.M.E., Etcheto, J., Knott, K., Pedersen, A., Wrenn, G., and Young, D.T. 1978b, Space Sci. Rev., 22, 633.
Etcheto, J., and Bloch, J.J. 1978, Space Sci. Rev., 22, 597.
Geiss, J., Balsiger, H., Eberhardt, P., Walker, H.P., Weber, L., Young, D.T., and Rosenbauer, M. 1978, Space Sci. Rev., 22, 537.
Gendrin, R. 1970, Space Sci. Rev., 11, 54.
Gendrin, R. 1975, Space Sci. Rev., 18, 145.
Gendrin, R., Lacourly, S., Gokhberg, M.V., and Troitskaya, V.A. 1966, Ann. Geophys., 22, 329.
Gendrin, R., Perraut, S., Fargetton, H., Glangeaud, F., and Lacoume, J.L. 1978, Space Sci. Rev., 22, 433.
Gendrin, R., and Roux, A. 1980, J. Geophys. Res., to be published
Gul'elmi, A.V., Klaine, B.I., and Potapov, A.S. 1975, Planet. Space Sci., 23, 279.
Gurnett, D.A. 1976, J. Geophys. Res., 81, 2765.
Gurnett, D.A., Shawhan, S.D., Brice, N.M., and Smith, R.L. 1965, J. Geophys. Res., 70, 1665.
Higel, B. 1978, Space Sci. Rev., 22, 611.
Johnson, J.F.E., Sojka, J.J., and Wrenn, G.L. 1978, Space Sci. Rev., 22, 567.
Jones, D. (for GEOS S-300 Experimenters) 1978, Space Sci. Rev., 22, 327.
Kimura, I. 1974, Space Sci. Rev., 16, 389.
Knott, K. 1975, ESA/ASE Scient. Techn. Rev., 1, 173.
Knott, K. 1978, Space Sci. Rev., 22, 321.
Kodera, K., Gendrin, R., and de Villedary, C. 1977, J. Geophys. Res., 82, 1245.
Märk, E. 1974, J. Geophys. Res., 79, 3218.
Mende, S.B., Arnoldy, R.L., Cahill, J. Jr., Doolittle, J.H., Armstrong, W.C., and Fraser-Smith, A.C. 1980, J. Geophys. Res., 85, 1194.
Paschmann, G. 1979, p. 25 in Magnetospheric Boundary Layers (Ed. B. Battrick), ESA-SP/148, Paris.
Paschmann, G., Sonnerup, B.U.O., Papamastorakis, I., Schopke, N.,

Haerendel, G., Bame, S.J., Asbridge, J.R., Goshing, J.T., Russell, C.T., and Elphic, R.C. 1979, Nature, 282, 243.
Perraut, S. 1974, Thesis, Paris.
Perraut, S., Gendrin, R., Robert, P., Roux, A., and de Villedary, C. 1978, Space Sci. Rev., 22, 347.
Perraut, S., Gendrin, R., Robert, P., and Roux, A. 1979, p. 113 in Magnetospheric Boundary Layers (Ed. B. Battrick), ESA-SP/148, Paris.
Prangé, R. 1978, Ann. Geophys., 34, 187.
Richter, A.K., Keppler, E., Axford, W.I., and Denskat, K.U. 1979, J. Geophys. Res., 84, 1453.
Robert, P., Kodera, K., Perraut, S., Gendrin, R., and de Villedary, C. 1979, Ann. Telecomm., 34, 179.
Russell, C.T., and Elphic, R.C. 1979, Geophys. Res. Letters, 6,33.
Shepherd, G.G., Boström, R., Derblom, H., Fälthammar, C.G., Gendrin, R., Kaila, K., Korth, A., Pedersen, A., Pellinen, R., and Wrenn, G. 1980, J. Geophys. Res., to be published.
Stéfant, R. 1963, Ann. Geophys., 19, 250.
S-300 Experimenters 1979, Planet. Space Sci., 27, 317.
Taylor, W.W., and Lyons, L.R. 1976, J. Geophys. Res., 81, 6177.
Ungstrup, E., Neubert, T., and Bahnsen, A. 1978, Space Sci. Rev., 22, 453.
Wilken, B., Fritz, T.A., Korth, A., and Kremser, G. 1978, Space Sci. Rev., 22, 647.
Young, D.T., de Villedary, C., Gendrin, R., Perraut, S., Roux, A., Jones, D., Korth, A., and Kremser, G. 1979, EOS, 60, 598.
Young, D.T., Perraut, S., Roux, A., de Villedary, C., Gendrin, R., Korth, A., Kremser, G., and Jones, D. 1980, J. Geophys. Res., submitted to.

ELECTRON CYCLOTRON WAVES IN THE EARTH'S MAGNETOSPHERE

P.J. Christiansen, M.P. Gough and *K. Rönnmark

Plasma and Space Physics Group, University of Sussex,
Brighton, Sussex, BN1 9QH, U.K.

*Kiruna Geophysical Institute, University of Umea,
Umea, Sweden.

INTRODUCTION

Observations of electrostatic waves in the Earth's magnetosphere which exhibited complex frequency banding related to harmonics of the electron gyrofrequency ("3/2 f_{ce} emissions") were first reported by Kennel et al (1). The authors also pointed out that, given the large amplitudes often attained by the waves ($\gtrsim$ 1 mV/m), electrons would be rapidly diffused in velocity space and the resulting precipitated fluxes could provide a significant contribution to the diffuse component of auroral precipitation.

This first report has stimulated considerable efforts during the past decade by theorists and spacecraft experimentalists to understand both the basic plasma physics of these waves and the role that they play in magnetospheric processes, efforts which have recently been extended to similar phenomena occurring in the Jovian magnetosphere.

By virtue of familiarity, this brief review will be illustrated by examples of data from the GEOS satellites (Knott (2)) and particularly from the S-300 wave experiment (S-300 experimenters (3)). The interpretation of many of these data relies on the use of novel diagnostic experiments - the S-301 resonance sounder (Etcheto et al (4)), the mutual impedance probe (Decreau et al (5)) and data from the particle experiments S-302 (Johnson et al (6)) and S-310 (Borg et al (7)).

C. S. Deehr and J. A. Holtet (eds.), Exploration of the Polar Upper Atmosphere, 355–366.

SOME THEORETICAL ASPECTS

A magnetised plasma is enormously rich in wave phenomena. The electrostatic electron cyclotron harmonic (ECH or Bernstein) modes which are important to this review were first discussed by Bernstein (8), and the general linear dispersion relation of these waves, which propagate almost at right angles to the ambient magnetic field, by Harris (9). There followed an extensive literature mainly with reference to laboratory plasmas which is reviewed by Mikhailovskii (10).

It was with the paper by Fredericks (11) that the specific problem of magnetospheric ECH instabilities began to be tackled, and in the subsequent contributions of Young et al (12), Karpman et al (13), Ashour-Abdalla et al (14), Ashour-Abdalla and Kennel (15, 16) the basis of a reasonable (in the sense that plausible velocity distributions are used) linear stability theory has emerged. The consensus view appears to be that a plasma containing at least two components should exist - a hot component, (density n_h, temperature T_h) and a cool component (n_c, T_c). The cool component, of ionospheric origin, controls to a large degree the dispersion of the waves, and aids destabilisation. The hot component of plasma sheet origin, has a distribution function $f(v)$ containing regions of velocity space where $\partial f/dv_{\perp} > 0$, developed as a result of inward convection of $\sim$ keV electrons from the geomagnetic tail (Ashour-Abdalla and Cowley (17). This provides the 'free energy' to drive the instabilities. The cool plasma upper hybrid frequency $f_{uhc} = (f_{pc}^2 + f_{ce}^2)^{\frac{1}{2}}$ (f_p is the electron plasma frequency), is important over a large range of the ratio n_c/n_h, since in its vicinity wave modes with low group velocities exist and may give rise to strong convective or even absolute instabilities. The emphasis of much of this early work, guided by the existing experimental results, was on waves in the lowest gyroharmonic bands, but has more recently been extended to cover instabilities observed in higher harmonic bands by Hubbard and Birmingham (18) and Ronnmark et al (19).

The dielectric constant $\varepsilon(\omega,\underline{k})$ for electrostatic waves $(\underline{k}//\underline{E})$ with complex frequency $\omega = \omega_r + i\overline{\gamma}$ and real wave vector $\underline{k}$ (components $k_{\perp}$ perpendicular and $k_{\parallel}$ parallel to the ambient magnetic field) is given by equation [1] below:

$$\varepsilon(\omega,\underline{k}) = 1 - \frac{\omega_p^2}{k^2} \sum_{n=-\infty}^{\infty} \int d\underline{v} \frac{J_n^2(kv_{\perp}/\omega_{ce})}{k_{\parallel}v_{\parallel}-\omega-n\omega_{ce}} \left[\frac{n\omega_{ce}}{v_{\perp}} \frac{\partial f}{\partial v_{\perp}} + k_{\parallel} \frac{\partial f}{\partial v_{\parallel}}\right]$$

where ω_p is the plasma frequency, J_n Bessel function of order n, f is the distribution function; $v_{\parallel}$, $v_{\perp}$ components of the particle velocity. For $\gamma \ll \omega_r$ the real dispersion and growth rate can be calculated from

$$\varepsilon_r(\omega,\underline{k}) = 0; \qquad \gamma = -\varepsilon_i/(\partial\varepsilon_r/\partial\omega) \qquad [2]$$

where ε_r is evaluated from 1 using the principal value of the integral and

$$\varepsilon_i\ (\omega,k) = -\frac{\omega_p^2}{k^2|k\ |}\sum_n \int 2\pi\ v_\perp dv_\perp J_n^2\left(\frac{k_\perp v_\perp}{\omega ce}\right).$$

$$\left[\frac{n\omega_{ce}}{v_\perp}\ \frac{\partial f}{\partial v_\perp} + k_{\shortparallel}\ \frac{\partial f}{\partial v_{\shortparallel}}\right] \quad v_{\shortparallel} = (\omega_r - n\omega_c)/k_{\shortparallel} \qquad [3]$$

In a two component plasma the low energy component will often determine the real part of the dispersion (Young et al (12)) so in Figure 1(a) we show the perpendicular dispersion relations for ECH waves for three values of the ratio ω_p/ω_{ce}, corresponding to wave propagation at three values of n_c. In a multicomponent plasma, corrections due to the hotter component must be introduced at small $k_\perp$. Note the low perpendicular group velocity modes ($vg_\perp = \partial\omega/\partial k_\perp \to 0$) in the bands containing and above the respective upper hybrid frequencies (f_q's).

Restricting ourselves to positive energy waves ($\partial\varepsilon_r/\partial\omega > 0$) (10) with $\gamma_k = -\varepsilon_i/(\partial\varepsilon_r/\partial\omega > 0$ for instability, then from [3] the destabilising contribution is proportional to $\sum_n J_n(\tilde{k})\,\partial f/\partial_\perp|v_{\shortparallel} = (\omega_k - n\omega_{ce})/k_{\shortparallel}$ since in the loss cone region $\partial f/\partial v_\perp > 0$ and $\partial f/\partial v_{\shortparallel}$ is negative. The reader unfamiliar with the field may well wonder how the frequencies and wavenumbers of the unstable waves are related to the destabilising distribution function. Mapping from $(\omega,\underline{k})$ space to $\underline{v}$ space can be performed by a graphical technique (Ronnmark et al (19)) but in Figures 1(b) and 1(c) we give a (rougher) illustration for waves in the band $6 < \omega/\omega_{ce} < 7$, for the $n = 6$ resonance, and for two values of $v_{\shortparallel}$, obeying the resonance condition $\omega - k_{\shortparallel}v_{\shortparallel} = 6\omega_{ce}$.

Finally we note that for computational simplicity the frequency was assumed complex; $\gamma > 0$ does not imply that the waves grow at this rate at all points in space (absolute instability). This is only true if the group velocity is also zero, otherwise the waves exponentiate as they convect in space at a rate

$$k_i = -\gamma/\left[(\partial\omega/\partial k_{\shortparallel})^2 + (\partial\omega/\partial k_\perp)^2\right]^{\frac{1}{2}},$$

thus emphasing the importance of the group velocity (15).

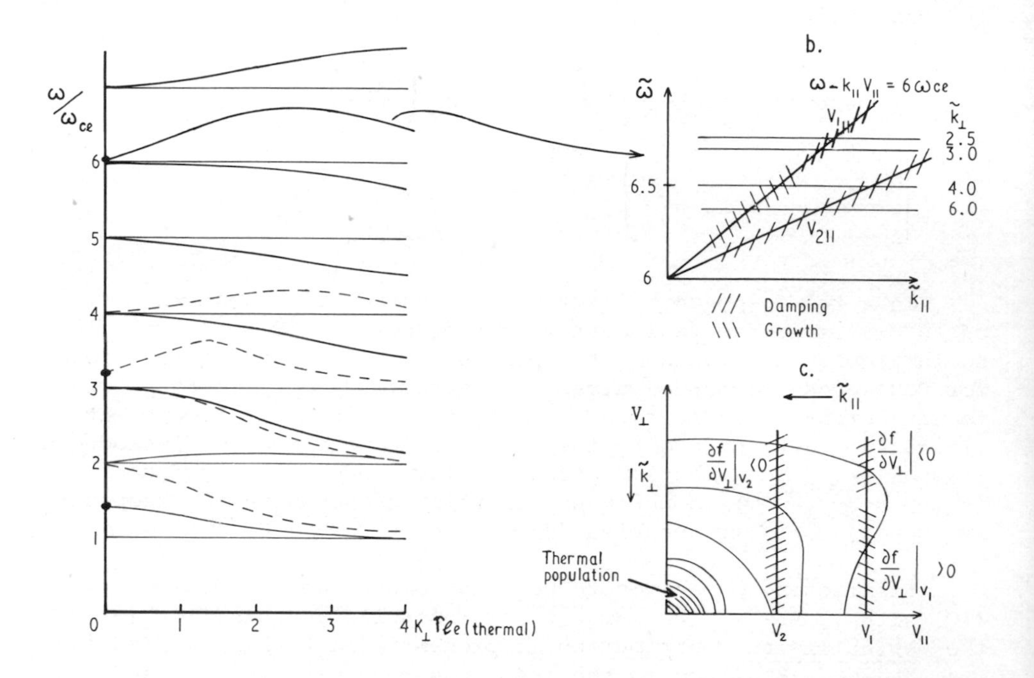

Fig.1 (a). ECH wave dispersion relations in maxwellian plasma, for several values of ω_p/ω_{ce}, and perpendicular propagation (gryoradius $r_{\ell e} = v_{th}/\omega_{ce}$). Fine lines ($\omega_p/\omega_{ce} = 1$); dashed lines (= 3); heavy lines (= 6); dots denote respective upper hybrid frequencies. Note low v_{gl} modes in these bands and in those above (f_q frequencies).
(b) Sketch of parallel dispersion in n = 6 band ($\omega_p/\omega_{ce} = 6$), demonstrating resonance condition $\omega - k_\parallel v_\parallel = 6\omega_{ce}$ for two parallel velocities $v_1 > v_2$. Only curves for $k_1 > 2.5$ are plotted to avoid confusion from double valued nature of the dispersion relation.
(c) Equivalent plot in velocity space, for distribution function with cool component and weak loss cone (contours of constant f are shown). Hatched regions correspond to stabilizing and destabilizing n = 6 contributions to integral in eqn. [3] .

OBSERVATIONS

The early reports of natural electrostatic emissions related to the electron gyrofrequency and its harmonics (Kennel et al (1), Scarf and Fredericks (21), Shaw and Gurnett (22), Anderson and Maeda (23)) showed enormous spectral variations ranging from single features just above (and in a few cases below) f_{ce}, multiplets at low gyroharmonics, to, at that stage, somewhat strange high frequency features at $\sim nf_{ce}$, $n >> 1$ (22). Their amplitudes also showed spectacular variation, ranging from $\sim 1\ \mu V/m$ to greater than 10 mV/m. The reconciliation of these results with theoretical models required detailed measurements of critical plasma parameters - for example, that of the cold plasma density in order to explain the dominant emission frequency. For $n \gtrsim 0.5\ cm^{-3}$, the parameter is measured routinely by the GEOS diagnostics (4,5) and identifications of the high frequency modes are described by Christiansen et al (24, 25). In Figure 2 we show a typical comparison of low group velocity modes stimulated by the GEOS resonance sounder (4) with high frequency natural emissions in the region of f_{uhc}. The spectral variations of ECH waves encountered during an orbit of the GEOS II satellite (Gough et al (26)) are demonstrated in Figure 3(a), while in Figure 3(b) characteristic spectra in the classification scheme of (26) (which differs only in detail from that of (18)) can be seen. These measurements, together with corroborative polarization studies (Kurth et al (27)) confirm the identification and controlling role of f_{uhc} for waves in the classes 3,4 ($n_c > n_h$, $\omega_{pc} > \omega_{ce}$). Diagnostic difficulties are considerably greater for class 1,2 emissions; for these, Hubbard et al (28) have presented an initial study using a combination of continuum cut off and medium energy particle measurements and conclude that $n_c/n_h < 1$, $\omega_{pc} \lesssim \omega_{ce}$.

EXPERIMENTAL DISTRIBUTION FUNCTIONS AND WAVE AMPLIFICATION

Explanation of the variations in amplitudes in ECH emissions require diagnostics of a higher order than those discussed above. With the launching of the advanced satellites in the GEOS and ISEE series measurements of the electron distribution function itself, and the identification of free energy features has been seen as an important first step. Anisotropic distributions with weak loss cones above $\sim$ 200 eV, and associated with strong, sporadic dayside class 3 emissions have been observed on GEOS I (19, 20), Horne et al (29)), but the ISEE results are at present less clear cut (Sentman et al (30)). In (19,20,29) the authors haveused models of the measured distribution function (in the range 0.5 eV - 25 keV) as direct inputs for linear stability calculations. As the waves are convectively unstable one must assume that amplification occurs in a spatially limited region, the dimensions of which, though at present somewhat uncertain, can be estimated roughly (15,20,29,31)

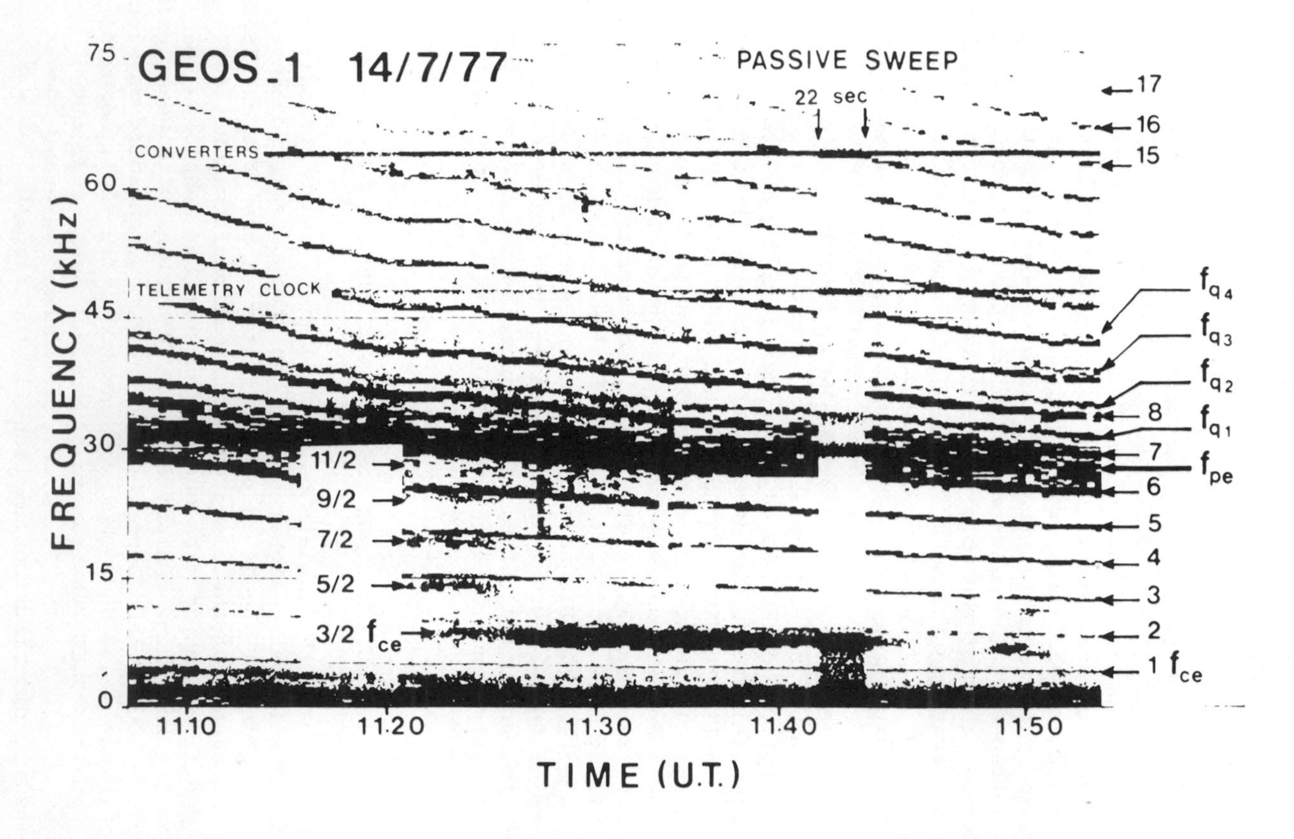

Fig. 2. Comparison of GEOS active resonances and passive measurements, grey scaled spectrogram. Note close intercomparison of emissions around f_{pe} ($\approx f_{uh}$) and f_q frequencies. The "$n + \frac{1}{2}$" f_{ce} signals are natural instabilities detected during sounder reception.

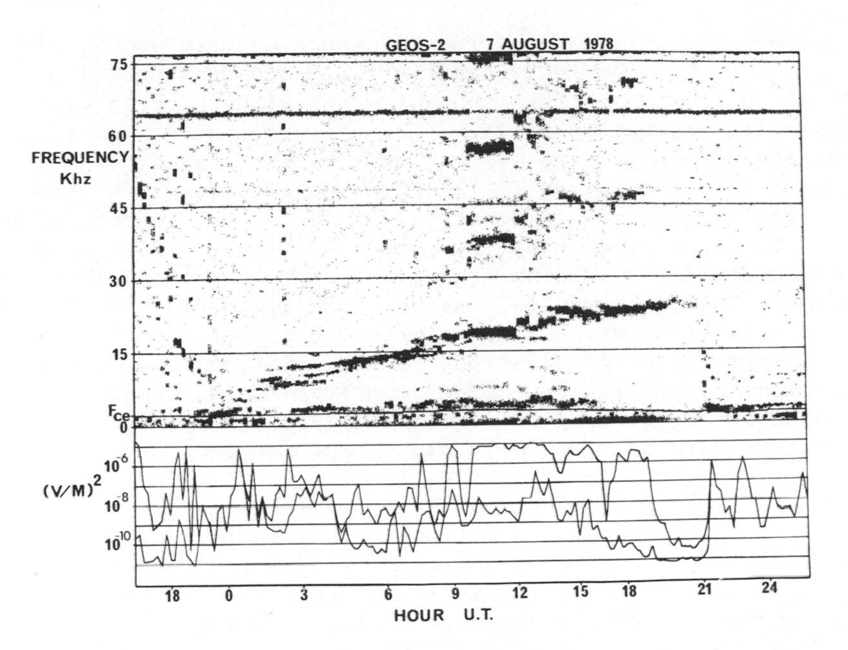

Fig. 3(a) Grey scaled GEOS 2 electric field data emphasing ECH emissions over > 24 hours at 6.6 Re. LT = UT + 1hr. Traces below show integrated spectral densities for $f > f_{ce}$ (upper plot) and $1 < f/f_{ce} < 2$ (lower plot) thus identifying dominant emission, and local time evolution of f_{pc}. The harmonic generation (9-18UT) is a preamplifier saturation effect.

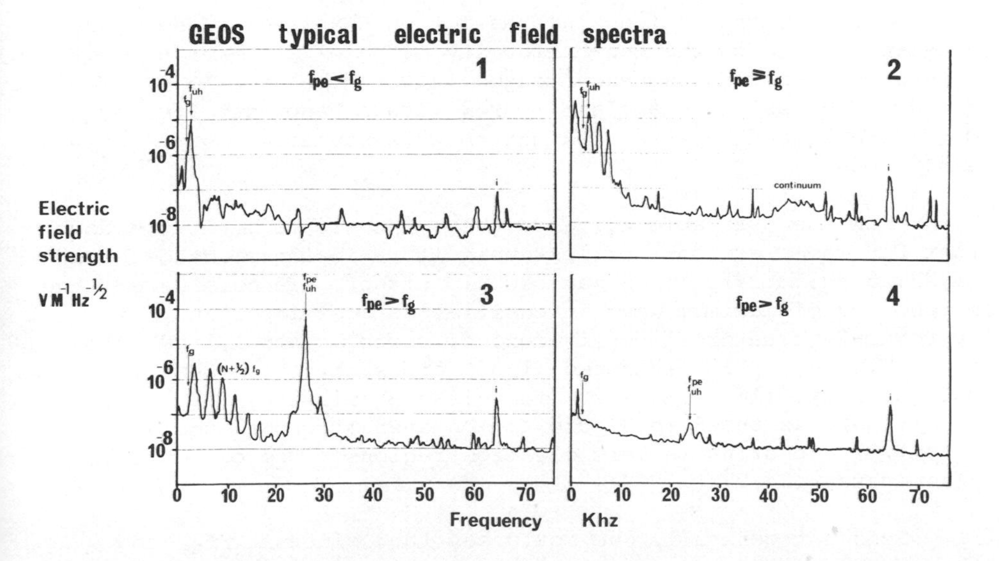

Fig. 3(b) Classification of ECH wave emissions. Typical spectra.

since the waves will detune from resonance as they convect in regions of varying ambient magnetic field and density. The predictions show an encouraging agreement with observed wave spectra with respect to relative harmonic amplitudes and temporal variations within a given event. Weak class 4 emissions have also been modelled in this way (29). In cases studied so far the properties of the cool component (T_c < 10 ev) are important, and support the contention that low group velocities and lack of cyclotron damping resulting from a very cool component ($T \sim 1$ eV) lead to high growth rates. Searches for free energy features, with more limited particle diagnostics, are also reported in (28), while less detailed observations of the relationships between enhancement of moderately energetic particles and class 1,2 emissions are noted in (23) and by Scarf et al (21) in the somewhat neglected area of ECH wave occurrence in magnetospheric substorms.

The weakest waves observed (≤ 1 µV/m) may result, not from instabilities at all but from fluctuations induced by stable distributions of suprathermal electrons (23) and the strongest waves (≥ 3 mV/m) are almost certainly amplitude limited by quasi-linear and non-linear processes. Distributions other than $\sim$keV loss cones have been proposed as instability sources, e.g. $\sim$10 keV "shoulder" features (Curtis and Wu (32)), regions of $\partial f/\partial v_\perp > 0$ at large pitch angles (12), and temperature anisotropies. Observational evidence is at present very limited, and in the last case much spectral observation contradicts theoretical predictions (12).

MORPHOLOGY

General morphological information on ECH waves is somewhat elusive because of the non-uniformity of satellite orbital coverage; because of limited frequency range (it would be easy to mis-identify a class 3 emission as a class 1 or 2 with a bandwidth limited to 0 - 10 kHz); and because of a dearth of systematic reports of wave amplitudes.

From the available evidence there appears to be little doubt that ECH waves are more or less ubiquitous in the magnetosphere (1,22,25,28,33,34). When macroscopic effects are considered, the morphology of maximum wave intensities is most important. The approximate equatorial confinement of strong class 1,2 emissions to within $\lambda m \approx \pm 10^\circ$ is noted (1,33) and such confinement is rather more dramatic ($\lambda m = \pm 1 - 2^\circ$) for class 3 emissions ((25,26), Gough et al (35)) as shown in Figure 4. An even stronger subclass ($\geq$ 3mV/m) appearing in cold electron regions close to the dayside plasmapause (24,27).

Somewhat generally one could say that since f_{pc}/f_{ce} and therefore f_{uhc}/f_{ce} decreases with radius, so should the dominant dimensionless emission frequency, with the spectral type evolving with increasing radial distance, from class 3,4 to class 1. Strong instabilities arise when the hot "free energy" electrons reach regions of significant cool plasma density. Several examples have been reported ((25), and Gurnett et al (36)). With respect to the geostationary orbit, local time asymmetries (see Figure 3a) can be related to the well known asymmetries in the plasmapause and the presence of the source of "fresh", hot, sheet plasma in the $\sim$ 22-02 hrs. L.T. sector.

WAVE-PARTICLE EFFECTS

For the proceedings of an institute of this kind it would be pleasant to report that definitive studies of diffuse auroral zone precipitation caused by ECH wave-related velocity space diffusion, have followed the work of Lyons (37). However these difficult measurements have not yet been made. Lyons showed that for class 1 emissions, and reasonable assumptions about the temperature of the electron component and wave number distribution, strong diffusion of keV electrons could be maintained by wave field of $\sim$ 1 mV/m, which are regularly observed at 6.6 R_E (see Figure 3(a)). The principle of experimental tests is clear. Assuming rough equatorial confinement, the differential fluxes in the loss cone (a conserved quantity in the absence of parallel electric fields) recorded both on the geomagnetic equator and at the conjugate point above the auroral zone can be compared, and the driving source sought in the magnetospheric wave spectrum. Only a few measurements have been attempted, by Meng et al (38), without, however, wave information.

Some evidence for local magnetospheric effects does exist. For example, the rather undramatic nature of the "free energy" distribution functions associated with strong dayside emissions (19,20,30), is probably indicative of rapid quasi-linear velocity space smoothing (20,30). More explicitly, it has been inferred (26) that the occurrence of anisotropic features in the supra-thermal electron distribution (the 100 eV "pancakes" of Wrenn et al (39) which evolve during dayside convection of the particles, is due to an absence of quasi-linear diffusion at large pitch angles (37). In order for such resonant interactions to occur, the conditions $\omega - k_{\parallel}v_{\parallel} = n\omega_{ce}$, $k_{\perp} \sim c_n/v_{\perp}$ imply such large wavenumbers, $k_{\perp}$ and $k_{\parallel}$, that the waves would be heavily cyclotron damped in normal circumstances. Extreme distributions, e.g. hot perpendicular rings might be invoked, but they themselves would be very rapidly destroyed by wave-particle processes. Non resonant diffusion and perpendicular heating of cool electrons has been suggested (17) as a mechanism by which initially absolute instabilities may rapidly become convectively unstable. Such a proposal might best be tested in nightside class 1-2 regions, where high

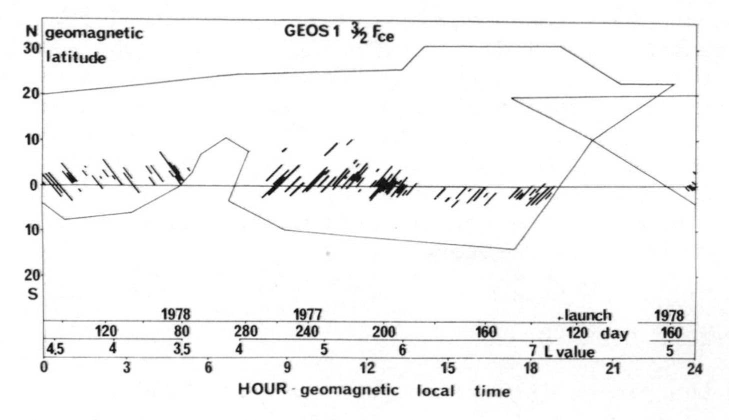

Fig. 4. Geos I observations (excluding storm periods) of class 3 emissions, as a function of geomagnetic latitude, local time and L value. The equatorial confinement is clear, and well within limits of observation (closed lines).

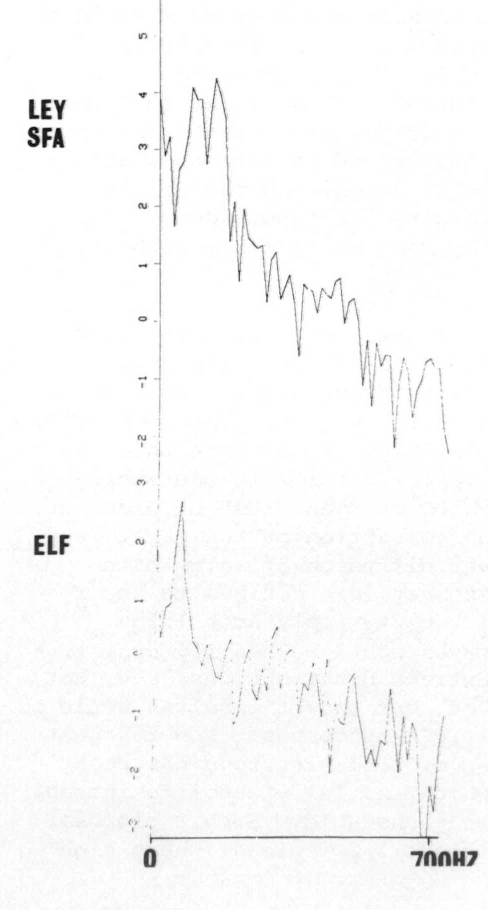

Fig. 5. Simultaneously recorded power spectra (GEOS II) in 700 Hz bands demonstrating 3 wave interactions. Upper trace is centred on f_{uh} emission at 22.8 Hz, and has two spectra features separated by ∼ 70 Hz. Lower trace (0-700 Hz) shows corresponding feature in vicinity of lower hybrid frequency at ∼ 70 Hz.

growth rates and low T_c/T_h ratios should exist. There is little evidence of such processes on the dayside given the approximate constancy of perpendicular suprathermal fluxes (Wrenn, personal communication) and the existence of low energy, quasi-field-aligned features (6,29).

NON-LINEAR EFFECTS

We do not wish to venture far into this domain, which is in a state of infantile disorder as far as (ECH) waves are concerned. Instead we point briefly to two areas in which non-linear wave couplings may be important. The first concerns the amplitude limitation of waves where linear theory appears to predict convection limited amplitudes in excess of those observed. The transfer of energy from strong ECH waves to low frequency ion waves or slow particles may help to accomplish this. The simplest processes in such schemes are 3 wave couplings (for coherent waves) and non-linear damping (both processes are of the same order in the random phase approximation). Three wave processes obeying the conditions $\omega_1 = \omega_2 + \omega_3$, $\underline{k}_1 = \underline{k}_2 + \underline{k}_3$ have been proposed in (25,35) and offer the greatest ease of detection. An example of the coupling of two ECH waves and an electrostatic wave in the vicinity of the lower hybrid frequency recorded simultaneously on the GEOS II long electric antenna is shown in Figure 5, but such results await a detailed comparison with existing theory (20,40). (In order to guard against false optimism the reader is urged to consult recent reviews (41,42) and a variety of texts, for example (43,44,45)).

The second area concerns the production of electromagnetic continuum radiation (Gurnett (46)). It is strongly suspected that at least some of this radiation emanates from regions of powerful class 3,4 ECH wave emissions and strong density gradients (27,47). The non-linear coalescence (3-wave interaction) of ECH and ion waves is a possible way of producing such radiation (Melrose (48)), but careful comparison of this process with linear conversation mechanisms must be made before definite conclusions can be drawn.

REFERENCES

(1) Kennel, C.F., Scarf, F.L., Fredricks, R.W., McGhee, J.H. and Coroniti, F.V.:J.Geophys. Res. 75, p.6136.
(2) Knott,K.:1978. Space Sci, Rev. 22, p. 321.
(3) S-300 Experimenters: 1979, Planet,Space Sci. 27, p. 317.
(4) Etcheto, J. and Bloch, J.J.:1978, Space Sci. Rev., p. 597.
(5) Decreau, P.M.E.,Behgin,C. and Parrot, M.: 1978, Space Sci. Rev., 22, p.581.
(6) Johnson, J.F.E., Sojka, J.J. and Wrenn, G.L.:1978,Space Sci. Rev., 22, p.567.
(7) Borg, H., Holmgren, L.A., Hultquist, B., Cambou,F.,Reme, H., Bahnsen, A., Kremser, G.: 1978, Space Sci. Rev., 22, p. 511.
(8) Bernstein, I.B.: 1958, Phys. Rev., 109, p.10.
(9) Harris, E.G.: 1959, Phys. Rev. Lett., 2, p.34.
(10) Milkhailovskii, A.B.: 1974,"*Theory of Plasma Instabilities*", Consultants Bureau, N.Y.
(11) Fredricks, R.W.: 1971, J. Geophys. Res., 76, p.5344.
(12) Young, T.S., Callen, J.D. and McCune, J.E.: 1973, J. Geophys. Res., 78, p.1082.
(13) Karpman, V.I., Alekhin, Ju.K., Borisov, N.D. and Rajabova, M.A.: 1975, Plasma Phys., p.34.
(14) Ashour-Abdalla,M., Chanteur, G. and Pellat, R.: 1975, J. Geophys. Res. 80, p.2775.
(15) Ashour-Abdalla, M. and Kennel, C.F.: in B.M. McCormac (ed.),"*Magnetosphere Particles and Fields*", D. Reidel Publ. Co., Dordrecht, p.181.
(16) Ashour-Abdalla,M. and Kennel, C.F.: 1978, J. Geophys. Res., 83, p.1531.
(17) Ashour-Abdalla,M. and Cowley, S.W.H.: in B.M. McCormac (ed.),"*Magnetosphere Particles and Fields*", D. Reidel Publ. Co., Dordrecht, p.421.
(18) Hubbard, R.F., and Birmingham, T.J.: 1978, J. Geophys. Res., 83, p.4837.
(19) Ronnmark, K., Borg, H., Christiansen, P.J., Gough M.P. and Jones, D.: Space Sci.Rev., 22, p.401.
(20) Ronnmark, K.: 1979, Doctoral Thesis, University of Unea, Sweden.
(21) Scarf, F.L., Fredricks, R.W., Kennel, C.F. and Coroniti, F.V.:1973, J.Geophys.Res.,78,p.3119,
(22) Shaw, R.R. and Gurnett, D.A.: 1975, J. Geophys. Res., 80, p.4259.
(23) Anderson, R.R. and Maeda, K.: 1977, J. Geophys. Res., 82, p.135.
(24) Christiansen, P., Gough, P., Martelli, G.,Bloch, J.J., Cornilleau, N., Etcheto, J., Gendrin, R., Beghin, C., Decreau, P. and Jones, D.: 1978, Nature, 272, p.682.
(25) Christiansen, P., Gough, P., Martelli, G., Bloch, J.J., Cornilleau, N., Etcheto, J., Gendrin, R., Beghin, C., Decreau, P. and Jones, D.: 1978, Space Sci. Rev. p. 383.
(26) Gough, M.P., Christiansen, P.J., Martelli, G. and Gershuny, E.J.:1979, Nature, 279, p.515.
(27) Kurth, W.S., Craven, J.D., Frank, L.A. and Gurnett, D.A.:1979, J. Geophys. Res.,84,p.4145.
(28) Hubbard, R.F., Birmingham, T.H. and Hones, E.W.: 1979, J. Geophys. Res., 84, p.5828.
(29) Horne, R.A., Christiansen, P.J., Gough, M.P., Ronnmark, K., Johnson, J.F.E., Sojka, J. and Wrenn, G.L.: in"*Advances in Space Research*", Vol. I. (K.Knott ed.), Pergamon (in press).
(30) Sentman, D.D., Frank, L.A., Kennel, C.F., Gurnett, D.A. and Kurth, W.S.: 1979, Geophys. Res. Letts., 6, p.781.
(31) Barbosa, D.D. and Kurth, W.S.: 1979, University of Iowa, preprint, No. 79-53.
(32) Curtis, S.A. and Wu, C.S.: 1979, J. Geophys. Res., 84, p. 2057.
(33) Fredricks, R.W. and Scarf, F.L.: 1973, J. Geophys. Res., 78, p.310.
(34) Mosier, S.R., Kaiser, M.L. and Brown, L.W.: 1973, J. Geophys. Res., 78, p.1673.
(35) Gough, M.P., Christiansen, P.J. and Gershuny, E.J.:"*Advances in Space Research*",Vol. I. (K.Knott ed.), Pergamon, (in press).
(36) Gurnett, D.A., Anderson, R.R., Scarf, F.L., Fredricks, R.W. and Smith, E.J.: 1978, Space Sci. Rev., 23, p. 103.
(37) Lyons, L.R.: 1974, J. Geophys. Res. 79, p.575.
(38) Meng, C.I., Mauk, B. and McIlwain, C.E.: 1979, J. Geophys. Res., 84, p.2545.
(39) Wrenn, G.L., Johnson, .J.F.E. and Sojka, J.: 1979, Nature, 279, p. 512.
(40) Ronnmark, K.: 1977, Planet Space Sci., 25, p. 149.
(41) Franklin, R.N.: 1977, Rep. Prog. Phys., 40, p.1369.
(42) Thornhill, S.G. and ter Haar, D.: 1978, Phys. Rep., 43C, p.43.
(43) Davidson, R.C.; 1972,"*Methods in Nonlinear Plasma Theory*", Academic Press, N.Y.
(44) Tsytovich. V.N.: 1977,"*Theory of Turbulent Plasma*",Consultants Bureau, N.Y.
(45) Akhiezer, A.I., Akhiezer, I.A., Palovin, R.V., Sitenko, A.G. and Stepanov, K.N.: 1975, *Plasma Electrodynamics*",Pergamon, Oxford.
(46) Gurnett, D.A.: 1975, J.Geophys. Res., 80, p.2751.
(47) Christiansen, P.J., Gough, M.P., Etcheto, J., Trotignon, J.G., Jones, D., Belmont, G.,Roux, A.: 1979, XVII I.U.G.G. General Assembly, IAGA Bulletin No. 43, p.312.
(48) Melrose, D.B.: 1979, U. of Sydney, Theoretical Physics Preprint (to be published).

MAGNETOSPHERIC HOT PLASMA MEASUREMENTS IN RELATION TO WAVE-PARTICLE INTERACTIONS ON HIGH-LATITUDE MAGNETIC FIELD LINES

Bengt Hultqvist

Kiruna Geophysical Institute

P.O. Box 704, S-981 27 Kiruna, Sweden

ABSTRACT

This paper summarizes what the hot plasma observations in the upper ionosphere and near the equatorial plane on high latitude magnetic field lines tell about magnetospheric wave particle interactions. The information about the spatial and temporal distributions of wave particle interactions within the magnetosphere that is obtained from particle measurements is dealt with. Some effects of the cold plasma on the hot plasma turbulence are described and discussed. Relations between electron and ion scattering processes obtained from plasma measurements are summarized. Finally, the identities of the specific wave particle resonance effects are discussed.

1. INTRODUCTION

Collective plasma processes are known, believed, or suspected to play a large part in many important phenomena in the magnetosphere: in the feeding of the plasma into the magnetosphere, in the loss processes of the hot magnetospheric plasma, in the spatial redistribution of the hot plasma within the magnetosphere, in transforming the energy distribution of the auroral particles, in initiating substorms, in producing anomalous resistance and double layers which decouple the ionosphere electrically from the outer magnetosphere and accelerate auroral particles, in coupling the precipitation of auroral electrons with that of auroral ions, in making the Earth a strong cosmic radio emission source, etcetera. Significant transfer of mass, momentum and energy between various parts of the magnetosphere/ionosphere system characterizes many of these processes.

C. S. Deehr and J. A. Holtet (eds.), Exploration of the Polar Upper Atmosphere, 367–380.

The wave particle interactions, the effects of which can be seen most easily in the distribution functions of the hot magnetospheric electrons and ions, are those which redistribute the particles in pitch angle. Pitch angle diffusion of particles into the atmospheric loss cone is one of the main loss mechanisms for the hot plasma in the magnetosphere. A large part of this report deals with the question what observations of the hot plasma in and near the loss cone may tell us about the pitch angle diffusion mechanisms and thus about the wave particle interactions causing this diffusion.

2. CONCLUSIONS ABOUT WAVE PARTICLE INTERACTIONS ON AURORAL FIELD LINES DERIVED FROM OBSERVATIONS OF PITCH ANGLE DIFFUSION

As a basis for the considerations in this report we list here a number of conclusions drawn from measurements of pitch angle distributions at various altitudes along auroral field lines. The basis for the various conclusions are summarized and discussed in separate sections together with general discussions of the physical implications.

The observational results to be treated below are the following:

- The hot plasma is permanently affected by wave particle interactions which give rise to pitch angle diffusion in a wide region outside the plasmapause.
- The wave particle interactions(s) which scatter(s) keV ions into the loss cone usually occur(s) in a region which closely coincides with the region where keV electron pitch angle diffusion occurs. The electron and ion precipitation outside the plasmapause seems to be coupled in complex and variable ways.
- The wave particle interactions scattering keV ions do not depend on the existence of wave electron interactions.
- Cold plasma appears to inhibit the turbulence outside the plasmapause rather than to enhance it and to determine thereby the inner boundary of the pitch angle scattering region.
- There occur wave particle interactions not only near the equatorial plane but also along the auroral field lines between a few thousand kilometers altitude and magnetic latitudes of a few tens of degrees.
- The wave particle interactions affecting keV ions and electrons outside the plasmapause are mostly so strong that less than a quarter bounce is required to scatter the pitch angle over a much wider angle than that of the loss cone.
- The electromagnetic ion cyclotron instability is not likely to be of importance for the strong pitch angle diffusion of keV ions outside the plasmapause.

3. PERMANENTLY TURBULENT REGION OUTSIDE THE PLASMAPAUSE

That the hot plasma in the magnetosphere is permanently in a turbulent state follows from the observation of keV ion pitch angle distributions associated with diffuse aurora. These are characterized by the absence of any peaks in the pitch angle distribution at angles less than 90°. Field aligned pitch angle distributions resulting from acceleration of ions along the magnetic field lines are also found but they are generally more limited in spatial extent and transient in time. Their existence does not affect the general conclusions drawn here.

In the auroral zone at altitudes below a couple of thousand kilometers the hot ion distributions have mostly been found to be roughly isotropic in most of the precipitation region at keV energies and in a part of the region at hundreds of keV. With "roughly isotropic" is then meant that the fluxes near 0° pitch angle are of the same order of magnitude as the fluxes near 90°. This has been shown for keV energies by Hultqvist et al., (1971, 1974), Bernstein et al., (1974), and Galperin et al., (1976) and for hundreds of keV energies by Søraas (1972), Søraas and Berg (1974), and Mizera (1974). These observations are illustrated in Fig 1 by the lowermost frame where the count rates of 6 keV ions at 10° (upward pointing triangles) and 80° pitch angle (downward pointing triangles) are given for an ESRO 1A pass from dawn at subauroral latitudes through the auroral zone in early morning hours over the polar cap through the evening auroral zone. Equal count rates for the 10° and 80° ion detectors means a factor of two higher flux at 80° than at 10°. We thus see that 6 keV ions are roughly isotropic in most parts of the precipitation regions (except at the low latitude edge of the auroral zone where the 80° fluxes extend to lower latitudes than do the 10° fluxes). Two examples of observations of ions at hundreds of keV energy are shown in Figs 2 and 3. The region with roughly isotropic pitch angle distribution stands out clearly.

This roughly isotropic condition, indicating a fairly strong to quite strong pitch angle diffusion is observed in the central part of the auroral precipitation zone practically always, at all local times, and for all levels of magnetospheric disturbance. Only on the central dayside equatorward of the cusp a deep loss cone is sometimes seen in the low altitude data.

Under quiet conditions the region with strong diffusion is in the night located between about 67° and 69° invariant latitude. The latitudinal width shows a marked K_p dependence with an equatorward broadening with increasing disturbance level.

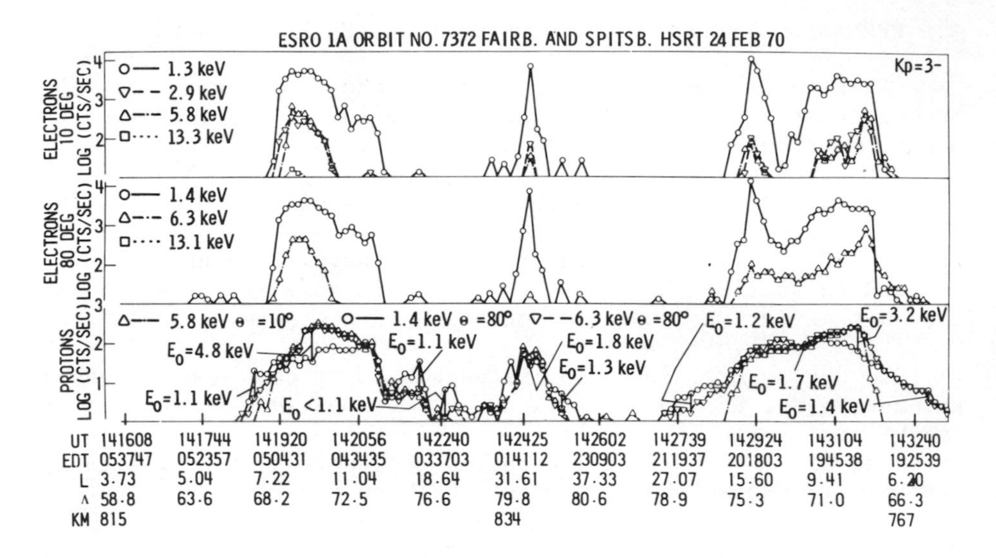

Figure 1: KeV electron and ion data from ESRO 1A orbit No. 7372, EDT means excentric dipole time and Λ invariant latitude (after Hultqvist et al., 1974). The calibration factors for transforming count rates into particles $cm^{-2} s^{-1} sr^{-1} keV^{-1}$ are the following for the various detectors:

10°, p^+,	5.8 keV:	0.37×10^4	80°, p^+, 1.4 keV: 13.5×10^4
el,	1.3 keV:	12.7×10^4	p^+, 6.3 keV: 0.74×10^4
el,	2.9 keV:	50×10^4	el, 1.4 keV: 10.6×10^4
el,	2.9 keV:	0.48×10^4	el, 6.3 keV: 0.94×10^4
el,	13.3 keV:	0.50×10^4	el, 13.1 keV: 4.5×10^4

The roughly isotropic ion precipitation zone in general extends to somewhat lower latitudes at hundreds of keV than in the keV range (Søraas et al., 1977). In L-value the difference is mostly a few tenths of an L unit (0.2 - 0.7).

Fig 2 relates the location of the roughly isotropic region to the plasmapause. As can be seen there, the precipitation region is located well outside the plasmapause. The same spatial relation is demonstrated for keV ions in Fig 4, which shows the dependence of the latitudinal location (L value) of the point where the 6 keV precipitated (10° pitch angle) ion flux, as observed by means of ESRO 1A, falls below the detection threshold on geomagnetic acitivity (K_p). Also shown are plasmapause locations determined by Chappell et al. (1970) with OGO 5. The ion observations

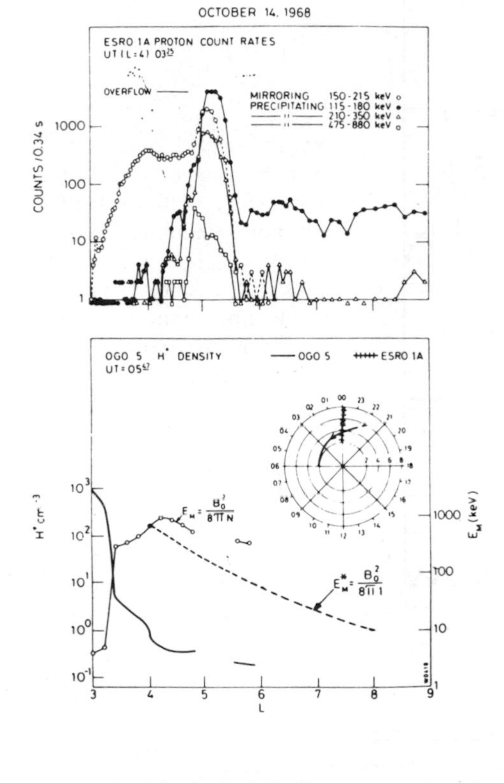

Figure 2: The upper half of the figure gives the count rate of locally mirroring and precipitating protons plotted versus L. The lower half gives the measured plasma density and calculated values of the magnetic energy per particle both plotted versus L. The orbit segments of the satellites are indicated in the L versus local time polar diagram. Counts can be converted to protons $(cm^2\ s\ sr\ keV)^{-1}$ by multiplying with the following figures: for precipitating protons in the range 115-180 keV, 1.2; 210-350 keV, 0.54; 475-880 keV, 0.19; and for mirroring protons in the range 150-215 keV, 1.2 (after Søraas and Berg, 1974).

were made in the evening sector whereas the plasmapause determinations were made in the early morning sector. If these differences in local time are taken into account the two sets of points will overlap even more. In spite of the fact that the two kind of data originate in different spacecraft they clearly demonstrate that the keV ion precipitation takes place outside the plasmapause.

4. COINCIDENCE OF REGIONS OF PITCH ANGLE DIFFUSION FOR KEV ELECTRONS AND IONS AND RELATIONS BETWEEN ELECTRON AND ION PRECIPITATION

KeV electrons and ions are found in and close to the atmospheric loss cone in closely coinciding zones (except sometimes in the afternoon sector where keV ions without any associated keV electrons often are found). This is illustrated by Fig 1 and also by Figs 5 and 6. Fig 5 has been included to show the very close coincidence of the boundaries of the main electron and ion regions

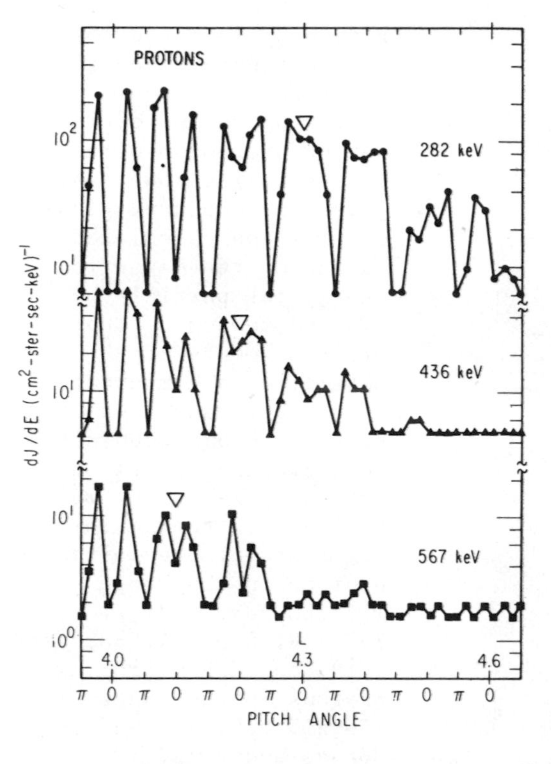

Figure 3: Flux versus pitch angle and L plots for 282, 436 and 567 keV protons measured by OV 1-19 on March 20, 1969 at 0635 UT. The triangles mark the place where maximum proton precipitation occurs in the 100 km loss cone, the width of which is illustrated by the width of the symbols (after Mizera, 1974).

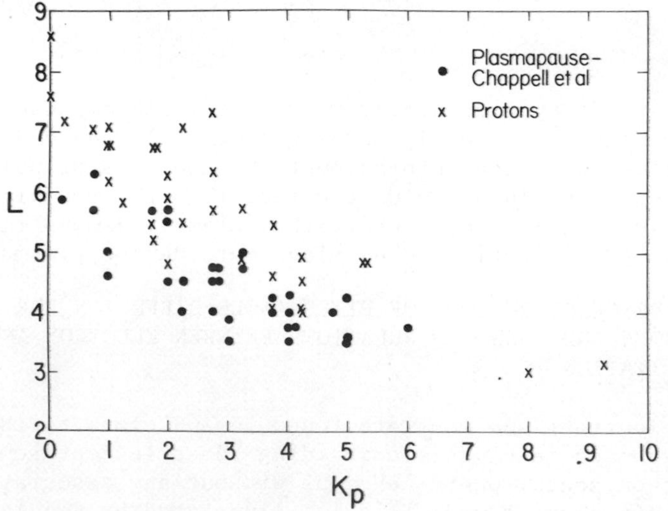

Figure 4: Dependence of the latitudial location (L) of the point where the 6 keV precipitating (10°) ion flux falls below the detection threshold on geomagnetic activity (K_p). Also shown are plasmapause locations determined by Chappell et al. (1970) for equivalent geomagnetic conditions (after Bernstein et al., 1974).

that sometimes is found. Poleward of the 40 keV trapping boundary there are no measurable fluxes of keV particles. In Fig 6, on the countrary, there are higher keV electron fluxes poleward of the trapping boundary than equatorward of it, while the keV ions show the opposite relation. But still, the keV ions and electrons are generally found in closely coinciding regions.

The magnetospheric ion population is mostly more nearly isotropic in the entire energy range from one keV to hundreds of keV than are the electrons in the same energy range. This is illustrated for instance by Fig 1. At a few keV energy also the electrons show mostly a fairly well filled loss cone but the degree of variability in this respect is higher for the electrons than for the positive ions even in diffuse auroral conditions. For auroral electron energies above 10 keV, say, the pitch angle distribution approaches isotropy only when the precipitation intensity is very high, as found already by O'Brien (1964) with the Injun 3 satellite. Most common for these electrons is a

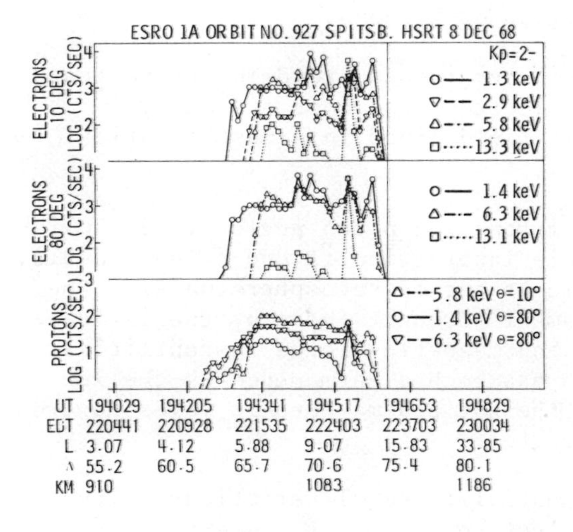

Figure 5: KeV electron and ion data from an ESRO 1A pass showing closely similar spatial distribution of electron and ion precipitation. Each data point representes measurements during 8 sec. The center energy and approximate pitch angle of the various detectors are shown in the figure. EDT means excentric dipole time, Λ is the invariant latitude and the satellite altitude is given in km. The straight vertical lines over some of the diagrams indicate the location of a trapping boundary for >40 keV electrons as determined on ESRO 1A by Page and Shaw (1972). The factors for conversion from count rates to differential fluxes are given in the caption of Fig 1.

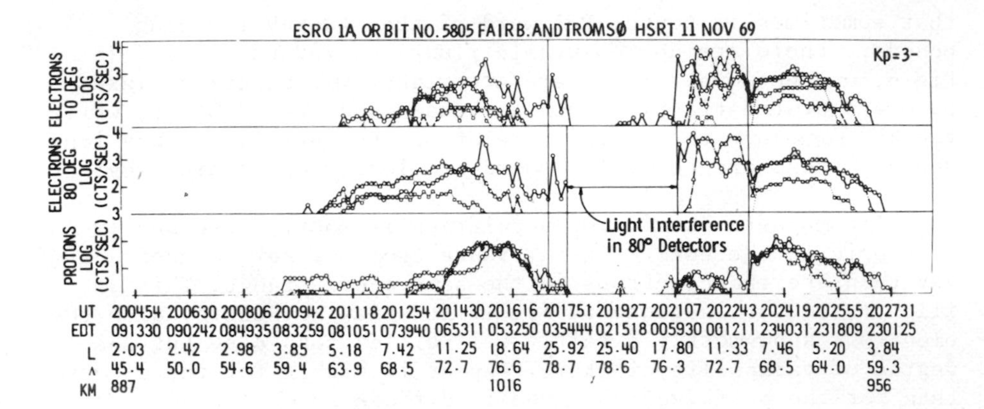

Figure 6: KeV electron and ion data from an ESRO 1A passage illustrating the coincidence of the regions in which keV electrons and ions are found in the upper ionosphere. For the meaning of the symbols as well as about the vertical straight lines and the values of the factors for conversion from count rate to differential flux the reader is referred to Figs 1 and 5.

pronounced loss cone distribution. It is of interest to note that in velocity space the size of the region where strong pitch angle diffusion occurs for electrons appears to be well comparable with that of the corresponding region of protons.

Thus, the observations in the ionosphere demonstrate that processes precipitating keV electrons and ions are active simultaneously along magnetic field lines in almost identical regions of the magnetosphere.

The persistence of the strong ion pitch angle diffusion indicates that the wave particle interaction is due to some basic, always present characteristic of the magnetosphere outside the plasmapause. As soon as plasma is brought there by the influence of the large scale electric and magnetic fields instabilities are excited. A process which has such a permanence is the increase of the anisotropy of the hot plasma when it moves towards the Earth. There may be other too.

The relations between spatial/temporal variations in the precipitation rates of keV electrons and ions are extremely variable as can be seen already by studying Figs 1, 5, and 6. Practically all kinds of relations can be found in the data. The association of electron spectrum hardening in the keV range and ion spectrum softening with enhanced fluxes of both ions and electrons was the most commonly seen combination in the ESRO 1 data (Hultqvist et al., 1974).

The observational result that the strong pitch angle diffusion regions associated with diffuse aurora largely coincide for keV ions and electrons outside the plasmapause, in combination with the mentioned wide variety of relations between spatial/temporal variations of keV ion and electron characteristics, indicates that no single basic physical process that affects only a small number of variables of the magnetospheric plasma is the cause of the pitch angle diffusion. Instead, a complex set of physical processes seems to affect simultaneously both ions of energies up to hundreds of keV and keV electrons. The processes affecting ions and electrons are strongly coupled but appear to be so in complex and variable ways. A well developed turbulence with strong coupling between several kinds of waves, permanently present in a region outside the plasmapause, is consistent with the observations.

In the afternoon local time sector keV ion precipittion is frequently seen without any keV electrons being precipitated. An example is shown in Fig 7 where the measurements in the left hand part were taken in the afternoon. As can be seen there, no electrons at al were found above 1 keV energy, whereas good fluxes of keV ions were recorded.

Hultqvist et al. (1980) have recently demonstrated that the absence of precipitated electrons in the afternoon is due to the electrons having been convected away from this region of the magnetosphere by an enhanced large scale electric field associated with a fairly strong magnetospheric disturbance.

Observational results like those shown in Fig 7 thus demonstrate that strong pitch angle diffusion may occur also in the absence of hot electrons. The wave particle interaction(s) that give rise to the ion scattering thus can exist without hot electrons present and thus without wave particle interactions involving hot electrons.

5. DEPENDENCE OF WAVE PARTICLE INTERACTIONS ON COLD PLASMA DENSITY

The observation that the strong pitch angle diffusion of ions and keV electrons occurs outside the plasmapause may be due either to the plasmapause formation and the hot plasma distribution being determined by the same process(es) or to the cold plasma in the plasmapause region determining the distribution of the hot plasma turbulence.

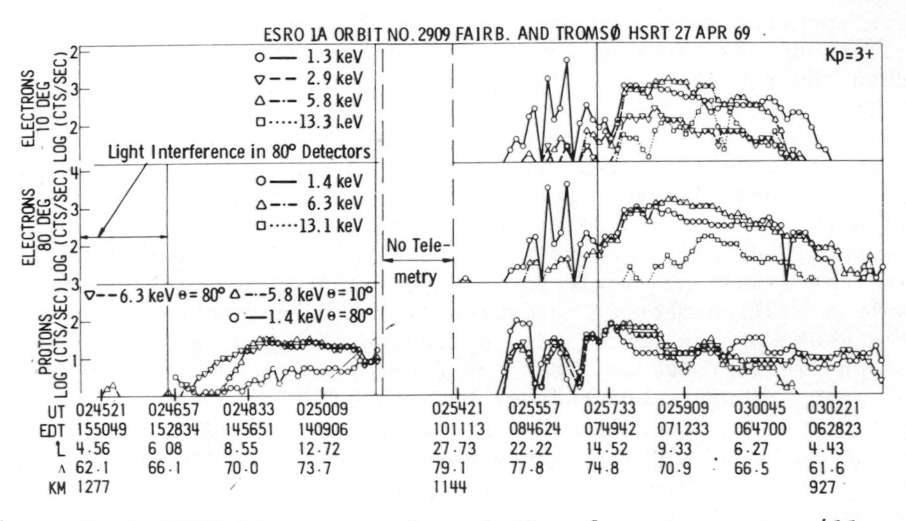

Figure 7: An ESRO 1A passage through the afternoon sector illustrating "pure" ion precipitation. About the vertical straight lines and values of conversion factors, see captions of Figs 1 and 5.

Measurements show that the ring current ions reach well within the plasmapause in most cases (Frank, 1971; Williams and Lyons, 1974; see also Hultqvist, 1975b for a discussion). The ions are thus present at and within the plasmapause but are not being scattered into the loss cone. The observations are thus not consistent with the hypothesis that the inner boundary of the region containing the hot plasma and the outer boundary of the cold plasma region - the plasmapause - coincide and are determined by the same process. Instead, they indicate that the cold plasma has a strong quenching effect on those instabilities which give rise to the pitch angle diffusion of ions and electrons outside the plasmapause.

6. ON THE ALTITUDE DISTRIBUTION OF THE WAVE PARTICLE INTERACTION PROCESSES AFFECTING THE KEV IONS

The field aligned ion beams ejected into the magnetosphere from the hot-cold plasma interaction region at several thousand kilometers altitude on auroral field lines that were first observed by means of the S3-3 satellite by Shelley et al. (1976), Ghielmetti et al. (1977, 1978), Fennell et al. (1977), Mizera and Fennell (1977), and Mizera et al. (1977) have been found to be frequently quite narrow (a few tens of degrees) at the altitude of observations by S3-3 (4000 - 8000 km). They are therefore expected to be very narrow when they reach the vicinity of the equatorial plane. The overall probability to observe outward flowing keV ions (UFI) in an S3-3 pass through the auroral zone at altitudes above 4000 km has been found to be about 60% (which is somewhat less than the probability of seeing discrete aurora, probably because of measurement limitations).

Measurements at magnetic latitudes below 30° on auroral field lines by means of the GEOS 1 satellite have not revealed a single case of extremely well-collimated ions. It is true that a systematic investigations of all GEOS 1 measurements very close to magnetic field lines has not yet been carried out. The existence under some conditions of very narrow ion beams near the equator can, therefore, not be excluded. However, it seems quite safe to conclude from the GEOS 1 data studied hitherto that normally the well collimated ion beams observed by S3-3 are scattered over pitch angle ranges of the order of tens of degrees when they reach the vicinity of the equatorial plane. This is also indicated by the general absence of observations by S3-3 of strongly field aligned beams returning from the opposite hemisphere. Important wave particle interactions thus appear to take place along the geomagnetic field lines between an altitude of about an earth radius and the vicinity of the equatorial plane. As GEOS 1 went up to a maximum magnetic latitude of about 30° the effective wave particle interactions seem to occur also fairly far from the equatorial plane. They may even take place close to the altitude range of S3-3. This does, of course, not mean that there is no scattering taking place close to the equator.

More detailed observational results related to the above considerations are expected to be available within the next year or two.

7. ON THE IDENTITIES OF THE INSTABILITIES AFFECTING KEV IONS AND ELECTRONS

Cornwall et al. (1971) have suggested that the electromagnetic ion cyclotron instability driven by an anisotropic particle distribution in the equatorial plane plays a role in determining the spatial distribution of the pitch angle diffusion. Their theoretical model suggests that the ring current ions will be subject to enhanced pitch angle and energy diffusion just within the plasmapause and at high L shells where the magnetic field intensity is small. At intermediate locations the ring current should be stable because the particle energy required for resonance with the cyclotron waves exceeds that present in the ring curren distribution. According to the theoretical model the difference in the position of the low L limit for strong diffusion between 6 and 115-180 keV protons is expected to be several units of L (when the plasma density and energetic particle anisotropy are assumed not to decrease with decreasing L). This contrasts with the observed difference of a fraction of an L unit. Quite an unrealistic variation of plasma density and/or anisotropy with L is required to interpret the observations in terms of the electromagnetic ion cyclotron instability. It may, therefore, be concluded that this instability is not likely to be an important constituent in the turbulence that affects the keV ions on auroral zone magnetic field lines.

Coroniti et al. (1972) have demonstrated that an electrostatic loss cone instability may be excited in the ring current throughout the low density region beyond the plasmapause. The growth rate for this instability decreases with increasing cold plasma density. It is a strong candidate as a constituent in the turbulence causing the diffusion of the ions.

Ashour-Abdalla and Thorne (1977, 1978) have proposed that electrostatic ion cyclotron instability is excited at frequencies confined to bands centered between multiples of the ion gyrofrequency even under quiet conditions, due to the free energy associated with the loss cone distribution of the hot ions near the equatorial plane. The instability is, however, severly suppressed by thermal electron Landau damping. Heating of the electrons to the order of 100 eV temperature in the turbulent region is, therefore, a requirement for the instability to be of importance. Whether is happens or not is not known. In any case the instability is not likely to be important on the dayside and should be quenched completely within the plasmasphere.

There is no a priori reason for the ion precipitation to be confined to the same L range where the energetic electrons are precipitated when resonant wave-particle interactions are considered. Ashour-Abdalla and Thorne (1978) suggest that the coupling between electron and ion precipitation is due to the requirement that the cold electrons be heated to the 100 eV range in order not to quench the ion cyclotron turbulence. Such a heating may occur only in the region of electron precipitation according to Ashour-Abdalla and Thorne (1978) and, therefore, ion precipitation occurs mainly in the electron precipitation region. However, the ion precipitation zone frequently extends farther equatorwards than the electron precipitation and in the afternoon sector sometimes only ions are precipitated (Hultqvist et al., 1974). Measurements of the low energy electron component together with the ions are needed before it will be possible to judge whether or not the mechanism for the coupling between the electron and ion precipitation proposed by Ashour-Abdalla and Thorne plays a significant role for the simultaneous ion and electron-precipitation.

Nambu (1975), Rönnmark and Stenflo (1976), and Rönnmark (1977) have investigated the nonlinear generation of electrostatic ion loss cone waves by electrostatic turbulence near three halves of the electron cyclotron frequency and of high frequency waves by low frequency electrostatic turbulence. The investigations show that such nonlinear wave coupling may be of importance in the magnetosphere under some conditions. These or similar nonlinear processes offer alternative possibilities to the one suggested by Ashour-Abdalla and Thorne (1978) of understanding the simultaneous precipitation of hot ions and electrons into the atmosphere and have to be studied further.

REFERENCES

Ashour-Abdalla, M., and Thorne, R.M. 1977, Geophys. Res. Lett. 4, 45.
Ashour-Abdalla, M., and Thorne, R.M. 1978, Preprint PPG-352. Center Plasma Phys. and Fusion Engineering, UCLA.
Bernstein, W., Hultqvist, B., and Borg, H. 1974, Planet. Space Sci. 22, 767.
Chappell, C.R., Harris, K.K., and Sharp, G.W. 1970, J. Geophys. Res. 75, 50.
Cornwall, J.M., Coroniti, F.V., and Thorne, R.M. 1971, J. Geophys. Res. 76, 4428.
Coroniti, F.V., Fredericks, R.W., and White, R. 1972, J. Geophys. Res. 77, 6243.
Fennell, J.P., Mizera, P.F., and Croley, D.R. 1977, Paper GA - 124, IAGA/IAMAP Joint Assembly Seattle, August 22 - Sept. 3.
Frank, L.A. 1971, J. Geophys. Res. 76, 2265.
Galperin, Yu.I., Kovrazhkin, R.A., Ponomarev, Yu.N., Crasnier, J. and Sauvaud, J.A. 1976, Ann. Geophys. 32, 109.
Ghielmetti, A.G., Shelley, E.G., Johnson, R.G., and Sharp, R.D. 1977, Paper GA - 126, IAGA/IAMAP Joint Assembly Seattle, August 22 - Sept. 3.
Ghielmetti, A.G., Johnson, R.G., Sharp, R.D., and Shelley, E.G. 1978, Geophys. Res. Lett. 5, 59.
Hultqvist, B., Borg, H., Riedler, W., and Christophersen, P. 1971, Planet. Space Sci. 19, 279.
Hultqvist, B., Borg, H., Christophersen, P., Riedler, W., and Bernstein, W. 1974, NOAA Techn. Rept. ERL 305-SEL 29.
Hultqvist, B. 1975, p. 291 in Physics of the Hot Plasma in the Magnetosphere, Eds B. Hultqvist and L. Stenflo, Plenum Press, New York.
Hultqvist, B., Aparicio, B., Borg, H., Arnoldy, R., and Moore, T.E. 1980, Decrease of keV electron and ion fluxes in the dayside magnetosphere during the early phase of magnetospheric disturbances, Submitted to Planet. Space Sci.
Mizera, P.F. 1974, J. Geophys. Res. 79, 581.
Mizera, P.F., and Fennell, J.F. 1977, Geophys. Res. Lett. 4, 311.

Mizera, P., Fennell, J.F., and Vampola, A.L. 1977, EOS 58, 716.
Nambu, M. 1975, Phys. Rev. Lett. 34, 387.
O'Brien, B.J. 1964, J. Geophys. Res. 69, 13.
Rönnmark, K., and Stenflo, L. 1976, Planet. Space Sci. 24, 904.
Rönnmark, K. 1977, Planet. Space Sci. 25, 149.
Shelley, E.G., Sharp, R.D., and Johnson, R.G. 1976, Geophys. Res. Lett. 3, 654.
Søraas, R. 1972, p. 120 in Earth's Magnetospheric Processes, (Ed. B.M. McCormac), D. Reidel Publ. Co., Dordrecht, Holland.
Søraas, F., Lundblad, J.Å., and Hultqvist, B. 1977, Planet. Space Sci. 25, 757.
Williams, D.J., and Lyons, L.R. 1974, J. Geophys. Res. 79, 4195.

MECHANISMS FOR INTENSE RELATIVISTIC ELECTRON PRECIPITATION

Richard M. Thorne and Leo J. Andreoli

Department of Atmospheric Sciences, University of California, Los Angeles, California 90024 U.S.A.

ABSTRACT: From an indepth analysis of over fourteen months of data from the S3-3 satellite, 313 relativistic electron precipitation events have been detected with essentially isotropic flux over the upward looking hemisphere. The majority of events occur at night in a narrow latitudinal zone embedded within a broader region of intense energetic ion precipitation. Three distinct classes of precipitation have been identified each associated with strong diffusion resonant scattering by known magnetospheric plasma waves. The intense electron energy deposition is a major source of middle atmospheric odd hydrogen and odd nitrogen molecules at sub-auroral latitudes; this can lead to observable catalytic destruction of mesospheric ozone.

1. INTRODUCTION

During geomagnetically disturbed periods the precipitation of energetic electrons from the Earth's radiation belts can provide the dominant source of ionizing energy input for the middle latitude mesosphere. One particularly intense class of precipitation is the relativistic electron precipitation (REP) event which is characterized by extremely high energy (>100 keV) electrons near the upper limit imposed by strong pitch-angle diffusion. Such events were first identified through the absorption of forward scattered radio signals associated with a pronounced increase in D-region ionization (Bailey and Pomerantz 1965; Bailey 1968). Subsequent radio wave studies identified the general morphology of such events and clearly indicated a direct association with substorm activity (e.g. Rosenberg et al. 1972; Thorne and Larsen 1976). However, a quantitative

C. S. Deehr and J. A. Holtet (eds.), Exploration of the Polar Upper Atmosphere, 381–394.

assessment of the electron energy deposition and its effect on the middle atmosphere requires direct satellite or rocket observations (e.g. Vampola 1971; Mathews and Simons 1973). Only very recently have detectors been flown which are capable of measuring the electron energy spectrum and pitch-angle distribution with the resolution required for accurate modeling studies of the atmospheric response (e.g. Reagan 1977; Thorne, 1978). The results presented here provide the first comprehensive survey of intense REP events obtained from an indepth study of 14 months of S3-3 satellite data (courtesy A. Vampola). Although emphasis here is placed on identifying the mechanisms responsible for the observed strong diffusion electron precipitation, a brief assessment is also made (Section 4) of the impact of such intense precipitation on the chemistry of the middle atmosphere.

The characteristic signature of strong diffusion scattering, which requires stochastic pitch-angle diffusion across the size of the atmospheric loss cone within a particle transit time along the magnetospheric flux tube, is an essentially isotropic pitch-angle distribution over the upward looking hemisphere. Studies using low altitude polar orbiting satellite data have previously shown that such "isotropic" distributions are characteristic of the diffuse auroral zone with a general coincidence between the regions of low energy (~keV) ion and electron precipitation (e.g. Hultqvist et al. 1974). An important new result from the S3-3 analysis is that high energy (>100 keV) electrons frequently exhibit a more confined zone of strong diffusion scattering which is also generally coincident with strong diffusion ion precipitation. All but seven out of a total of 313 individual strong diffusion REP events observed on S3-3 exhibit such a correlation. This feature suggests a common scattering process which has a direct bearing on the potential electron precipitation mechanisms discussed in Section 2. Three distinct classes of high energy electron precipitation have been identified and examples of each are presented in Section 3. Each class can be related to the anticipated scattering by known magnetospheric plasma waves.

2. QUASI-LINEAR SCATTERING BY PLASMA WAVES

Under adiabatic conditions energetic electrons can execute bounce motion between magnetic "mirror" points in the inhomogenous geomagnetic field. The condition for such trapping may be obtained from conservation of the first adiabatic invariant $p_\perp^2/\gamma mB$, where $p_\perp = p \sin\alpha$ is the particle's momentum perpendicular to the geomagnetic field of strength B, m is the rest mass and $\gamma = (1-v^2/c^2)^{-1/2}$; this mandates that $\sin^2\alpha/B$ is conserved during the bounce motion. Particles with $\alpha_L < \alpha < \pi - \alpha_L$ where

$$\sin^2\alpha_L = B/B_A \qquad (1)$$

will mirror above the top of the atmosphere (where $B = B_A$) and thus be trapped. Those within the "loss-cone" ($|\alpha| < \alpha_L$) will precipitate into the atmosphere and be removed. Processes which violate the first adiabatic invariant can therefore provide stochastic pitch-angle diffusion and eventual precipitation loss from the trapped particle environment. For high energy ions and electrons in the outer radiated belt the most effective pitch-angle scattering mechanism involves resonant interactions with natural magnetospheric plasma waves which appear Doppler shifted in frequency to some multiple of the relativistic gyrofrequency:

$$\omega - k_{\parallel} v_{\parallel} = n\Omega_{\pm}/\gamma_{\pm} \;. \tag{2}$$

Here ω is the wave frequency, $k_{\parallel}$ and $v_{\parallel}$ are the wave propagation vector and particle velocity along the field direction, and n is an integer.

2.1 Wave Amplitudes Required For Strong Diffusion

During the resonant interaction between particles and electromagnetic waves the rate of stocastic pitch-angle scattering is approximately (Kennel and Petschek 1966)

$$D^{\pm}_{\alpha\alpha} \approx \frac{\langle\Delta\alpha\rangle^2}{2\Delta t} \sim \frac{\Omega_{\pm}}{\gamma_{\pm}} \left(\frac{B'}{B}\right)^2 \tag{3}$$

where B' is the total wide band amplitude of the fluctuating magnetic field. Under strong pitch-angle diffusion one requires

$$D_{\alpha\alpha} \geq D_{SD} \approx \alpha_L^2/2\tau_{1/4B} \;; \tag{4}$$

namely particles must be scattered over the dimension of the loss cone α_L within the quarter-bounce time $\tau_{1/4B} \sim L\, R_e/V$, where L is the radial extent of the field line measured in Earth radii R_e and V the particle velocity. Assuming that particles spend a fraction f of their orbit in resonance with waves near the equator, where $\alpha_L^2 \sim 1/2L^3$, one can combine (3) and (4) to obtain the minimum wave amplitude for strong diffusion scattering

$$B'_{SD^{\pm}} \approx (\gamma_{\pm}^2-1)^{1/4} \left\{\frac{cB^2}{4fL^4 R_e \Omega_{\pm}}\right\}^{1/2} \tag{5}$$

For a dipole field with $f \sim 1/5$ this has a numerical value

$$B'_{SD^-} \approx 10^2(\gamma_-^2-1)^{1/4}/L^{7/2}, \text{ gammas} \tag{5a}$$

for relativistic electrons and

$$B'_{SD^+} \approx 3 \times 10^2 E_+^{1/4}/L^{7/2}, \text{ gammas} \tag{5b}$$

for non-relativistic protons where E_+ is the proton kinetic energy measured in keV.

Because the electrostatic waves of interest to this study are polarized with $k_\perp >> k_\parallel$ the resonant diffusion occurs primarily in the particle velocity perpendicular to $\underline{B}$. The appropriate diffusion coefficient

$$D_{\perp\perp}^{\pm} = \frac{<\Delta v_\perp>^2}{2\Delta t} \quad (\frac{cE'}{B})^2 \quad \frac{\Omega_\pm}{\gamma_\pm} \tag{6}$$

where E' is the total wide-band fluctuating electric field. For strong diffusion scattering near the loss-cone one requires

$$D_{\perp\perp} \geq D_{SD} \approx v_{res}^2 \alpha_L^2 / 2\tau_{1/4B} \tag{7}$$

Combining (6) and (7) yields the required minimum fluctuating electron field for strong diffusion scattering

$$E'_{SD^\pm} \sim \beta_\pm B'_{SD^\pm} \tag{8}$$

where $\beta_\pm = v_{res}^{\pm}/c$.

For relativistic electrons ($\beta_- \to 1$)

$$E'_{SD^+} \approx 3 \times 10^2 (\gamma_-^2 - 1)^{3/4} / \gamma_- L^{7/2} \quad mv/m \tag{8a}$$

While for non-relativistic protons

$$E'_{SD^+} \sim 10^2 E_+^{3/4} / L^{7/2} \quad mv/m \tag{8b}$$

where again E_+ is measured in keV.

A comparison of the fluctuating wave amplitudes required for strong diffusion at L = 6 is shown in Figure 1 as a function of ion or electron energy. For a given power spectral density it is clear that strong diffusion is most likely at high L and electrostatic waves are more effective in scattering resonant particles particularly in the case of ions.

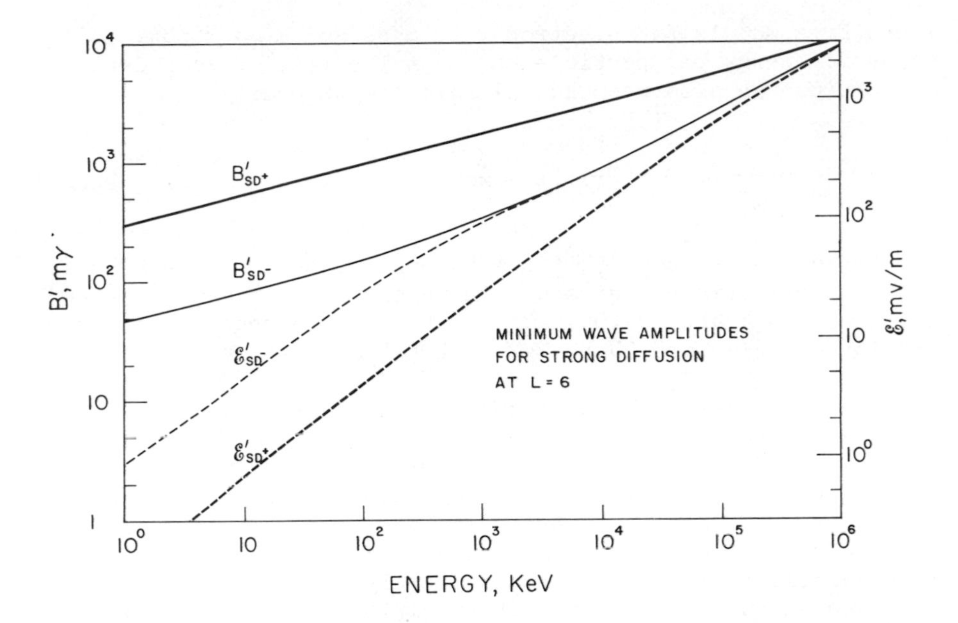

Figure 1. Minimum fluctuating electric or magnetic wave amplitudes required to scatter protons (+) or electrons (-) on strong diffusion at L = 6. At other locations the required amplitudes scale as $L^{-7/2}$.

2.2 Potential Scattering Waves and Resonant Electron Energies

There are four basic classes of observed magnetospheric plasma waves capable of resonating with outer radiation blet electrons.

a. Whistler-Mode Chorus. These are electromagnetic waves with frequency $\omega = (0.1\text{-}0.5)\Omega_-$ which are generated near the geomagnetic equator (but outside the plasmapause) during substorms (e.g. Tsurutani and Smith 1974). For parallel propagation the waves are right-hand circularly polarized and the refractive index

$$\mu_{\parallel} = \frac{k_{\parallel} c}{\omega} = \left\{ \frac{\omega_p^2}{\omega\Omega_-} \cdot \frac{1}{(1 - \frac{\omega}{\Omega_-})} \right\}^{1/2} \tag{9}$$

Using (2) and (9) the resonant electron kinetic energy

$$E^-_{res} = \left\{ \left[\frac{2E_M}{E^-_o} \cdot \frac{\Omega_-}{\omega} \left(1 - \frac{\omega}{\Omega_-}\right)^3 + 1 \right]^{1/2} - 1 \right\} E^-_o \tag{10}$$

where $E_o^- = mc^2$ is the electron rest mass and $E_M = B^2/8\pi N$ is the magnetic energy per particle based on the total plasma density N. Chorus can also resonant with ions with energy

$$E^+_{res} = \frac{m^+}{m^-} \cdot E_M \left(\frac{\omega}{\Omega_-}\right) \left(1 - \frac{\omega}{\Omega_-}\right). \tag{11}$$

For typical outer zone $(8 \geq L \geq 4)$ conditions $E_M \sim 10$-100 keV. Resonant electron energies are therefore in the range (25-500) keV near L = 4 and (2-100)keV near L = 8. Corresponding resonant ion energies are 20-50 MeV at L = 4 and 2-5 MeV at L = 8.

Magnetospheric chorus occurrence typically peaks in the midnight and late morning local time sectors with maximum wide-band wave amplitude in the range 10-100 mγ (Tsurutani and Smith 1977). From Figure 1 it is clear that such waves are capable of providing strong diffusion scattering for the lower energy resonant electrons but the scattering will generally become weaker at relativistic energies. Chorus is rarely intense enough to excite REP events but when it can such precipitation will most likely be confined to the higher L regions. No significant ion precipitation is expected during such events.

b. Electromagnetic Ion-Cyclotron Waves. These are low frequency $(\omega \sim (0.1$-$0.5)\Omega_+)$ waves sometimes referred to as Pc1 or IPDP pulsations. For a single ion plasma the waves are basically left-hand circularly polarized with a refractive index

$$\mu_{L} = \left\{\frac{\omega_p^2}{\Omega_- \Omega_+} \cdot \frac{1}{1 - \frac{\omega}{\Omega_+}}\right\}^{1/2} \tag{12}$$

Resonance with electrons requires large Doppler shift (or high velocity electrons) with energy

$$E^-_{res} = \left\{\left[\frac{2E_M}{E_o} \cdot \left(\frac{m^+}{m^-}\right) \left(\frac{\Omega_+}{\omega}\right)^2 \left(1 - \frac{\omega}{\Omega_+}\right) + 1\right]^{1/2} - 1\right\} E_o^- \tag{13}$$

Resonant ions have an energy

$$E^+_{res} = E_M \left(\frac{\Omega_+}{\omega}\right)^2 \left(1 - \frac{\omega}{\Omega_+}\right) \tag{14}$$

Theoretical arguments suggest that ion-cyclotron waves are most easily generated inside the high density plasmasphere (Cornwall et al. 1970) or within detached plasma regions (Thorne 1974) where $E_M \sim (0.3 - 10)$keV. Resonant electron energies are therefore in the ultra relativistic range (0.5 - 50)MeV (Thorne and Kennel 1971) while resonant ion energies are $(1\text{-}10^3)$keV.

Ion-cyclotron waves have been observed primarily in the dusk sector of magnetosphere with typical peak amplitude comparable to a few gamma (Bossen et al. 1976). Using Figure 2 we note that this is sufficient to drive both resonant ions and high energy electrons onto strong diffusion. However, unless the plasma density is very high (or E_m very low) the resonant electron energies will generally be well above 1 MeV, making such events difficult to detect.

c. Electrostatic Electron Cyclotron Waves. Purely electrostatic waves have been observed in the outer magnetosphere in frequency bands centered between harmonics of the electron gyrofrequency. These are often referred to as $(n + 1/2)\Omega_-$ or upper hybrid waves. The waves are typically polarized with $\bar{k}$ almost perpendicular to $\bar{B}$ ($k_\perp/k_\parallel \sim 0(10)$) and the parallel propagation vector $k_\parallel$ is comparable to the inverse Larmor radius of the hot plasmasheet (~keV) electrons; $k_\parallel \sim 0(\Omega_-/V_-)$ Ashour-Abdalla and Kennel 1978). Typical electron resonant energies are therefore comparable to a few keV. In fact, Lyons (1974) has explicitly demonstrated that these waves are ineffective in scattering high energy electrons. They will therefore not be considered further.

d. Electrostatic Ion-Cyclotron Waves. In direct analogy with electron electrostatic waves one can anticipate the presence of electrostatic ion harmonic waves in bands centered between the ion gyroharmonics. Such electrostatic ion-cyclotron waves have been observed in the topside ionosphere (Kelley et al. 1975) and broad-band ($\Omega_+ < \omega < \omega_{LH}$) electrostatic waves appear to be a permanent feature of the diffuse auroral field lines (Gurnett and Frank 1977). The waves are obliquely polarized ($k_\perp/k_\parallel \sim 0(10)$) with a parallel propagation vector typically comparable to the inverse Larmor radius ($k_\parallel\rho_+ \sim 0(1)$) of plasmasheet ions (Ashour-Abdalla and Thorne 1978). Because of the wide frequency band (involving many harmonics of the ion gyro-frequency) resonant ions can extend from the thermal (E^+_{th}) plasmasheet energies (~keV) up to seven hundred keV. Resonant electrons energies are typically

$$E^-_{res} \approx \{[\frac{2E^+_{th}}{E^+_o} \cdot (\frac{m^+}{m^-})^2 (\frac{1}{k_\parallel\rho_+})^2 + 1\,]^{1/2} - 1\}\, E^-_o \qquad (15)$$

where E^+_o is the ion rest mass. With the observed range of ion thermal energies (100 eV → several keV) and the theoretically anticipated values of $k_\parallel$ for unstable waves ($k_\parallel\,\rho_+ \sim 1/2 \to 2$,

Ashour-Abdalla and Thorne (1978)) resonant electron energies are typically between tens of keV to several Mev.

The broad-band electrostatic waves reported by Gurnett and Frank (1977) have peak wide-band wave amplitude between 10-100 mV/m. Referring to Figure 1, it is clear that such waves have ample intensity to scatter all resonant ions up to several hundred keV on strong diffusion and thus account for the persistent diffuse auroral ion precipitation (Ashour-Abdalla and Thorne 1978). The more intense waves can also scatter realtivistic electrons on strong diffusion particularly at higher L but in general the rate of scattering will be higher at lower electron energies.

3. S3-3 OBSERVATION OF REP EVENTS

In the previous section theoretical arguments were presented for three specific mechanisms capable of providing strong diffusion scattering for high energy electrons. All REP events identified during the analysis of S3-3 data can be uniquely described by one of these mechanisms, and an example of each is illustrated below.

3.1 Case Study Analysis

a. Whistler-mode event. Our analysis was initially restricted to a search for strong diffusion scattering of electrons above 235 keV. Probably because of these stringent criteria only seven events were identified which could be linked to whistler-mode scattering. All were found in the late morning sector where chorus occurrence is maximum. One example is illustrated in Figure 2. Scattering is most intense in the lowest (33 keV) energy channel, becoming progressively weaker at higher energy. No concomitant ion precipitation is perceptable.

b. Electromagnetic ion-cyclotron wave event. From over 14 months of data only four events fit the expected criteria for scattering by electromagnetic ion-cyclotron waves. All occurred at low L near the dusk meridian where such waves are preferentially excited. An example of such an event occurring near L = 4.3 is shown in Figure 3. No perceptible electron precipitation occurred below 160 keV. At 235 keV there is evidence of a partially filled loss cone. As the electron energy increases the rate of scattering becomes progressively more intense reaching the strong diffusion level above 850 keV. The restricted latitude range of relativistic electron precipitation is embedded within a broader zone of strong diffusion ion precipitation.

c. Electrostatic ion-cyclotron wave event. The majority (302) of detected events had the characteristic signature of

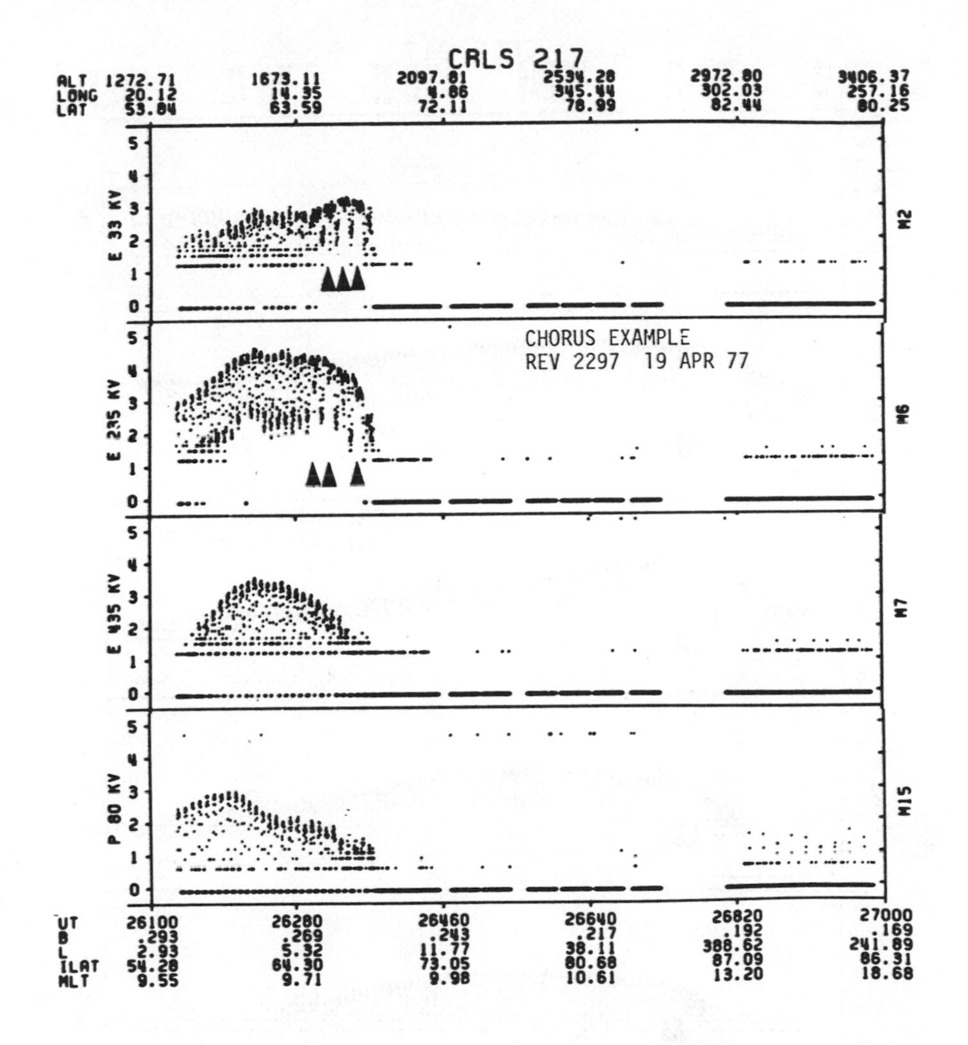

Figure 2. Electron counting rate at 33, 235 and 435 keV observed by the low altitude polar orbiting satellite S3-3 on 19 April 1977. All pitch-angles are sampled within the 20 second spin period of the satellite. At low L the electron flux exhibits a spin modulation consistent with a normal loss-cone distribution. Near the outer limit of electron trapping the lower energy electron flux becomes isotropic over the upward looking hemisphere, indicating strong diffusion scattering. The absence of any perceptible ion precipitation (lower panel) suggests whistler-mode scattering.

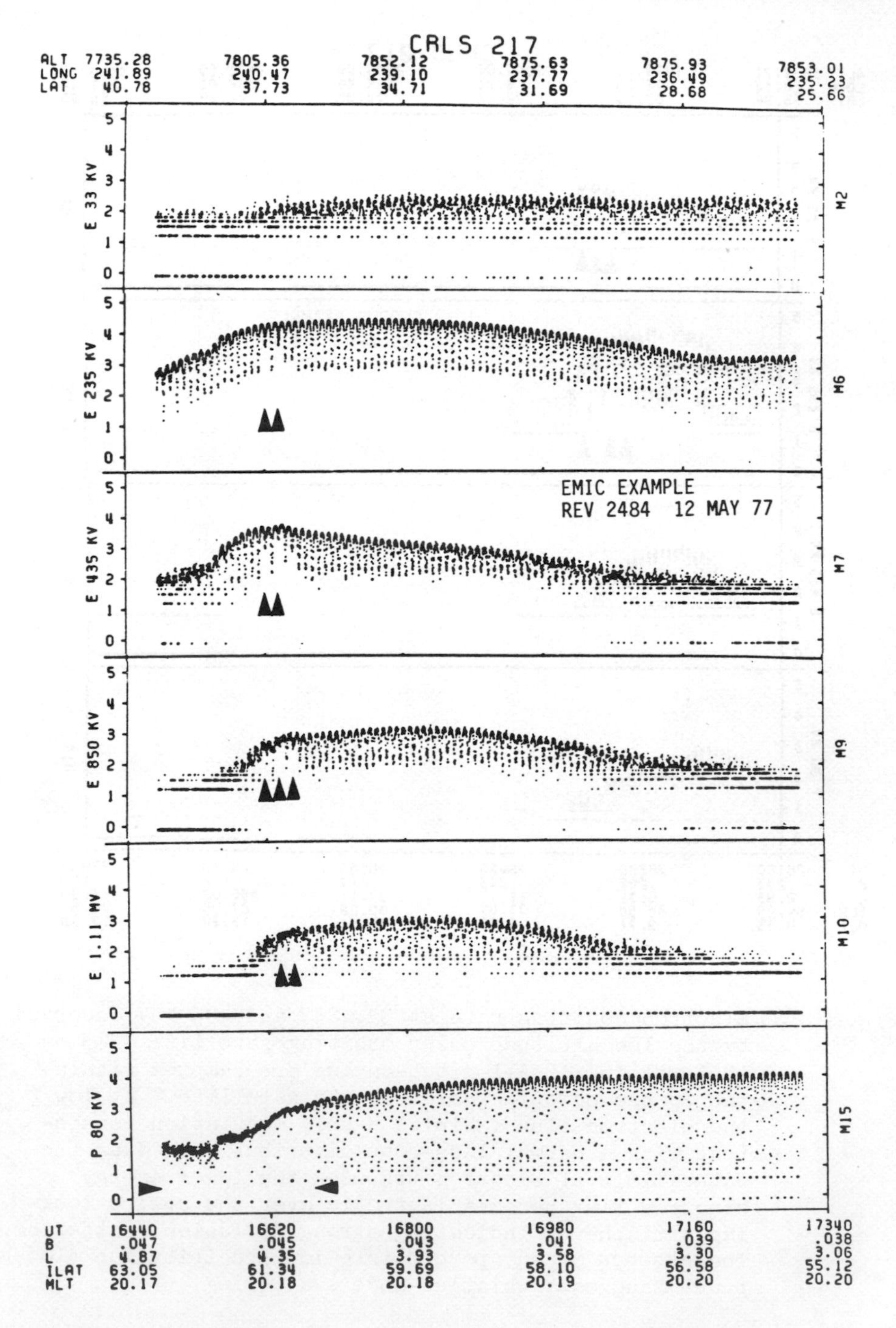

Figure 3. An example of scattering by electromagnetic ion-cyclotron waves observed on S3-3 on 12 May 1977.

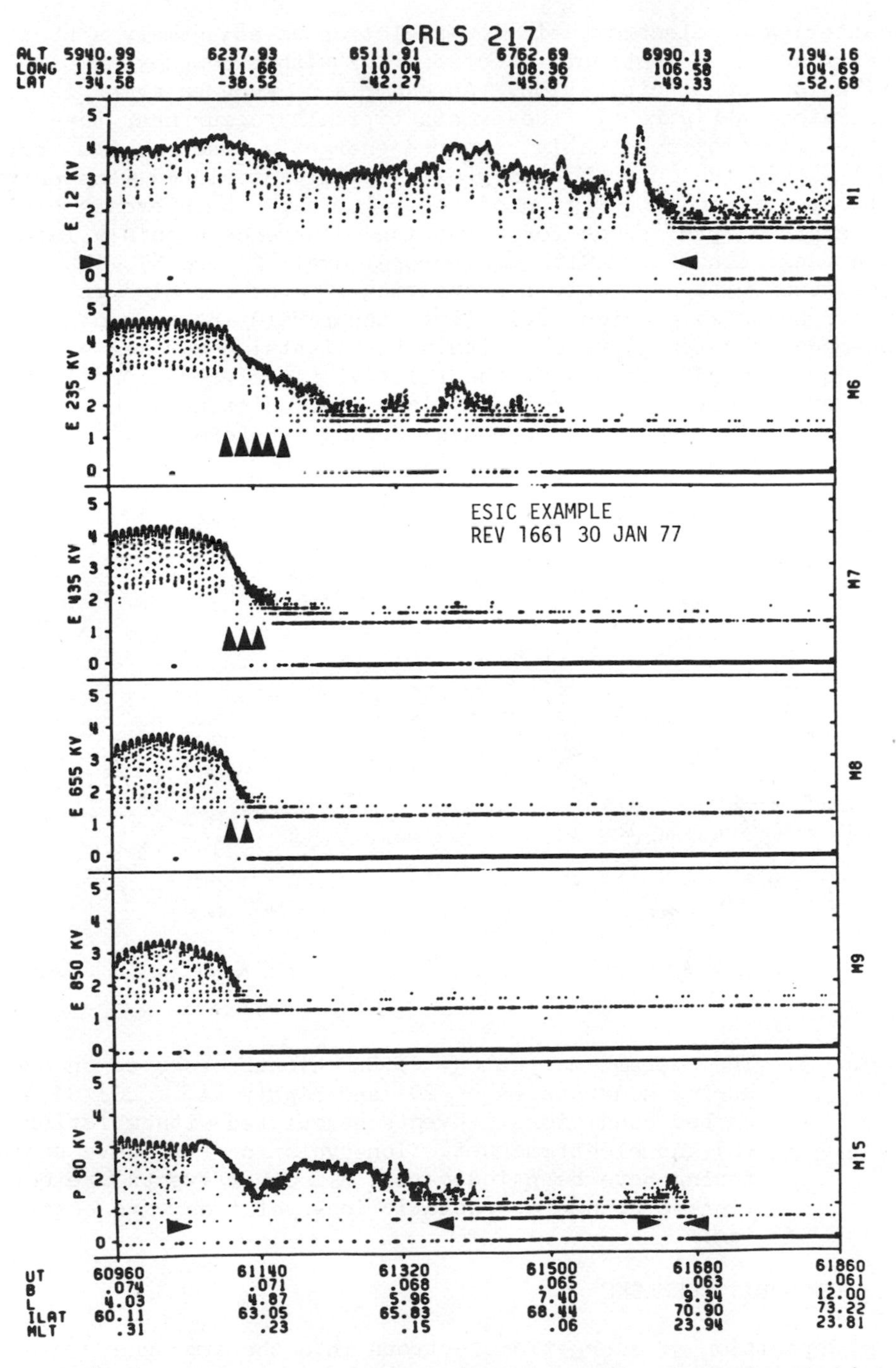

Figure 4. An example of scattering by broadband electrostatic ion-cyclotron waves observed on S3-3 on 30 January 1977

scattering by electrostatic ion-cyclotron waves; namely a broad energy range of isotropic electron flux with accompanying strong diffusion ion precipitation. An example of such an event is illustrated in Figure 4. The events typically occur near the high-L limit of substantial trapped energetic electron flux but well within the outer boundary of trapping associated with entry into open field line region of the polar cap. Such events exhibit a strong preference for nightside occurrence within a latitude range associated with the auroral oval (Figure 5). This is consistent with a parasitic scattering process for high energy electrons which gradient drift into the persistent zone of strong ion-mode turbulence present within the nightside auroral oval. The equatorward movement of the relativistic electron precipitation during more disturbed conditions is consistent with the anticipated displacement of the oval during substorms.

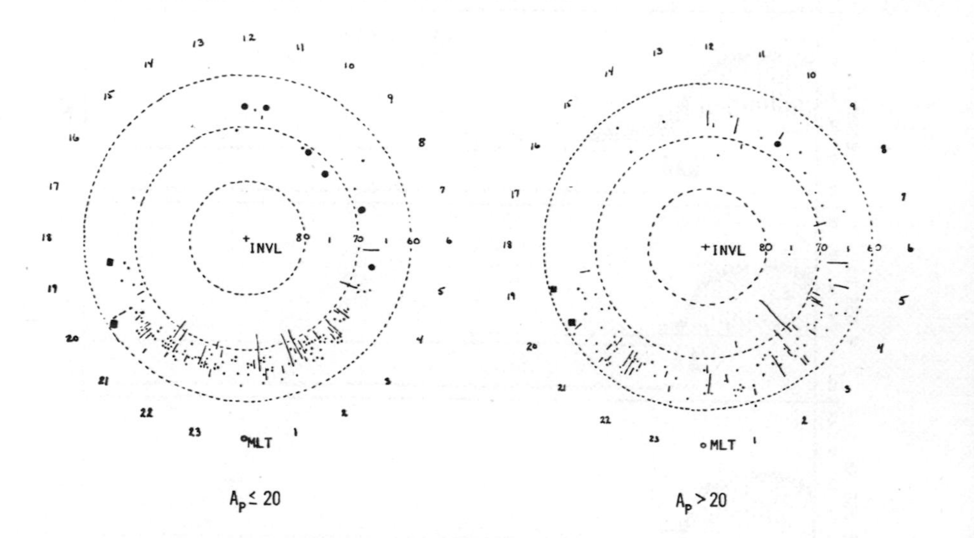

Figure 5. The overall morphology of REP events observed on S3-3 during moderate ($A_p \leq 20$) and highly ($A_p > 20$) disturbed conditions. Events associated with whistler (o) and electromagnetic ion-cyclotron (■) wave scattering have been indicated; all other events are consistent with electrostatic ion-cyclotron wave scattering.

4. CONCLUDING REMARKS

Deposition of energetic electrons into the atmosphere can both provide a major D-region ionization source (Reagan 1977; Thorne 1978) and it can lead to the production of odd hydrogen and odd nitrogen molecules which are known to be catalytic destruction agents for middle atmospheric ozone (Johnston and Podolske 1978). As a result of enhanced odd hydrogen production the extremely hard and intense REP events described above should yield measurable (≈ 30%) localized depletion of mesospheric ozone at sub-auroral latitudes (Thorne 1980). With the observed 5-10% occurrence frequency (Andreoli 1980), such events also constitute the major annual source of nitric oxide throughout the sub-auroral mesosphere and their effect may even be important in the upper stratosphere (Thorne 1977). Furthermore, because our S3-3 analysis was restricted to a limited class of events exhibiting the signature of strong diffusion at relativistic (>230 keV) energies, the above estimates of concomitant atmospheric effects are conservative. Weak diffusion precipitation events and those confined to lower energy electrons are far more frequent and their accumulative effects could therefore be important, but this conjecture must await future experimental verification.

ACKNOWLEDGEMENTS

We express sincere thanks to A. Vampola for making the S3-3 data available to us and to J. DeVore for his help in modelling the chemical response of the atmosphere during REP events. This work was supported in part by NASA grant NSG 5190 and NSF grant ATM 77-24843.

REFERENCES

Andreoli, L. J., 1980, Relativistic Electron Precipitation: An Observational Study, Ph.D. Thesis, Department of Atmospheric Sciences, UCLA.

Ashour-Abdalla, M., and Kennel, C. F., 1978, J. Geophys. Res., 83, 1531.

Ashour-Abdalla, M., and Thorne, R. M., 1978, J. Geophys Res., 83, 4755.

Bailey, D. K., 1968, Rev. Geophys., 6, 289.

Bailey, D. K., and Pomerantz, M. A., 1865, J. Geophys. Res., 70, 5823.

Bossen, M., McPherron, R. L., and Russell, C. T., 1976, J. Geophys. Res., 81, 6083.

Cornwall, J. M., Coroniti, F. V., and Thorne, R. M., 1970, J. Geophys. Res., 75, 4699.

Gurnett, D. A., and Frank, L. A., 1977, J. Geophys. Res., 82, 1031.

Heaps, M. G., 1978, U.S. Army Atmospheric Sciences Laboratory Report ASL-TR-0012.

Hultqvist, B., Borg, H., Christophersen, P., Riedler, W., and Bernstein, W., 1974, NOAA Technical Rept.; ERL 305-SEL29.

Kelley, M. C., Bering, E. A., and Mozer, F. S., 1975, Phys. Fluids, 18, 1590.

Kennel, C. F., and Petschek, H. E., 1966, J. Geophys. Res., 71, 1.
Lyons, L. R., 1974, J. Geophys. Res., 79, 575.
Mathews, D. L., and Simons, D. J., 1973, J. Geophys. Res., 78, 7539.
Reagan, J. B., 1977, p. 145 in Dynamical and Chemical Coupling of Neutral and Ionized Atmosphere (Eds. B. Grandal and J. A. Holtet) Reidel Publishing, Dordrecht, Holland.
Rosenberg, T. J., Lanzerotti, L. J., Bailey, D. K., and Pierson, J. D., 1972, J. Atmos. Terr. Phys., 34, 1977.
Thorne, R. M., 1974, J. Atmos. Terr. Phys., 36, 635.
Thorne, R. M., 1977, Science, 21, 287.
Thorne, R. M., 1978, in proceedings of joint IAGA/IAMAP Assembly in Seattle, published by NCAR.
Thorne, R. M., 1980, P. Ap. Geophys., 118, 128.
Thorne, R. M., and Kennel, C. F., 1971, J. Geophys. Res., 76, 4446.
Thorne, R. M., and Larsen, T. R., 1976, J. Geophys. Res., 81, 5501.
Tsurutani, B. T., and Smith, E. J., 1974, J. Geophys. Res., 79, 118.
Tsurutani, B. T., and Smith, E. J., 1977, J. Geophys. Res., 82, 5112.
Vampola, A. L., 1971, J. Geophys. Res., 76, 4685.

ELECTROSTATIC WAVES IN THE IONOSPHERE

Eigil Ungstrup

Danish Space Research Institute, Lyngby, Denmark

ABSTRACT

The study of electrostatic waves is important for the understanding of the fine structure of the ionosphere. Although a number of different instabilities occur in the ionosphere we shall limit ourselves to a discussion of two important instabilities that occur in the E-region of the ionosphere. These instabilities are known as the Farley-Buneman two-stream instability and the gradient drift instability. The instabilities will be discussed from in situ rocket observations in the ionosphere.

1. INTRODUCTION

Electrostatic waves play an important role for the fine-structure of the ionosphere and magnetosphere. Before instruments were carried into the ionosphere by rockets and satellites we only had indirect evidence of the existence of such waves because electrostatic waves occur only in plasmas. Our indirect evidence came from radio aurora and from scattered reflections from the E and F region of the ionosphere observed with ionsondes (e.g. slant sporadic E and spread F).

With the development of rockets and satellites it became possible to carry instruments into the ionosphere and directly observe the electrostatic waves. During the last fifteen years a large number of such experiments have been carried out (Shawhan and Gurnett 1968, Gurnett and Mosier 1969, Kelly et al. 1970, Kelly and Mozer 1973, Kelly et al. 1975, Ungstrup 1975).

C. S. Deehr and J. A. Holtet (eds.), Exploration of the Polar Upper Atmosphere, 395–406.

An important property of electrostatic waves is that they seldom propagate over large distances. Usually their speeds are comparable to thermal speeds of ions or electrons and therefore they experience strong Landau or cyclotron damping. Consequently they are damped out after propagating only a short distance away from the unstable region where they are generated. One therefore has to pass through or very close to an unstable region in order to observe electrostatic waves.

In this paper we shall limit ourselves to instabilities that are rather well understood and which are of great importance in the ionosphere. We shall especially treat the Farley-Buneman or Hall-current two-stream instability, and the gradient-drift instability. Other instabilities of importance in the ionosphere are the Post-Rosenblutn instability (Ott and Farley 1975, St.-Maurice 1978) and the electrostatic ion-cyclotron instability (Kindel and Kennel 1971, Ungstrup et al. 1979) but it is not possible to include a discussion of these instabilities in this short paper.

2. THE FARLEY-BUNEMAN TWO-STREAM INSTABILITY

In 1960 to 1963 it became clear from Doppler measurements on radar reflections from polar auroras and from the equatorial electrojet, that strong irregularities in the ionospheric E-region moved with phase velocities of the order of the sound speed. These observations were interpreted by Farley (1963 a,b) and Buneman (1963) in term of a two-stream instability that created field aligned irregularities moving through the ionosphere with the ion sound speed. Later Rogister and D'Angelo (1970) presented a unified treatment of the two-stream instability and the gradient-drift instability based on fluid theory. Here we shall present a short derivation of the two-stream instability based on a two-fluid theory of plasmas.

In the altitude range 90-115 km we have $\nu_{en} << \Omega_e$ and $\nu_{in} \gtrsim \Omega_i$, where $\nu_{en,in}$ is respectively the electron and ion collision frequency with the neutrals and $\Omega_{e,i}$ is the gyro frequency of the electrons and ions. Because of these conditions the electrons drift through the plasma with a drift velocity $\bar{v}_D = \bar{E} \times \bar{B}/B^2$, whereas the ions are impeded in their motion by the neutrals. We shall show that if the electron drift velocity relative to the ion background exceeds the local sound speed then longitudinal plasma waves will grow.

We start from the fluid equations

$$\frac{\partial n_j}{\partial t} + \nabla \cdot (n_j \bar{v}_j) = 0 \qquad (1)$$

$$n_j m_j \left(\frac{\partial \bar{v}_j}{\partial t} + (\bar{v}_j \cdot \nabla)\bar{v}_j\right) = -\nabla(n_j \kappa T_j) + n_j q_j(\bar{E} + \bar{v}_j X \bar{B}) - \nu_j n_j m_j \bar{v}_j \qquad (2)$$

where m_j and q_j are the mass and charge of the j-component particle, while n_j is the particle density, $\bar{v}_j$ the fluid velocity, T_j the temperature of the j-component fluid, ν_j is the collission frequency between the j-component particle and neutrals, and κ is the Boltzmann constant. Further we assume the plasma to be uniform, that there is quasi-neutrality ($n_e \simeq n_i$), that $T_e = T_i$, and that the E-field is in the X-direction and that the B-field is in the Z-direction.

The zero-order solution is found from the fluid equations which reduces to

$$n_e q_e(\bar{E} + \bar{v}_e \; X \; \bar{B}) = \nu_e n_e m_e \bar{v}_e \; ; \; n_i q_i(\bar{E} + \bar{v}_i \; X \; \bar{B}) = \nu_i n_i m_i \bar{v}_i \qquad (3\ a,b)$$

from which equations we find the Hall and Pedersen drifts for electrons and ions

$$V_{oye} = - \frac{\Omega_e^{\,2}}{\Omega_e^{\,2} + \nu_e^{\,2}} \frac{E}{B} \simeq - \frac{E}{B} \; ; \; V_{oyi} = - \frac{\Omega_i^{\,2}}{\Omega_i^{\,2} + \nu_i^{\,2}} \frac{E}{B} \qquad (4\ a,b)$$

$$V_{oxe} = - \frac{\Omega_e \nu_e}{\Omega_e^{\,2} + \nu_e^{\,2}} \frac{E}{B} \simeq 0 \; ; \; V_{oxi} = \frac{\Omega_i \nu_i}{\Omega_i^{\,2} + \nu_i^{\,2}} \frac{E}{B} \qquad (5\ a,b)$$

We now linearize the fluid equations with the following conditions

$$n = n_o + n_1(y,t) \qquad n_e \simeq n_i \simeq n$$

$$\phi = \phi_o(x) + \phi_1(y,t) \qquad \Omega_e >> \nu_e$$

$$\bar{v} = \bar{v}_o + \bar{v}_1(y,t) \qquad \Omega_i \simeq \nu_i$$

$$v_{oxe} \simeq 0$$

and assume the first order perturbations to be travelling waves of the form exp (iky − iωt) where ω is complex.

By linearizing equation (1) and (2) we get, when we reglect the electron inertia,

$$(-i\omega + ikv_{oe})\frac{n_1}{n_o} + ikv_{1ye} = 0 \tag{6}$$

$$-ik\kappa T_e\frac{n_1}{n_o} + eBV_{1xe} - \nu_e m_e v_{1ye} + ike\phi_1 = 0 \tag{7}$$

$$-\nu_e m_e v_{1xe} - eBV_{1ye} = 0 \tag{8}$$

$$(-i\omega + ikv_{oyi})\frac{n_1}{n_o} + ikv_{1yi} = 0 \tag{9}$$

$$ik\kappa T_i\frac{n_1}{n_o} + ike\phi_1 + eBv_{1xi} + (-i\omega + ikv_{oyi} + \nu_i)m_i v_{1yi} = 0 \tag{10}$$

$$(-i\omega + ikv_{oyi} + \nu_i)\, m_i v_{1xi} - eBv_{1yi} = 0 \tag{11}$$

This set of homogeneous equations has a non-trivial solution when the determinant is equal to zero. Setting the determinant to zero results in the following dispersion equation

$$(-i\omega + ikv_{oyi} + \nu_i)\,\Big[\nu_e(-i\omega + ikv_{oyi})(-i\omega + ikv_{oyi} + \nu_i) +$$

$$\nu_e k^2 c_s^2 + (-i\omega + ikv_{oe})\Big(\nu_e^2\frac{m_e}{m_i} + \Omega_e\Omega_i\Big)\Big] + \nu_e\Omega_i^2(-i\omega + ikv_{oyi}) = 0 \tag{12}$$

where $c_s^2 = \frac{\kappa(T_e+T_i)}{m_i}$ is the ion sound speed and $\Omega_j = \frac{eB}{m_j}$ the gyrofrequency for the j-component particle. We can neglect $\nu_e^2\frac{m_e}{m_i}$ compared to $\Omega_e\Omega_i$ because $\nu_e \ll \Omega_e$.

Setting $\omega = \omega_k + i\gamma_k$ with both ω_k and γ_k real and considering the case where

$$|\gamma_k| \ll |\omega_k| \text{ and } (v_{oye} - v_{oyi})^2 \gg \frac{\nu_e\nu_i}{\Omega_e\Omega_i}C_s^2 = \psi C_s^2 \tag{13}$$

where

$$\psi = \frac{\nu_e\nu_i}{\Omega_e\Omega_i} \ll 1,\text{ we find } \omega_k = \frac{kv_{oye}}{1+2\psi} + \frac{4\psi kv_{oyi}}{1+2\psi} \simeq kv_{oye} \tag{14}$$

When $\omega_k = kv_{oye}$ is introduced into the equation for the imaginary part we find

$$\gamma_k = \frac{\nu_e k^2}{\Omega_e \Omega_i}((v_{oye}-v_{oyi})^2 - C_s{}^2)\frac{1}{1+\psi} \simeq \frac{\nu_e k^2}{\Omega_e \Omega_i}((v_{oye}-v_{oyi})^2 - C_s{}^2) \qquad (15)$$

If we substitude $\omega_k = kV_{oiy}$ in stead we find

$$\gamma_k = -\nu_e \frac{k^2 C_s^2}{\Omega_e \Omega_i} \qquad (16)$$

We thus see that only waves with phase veloicity equal to the electron drift velocity can grow, while waves with phase velocity equal to the ion drift velocity are always damped.

For higher collision frequencies ($\nu_i >> \Omega_i$) this simple derivation does not hold (see eq. 13). This case has been considered by Sudan et al. (1973) and by Rogister and D'Angelo (1970) and their result is essentially the same as we found in equations (14) and (15).

From equation (15) it follows that the difference in drift velocity for the electrons and ions must exceed the ion sound speed for the instability to be excited. This limits the altitude range in which the instability can be excited in the upward direction since the ions will have the same drift velocity as the electrons when the ion-neutral collision frequency becomes neglible. This occurs typically around 115 km altitude.

In the bottom of the E-region the collision frequencies increase as we go down in altitude and this will increase ψ so that the final approximation in (14) and (15) becomes invalid. The effect of the increased collision frequency is to reduce the phase velocity of the waves as well as the growth rate and for sufficiently high collision frequency the instability may not be excited even though $v_{oe} > c_s$.

3. OBSERVATIONS OF THE FARLEY-BUNEMAN TWO-STREAM INSTABILITY

Detectors for electrostatic waves have been flown by the Danish Space Research Institute on a number of rocket payloads. During disturbed geophysical conditions electrostatic waves are usually observed in the E-region of the ionosphere and in most cases these waves can be attributed to the two-stream instability.

In figure 1 we show the VLF spectrum from 1.2 to 10.0 kHz observed on payload S 70/1 launched from ESRANGE, Sweden, with a 3 m dipole antenna. The strong broad-band electrostatic signals observed from 85 to 102 s and again from 355 to 370 s are signatures of the two-stream instability.

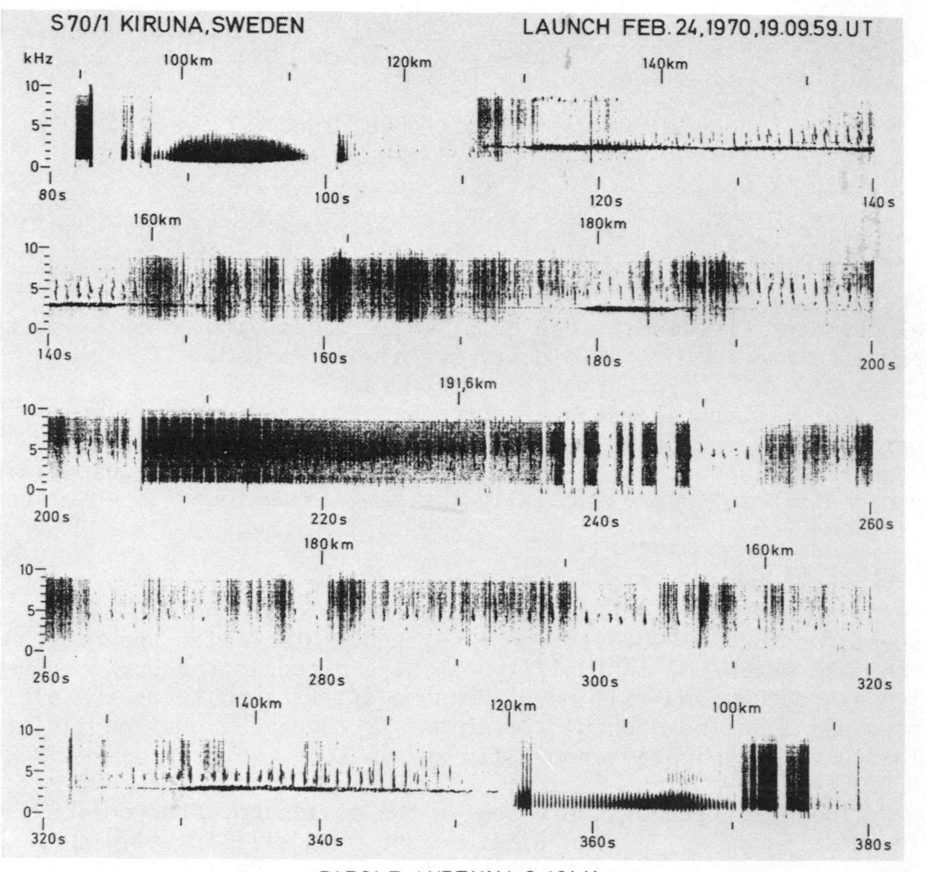

Figure 1. Observations of the spectrum of electrostatic and electromagnetic waves in the ionosphere on payload S 70/1 launched from ESRANGE. Time after launch is shown at the bottom of the spectrum and the altitude of the payload at the top of the spectrum. The spectrum is shown for the frequency range 1.2 to 10 kHz. The blackening is proportional to signal strength except that gain changes of 20 dB are automatically introduced at irregular intervals to keep the signal level within the dynamic range of the telemetry. The strong signals observed between 85 and 102 s and again from 355 to 370 s in the frequency range 1 to 4 kHz are electrostatic waves caused by the two-stream instability in the E-region of the ionosphere.

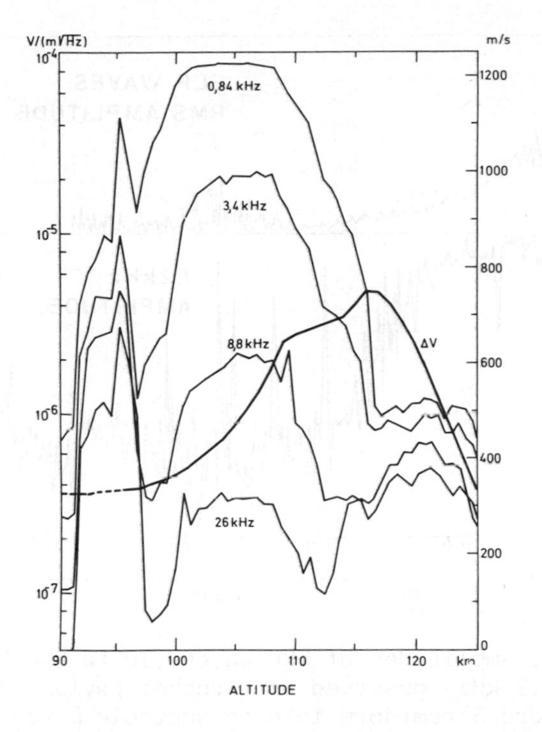

Figure 2. Intensity of the electrostatic signals on four different frequencies observed in the altitude range 90 to 125 km together with drift velocity of electrons with respect to ions computed from the measured DC electric field.

The amplitudes of the electrostatic waves on four different frequencies are shown in figure 2 together with the relative drift of electrons with respect to ions. The drift velocity exceeds the ion-acoustic wave speed from 100 to 120 km. However, the instability occurs in the altitude range 96 to 116 km. There are several possible causes for this discrepancy of 4 km. The collision frequency model used in the computations is derived from Thrane and Piggot (1966) and from Thrane (1968). Comparision with Dalgarno (1961) seems to indicate that our collision frequencies are possibly too high. A reduction of the collision frequency by a factor of 2 to 2.5 gives very good agreement between the expected and observed altitudes for the two-stream instability. We must, however, point out that other factors than the collision frequency affect the altitudes of this instability. We shall mention the increase in ion sound speed caused by the increase in temperature with altitude and the unknown effects of a possible neutral wind pulling the ions along.

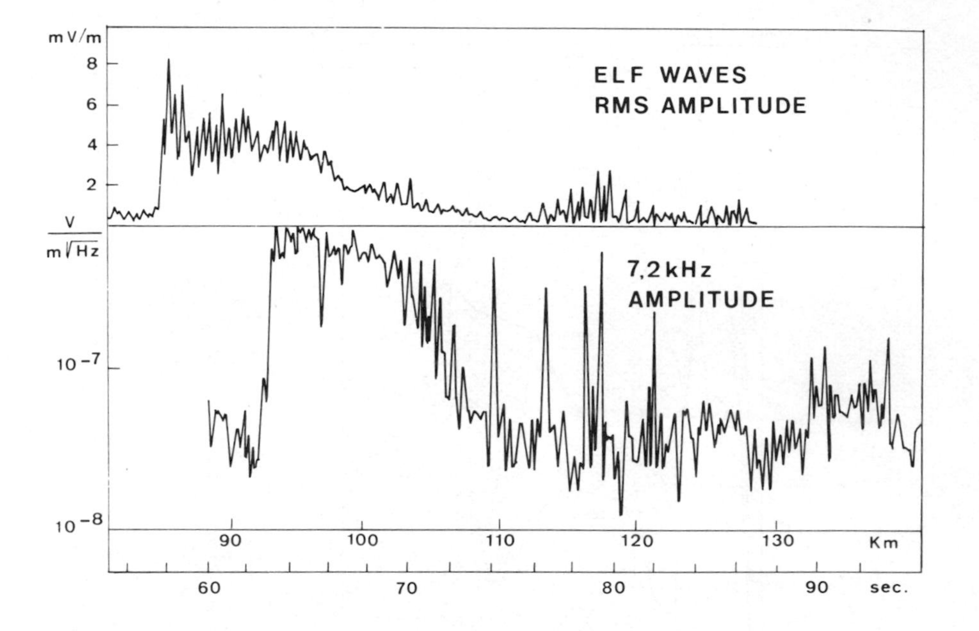

Figure 3. Amplitudes of ELF waves (10 to 200 Hz) and VLF waves (7.2 kHz) observed on a rocket payload launched from Søndre Strømfjord into an unstable E-region.

There is an important variation in the spectrum of the electrostatic waves as function of altitude. This is shown in figure 3, (Olesen et al. 1976, Bahnsen et al. 1978) where we show the amplitude of ELF waves below 200 Hz compared to the amplitude of VLF waves on 7.2 kHz as a function of altitude. The ELF waves occur in the altitudes 85 to 98 km whereas the VLF waves occur between 92 and 105 km. This change in the spectrum with altitude seems to be in agreement with theory as the unstable frequencies decrease with increasing collision frequency (see equation 14).

The two-stream instability of Farley and Buneman is a very important instability in the high latitude ionosphere where it is responsible for most observations of radio aurora as well as for the slant sporadic E observations observed by ionsondes. These phenomena are also associated with rather poor conditions for HF voice communication in the auroral zone and polar cap where the readability is reduced by flutter fading.

4. THE GRADIENT-DRIFT INSTABILITY

The gradient-drift instability was first investigated theoretically by Simon (1963) and by Hoh (1963). It was applied to the equatorial electrojet by Maeda et al. (1963), Knox (1964),

Reid (1968), Rogister and D'Angelo (1970), Whitehead (1971), Farley and Balsley (1973), and Sudan et al. (1973), to mid-latitude sporadic-E by Tsuda et al. (1966) and to radio aurora by Unwin and Knox (1971).

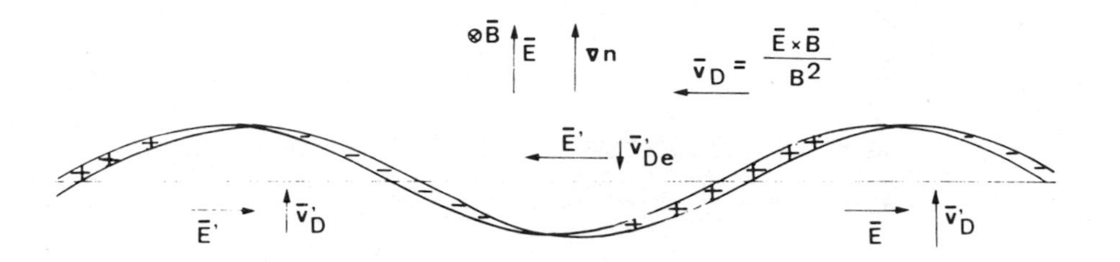

Figure 4. Principles for the gradient-drift instability

The operation of one mode of the instability can be understood from figure 4. The density gradient and electric field are pointing upwards and the magnetic field is horizontal. If the density contours are pertubed the $\bar{E} \times \bar{B}$ drift of the electrons will set up space charges as shown. (The ions are assumed to be inhibited in their motion by the neutrals). The space charges act to set up secondary electric fields that act together with the magnetic field to carry enhanced regions downward and depleted regions upward so that they both appear to grow against the background density. If we reverse either the gradient or the electric field the perturbations tend to disappear.

For the instability to work at high latitudes it is necessary that the angle between the density gradient and the electric field is less than 90 degrees, but this condition can easily be satisfied in auroral forms and also a horizontally stratified medium if the inclination of the magnetic field is not too high and the DC electric field in the northern hemisphere is directed northward perpendicular to the magnetic field.

In figure 5 we show observations from payload S 70/2 launched from ESRANGE, Sweden. We see that the amplitude of the electrostatic waves has a narrow peak between 101 and 107 km on 0.84 kHz. The spectrum is very steep being a factor of 12 down on 3.4 kHz corresponding to an amplitude fall off as $f^{-1.8}$.

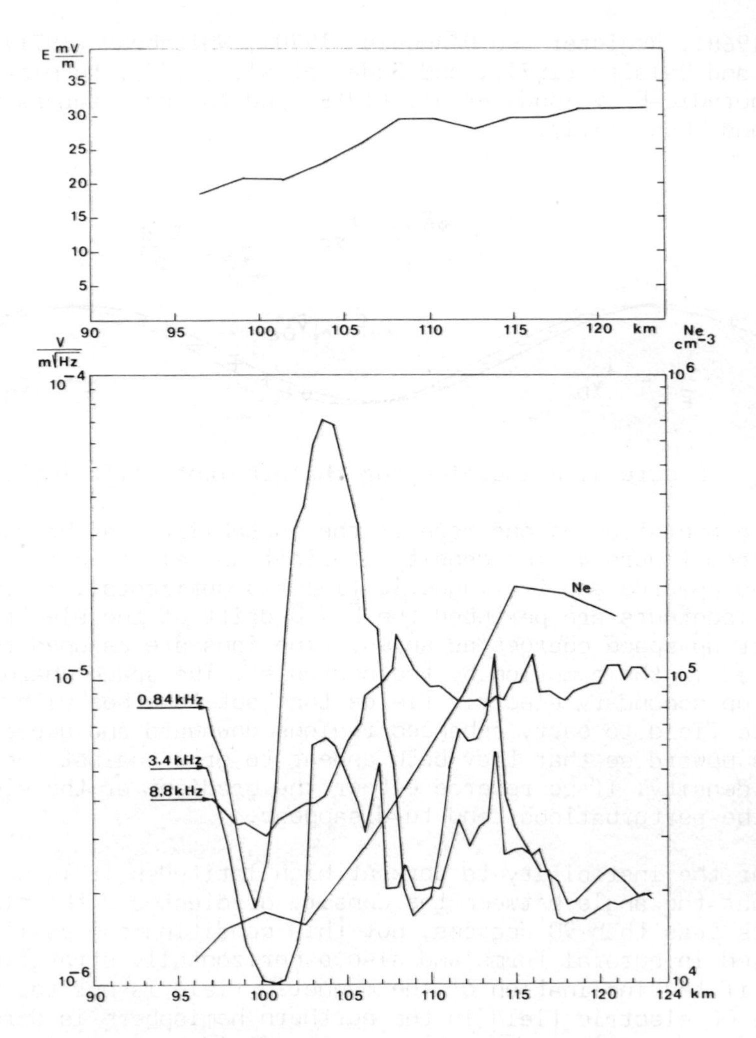

Figure 5. DC electric field, electron density, and VLF amplitudes measured on payload S 70/2 launched from ESRANGE.

The electric field increases from 20.0 to 29.5 mV/m between 101 and 108 km and remains around 30 mV/m until the rocket is out of the E-region. One expects the medium to go unstable to the two-stream instability when the DC electric field exceeds 20 to 25 mV/m if the temperature in the E-region is normal. At 104 km where we have the amplitude peak the medium is thus only on the limit of being unstable to this instability and if it was responsible for the waves we expect an unstable region between 104 and 115 km with peak amplitudes at 108 to 110 km, which is not in agreement with observations.

The electron density profile in figure 5 shows a density gradient in the altitude range 102.5 to 115 km. The e-folding length of this density gradient is $L \simeq 4$ km, but the altitude resolution on the Langmuir probe electron density measurements is rather poor because of a slow sweep rate and therefore we can only derive an average density gradient. It is possible that the gradient locally may be larger. From the average gradient one expects the instability to occur in the altitude range of 102 to 114 km. The direction of the observed electric field is about 10 degrees East of North and thus in the right direction for the gradient-drift instability to occur.

The growth rate for the gradient-drift instability is (Rogester and D'Angelo 1970)

$$\gamma_k = \frac{1}{1+\psi} \frac{\nu_i}{\Omega_i} \frac{\omega_k}{k} \frac{1}{n_o} \frac{\partial n_o}{\partial z} \tag{17}$$

from which we see that we have no threshold for this instability and that large wavelength waves (with low wave number, k) will grow fastest.

The gradient-drift instability therefore will generate irregularities with wavelengths of the order of 10^2 meters or more. These waves are not observed directly by the AC electric field on 840 Hz or higher frequencies, but the primary waves cause secondary gradients that have a smaller e-folding length than the ~ 4 km which follow from the observed electron density profile. In these secondary gradients small scale turbulence is generated in the same way as discussed for the equatorial electrojet by Sudan et al. (1973) and Farley and Balsley (1973). This cascading accounts for the steep spectrum discussed earlier.

It is suggested that the electrostatic waves observed on payload S 70/2 are caused by the gradient drift instability in stead of the two-stream instability. The observations do not agree with the criteria for the two-stream instability as this instability would have caused the electrostatic waves to occur over a broader altitude range and to peak at a greater altitude.

The narrow peak in wave amplitude can be understood in terms of the gradient drift instability if we have a stronger gradient in part of the altitude range. The steep spectrum also points to the gradient-drift instability. Since this instability generates longer wavelengths than the two-stream instability we will find the transition from a flat spectrum to a steep spectrum at lower frequencies than for the two-stream instability. On payload S 70/1 where we assume the waves to be generated by the two-stream instability we have the transition to a steep spectrum in the range

3.4 to 8.8 kHz but on payload S 70/2 this transition occurs below 840 Hz.

REFERENCES

Bahnsen, A.,Ungstrup,E., Fälthammar, C.-G., Fahleson, U., Olesen, J.K., Primdahl, F., Spangslev, F., and Pedersen,A, 1978, J. Geophys.Res., 83, 5191
Buneman, O. 1963, Phys. Rev. Letters, 10, 285
Dalgarno, A. 1961. Ann. Geophys., 17, 16
Farley, D.T. 1963, Phys Rev. Letters, 10, 279
Farley, D.T. 1963, J.Geophys. Res., 68, 6083
Farley, D.T., and Balsley, B.B. 1973, J.Geophys. Res., 78, 227
Gurnett, D.A. and Mosier, S.R. 1969, J.Geophys. Res., 74, 3979
Hoh, F.C. 1963, Phys. Fluids, 6, 1184
Kelley, M.C., Mozer, F.S., and Fahleson, U. 1970, Planet Space Sci., 18, 847
Kelly, M.C., and Mozer, F.S. 1973, J.Geophys.Res., 78, 2214
Kelly, M.C., Bering, E.A., and Mozer, F.S. 1975, Phys.Fluids, 18, 1590
Kindel, J.M., and Kennel, C.F. 1971, J.Geophys.Res., 76, 3055
Knox, F.B. 1964, J.Atm.Terr. Phys., 26, 239
Maeda, K.-I., Tsuda, T., and Maeda, H. 1963, Rep. Ionos.Res.Japan, 17, 147
Olesen, J.K., Primdahl, F., Spangslev, F., Ungstrup, E., Bahnsen, A., Fahleson, U., Fälthammar, C.-G., and Pedersen, A. 1976, Geophys. Res. Letters, 3, 711
Ott, E., and Farley, D.T. 1975, J.Geophys. Res., 80, 4599
Reid, G.C. 1968, J.Geophys. Res., 73, 1627
Rogister, A., and D'Angelo, N. 1970, J.Geophys.Res., 75, 3879
St.-Maurice, J.P. 1978, Planet.Space Sci., 26, 801
Shawhan, S.D., and Gurnett, D.A. 1968, J.Geophys.Res., 73, 5649
Simon, A. 1963, Phys.Fluids, 6, 382
Sudan, R.N., Akinrimisi, J., and Farley, D.T. 1973, J.Geophys. Res., 78, 240
Thrane, E.V., and Piggott, W.R. 1966 J.Atm.Terr.Phys., 28, 721
Thrane, E.V. 1968, Ionospheric Radio Communications, Plenum Press, 63.
Tsuda, T., Sato, T., and Maeda, K.I. 1966, Radio Sci., 1, 212
Ungstrup, E. 1975, J.Geophys.Res., 80, 4272
Ungstrup, E., Klumpar, D.M., and Heikkila, W.J. 1979, J.Geophys. Res., 84, 4289
Unwin, R.S., and Knox, F.B. 1971, Radio Science, 6, 1061
Whitehead, J.D. 1971, J.Geophys.Res., 76, 3116

THE CHANGING AURORA OF THE PAST THREE CENTURIES

S. M. Silverman

Air Force Geophysics Laboratory, Hanscom AFB, MA 01731

J. Feynman

Dept of Physics, Boston College, Chestnut Hill, MA 02167

INTRODUCTION

Studies of the aurora are of interest not only for their own sake but also for the information that can be inferred from them on the solar wind, solar variability, magnetospheric dynamics, and the dynamics of the upper atmosphere.

In this paper we review some of the changes that have taken place on the time scale of from two to ten decades in the frequency and latitude of occurrence of the aurora in the region near and south of the present auroral oval. We relate these changes, qualitatively to solar wind parameters. The review is in two sections based on the primary data source used. Swedish auroral observations from 1721 to 1876 have been catalogued by Rubenson (1879, 1882) and forms the major data base for that period. For the period from 1876 to at least 1943 auroral observations are available but not yet completely catalogued. We report on some observations from individual stations for this period. The major data base for the period before 1876, however, is the aa indices derived from magnetic data by Mayaud (1973), which can be used as proxy data for auroral observations because of the close connection between the two phenomena, the recognition of changes of oval position had been made by Tromholt (1880). Basing his conclusions on data from Greenland, Tromholt stated (translation by S.S.) "...during the course of

C. S. Deehr and J. A. Holtet (eds.), Exploration of the Polar Upper Atmosphere, 407–420.

the 11 year period the auroral zone is displaced laterally in such a way that during sunspot minimum this zone is more to the north than during sunspot maximum... The auroral maximum which in temperate zones coincides with sunspot maximum arises from the fact that the auroral zone is then in its most southerly position; and this is why the polar regions have at the same epoch a minimum of auroras; on the other hand, the auroral zone has its most northerly position when the sunspot number is at a minimum, which leads to a minimum of auroras in temperate regions and a maximum of auroras in the polar zone". Tromholt also pointed out the oscillatory movements of the zone in the course of the day and seasonally. In short, the basic characteristics of oval behavior had been described by Tromholt in 1880.

Other well-known studies of the latitudes at which aurora occur are those of Loomis (1865) and Fritz (1881) in the latter part of the nineteenth century. These are statistical in nature and thus tend to disguise some interesting aspects of the problem. This will become clear from our discussion of the data from some specific locations. Much emphasis in past work has been on the location of the auroral oval with some apparent discrepancies in results. Vestine (1944) used data from the first and second International Polar Year, 1882-1883, and 1932-1933, to determine the position of the most frequent auroral sightings, and concluded that this position had not changed for the two periods, despite the fact that 1883 was a sunspot maximum year and 1933 was a sunspot minimum year. Note, however, that the level of geomagnetic activity measured by midlatitude indices (Mayaud, 1973) was about the same in these two periods in spite of the difference in the phases of the sunspot number cycle. In fact, a change in the position of the oval has been observed. Sandford (1968) studied the latitudes of the discrete auroras in the southern hemisphere during the 1958-1959 IGY and the 1963 IQSY. In his study of discrete midnight auroras as a function of the local disturbance index, K, he found that for the same value of K aurora were seen at more equatorward stations in 1958-1959 than in 1963. Feldstein and Starkov (1968) also compared the positions at which aurora were seen during the IGY and IQSY as a function of disturbance level. They state that the center line of the oval moved northward 1.5° between the IGY and the IQSY, in agreement with the Sandford results.

THE AURORA AND SOLAR ACTIVITY

It has been known since at least the middle of the nineteenth century that auroral frequency is correlated with solar activity (Loomis, 1865; Fritz, 1880). A more recent study by Silverman (1978) using the Rubenson catalog has shown that auroral frequency, unlike the sunspot cycle, has typically, two peaks during a solar cycle, as does the geomagnetic activity. It was also pointed out that auroral frequency is to be responsive to active regions near solar maximum and to coronal holes during the declining phase of the sunspot cycle.

In view of these connections we briefly review here the solar activity of the past three centuries. The Maunder minimum of sunspot activity [1645-1715] had ended by 1721, the date of the earliest auroral observations used in this study. By 1730 the number of auroras reported in Sweden had reached values typical of the number reported for periods of high solar activity for the next 150 years.

Sunspot activity shows two periods of considerably lessened activity around 1810 and 1900. These periods, taken together with the very low sunspot numbers around 1700, have led to the hypothesis that there is an 80-100 year cycle for sunspots. A considerably longer observational period would be required to establish this from sunspot data alone. Here we focus on auroral studies from 1720 to the present with particular emphasis on the periods around 1810 and 1900. In addition, we will describe changes in the latitudinal occurrence of Swedish auroras for time scales of the order of decades during the 18th century.

MAGNETIC INDICES

In a classic geomagnetic storm, the horizontal component (H) of the magnetic field increases at the storm sudden commencement, remains high for a variable amount of time and then a deep depression develops after which H gradually recovers. The general level of H is described by the hourly Dst index. Superposed on the decline of H there are irregular rises and declines generally corresponding to substorm activity. The range of these disturbances is described by of indices such as the K_p and aa indices. The K midlatitude indices are quasilogarithmically related to the a indices. The data set discussed here are the annual averages of the aa indices, <aa>. The advantage of <aa> is that a selfconsistent set of values due to Mayaud exists from 1868 to the present. For further discussion of indices see Allen and Feynman (1979) and the review by Berthelier in these proceedings.

The geographical location of the aurora is related to geomagnetic activity. Gussenhoven et al. (1980) have studied the equatorward position of the auroral oval defined by electron precipitation observed by DMSP and found that the average position of the equatorward boundary at a particular local time is a linear function of K_p, but that there is considerable spread in the positions for a given K_p. The average position

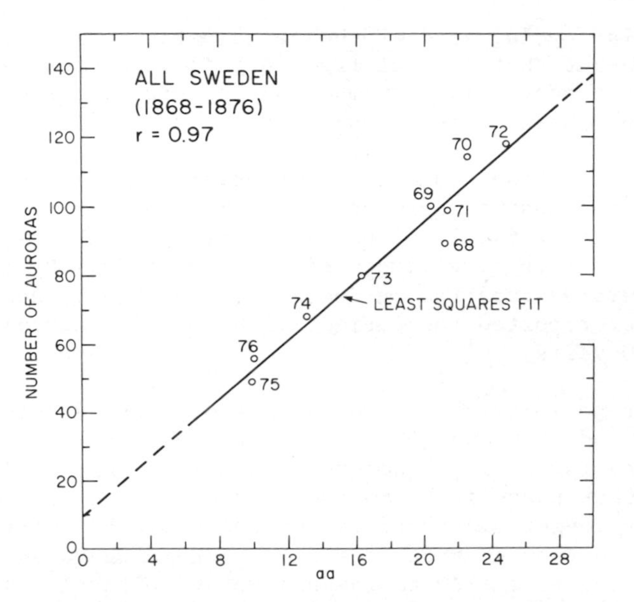

Figure 1. A comparison of the yearly averaged midlatitude geomagnetic index aa with the number of aurora reported for Sweden during the years when the data sets overlap.

of the equatorward boundary of auroral electron precipitation in the nighttime sector is north of Scandinavia for low activity ($K_p = 1$), for moderate activity ($K_p = 3$) the boundary within northern Scandinavia and for high activity ($K_p = 6$) it is in southern Scandinavia.

The Correlation Between Auroral Frequency and Magnetic Activity

Both the <aa> and the annual value of the number of auroras reported for Sweden are available from 1868 to 1877. Figure 1 shows that there is a close linear relationship between the two quantities, the correlation coefficient being 0.97. This indicates that <aa> can be used as proxy data for auroral frequency and that the Swedish data represent more than a strictly local quantity. If specific individual sites are used, however, caution must be exercised because of the possibility of changes in the latitudinal distribution of auroras, as is discussed below.

Magnetic Activity Behavior in the Twentieth Century

Geomagnetic activity in the 20th century shows secular changes which are related to the changes in the amplitude of the "long cycle" of sunspot number activity which was minimum near 1900. Figure 2 (from Feynman and Crooker 1978) compares the behavior of the sunspot number cycle with <aa> from 1900 to

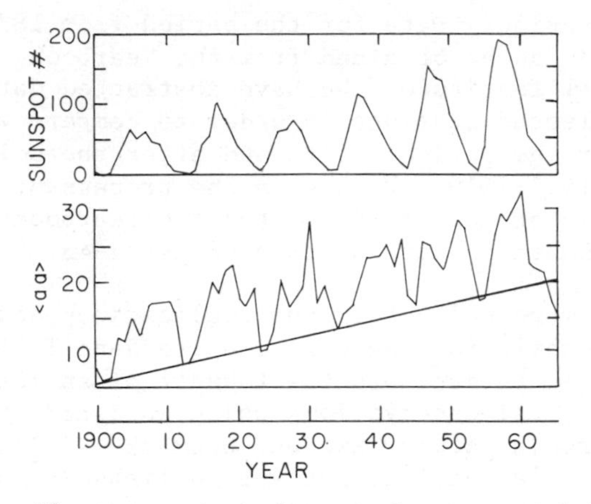

Figure 2. The sunspot number cycle compared to the geomagnetic disturbance cycle from 1900 to 1965.

1965. Note the difference of behavior of the two quantities. The amplitude of the sunspot number cycle increased and values at minimum remained relatively unchanged. With <aa> the values at minimum increased markedly, so that the maximum <aa> in the cycle beginning in 1901 was about the same as the minimum occurring in 1946. A straight line indicating the trend of the data has been drawn approximately through the minima of the <aa> for each solar cycle. If this trend is used as a base line it will be noted that the remaining part of the variation, which is the 11 year solar cycle contribution, is quite constant in amplitude (Feynman and Crooker 1978). Now <aa> has a correlation coefficient of 0.9 with either the solar wind parameters $\langle v^2 \rangle$ or $\langle |B_z| \rangle \langle v^2 \rangle$, where B_z is the z component of the of the interplanetary field in solar magnetospheric coordinates, and v the solar wind velocity. Auroral sightings are also related to geomagnetic activity, so that we may expect the aurora to show the same type of behavior, a long term trend with an extremum near 1900 and a superposed solar cycle variation.

THE AURORAL DATA BASE

Most auroral notations in Scandinavia, as in other parts of Europe, have been from the eighteenth century onwards. There are some systematic data earlier than this, such as the meteorological journal kept by Tycho Brahe in connection with his astronomical observations. From the early 1700's on a significant amount of material in both published and manuscript form exists. This material was collected by Robert Rubenson, director of the Central Meterological Institute of Sweden, and published as a catalog in 1879 and 1882 in the Proceedings of the Royal Swedish Academy of Science. Rubenson's catalog covering the period through 1877 can serve as one of the primary data bases for studies of this period. Data for the period from 1873 through at least 1943 can be obtained from the Yearbooks of the Swedish Meteorological Institute. We have abstracted data from these for several selected stations in order to compare with magnetic activity for the period before and after the solar activity minimum of 1900-1910. We are in the process of abstracting the remaining auroral data from the Yearbooks in order to extend the Rubenson catalog to at least 1943.

Data for Norway were collected from publications and manuscripts by Sophus Tromholt for the period up to June 1878. Tromholt, who taught in Bergen, was the founder of an international system of auroral observations which enlisted the aid of 1000-2000 observers in Norway, Sweden, Denmark, Finland, England and Iceland. His catalog, not yet published at his death at the age of 44 in 1896, was edited by Schroeter and published in 1902.

Several authors have used the combined results for Norway and Sweden given in this catalog. Unfortunately, this can not readily be done. The Tromholt catalog contains almost no data prior to 1761, so that the period from 1721 - 1760 is simply the Swedish figure. The Tromholt catalog also, despite the extensive sources used, contains unexpected gaps. Thus the period for several years following 1788, which was aurorally rich in Sweden and the rest of Europe, contains no reports from Norway. The Tromholt catalog must therefore be used with caution.

Similar problems exist with the auroral reports in the Yearbooks of the Norwegian Meteorological Institute. Many auroras have simply gone unreported by the observers, as noted by the editor of the Yearbooks. This can be seen clearly in, for example, the data from Alten, Lat. 69°58', where we might anticipate reasonably frequent auroral occurrence. In 1874, 30 auroras are reported, but in 1875, 1876 and 1877, only one aurora per year. Similarly, from 1904 to 1909 no auroras at all are reported, but in 1910 and 1911, there are 60 and 62 auroras reported, respectively. We are not in a position to evaluate these apparent inconsistencies, but feel that until Norwegian auroral data are critically re-evaluated they must be treated with caution. The Rubenson catalog of Swedish aurora and data found in the Yearbooks of the Swedish Meteorological Institute appears to be the more reliable data set. To show that the results obtained are not simply a peculiarity of the Swedish observations they may be compared with a tabulation of auroras seen at New Haven and Boston (Loomis, 1865). These data were compiled by Lovering and by Loomis largely from manuscript observations by several careful observers in those areas during the eighteenth and nineteenth centuries.

Auroral Frequency at Individual Stations, 1873-1943

The discussion up to this point has focused data summed on a regional basis. We may anticipate that the auroral frequency at a particular station will vary with solar cycle in a manner similar to that of the region as a whole and that this frequency will be strongly correlated with aa. However, observational problems may obscure this behavior at a given station. We have examined the behavior of auroral frequency for several stations in Sweden for the period from 1873-1943, which spans the 1900 minimum. Preliminary results indicate that, in general, the auroral frequency at a given station will follow the solar cycle variation, and that this frequency can be correlated with aa over a limited time period. For more extended time periods, however, the correlation will change at times when latitudinal shifts of auroral frequency occur. One station, Karesuando, (Geog. Lat. 68.4°N, 22.5°E; Geomag. (corr) lat. 65.2°) has a remarkable behavior pattern. Data for this station, as well as for Haparanda (Geo. Lat. 65.8°N, 24.2°E; Geomagn. (corr.) Lat. 62.6°), are shown in Figure 3.

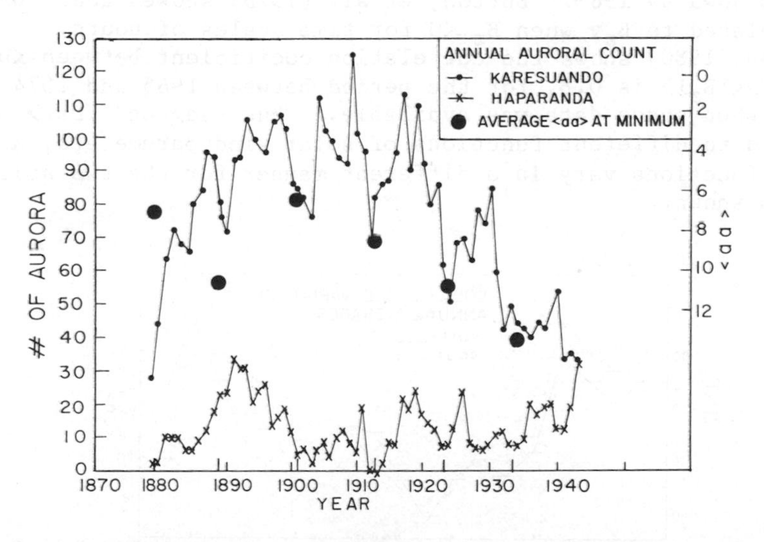

Figure 3. The left hand scale shows the number of auroras reported for two Swedish stations, Karesuando, now at 65.2° geomagnetic latitude, and Haparanda, now at 62.6° geomagnetic latitude. The right hand scale is inversely proportional to <aa>. The <aa> at minimums are shown as black dots.

We note first that the number of auroras at Karesuando is relatively large and ranges from three to ten times the number reported for Haparanda. This is presumably due to its northerly position near the auroral oval. There is a clear 11 year solar cycle variation in the Karesuando observations, with the relative minima occurring close to the sunspot number minima shown by arrows on the abscissa. The data also show a very pronounced secular variation centering on about 1903. For comparison with the secular trend of the magnetic variaiton we have shown the minimum <aa> for each cycle, denoted by large black dots. For these the scale is on th right of the figure and decreases upward. Except for 1879, the trends are clearly parallel. We might, however, have expected the inverse correlation, that is, that the auroral frequency would increase with increasing magnetic activity, as is, in fact, the case for the shorter term solar cycle variation, and for the Haparanda data.

Both authors agree that a partial explanation of this apparent anomaly may be provided by considering the relationships of <aa> and the ring current index <Dst> . Figure 4 (Feynman 1980) shows a comparison of <aa> and <Dst> from 1957 to 1974. We recall that <aa> and <Dst> are measures of different aspects of magnetospheric processes. From 1957 (the earliest year for which Dst is available) to perhaps 1968, <aa> and <Dst> appear to be linearly related to one another, but this relationship breaks down by 1969. Burton, et al. (1975) showed that Dst was related to $B_z v$ when $B_z < 0$ for time scales of hours. Feynman,(1980) shows the correlation coefficient between <Dst> and $\langle V \rangle \langle |B_z| \rangle$ is 0.8. for the period between 1965 and 1974, i.e. when space data was available. Thus <aa> and <Dst> are related to different functions of solar wind parameters, and these functions vary in a different manner for the two solar cycles shown.

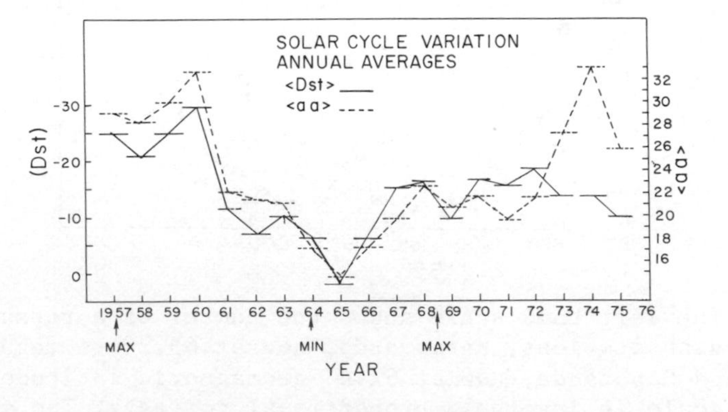

Figure 4. The ring current index <DST> is not simply related to the midlatitude index <aa>, as shown in the figure above.

A similar effect is found by Deehr et al. (1971) who separated the effects of Dst and aa and showed that the emission region narrows with increasing Dst and brightens with increasing K_p. Siscoe (1979) has discussed theoretically the influence of the ring current on the location at which auroras are seen. We propose, then, that the overall frequency of auroras will be functionally related to the solar wind velocity, since it is related to <aa>, but that the statistical location of auroral occurrence will depend on a different function of the solar wind parameters, as does the <Dst>. One of the authors (S.S.), (the other (J.F.) remaining somewhat unconvinced after long discussion), now proposes the following explanation for the inverse correlation of auroral frequency and <aa> over secular time periods for the Karesuando data. Akasofu and Chapman (1963) have shown that the lower latitude limit of quiet auroral arcs in the U.S. sector shows a quasi-hyperbolic variation with Dst. Thus small changes in Dst in the lower ranges will produce relatively large changes in latitude, while for larger values of Dst changes in this parameter produce relatively small changes in latitude. A theoretical model describing these results has been given by Siscoe (1979a). We may distinguish stations in the peri-oval region, that is near the southern edge of the oval, such as Karesuando, from those at some distance away, such as Haparanda. During periods of very low solar activity, as at the turn of the century, we assume that the statistical distribution of Dst will be centered at very low values, and auroras will be visible at the peri-oval station, but much less likely at the more distant station. As the Dst increases to even moderate levels, the aurora will occur much further south, because of the initial steep slope of the latitude-Dst curve. Consequently the peri-oval station will see fewer aurora, while the more distant number of auroras is likely to be directly related to Dst and no auroras will be seen unless the Dst has reached some threshold level. Thus for Karesuando the two solar wind parameters work to produce an inverse correlation for secular time scales, while for more southerly stations they produce a direct correlation for those periods in which a latitudinal shift (discussed elsewhere in this paper) has not occurred.

Eighteenth and Nineteenth Century Auroras

The yearly auroral frequency (number of nights per year aurora were reported) for all of Sweden for the years 1721-1876 is shown in Figure 5 (Feynman and Silverman, 1980). The year is taken from July to June of the following year since there are few reports in the summer months when there is little darkness.

The minimum in the long cycle of sunspots occurred near the beginning of the 19th century. Consistent with this very few aurora were reported for the period from 1809 to 1815. In the entire 156 years shown in Figure 5 only 14 had fewer than 10 auroras, and of these six occurred during the period of seven years from 1809-1815.

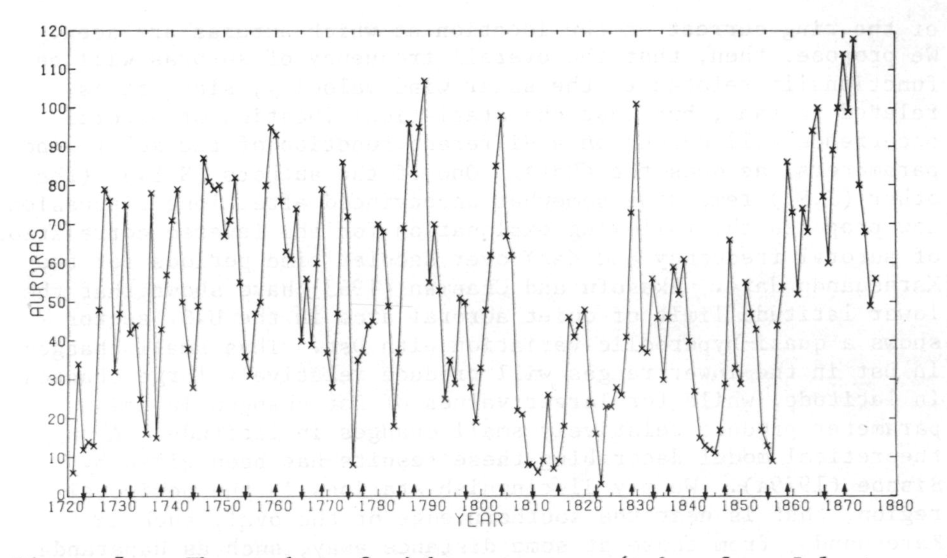

Figure 5. The number of nights per year (taken from July to June) aurora were reported in Sweden from 1721 - 1876.

Auroral frequencies for different regions of Sweden have also been examined (Feynman and Silverman, 1980). Rubenson divided Sweden into four geographic latitude regions, from 55°N to 58°30'N to 61°30'N, from 61°30'N to 65°N, and north of 65°. If an aurora is extensive it may be seen in more than one region and will be reported in each. Thus the sum of the reported auroras for the different districts may be greater than the number given for all Sweden. To provide better statistics we have nevertheless taken the two northern districts together and the two southern districts together. While some error is present due to the same aurora being reported in more than one district occasionally we feel that this is outweighed by the benefit of better statistics.

During the period from 1809 to 1815 a total of 73 auroras are listed in Rubenson's catalog for all Sweden, 72 auroras were seen in the Northern district, and only 7 in the southern district. This behavior is consistent with that seen during the period of the long cycle minimum of 1900.

Latitudinal shifts in auroral occurrence are found on several occasions during the period from 1721-1876. Figure 6 shows the ratio of the difference between the number of northern and southern auroras to the sum. Plus one indicates that aurora were seen only in the north and minus one indicates they occurred only in the south. A remarkable pattern is apparent in the figure. For 42 of the 44 years between 1793 and 1837 the aurora appeared predominantly in the north. This tendency was especially

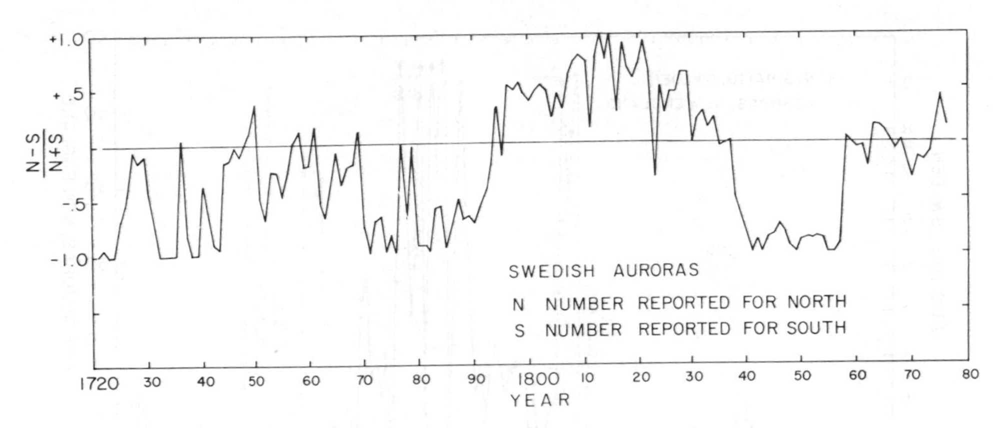

Figure 6. A comparison of the number of auroras listed by Rubenson for the two northernmost Swedish districts to the number listed for the two southernmost districts.

marked from 1807 to 1822, which is added evidence that the long cycle minimum occurred about 1810. During the twenty three years before 1793 and the twenty one years after 1837, almost all the aurora that were seen appeared only in the south of Sweden. From 1858 to the end of the series auroras occurred about equally in the north and south. From 1771 to 1876 there was, then, a period of 106 years, or about 9 1/2 solar cycles during which the Swedish aurora underwent a well defined systematic change in the pattern of occurrence, appearing first only in the south, moving north, and then appearing only in the south again, and finally moving to midSweden. The changes in geographic distribution took place fairly suddenly, typically over a period of 2 or 3 years (for example, circa 1793, 1839 and 1858). The times of these changes have no apparent relation to the phases of the 11 year sunspot cycle.

A change in the pattern of observed occurrence of aurora can be due to several factors including a change in the earth's dipole field, a change in the number or character of observers, or a change in the solar wind which drives the aurora. A change in the earth's dipole of the type required can be ruled out. Barraclough (1974) calculated the latitudes and longitudes of the northern geomagnetic pole of the centered dipole field as 79.9° lat. and 307.2° long. in 1750, 78.7° lat. and 296.0° long. in 1850, and 78.7° lat. and 294.8° long. in 1890. Thus a point in Sweden at about 60°N geographic would shift about 1° geomagnetically from 1750 to 1850 which appears to be insufficient to cause the effect. In addition, the dipole drifted in the same direction throughout the period and was not cyclic as would be required to produce the observed change in occurrence patterns. Calculations involving higher moments of the earth's field (Roederer, 1974) leave this conclusion unchanged.

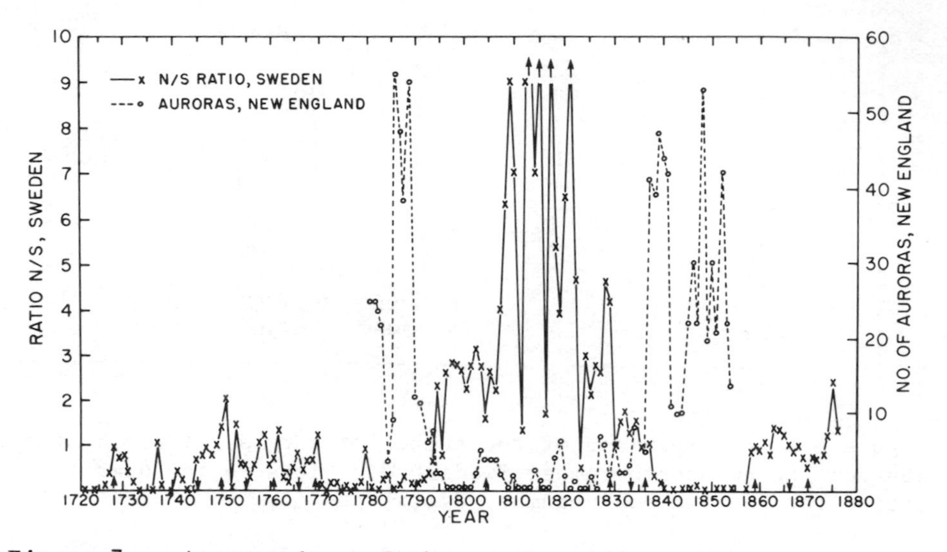

Figure 7. A comparison of the northern to southern aurora in Sweden and the number of auroras reported for the Boston-New Haven area.

The non-local nature of the change in auroral occurrence frequency was confirmed by comparing the Swedish data with that from the northeastern United States (Loomis 1865). The number of aurora was reported annually from the Boston -New Haven region (Loomis, 1866) . From about 1792 to about 1836 almost no auroras were seen but before and after that date auroras were very frequent. The comparison with the Swedish data is facilitated by plotting the ratio of northern to southern auroras in Sweden as a solid line, Figure 7, and comparing it with the Boston-New Haven data superimposed on it. The agreement between the two sets of data is very close. In the late 18th century when the Swedish auroras appear primarily in the south, auroras were seen in Boston-New Haven, and when the Swedish auroras move to the north, few auroras are seen in Boston-New Haven. The southward movement of Swedish auroras is contemporaneous with the reappearance of Boston-New Haven auroras. Thus the latitude effect, which is non-local in nature, must be caused by a combination of solar wind parameter changes. Unfortunately it is not possible to define these more precisely at this time since studies relating the auroral latitudes or brightness with interplanetary parameters are not yet available (Feynman and Silverman 1980).

Conclusions (Both authors agreeing)

A long cycle of 80-100 years occurs in sunspot numbers, mid latitude geomagnetic disturbance data, and in auroral frequency in regions sufficiently south of the auroral oval, with auroral occurrence displaying a minimum at the time of the long trend minimum in <aa>. We recall that geomagmetic activity and solar wind properties are not directly related to the number of sunspots. Hence, the long solar cycle expresses itself in at least two seperate aspects, first, the amplitude of the 11 year cycle in the number of sunspots and second in the long cycle variation in aurora and magnetic activity. The peri-oval station, Karesuando, however, shows a contrasting inverse trend to the magnetic activity on a secular basis (1879-1943) even though it shows a direct relationship for times of the order of a solar cycle. The auroral frequency range for a solar cycle at Karesuando varies much less than the threefold variation over the entire period. The results for Karesuando should be explicable in terms of the long cycle and the eleven year cycle varying as two different functions of the solar wind parameters, including velocity and the southward component of the interplanetary field, and perhaps some others, such as density. Also, the latitude at which aurora occurred underwent a clearly patterned change during the 106 year interval between 1771 and 1876, and there is evidence for such changes from 1720 to 1771.

We conclude, therefore, that the variation of the configuration of the magnetosphere during the last three centuries has been much greater than has been observed to date since in situ space observations began. These changes can be expected to alter both the energy input into the thermosphere and the latitude of that input. Additional studies of the interrelationship of the interplanetary parameters, and magnetospheric configuration with auroral frequency and positions of occurrence are clearly indicated.

Acknowledgment: One of us (J.F.) was supported by The Air Force Geophysics Laboratory, Contract F19628-79-C-003.

REFERENCES

Akasofu, S.I. and Chapman, S. 1963, J. Atmos. Terr. Phys. 25, 9
Allen, J.H. and Feynman, J. 1980, in Proceedings of the International Symposium on Solar Terrestrial Predictions, Ed. R.F. Donnelly, NOAA Washington: Government Printing Office
Barraclough, D.R. 1974, Geophys. J. R. Astr. Soc. 36, 497
Burton, R.K., McPherron, R.L. and Russell, C.T. 1975, J. Geophys. Res. 80, 4204
Deehr, C.S., Egeland, A. and Søraas, F. 1971, in The Radiating Atmosphere Ed. B. M. McCormac, Dordrecht-Holland, D. Reidel
Eddy, J.A. 1976, Science, 192, 1189
Feynman, J. and Crooker, N.U. 1978, Nature, 275, 626
Feynman, J. and Silverman, S.M. 1980, J. Geophys. Res., in press
Feynman, J. 1980, paper in preparation

Feldstein, Y.I. and Starkov, G.V. 1968, Planet. Space Sci., 16, 129
Fritz, H. 1881 Das Polarlicht, Leipzig: F.A. Brockhaus
Gussenhoven, M.S., Hardy, D.A. and Burke, W.J. 1980 J. Geophys. Res. in press
Loomis, E. 1866 Annual Report, Smithsonian Institution, 13, 1
Mayaud, P.N. 1973 IAGA Bulletin, 33, 262
Roederer, J.G. 1974, Geophys. Res. Let. 1, 367
Rubenson, R. 1879, 1882 Kongl. Svenska Vetenskap - Akademiens Handlingar 15 No. 5, 18 No. 1.
Sandford, B.P. 1968, J. Atmos. Terr. Phys., 30, 1921
Silverman, S.M. 1978, Presented at Solar-Terrestrial Conference, Yosemite, Cal.
Siscoe, G.L. 1979a, Planet. Space Sci. 27, 285
Siscoe, G.L. 1979b, Planet. Space Sci. 27, 997
Tromholt, S. 1882, Danish Meteorol. Inst. Yearbook for 1880
Tromholt, S. 1902, Catalog der in Norwegen bis June 1878 beobachteten Nordlichter. (Ed. J. Fr. Schroeter) Kristiania: Jacob Dybwad
Vestine, E.H. 1944, Terr. Magn. and Atmos. Elect. 49, 77

AURORAE, SUNSPOTS AND WEATHER, MAINLY SINCE A.D. 1200

D. J. Schove

St. David's College
Beckenham, Kent, BR3 3BQ, U. K.

ABSTRACT

Auroral records received for the Spectrum of Time project were used in 1955 to estimate sunspot activity and the dates maxima and minima back to 649 B.C. An additional set of rules has been developed (Schove 1979a) (especially the so-called X + 2 and X + 3 rules) and has made possible further improvements utilizing the separate auroral maxima associated with flares and coronal holes on the sun. A further set can now be given. 1) the time between sunspot maxima depends especially on the ratio of the amplitudes: the time between minima is high if the next cycle is very weak and low when the two consecutive cycles are both strong. 2) The time of rise is usually dependent on the strength of the next maxima, and the time of fall is low when a moderate cycle is followed by a strong one.

Sun-weather relationships can be investigated using long time-series of ice-core, tree-ring or varve data if sunspot cycle amplitude classes are taken into account. The sun affects the weather indirectly through the "Atmospheric Pressure Parameter" which responds differently to strong and weak sunspot cycles.

The 2.2 year stratospheric cycle responds to solar activity in such a way that 2- and 3-year Weather Cycles are useful clues to sunspot activity.

The 200-year solar cycle is confirmed in prehistoric times by radiocarbon. Satisfactory proof of the existence or length of the 11-year sunspot cycle before 200 B.C. is still lacking.

C. S. Deehr and J. A. Holtet (eds.), Exploration of the Polar Upper Atmosphere, 421–430.

Historical exploration relevant to Sunspot Cycles can now be undertaken backwards through time in four ways:

I Aurorae and Sunspots to 200 B.C. (cf. Schove 1981)
II Ice-cores to A.D. 1200 (cf. Hammer et al, 1979) and soon, probably to 43 B.C.
III Tree-rings in Turkey can be dated absolutely to c.1500 B.C. (Schove 1978, 1979b) and in Europe may soon go back to c.9000 B.C.
IV Varves in the USSR, if corrected by 145 years, are absolutely dated c.1500/1200 B.C.

I AURORAE AND SUNSPOTS

A large unpublished collection of aurorae received for the so-called Spectrum of Time project was used to estimate the dates of the 11-year cycle back to 649 B.C. (Schove 1955). Some of the descriptions enable us to determine when the direction of magnetic north was different from its present value: one twelfth-century lady thus walked all through the night following an aurora on the horizon slightly E of geog. N and founded a convent at the spot to which she was led. The auroral oval must have been nearer the Mediterranean in e.g. the 2nd century B.C. and nearer China in the first millenium A.D. In my estimates of auroral intensity (Schove 1962) no corrections were made for changes either in the mean Europe-China geomagnetic latitude or in the earth's magnetic intensity, and suggestions would be appreciated.

In the following diagrams of auroral occurance frequency in relation to sunspot cycle, X and N refer to the years of Zurich maximum and minimum sunspot number respectively; T is the time of rise from N to X and U the time of fall from X to the following N ($=N_2$). S,M,W relate to the intensity of the cycle: Strong, Medium and Weak.

Figures 1-5 show the distribution in percent of aurorae per year among the 11 years of the solar cycle (X-4 to X+6) (data from Schove 1979a Table I) for the years from 1710 to the present.

Auroral occurrence for all classes of sunspot cycles from this period is shown in fig. 1. The main maximum of auroral activity, associated especially with solar flares, occurs in year X at lower latitudes, X+1 in Holland and X+1.5 in Scandinavia. In central Europe, for instance, 24% occur in year X. Thus, among medieval aurorae, recorded mostly in central Europe, 1 in 4 should belong to years of maximum. The secondary maximum, associated especially with coronal holes (Deehr, 1980), is weak in the Mediterranean but strong in the Netherlands, where it occurs at X+3 but is further delayed in Scandinavia. Deehr

(1980) has shown that blood-red upper border type A aurorae are associated with the main maximum and type B (green with magenta or red lower borders) with the secondary maximum.

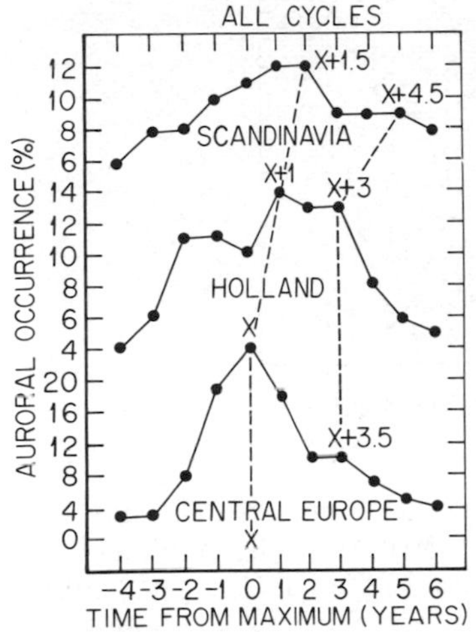

Fig. 1 Auroral occurrence-frequency percentages for each year of the 11-year period X - 4 to X + 6 for all cycles from 1710 to present.

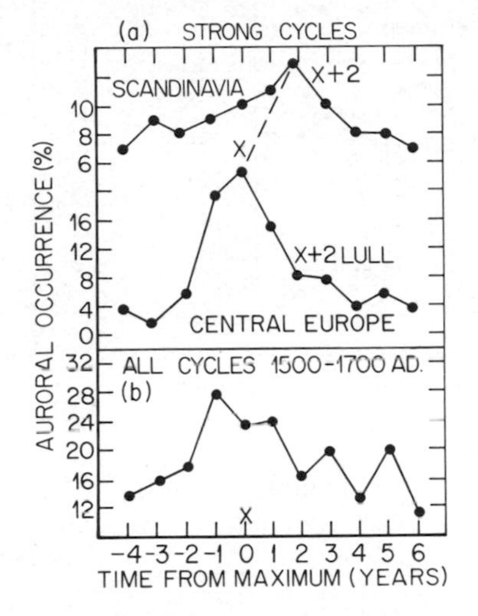

Fig. 2a Auroral frequency as in fig. 1 for strong sunspot cycles. (R > 140) (b) auroral frequency 1540-1710 with indications of redness, corona or movement (Link, 1962, 1964).

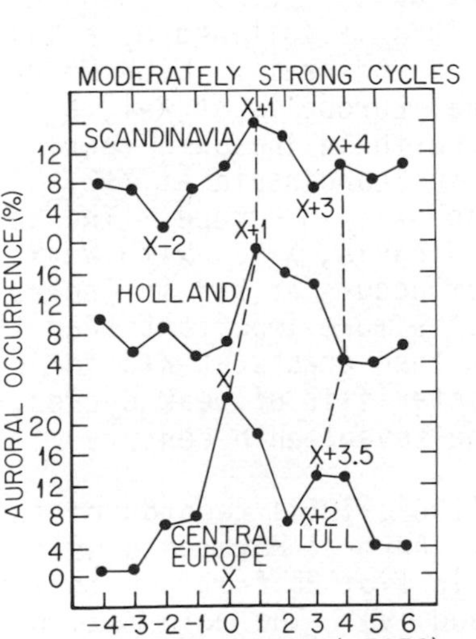

Fig. 3 Auroral frequency as fig. 1 for Moderately Strong cycles. (140 > R > 100)

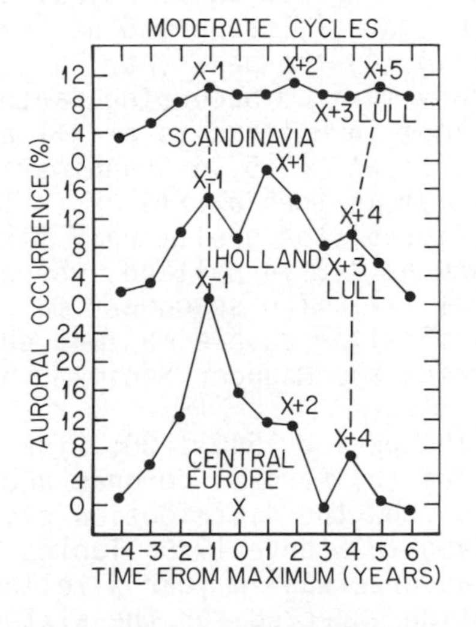

Fig. 4 Auroral frequency as in fig. 1 for Moderate cycles (100 > R > 87)

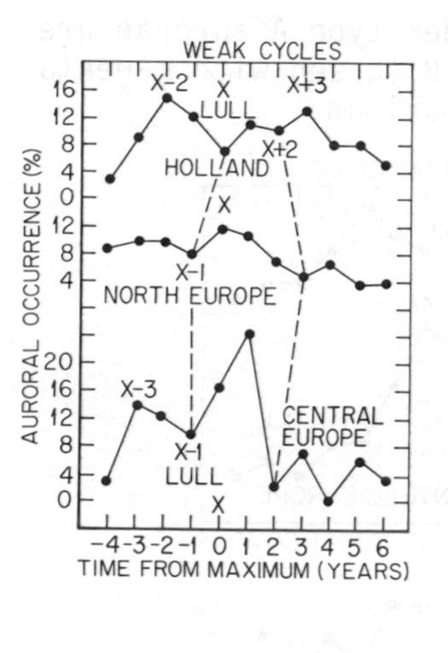

Fig. 5 Auroral frequency as fig. 1 for Weak cycles (R < 87).

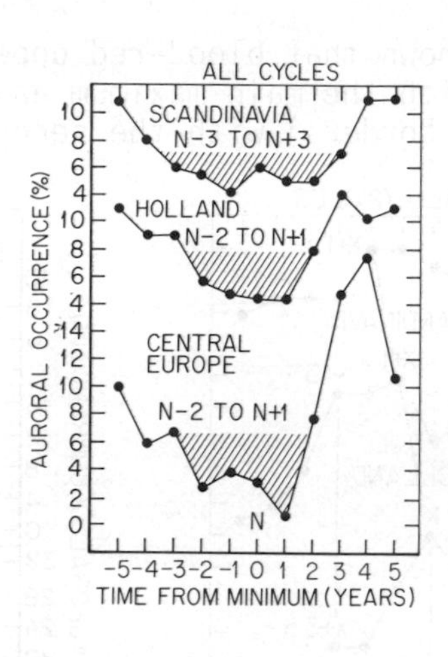

Fig. 6 Auroral frequencies relative to sunspot Minimum (N) for all cycles from 1710 to the present.

Strong (S), Moderately Strong (MS), Moderate (M) and Weak (W) cycles are considered separately in figs. 2 to 5. Strong cycles (fig. 2) show a single maximum, X in central Europe and X+2 in Scandinavia, so that, if the winter of 1979/80 is to be the time of the (smoothed) sunspot maximum, most aurorae should be seen in Scandinavia in winter 1982. With moderately strong cycles (fig. 3), the main maximum is at about X+0.25 in Central Europe and at X+1.25 in Scandinavia. This is followed by a lull at X+2 in central Europe and at X+3 in Scandinavia. With Moderate cycles (fig. 4) the peak year in central Europe is at X-1, a preliminary maximum occurring farther north in the same year. The main maximum in Holland is at X+1 and in Scandinavia at X+2. The lull is then at X+3.5 in Scandinavia and X+3 in Europe. The secondary maximum is then at X+4 or in Scandinavia, X+5. With Weak cycles (fig. 5) the preliminary maximum occurs at X-3 in Central Europe and at X-2 in Holland, where it is more important than the maximum of X+1. The secondary maximum then occurs at X+3 in Holland. Similar double maxima, characteristic of weak cycles, occur during the Maunder Minimum in the seventeenth century.

In the period 1540-1700, Link's (1962, 1964) records of (a) Red Aurorae (b) Auroral Coronae and (c) Auroral Movements were added and show the distribution given in fig. 2b from X-4 to X+6 (X as given in Schove 1979a Tables V and IV). The main maximum for red aurorae was in year X following the "strong cycle" distribution expected for the sixteenth century. The rest of the distribution shows characteristics of a "weak cycle" from the seventeenth century. These data 'should' be separated in future study.

Weak cycles, with a Gnevyshev gap (Gnevyshev, 1977) in large spots near year X, are clearly evident in the Maunder Minimum. In the light of Eddy's (1976) work, however, the proposed intensities have been reduced from those given in 1955. The auroral double cycles are very clear, however, and Brekke and Egeland (1979) include illustrations of aurorae from central Europe for 1663 and 1681.

The percentage distribution of aurorae relative to Zurich minimum, compared with the totals from N-5 to N+5 is illustrated in fig. 6. The little auroral maximum (cf. Schove 1979a p. 426) in year N in Scandinavia is characteristic even in 1740/70 when the oval lay less far north than in its poleward withdrawal found by Silverman and Feynman (1980) for the period 1800/40. In the period 1500/ 1700 the auroral minimum in central Europe was year N (not N+1), the three-year gap in auroral phenomena, except c.1554 and c.1679, being well-marked.

The next set of diagrams illustrates some rules based on the relationship between the intensity (R_M) of two adjacent cycles and the values of X-X, N-N, T and U as found in the Zurich data from 1749.

The time between two maxima (X-X) is dependent especially on the ratio of the intensities (cf. fig. 7) of the adjacent cycles, and varies considerably from 17 years between 1788 and 1805 when the ratio was 0.35 to 7 years between 1829 and 1837 when the ratio was 2.1.

The time between two minima (N - N fig. 8) depends on the weakness of the adjacent cycles, being under 10 if both cycles are strong and over 12 if the next cycle is to be very weak.

The time of rise (T in fig. 9) often foreshadows the strength of the coming maximum; the intensity can be greater but, since 1700, never less, than predicted in this way.

The time of fall (U in fig. 10) can be over 10 when a Very Strong (SS) cycle is followed by a Very Weak (WW) one, so that a minimum in 1986 would foreshadow another Strong maximum (S). Nevertheless, it is under 5 1/2 years only when a Moderate cycle (M) is followed by a Strong (S or SS) one and the 22-year cycle, persistent since c. 1840, favours a later minimum and a weaker maximum.

These empirical rules help us to improve slightly Wolf's turning point before 1720 and tentative sunspot numbers have been constructed back to 1600 (Schove, 1981).

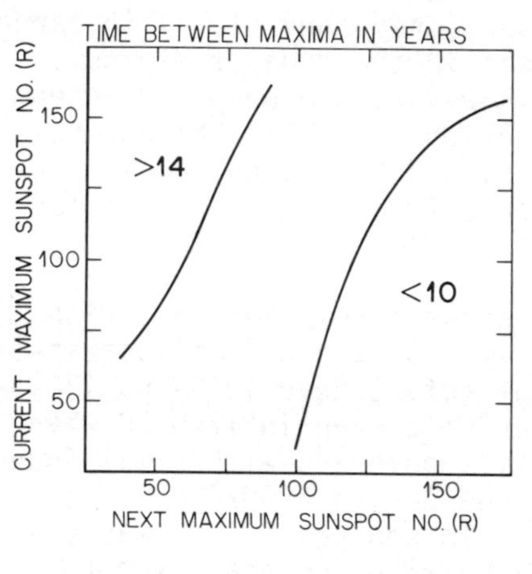

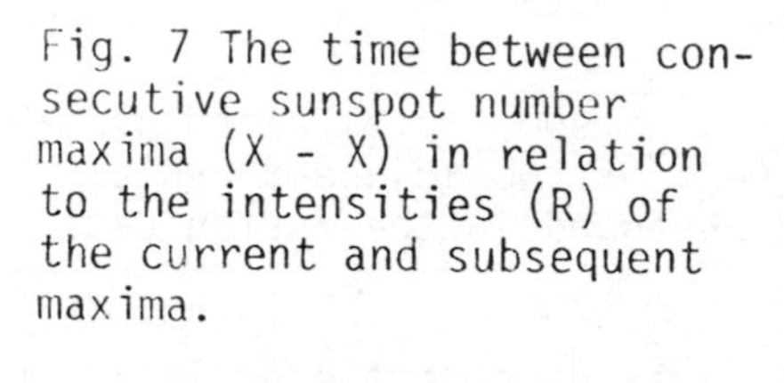

Fig. 7 The time between consecutive sunspot number maxima (X - X) in relation to the intensities (R) of the current and subsequent maxima.

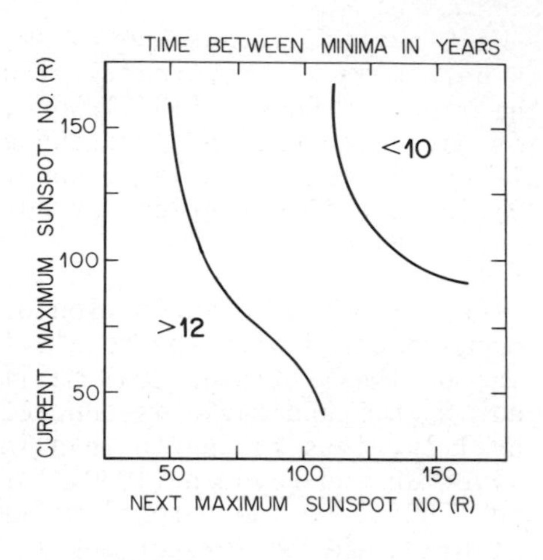

Fig. 8 The time between consecutive sunspot number minima (N - N) in relation to the intensities (R) of the current and subsequent maxima.

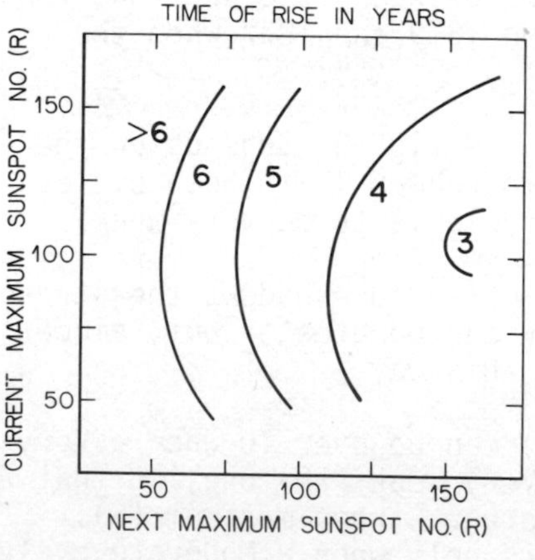

Fig. 9 The time of rise (X - N) in relation to the intensities (R) of the current and subsequent maxima.

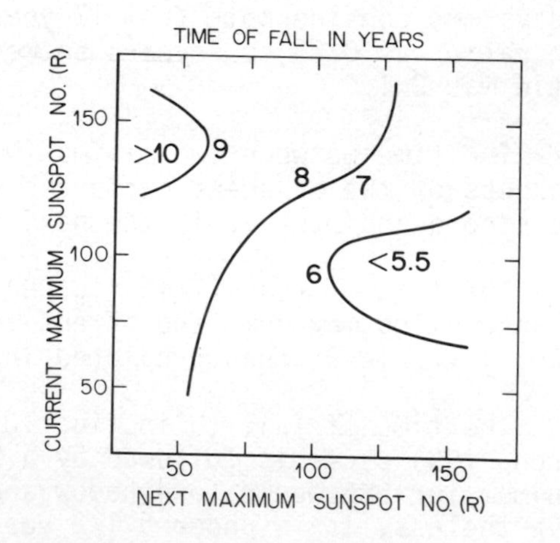

Fig. 10. The time of fall (N - X) in relation to the intensities (R) of the current and subsequent maxima.

SUNSPOTS AND WEATHER

The pressure parameter

Solar-Weather relationships can be investigated successfully only if a large number of solar cycles of each amplitude class is available. Strong and Weak cycles often have opposite effects, (Schove 1977 figure 4.6) although no magnetic mechanism is known and a 22-year cycle can be more important than the 11-year cycle, especially near the neutral lines of the Pressure Parameter map referred to below (fig. 11).

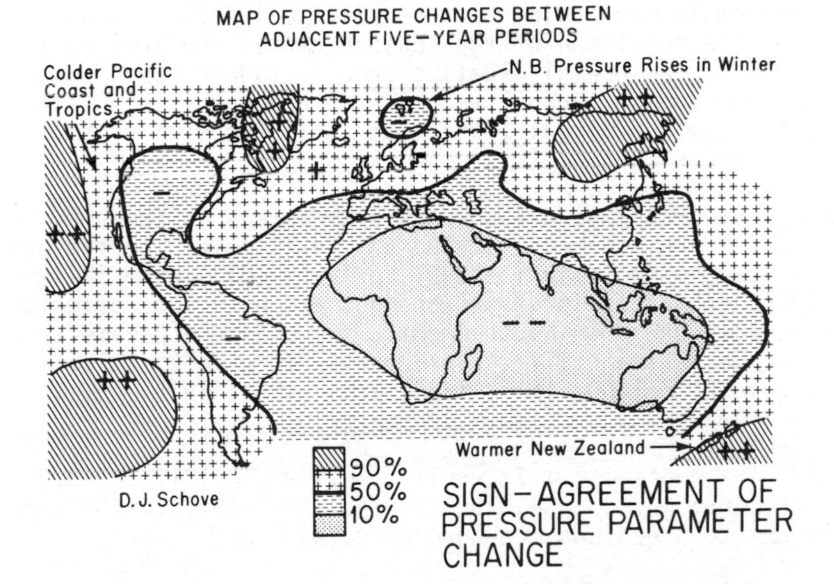

Fig. 11 The first eigenvector of pressure changes between consecutive five-year periods. Reprinted from Schove (1977) with permission from the Internat. African Institute.

The opposite effects of Strong and Weak cycles on the Pressure Parameter can at least be interpreted. The earth's atmosphere bulges over the Old World Tropics due to solar heating of the adjacent continents of Africa, Australia and Asia. When the five-year period is centered on a strong sunspot cycle, the Upper High at 200 mb erupts - a volcano of air - and its overflowing air causes a pressure rise in the Upper Trough regions, one of which is over the Davis Strait off W. Greenland. N. Scandinavia on the other hand has a Ridge in the Upper Troposphere from July to September and, at least in those months, the pressure falls there too.

The geographical pattern for the year as a whole (The summer pattern is very different) has been mapped (Schove 1977) and this, the First Eigenvector or Principal Component of Pressure Change between 5-year periods, is termed the Pressure Parameter pattern. The writer originally experimented with anomaly maps for 5-year periods and then there was no obvious eigenvector - it is especially when 5-year differences are plotted that the pattern emerges.

With a weak cycle the volcano of air erupts only over a limited area of the Indian Ocean and the mountain of air over the Upper High continues to grow upward, and the troughs, as over the Davis Strait, deepen slightly. All these effects are seen in pressure anomaly maps going back usually to 1841/5. In addition, the larger the magnitude of the sunspot cycle, the greater the area erupting, the normal negative area including Cape Town, S.E. Australia and India. Annual values have been published back to 1796 by Berlage and myself (1967) but it is the sign of the five-year differences that matters.

The West Greenland pressure usually agrees in tendency with the Pressure Parameter and is high in the five-year period centered on 1890, especially in 1894, in 1905, 1913, especially 1919, 1935, etc. These pressure changes are no doubt reflected in Greenland Ice Cores. In North America the pressure parameter effects reach maxima c.1777, c.1787, c.1797, c.1804, c.1825, c.1831, c.1850, c.1860, 1877, 1880, 1894, 1909, 1914, 1927, 1941...

Minima are dated c.1772, c.1782, c.1792, c.1813, c.1837, c.1857, c.1866, 1878, 1886, 1904, 1914, 1927, 1941... These may be relevant to Greenland Ice-Core results.

In the 18th century the Pressure Parameter, based mainly on lower latitude evidence, again over five-year periods, was

High 1707, 1715, 1731, 1740, 1752, weakly so in 1766 and 1755,
also 1787, 1803...
Low 1969, 1702, 1712, 1726, 1737, 1744, 1758, weakly in 1770,
very strong in 1783, 1793, and 1806.

For any meteorological element in a particular season we can compare the results obtained by using a Sunspot worksheet (copies available) or adopting these Pressure Parameter turning-points. Often the correlation is better if we use these turning-points instead of the observed sunspot maxima and minima.

Ice-cores, tree rings and varves all provide the long series of proxy data needed for investigation of sunspot parameter effects. Ice-cores reflect summer temperatures and are soon expected to be correctly dated back to 43 B.C. when Latin and Chinese evidence leads us to expect a volcanic dust layer. A preliminary investigation of summer and winter temperature in the Greenland ice-cores since A.D. 1200 (Hammer, et al. 1979) confirms that strong and weak cycles have opposite but significant effects.

Tree-rings are proving very useful as climatic indicators. N. European summer raininess (cf. Schove 1979c) is reflected in the width of oak rings and a similar effect of winter raininess is found in the well-dated 'Hittite' tree-rings of Turkey. High-level trees in the Western USA reflect summer temperature (Harris 1980); low level trees usually reflect winter rainfall (Schulman 1951). These series cover many centuries and can be used with estimated dates of sunspot maxima (Schove 1955) to determine the effects of strong and weak solar cycles.

Varves are annual layers of sediment with a double structure that enables each varve to be dated to the exact year. The thickness often depends on the rainfall. Varves in the USSR have been cross-dated (Schove 1979b) with the Turkish tree-rings in the period 1500/1000 B.C. In Lake Van, Kempe and Degens (1979) believe that the varves reflect the solar cycle throughout the Holocene. Moreover, if magnetic storms shake molecules until they align themselves with the magnetic field, we might expect a magnetic cycle in some varves, and Mörner (1979) is making investigations of an apparent 11-year cycle in varve-magnetism.

Quasi-biennial cycles are seen in a curious alternation of West and East winds in the Lower Stratosphere with a mean period of 2.17 years, just under one-fifth of the sunspot cycle. The amplitude and the frequency of this cycle may be linked to solar activity (Schove 1969, 1971). After delays, different for different seasons and places, the Tropospheric weather is affected. 2-year cycles reflect strong activity and 3-year cycles are associated with weak activity the contrast makes it possible to use ice-core, tree-ring and varve cycles as a clue not only to sunspot fluctuations but also to the phase of the solar cycle in prehistoric times.

The longer cycle of 203 years (Schove 1955) in solar activity can now be traced in prehistory as it is found by Suess (1980) in radiocarbon.

Ice-cores, tree-rings, and varves thus lead to proxy series that are long enough to use for Sun-Weather relations.

Tests are needed to check if deviations, squares of deviation or 'oddness indices' are significantly related with specific amplitude-classes of sunspot cycle or with the Pressure Parameter. At present, physical methods have failed to provide satisfactory evidence for the 11-year sunspot cycle before c.220 B.C. and most of the 'evidence' for its meteorological effects since 1749 does not become statistically significant until allowance is made for the different effects of strong and weak sunspot cycles.

REFERENCES

Brekke, A. and Egeland, 1979. Nordlyset, Grondahl, Oslo.
Deehr, C. S., 1980. This volume.
Eddy, J. A., 1976. Science, 192, 1189.
Gnevyshev, M. N., 1977. Solar Physics, 51, 175.
Hammer, C. U., et al., 1979. Jnl. Glaciol. 20, 12, fig. 4.
Harris, R. R., 1980. An Index of Summer Temperature: The Western U.S. AD 590-1960 BA Dissertation. Cambridge Univ.
Kempe, S., and Degens, E., 1979. in Moraines and Varves (ed. Ch..Schluchter) Balkema.
Link, F., 1962. Geofys. sborník, Praha, No. 172.
1964, Geofys. sbornik, Praha, No. 212.
Mörner, N.-A., 1979. private communication.
Schove, D. J., 1966. J. Geophys. Res. 60, 127-146.
1962. J. Brit. Astr. Ann., 72, 30-35.
1969. Weather, 24, 390.
1971. Weather, 26, 201.
1977. in Drought in Africa (ed. D. Dalby et al.,) Internat. Afr. Inst., London.
1979a. Solar Physics, 63, 423-432.
1979b. in Moraines and Varves (ed. Ch. Schluchter) Balkema, 319-325.
1979c. Med. Archaeol., XXIII, Notes.
1980. Solar Terrestrial Predictions Proceedings, 3, A-111 to A-120., Donnelly, Ed. NOAA Boulder.
1981. Sunspot Cycles (Benchmark Book) Dowden, Hutchinson & Ross, Inc. New York.
Schove, D. J. and Berlage, H. P., 1965. Pure and Applied Geophysics, Basle, 61, 219-231.
Schulman, E., 1951. Amer. Met. Soc. Publ., 1024.
Silverman, S., and Feynman, J., 1980. This volume.
Suess, H. E., 1980. Sun and Climate Conference, Toulouse.

ANCIENT NORWEGIAN LITERATURE IN RELATION TO THE AURORAL OVAL

Asgeir Brekke

Institute of Mathematical and Physical Sciences
University of Tromsø, Tromsø, Norway

Alv Egeland

Norwegian Institute of Cosmical Physics
University of Oslo, Blindern, Norway

ABSTRACT. "The Poetic Edda" and "The King's Mirror" are well preserved Norse documents from the period between 700 and 1300 A.D. The latter states that the aurora was known to people living in Greenland but probably not observed in Norway at about 1200 A.D. "The Poetic Edda" do not include any decisive evidence for the aurora being known to the Norse scalds in the Viking era. This is a rather surprising fact as the scalds were much inspired by natural phenomena, and in particular occupied by celestial gestalts. In a search for an explanation of this lack of inspiration from the northern lights among the Norse scalds it is maintained that the position and shape of the auroral oval was different in the Viking era from the present day auroral oval.

1. INTRODUCTION

In the last decade we have experienced an increasing interest in the historical variability of the Sun-Earth relationship. It has been realized that large variations in this relationship have probably taken place from time to time and thus resulting in severe consequences to the life on Earth. The interest in solar variability and its influence on the weather has in particular been offered attention from a large community of different scientists.

C. S. Deehr and J. A. Holtet (eds.), Exploration of the Polar Upper Atmosphere, 431–442.

Investigations of the ancient history of the Sun-Earth relationship is a problem of putting bits and pieces of information from many different areas of science together in order to get a comprehensive view of the problem. Auroral physicists have also taken part in this puzzle and focused their interest on the historical occurrence of the aurora. The auroral catalogues collected by people like Lycosthenes (1517), Frobesius (1739), Lovering (1868), Fritz (1873, 1881) and Tromholt (1902) have been pulled out of the dust and scrutinized in a hope of revealing evidences which have been overlooked in the past.

It is believed that major natural events such as eruptive volcanos, flooding rivers, solar and lunar eclipses and maybe dramatic auroral displays, may, when they occurred, have made very strong impressions in the local population from time to time. It is not uncommon that stories about such events could be related from generation to generation almost as a folklore. Sometimes such events could even move poets to write down a few words about the incident, words if they are dated properly, can help the descendants to determine the time for the event.

In this work we will delve into the ancient Norse literature in a search for such traces of auroral events in the distant past.

2. THE NORSE LITERATURE

In the Norse literature written before 1300 A.D. there are two outstanding parts, for those seeking knowledge about Norse tradition and belief, "The Poetic Edda" and "The King's Mirror". "The Poetic Edda" consists of ten celestial and twenty hero-idolizing poems - by unidentified authors. The celestial poems render ancient mythes concerned with the accomplishment of the various gods, edited according to their significance in the order of Odin, Tor, Frøy etc. The heroic poems deal with the tragic fate of a few major individuals. Here much wisdom and worldy experience come forth. The poems stayed alive more or less intact by oral tradition until they were written down about 1270. They were well preserved in Iceland after the emigration from Norway which began around 700 A.D. Some of them were probably made there as well.

An exact dating of the origin of "Edda" is extremely difficult. Most certainly non of the poems are composed before 700 A.D. and modern research in the field indicates that they most likely were drawn up as late as 1000-1100 A.D.

The term "Edda" is synonymous with "great-grandmother" in old Norwegian which probably indicates that the poems were put into writing sometimes in the distant past. Form and content are closely interconnected in the "Edda" epics and it is practically impossible to enunciate exactly the author's meaning in a translation to foreign languages, even translation into modern Norwegian entail great difficulties.

"The King's Mirror" is probably written before 1250 and is made as a conversation between a father and his son, probably a king and his successor. The son is being tought by his father about all possible matters that could be of any need to know for a being leader in Norway in the old days. During the conversation the father says the following:

> "But as to that matter you have often inquired about, what these lights can be which the Greenlanders call the northern lights, I have no clear knowledge. I have often met men who have spent a long time in Greenland, but they do not seem to know definitely what these lights are." (Ref. see Størmer 1955.)

It appears from this quotation that the author of "The King's Mirror" probably never saw the aurora and he expresses himself as if it was not known to him that it could occur in Norway at all. On the other hand he states that the aurora was known to occur in Greenland.

These few words from a more than 700 year old book have puzzled many scholars from time to time. In the literature several authors have been discussing this particular point and in fact suggested that the aurora was a phenomenon that did not occur with the same frequency at about 1200 A.D. as it did at later times. de Mairan (1733) was in favour of this hypothesis although he was strongly disputed by Norwegian writers like Spidberg (1750), Ramus (1745, 1747), and Barhow (1751).

Beside from this peculiarity in "The King's Mirror" the author is in fact the very first to mention the phenomenon with its original name Nordurljos or Northern Light.

3. THE EARLY TRANSLATION OF "EDDA" AND FRITZ's USE OF IT

The first complete translated version of "Edda" was published by the Icelander Magnusson in 1821. This translation was strongly inspired by mythological viewpoints mixed up in the prevailing national romanticism of the time. Thus Magnusson maintained that the aurora was referred to in "Edda" with such picturesque images as "reflections of the Valkyries' shields" and "The Gjallarbridge". The Valkyries were the young maids which in Norse mythology picked up the deads from the battle field, and "The Gjallarbridge" was the bridge over the river "Gjoll" which defines the border between the two worlds, of the living and the dead. Magnusson also claimed that the name of horses owned by deities or Valkyries and originating from phrases for "lightflash", "flames and fires", could as well be regarded as synonymous for the aurora.

This way of interpreting the "Edda" was fairly popular in Europe in the last century, and it is the influence of this school we can find in books like "Das Polarlicht" of Fritz from 1881 and in "Electrisität und Magnetismus im Alterthume" by Urbanitzky from 1887, and in fact in the fairly recent "Keoeit" of Petrie from 1963.

The first stanza quoted by Fritz (1881) is verse 155 from "The Sayings of Hár" (Håvamål). "The Sayings of Hár" is a celestial poem consisting of five parts. Stanza 155 is from the last section where Odin relates the 18 magic songs he is in possession of. The stanza reads as follows in Hollander's translation from 1969:

"That tenth I know, if night-hags
sporting I scan aloft in the sky"

It is presumably the word "night-hags" or witches in the sky which made Fritz to think of northern lights in this context. There is, however, nothing in this stanza which can indicate that the author has been inspired by northern lights except that the phenomenon occurs in the night sky.

Another phrase that is included in Fritz's book is stanza 33 in "The Say of Hyndla":

"A fire see I burn, flameth the earth"

This poem is not very well preserved and is therefore difficult to translate into modern Norwegian, no wonder that it has initiated the most diverse interpretations. Even if the aurora was often associated with terms like "fires" and "flames" during the Middle-Age, no scholar today will dare to maintain that this is necessarily a description of northern lights. A reasonable conclusion is rather that the poem, probably refers to a volcanic eruption. The fact that the poem probably is written in Iceland also makes this more likely.

The third quotation included by Fritz (1881) constitutes verse 49 in "Second Lay of Helgi the Hunding-Slayer". This is a poem about Helgi who killed King Hunding and later fell in battle thus arriving in Valhall to become Odin's righthand man. When morning dawned, Helgi arose and said:

"Along redding roads to ride I hie me,
on fallow steed aery paths to fly:
to the west shall I of Windhelm's bridge,
ere Valhall's warriors wakes Salgofnir"

It is "the reddening roads" that Fritz here referred to as the aurora. From the end of the stanza it is clear that the scene of the story is laid to the early morning hours since Salgofnir is the rooster of Valhall which wakes up the warriors. "The reddening road" is therefore more likely to be synonymous with the red light of dawn.

The conclusion must therefore be that Fritz's interpretation of a few stanzas in "The Poetic Edda" are not carefully investigated, and cannot be taken as proof for the aurora being described in one of the oldest historical documents of Norway.

4. A MODERN LOOK ON "EDDA" REVEALS FEW TRACES OF THE AURORA

From a closer look at "Edda" we find that there is no place in these poems where the aurora is directly named, but there are a few stanza where formulations occur which in a poetic manner may have been descriptions of the phenomenon. We shall not go into this in deep detail but mention the word "vafrlóga" which occurs in several stanzas of "Edda". "Vafrlóga" is composed of "vafr" which is a verb for throw and "loga" is the meaning of flames, "vafrlóga" therefore are flames which are thrown in the air, a beautiful poetic description of the aurora. Even more interesting is the name "Hålogaland" which is a landscape in the northern part of Norway. Here the prefix "Hå" means high in the air and "loga" is again flames, and "land" has the same meaning as in English. Translated to English the meaning should be "The land under the flames high in the air", or "The land under the aurora" (Bugge 1871; Koht 1920).

In particular the meaning of this last name has been discussed back and forth from time to time, and any proof for a correct translation can hardly be given. Therefore we must conclude that there is no definite evidence for a description of the aurora in the Norse literature except in "The King's Mirror". Furthermore in this book it is alluded that the aurora did not take place in Norway at the beginning of the 13th century.

5. THE AURORA IN NORWEGIAN TRADITION

By investigating oral tradition in a community one can find many traces of old belief concerning natural phenomena like weather and sea, vegetation and fauna etc. In the Scandinavian tradition one can also find relations to the aurora in almost any district of the country. The most common belief has been that the aurora was related to the weather, and weather marks of any type can be found such as: after aurora comes wind, stormy weather, calm, snow etc. (Storaker 1923, 1924). Traditions that have roots in the Norse mythology on the other hand are more difficult to find. There are, however, a relation from Sweden which tells that the aurora is the breath of the god Tor and the big giants, the Jotnes, when they are in battle. If the aurora is in the south it is the breath of Tor and his rams, while in the north it is due to the breath of the giants.

From the west coast of Norway there is a relation which says that the aurora is due to the dead maidens dancing. They also claimed in this part of the country when an unmarried girl grew very old, that she would be taken to the aurora after death. This tradition may indicate that the popular conception of the aurora being related to the Valkyries, are not quite as off the route as modern scholars tend to believe, because the Valkyries were themselves dead maidens.

Further search in the folklore and tradition of the Scandinavian countries does not yield much more than these few suggestions of the aurora being a matter of common knowledge among the Norsemen.

6. THE POSITION OF THE AURORAL OVAL IN THE VIKING ERA

The fact that the aurora seems to have been a very feeble source of inspiration for the Norse poets, folklore and mythology is much surprising. At least we know from the people living in the Arctic region of Greenland, North America and Siberia that the aurora played a more central part in their tradition and folklore. For the Ottawa indians in Canada moreover the aurora was even an omen from their benefactor Nanahboozko. In modern time we also know that poets are portraying the aurora in many ways. From our own lifetime we know almost 50 poets only in Norway which have been writing about the aurora. It is therefore a puzzle why the scalds (the Norse poets) in the Viking era which were so closely united with the nature, hardly mentioned this phenomenon that today is a very popular theme for artists.

The auroral research in modern time has revealed that the occurrence of the aurora is closely related to processes on the sun which vary strongly with time, and in such a way that it was almost absent on long periods in the Late Middle Age, viz. the Maunder Minimum (1650-1715) and the Spører Minimum (1400-1510) (Eddy 1976) (see Fig. 1).

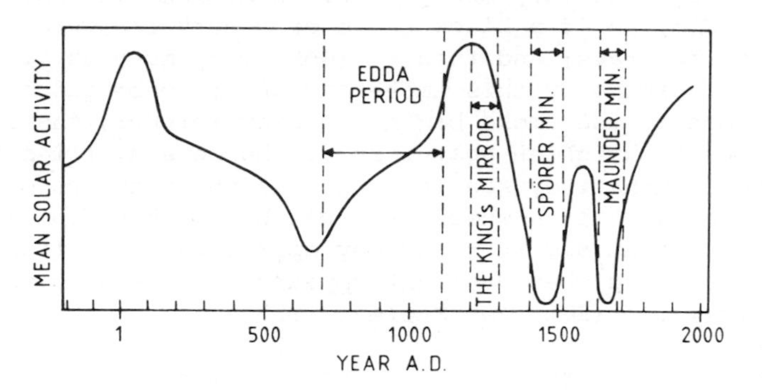

Fig. 1. The mean solar activity during the last two thousand years after Hughes (1977). Indicated are also the periods when "The Poetic Edda" and "The King's Mirror" were composed. The time periods of the Spörer- and Maunder-minima are also shown for references.

In the Maunder Minimum a well known poet, Petter Dass, lived on the coast of Hålogaland in Northern Norway. He wrote a very popular book called "Nordlands Trompet" at about 1675. The book which is a collection of poems describing the Northern Norway include a fairly large poem about the weather and climate in this part of Norway. Neither does the author mention the aurora by a single syllable in this book nor in any other of the many books he wrote. As for the scalds of the Viking era the aurora seemed to have escaped the mind of Petter Dass.

Recently Hughes (1977) based on C^{14} data derived a smoothed curve of solar activity from about 200 B.D. to 1980 (Fig. 1). From Hughes' result the Maunder and Spörer minima are quite evident, and also a strong minimum between 500 and 1000 A.D. In the last part of this period (1000-1100 A.D.) most likely "The Poetic Edda" was composed. Thus we can from the relatively low solar activity in this period understand to some extent why the authors of the poems probably never mentioned the aurora. The conflicting evidence stems from the strong maximum at 1200 A.D. just at the time "The King's Mirror" was written. As already indicated the author of this book probably never saw the aurora himself, and did not know of anybody who had seen it from Norway either. The author, which according to his book, was a very well educated man knew about the aurora only as a phenomenon occurring in Greenland. To further illustrate the author's detailed knowledge about the aurora it is illustrative to quote some more from the English translation of "The King's Mirror":

> "The light is very changeable. Sometimes it appears to grow dim, as if a black smoke or a dark fog were blown up among the rays; and then it looks very much as if the light were overcome by this smoke and about to be quenched. But as soon as the smoke begins to grow thinner, the light begins to brighten again; and it happens at times that people think they see large sparks shooting out of it as from glowing iron which has just been taken from the forge. But as night declines and day approaches, the light begins to fade; and when daylight appears, it seems to vanish entirely." (Ref. see Størmer 1955.)

We cannot understand these two conflicting facts only on the basis of variations in the solar activity, there must be some other factors involved in the determination of the occurrence of the aurora. We can imagine a few other such factors: variations in the position and the shape of the auroral oval and in the Earth's magnetic field strength. Geomagnetic recordings have been carried out only for the last 300 years (cf. e.g. Roederer 1974) but during this period, large secular variations in the Earth's magnetic field have been found. Thermoremanent magnetism of baked clay, however, in which the ancient geomagnetic field is fossilized, hasproven a useful technique to obtain knowledge of the Earth's magnetic field in the remote past. Relying on world-wide archomagnetic measurements Kawai and Hiroka (1967) have shown that the dipole axis has changed its position considerably from 700 A.D. until present time. This result is reproduced in Fig. 2 where an anticlockwise rotation of the Northern Geomagnetic Pole has been predominant in the time span studied. It is realized that these results are based on a very limited amount of data and certainly subject to large uncertainties. Even so Kawai and Hiroka's results have been used in Fig. 3 in order to indicate a possible location of the nightside auroral oval for the years 700, 1100, 1200 and 1970. An oval corresponding to the present oval for Q = 3 has been used. From this figure it is indicated that the auroral oval was far to the north of Norway at the time of the origin of "The King's Mirror". In spite of the solar activity being very high at this time, only in extreme cases would the aurora occur above the middle of Norway where the author of "The King's Mirror" presumably lived.

With respect to Greenland, however, the oval as shown here (Fig. 3) for 1200 A.D. passes too far to the north, because the Norwegian settlement on Greenland was at the southern end of the island, and from there the quotation from "The King's Mirror" implies that the aurora was seen. In our Figure we have used the form of an oval for Q = 3 as it occurs at the present time. Paleomagnetic studies indicate that non-dipolar fields are and have been typical in the Earth's magnetic field and that the non-dipolar terms at time can be quite dominant to the global magnetic field.

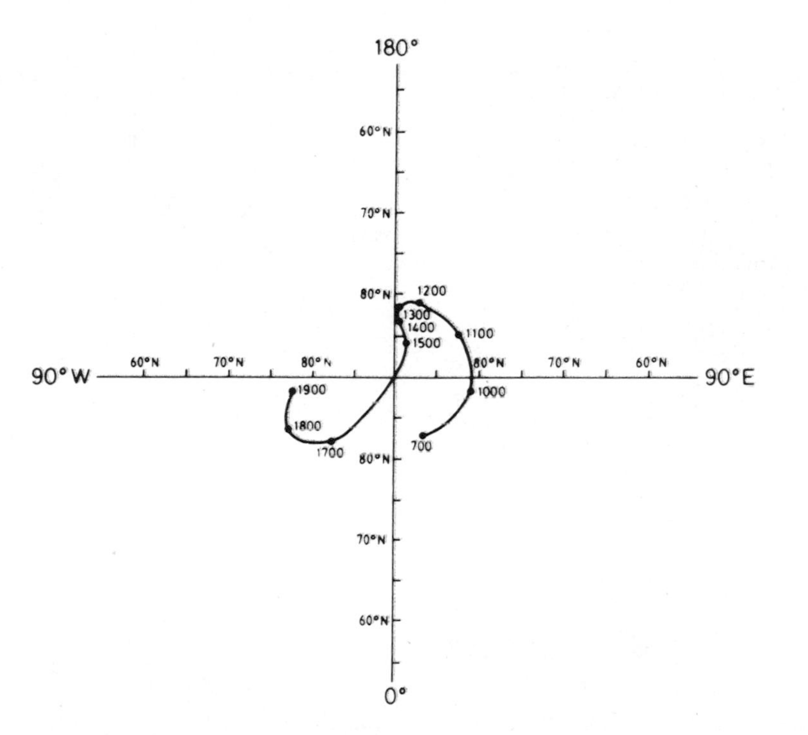

Fig. 2. The position of the geomagnetic pole in Northern Hemisphere as substantiated by Kawai and Hiroka (1965) based on their archaeomagnetic observations. The pole has on the average rotated counterclockwise until last century.

Siscoe and Siebeck (1979) have derived models for the auroral oval taking into account the contribution from such non-dipolar terms and shown that the oval becomes fairly much elongated compared to the present day oval when only modest contribution from the non-dipolar terms are present. We know at the moment very little about the status of the non-dipolar terms in historical time, but we have good reasons to believe that they in periods have contributed with different strength to the global field (Siscoe and Siebeck 1979). It then appears that in order to explain the statement in "The King's Mirror" one has to introduce non-dipolar components to the Earth's magnetic field.

At about 700 A.D. before the first poems in "Edda" was composed, the nightside oval (Fig. 3) was well in the south of Norway, but moved progressively northward until about 1100 A.D. just after the last poem in "Edda" probably was put together. During this period the solar activity increased from low to modest. If we

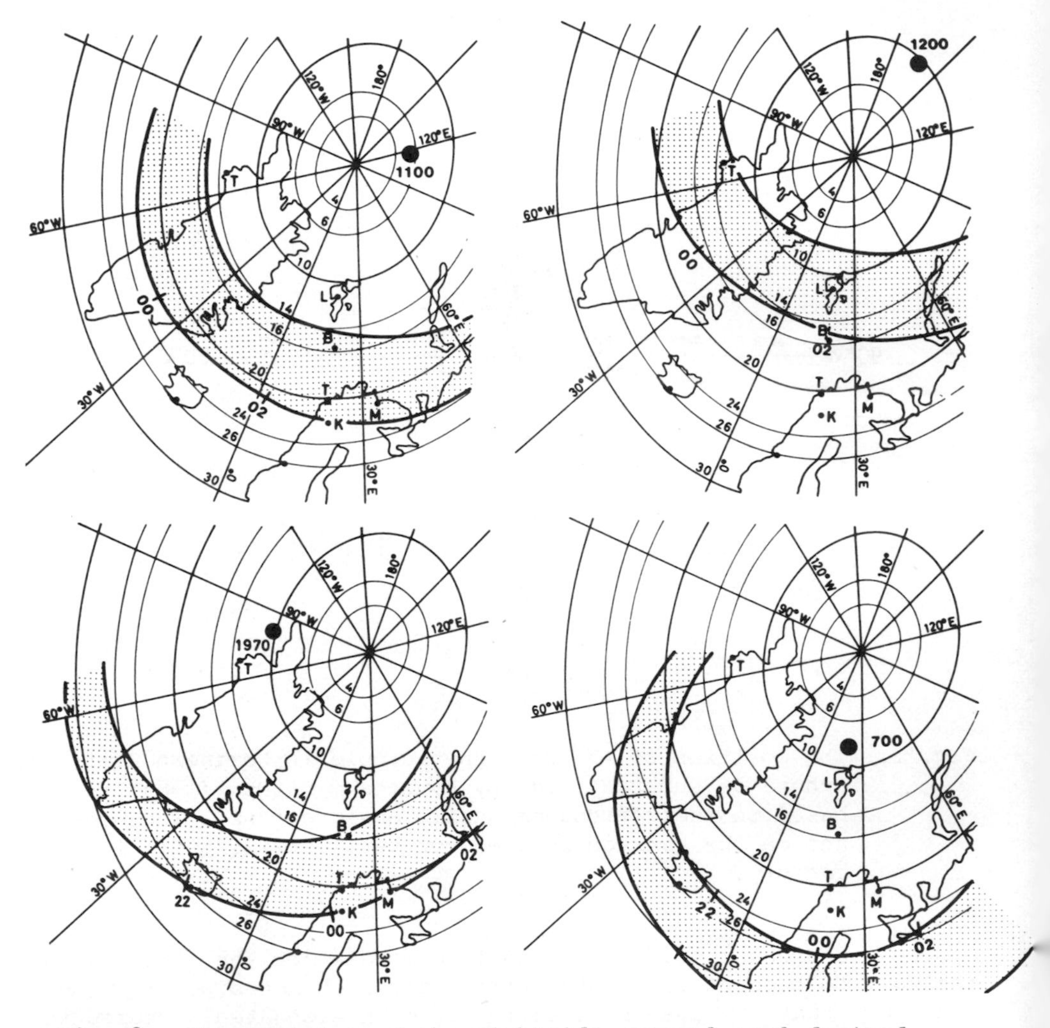

Fig. 3. The position of the nightside auroral oval derived in northern Europe for four different periods. The positions of the geomagnetic pole shown in Fig. 2 have been used together with an auroral oval corresponding the present Q = 3 activity.

put the "Edda" period somewhere between 1000 and 1100 A.D. as the latest research in the field tend to indicate, we see that the nightside oval is just touching the northern tip of Norway about in the same manner as at the present time. Since the authors of "Edda" lived in the southern part of Norway and in Iceland and may be also as far south as England, the oval would be so far north that the aurora could be a very rare phenomenon to the scalds. This becomes even more likely as the solar activity in this period was relatively low. Deehr et al. (1980) have recently pointed out that in particular observations of the red auroras together with knowledge of the solar activity can shed light on the behaviour of the magnetic field strength. If the dipole moment of the Earth's magnetic field is enhanced the aurora will tend to occur at higher latitudes unless the solar activity is also increased.

Our present knowledge of the strength of the Earth's magnetic field in the remote past is, however, very limited and further conclusions made on the basis of this must therefore be very fumbling.

7. CONCLUSION

"The Poetic Edda" and old Scandinavian folklore leave no conclusive evidence for the aurora being known to the Nordic scalds in the Viking era. "The King's Mirror" on the other hand states that the aurora was observed in Greenland and probably not in Norway at about 1200 A.D. Furthermore the author of this chronicle demonstrates detailed knowledge about the appearance of the phenomenon which clearly shows that he was well informed about its feature.

It is maintained that a correct interpretation of the lack of aurora in ancient literature in Scandinavia during a fairly long time span, can shed new light on our knowledge about solar activity, the Earth's magnetic field and the auroral oval. It is likely that the variability in the solar activity alone cannot explain the lack of aurora observed in Scandinavia in the Viking era, other factors as the strength of the Earth's magnetic field and the shape and position of the auroral oval must be taken into account.

REFERENCES

Barhow, L. 1751, Observationes von Phaenomeno Nord-Licht, Frankfurt und Leipzig.

Brekke, A. and Egeland A. 1979, Nordlyset - Fra Mytologi til Romforskning, Grøndahl & Søn A/S, Oslo.

Bugge, S. 1871, Tillægsbemerkning om navnet Hålogaland, Helgeland, Historisk Tidsskrift, Kristiania, 1, 136-140.

Deehr, C., Egeland, A., and Brekke, A. 1980, Historical Observations of the Aurora as a Measure of the Relative Effect of Solar Activity and Geomagnetic Field. Unpublished paper.

Eddy, J.A. 1976, The Maunder Minimum, Science, 192, pp. 1189-1202.

Fritz, H. 1873, Verzeichniss beobachteter Polarlicht, Wien, 1873.

Fritz, H. 1881, Das Polarlicht, Leipzig.
Frobesius, J.N. 1739, Nova et antiqua lumines atque Aurorae Borealis spectacula, Helmstadt.
Hollander, Lee M. 1969, The Poetic Edda, University of Dallas Press, Austin, Texas.
Hughes, D.W. 1977, The inconstant Sun, Nature, Vol. 226, pp. 405-406.
Kawai, N. and Hiroka, K. 1967, Wobbing Motion of the Geomagnetic Dipole Field in Historic Time During These 2000 Years, J. Geom. and Geol., Vol. 15, pp. 217-227.
Koht, H. 1920, Om namne Hålogaland, Haaløygminne, 3-11.
Lovering, J. 1868, On the Periodicity of the Aurora Borealis, Memoirs of the American Academy, Cambridge and Boston, Vol. X - Part I.
Lycosthenes, C. 1517, Prodigorum ac ostentorum, Chronicon, Basileae.
Magnusson, F. 1821, Den Ældre Edda, Vol. 1, p. 62, København, 1821.
de Mairan, J.J.D. 1733, Traité physique et historique de l'Aurore boréale, Paris.
Petrie, W. 1963, Keoeeit. The Story of the Aurora Borealis, Pergamon Press.
Ramus, J.F. 1745, Historiske og Physiske Beskrivelser over Nordlysets forunderlige Skikkelse, Natur og Oprindelse, Skrifter fra Det Kiøbenhavnske Selskap, 1, 317-396, Del I.
Ramus, J.F. 1747, Historiske og Physiske Beskrivelser over Nordlysets forunderlige Skikkelse, Natur og Oprindelse, Skrifter fra Det Kiøbenhavnske Selskap, 3, 147-212.
Roederer, J.G. 1974, Secular-invariant Relationships Among Internal Geomagnetic Field Coefficients, J. Geophys Res. Letter, pp. 367-370.
Siscoe, G.L. and Siebeck, O.G. 1979, Effects of Non-Dipole Components on Auroral Zone Configurations During Weak Dipole Field Epochs, Preprint.
Spidberg, J.C. 1750, In a letter to E. Pontoppidan mentioned in "Norges Naturlige Historie", by E. Pontoppidan, Kiøbenhavn 1752.
Storaker, J.Th. 1923, Rummet i Den Norske Folketro, Norsk Folkeminnelag VIII, Kristiania.
Storaker, J.Th. 1924, Elementene i Den Norske Folketro, Norsk Folkeminnelag X, Kristiania.
Størmer, C. 1955, The Polar Aurora, Oxford, p. 6.
Tromholt, S. 1902, Catalog der in Norwegen bis Juni 1878 beobachteten Nordlichter, Oslo.
Urbanitzky, A.R. von, 1887, Electrisität und Magnetismus in Alterthume, Wiesbaden, Martin Sanding.

ON THE LITERATURE OF THE AURORA IN NORDIC COUNTRIES

S.M. Silverman

Air Force Geophysics Laboratory, Bedford, MA 01731

Scandinavia is in a unique position among European regions for auroral observations. In the north it touches on the auroral oval. As magnetic activity increases the aurora is seen more and more to the south of Scandinavia. The study of auroral frequency in different latitudinal bands of the region thus provides useful and important information on the relationships between auroral frequency and magnetic activity and between these and the parameters related to the solar wind and solar variability (Feynman and Silverman, 1980; Silverman and Feynman, 1980). Furthermore the Norse settlements in Iceland and Greenland and voyages in other polar regions (as to Spitzbergen) have provided data and information on auroral occurrence in regions in and to the north of the auroral oval. The literature of the Nordic countries thus provides a rich data base for auroral studies. This literature is extensive from the early eighteenth century on. Here I provide an annotated bibliography of those publications which, with some exceptions, were personally accessible. For references which I have not myself seen I have drawn extensively on Eggers (1786) and Brekke and Egeland (1979). Additional references may be found in the auroral catalogs of Rubenson (1879, 1882) and of Tromholt (1902). A detailed and valuable discussion of much of this early literature is given by Brekke and Egeland (1979) in a recent book on the aurora (in Norwegian).

One of the more puzzling features of the literature before 1700 is the absence of mention of the aurora in the medieval Sagas. The Norse explored and settled in the Faeroes, Ireland, Scotland, Iceland, Greenland, and in North

C. S. Deehr and J. A. Holtet (eds.), Exploration of the Polar Upper Atmosphere, 443–448.

America (for a listing of voyages which touched Canada, see Cooke and Holland, 1970). In view of the extensive latitudinal coverage of these voyages and the known occurrence of aurora in Europe at latitudes a few degrees south of the Norse voyages, it seems very likely that auroras should have been observed. They were not, however, mentioned. I believe that the explanation is simply that the Sagas were of the nature of historical records and that the occurrence of an aurora was not relevant to their purpose. Where it was, as in the 13th century King's Mirror, meant for didactic purposes, the aurora was mentioned. A similar effect is seen in the reports of the voyages of exploration of, for example, Frobisher and Davis in the period from about 1570-1590. This period was extremely active aurorally and the aurora should have been noted at least on the voyages home, usually from mid-September to mid-October, yet they are not mentioned in the accounts of the voyages. It is only much later, when there was a heightened awareness of the importance of scientific observations on these voyages, that auroral notations occur in the accounts.

In Scandinavia, as elsewhere in Europe, the auroral literature grows rapidly from 1710, the end of the Maunder minimum, and in the full bloom of the renewed interest in science. This is reflected in the bibliography below, which is limited to the period before 1900. I would be grateful if errors and omissions are brought to my attention.

REFERENCES

Cooke, A. and Holland, C., 1970 The Polar Record, 15 169
Feynman, J. and Silverman, S.M., 1980 J. Geophys. Res., 85 2991
Silverman, S.M. and Feynman, J., 1980 Proceedings of this conference.
Tromholt, S., 1902 Catalog der in Norwegen bis Juni 1878 beobachteten Nordlichter, Edited by J. Fr. Schroeter, Kristiania.
Brekke, A. and Egeland, A., 1979 Nordlyset: Fra Mytologi til romforskning Oslo: Grøndahl and Søn.

BIBLIOGRAPHY

13th Century Kongesspeilet, Konungs Skuggsja. Translated into English as The King's Mirror, L.M. Larson, New York: The American Scandinavian Foundation, London: Oxford University Press, 1917. Apparently the earliest mention of the aurora in Scandinavian writings. The book is designed as a primer of all the knowledge that a gentleman would need. The King's Mirror describes the Greenland aurora and notes three theories for their origin: (1) Fire surrounds the earth, Greenland is on the outermost edge, and the aurora is the light of the fires; (2) they are occasional gleams of sunlight when the sun is far below the horizon; (3) the frost and the glaciers are so powerful that they radiate these flames.

1563 Beyer, Absalon Pedersson, Description of an aurora. An English translation of this description is in Oxaal, J., Mon. Weather Rev. 42 27-29 (1914), translated from Naturen (1913). It is reproduced in Silverman, S.M. and Tuan, T.F., Advances in Geophysics, Vol. 16, 1973, P. 155-266, at p.. 218. See also the more extended material in Brekke and Egeland (1979).

1582-1597 Brahe, Tyge [Tycho]. Brahe kept a meteorological journal during the years 15821597 at Uraniborg, Island of Hveen. This was published by the Danish Royal Academy of Sciences, Copenhagen, 1876.

1588 Olsen, Jorgen, Prognosticon, Stockholm. Mentions the aerial fiery flames often seen in the winter nights in the sixteenth century, and that especially in the years 1561-1568 they were seen on almost every clear night in the winter. Cited in Eggers (1786), p. 307-308, footnote o.

1596 Friis, Peder Clausson, Norriges Beskrivelse, includes discussion of the aurora taken from the King's Mirror. Cited in Brekke and Egeland (1979), also in Eggers (1786), p. 306.

1605 Linschoten, Jean Huyghens von, [Voyage] Amsterdam. Reprinted as Reizen naar het Noorden, 's Gravenhage: Martinus Nijhoff, 1914. See p. 198-199 of this edition. States that in the northern regions (off North Cape) in 1595 saw a glow in the sky on clear nights in all seasons, and which were very frequent in the winter nights. Cited in Eggers (1786), p.310-311.

1676 Wood John, [Voyage], Voyage to Norway and Nova Zembla. Noted that everyone who had been in Greenland knew about the aurora, which was frequent there. States that it was also seen in Norway, Iceland and other countries, and identifies it with the streaming, seen in the northern parts of England. Cited in Eggers (1786), p. 310.

1706 Torfaeus, Tormod, Grønlandia Antiqua, Torfaeus was the court historian and wrote histories of Norway, Greenland, Iceland and the Faeroes. He notes that the aurora was frequently seen in Greenland, Iceland and Norway and that, as a child in Iceland, he had seen it in resplendent glory, and he often noted it at his residence on the Island of Carmen. Cited in Eggers (1786), p. 308, and in Brekke and Egeland (1979).

1708 Arnelius, Suno, Exercitium Philosophicum de Chasmatibus, Discussion of the aurora. Cited in Brekke and Egeland (1979), p. 44.

1710 Römer, O. Miscellanea Berolini, 127-128. Description of the auroras of February 1, March 1, and 6 March, 1707 seen in or near Copenhagen.

1715 Ramus, Jonas, Norriges Beskrivelse. Discussion of auroral hypotheses. Relates the aurora to magnetism. Cited in Brekke and Egeland (1979), p. 45. Also cited in Eggers (1786).

1717 N.L. Reise-Beskrivelse til Island, Copenhagen. Notes the occurrence of the aurora in Iceland almost every night. The inhabitants considered it as providing as great a service as the moon. Cited in Eggers (1786), p. 312.

1723 Zorgdrager, C.G. Alte und Neue Grönländische Fischerei und Wallfischfang. This popular work on the whale fisheries (economically important at the time) went through many editions. In the Leipzig edition cited here the aurora is mentioned on pps. 38, 320, 322-332. In addition to noting the Greenland aurora he cites a number of other European observations. He notes that the most favorable months for aurora are February/March/April and September.

1723-1741 Burman, E.J., "Specimen Observationum Meteorologicum Upsalensium". Acta Literaria Sueciae, 1 387 f Meterological observations at Upsala. Auroral observations for 1722 on p.393. Observations for the year 1723 are at p. 513 f, with auroral observations at p. 518. Observations for 1724 are in 2 11f, auroras, p. 15; for 1725 in 2 139-141, auroras, p. 141; for 1726 in 2 254-257, auroras, p. 257; for 1728 in 2 513-515, auroras, p. 515. The observations for 1727 are not at the page listed in the index and I have not yet located them. Observations for 1729 and subsequently are by Anders Celsius and are found at: 1729, 2 610-611; 1730, 3 101-106, auroras, 106; 1731, 3 39-44, auroras, p. 44; 1739, 4 539-547, auroras, 542-547. Data For 1740 are in the successor to the Acta Literaria, the Acta Societatis Regiae Scientarum Upsalensis; for 1740 by Celsius, p. 38, auroral data p. 43-47. Data for 1741 are found in the Acta for the years 1744-1750 by O.P. Hiorter, p. 95, auroral data p. 100-111.

1724 Burman, E.J., "Observatio circa Lumen Boreale", Acta Literaria Sueciae, 1 566-570. Description of the aurora of 20 September 1717 at Upsala. [Reprinted in Phil. Trans. Roy. Soc. (London) 33 175-178 (1724-5).] A biography of Burman is at 3 114-120 (1730).

1724 Spidberg, Jens Christian, Historische Demonstration and Anmerkung über die Eigenschaften and Ursachen des sogenandten Nord-Lichts. Book on the characteristics and origin of the aurora. Cited in Brekke and Egeland (1979), p. 46.

1730 Spöring, Hermann D., [Meteorological Observations at Abo in 1730], Acta Literaria Sueciae 3 106-109. auroras noted at p.109.

1733 Celsius, Anders, CCCXVI observationes de lumine boreali ab a. 1716-1732, partim a se, partim ab aliis, in Suecia habitas, Norimbergae [Nuremberg]. A collection of 316 Swedish auroras from 1716-1732. Cited in Eggers (1786) and Rubenson (1882).

1734 Acta Literaria Sueciae, 3, end papers. Drawings of auroras at Copenhagen, Feb. 1,

March 1 and 6, 1707; Berlin, March 6, 1707; Giessen, November 26, 1710; Meissen, March 17, 1716; Eisleben, March 17, 1716; Halberstad, March 17, 1716; Helmstad, March 17, 1716; Danzig, March 17, 1716; Giessen, February 17 and March 1 1721; Brevillepont, France, September 26 and October 19, 1726, and September 26, 1731.

1736-1738 Sparschuch, Johann, [Meteorological observations at Linkoping for 1734-1738]. Acta Literaria Sueciae, 4 185-192, 249-254, 435-443. auroral notations passim.

1737 Waller, Nicolas, "Observationes nonnullae circa Lumen Nocturnam Boreale", Acta Literaria Sueciae, 4 220-229. Observations at Upsala, September 1736-February 1737.

1737 Celsius, Anders, "Observationes de Lumine Boreali", Acta Literaria Sueciae, 4 254 -262. Observations at different sites in Sweden.

1741 Heitman, Johan, Physiske Betaenkninger over Solens Varme, Luftens skarpe Kuld og Nordlyset, Copenhagen. Speculations on the aurora and its origin in chemical processes. Cited in Eggers (1786), p. 313-314, and Brekke and Egeland (1979), p. 47-48.

1741 Møller, Peter, Tanker om Nordlyset, Trondheim. A reaction to Heitman's book. Møller believed that the aurora originated in Sulfur and Saltpetre vapors emitted from Icelandic volcanoes and which were carried to southern countries by the North winds. Cited in Eggers (1786), p.316, and in Brekke and Egeland (1979), p. 49.

1745 Ramus, Joachim Freiderich, Historico-physica enarratio de stupendis luminis borealis phoenomenis, natura et origine, Copenhagen. From Scriptores Societatis Havniensis, Pars I, 317-394. Description and discussion of the aurora. Cited in Brekke and Egeland (1979), p. 49.

1746 Anderson, Johan, Nachrichten von Island, Grönland und der Strasse Davis, Hamburg. Includes discussion of aurora. Severely criticized by Horrebow (1752). Apparently contains many errors see the footnotes in Eggers (1786), p. 65-67.

1751 Barhow, Lars, Richtig angestellte und aufrichtig mitgeteilte Observationes vom Nordlicht, Leipzig. Observations and extensive discussion of the aurora. Cited in Eggers (1786), p. 247 and passim, and in Brekke and Egeland (1979), p. 53.

1751-53 Pontoppidan, Erich, Det første Forsøg paa Norges Naturlige Historie. Suggests the aurora is an electrical phenomenon. Cited in Eggers (1786), p. 317, and Brekke and Egeland (1979), p.55.

1752 Horrebow, Niels, Tilforladelige Efterretninger om Island. A description of Iceland, with severe criticism of the writings of J. Anderson. Includes daily meteorological observations from 1 August 1749 to 31 July 1751 with auroral notations. Notes that the aurora is very common, discusses its character, position, and origin. A French translation by Rousselot de Surgy and Meslin was published in Paris in 1764.

1760 Schøning, Gerhard, "Nordlysets Aelde beviist med gamle Skribenters Vidnesbyrd", Skrifter Kiøbenhavnske Selskab af Laerdoms og Videnskabers Elskere ...., p. 197-316. A discussion and descriptive catalog of auroral observations from ancient times to his own.

1762 Bergman, Thorbern, "Observations on Aurorae Borealis in Sweden" Phil. Trans. Roy. Soc. 52 479-486. Auroral descriptions.

1763 Jessen, Erich Johan, Det Kongrige Norge fremstillet efter dets naturlige og borgerlige Tilstand. Includes extensive discussion of the aurora in Chapter 7, p. 375-469. Cited in Eggers (1786), passim, and Brekke and Egeland (1979), p. 57-58.

1767 Bergman, Thorbern, "Aurorae Boreales", Nova Acta Reg. Sci, Upsala, 1 116-150. Descriptions of auroras observed from 1759-1762.

1774 Le Roy, Petr Ludovik, A Narrative of the Singular Adventures of Four Russian Sailors..., London. Description of the travails of four Russian sailors in East Spitzbergen, 17431749. On p. 96 there is a notation, "aurora borealis pretty frequent in winter".

1777 Wilcke, Johann Carl, [New results on magnetism] VetenskapsAkademien Handlingar. Noted that the aurora was along the magnetic axis. Cited in Brekke and Egeland (1979), p. 59.

1781 Fester, Diederich Christian, Matematiske og Physiske Betaenkninger over Nordlyset, Trondheim. Argued in support of Mairan's theory that the aurora is an extensiion of the zodiacal light into the earth's atmosphere. Cited in Eggers (1786) p. 317-318, and in Brekke and Egeland (1979), p. 60.

1786 Eggers, Christian Ulrich Detlev, Beschreibung von Island, Copenhagen. This description of Iceland includes a long (105 pages) discussion of the aurora, including all the European observations. It includes what appears to be a fairly complete collection of references on the aurora up to that date.

Late 18th Century Schiøtte, David, "Om de norske Bierge," Oecon. Magaz. 2 317 f. Discusses the different appearance and characteristics of the aurora in different regions of Norway. Cited in Eggers (1786) p. 322.

Late 18th Century Svendsen, Gudlaug, [Auroral observations] Observations of aurora in the western part of Iceland from 1771-1774. Cited in Eggers (1786) p 323, footnote.

1802 Olafsen, Eggert and Povelsen, Biarne, Voyage en Islande Paris. A description by regions of Iceland. Auroral material is found in Vol. 1, p. 13; Vol. 2. p. 172; and Vol 5, p. 112-118. Discusses a number of auroras in the period from 1753-1757 with particular emphasis on the meteorological conditions accompanying them.

1811 Hooker, William Jackson, Journal of a Tour in Iceland, Yarmouth. Description of the aurora, p. 284-285.

1815 Laing, John. An Account of a Voyage to Spitzbergen, London. Describes the aurora in the Shetlands (p. 23-25). Notes the aurora in Spitzbergen, probably a citation from someone else (p.83).

1818 Egede, Hans. A Description of Greenland, London. Egede was the first missionary to Greenland in modern times. Notes the occurrence of the aurora in the spring (p. 54-55).

1819 Henderson, Ebenezer. Iceland, or the Journal of a Residence in that Island During the Years 1814 and 1815, Edinburgh. Description of the aurora which he could see "almost every clear night the whole winter".(p.277-278).

1820 Scoresby, W., Jr. An Account of the Arctic Regions, Edinburgh. One of the more famous books on the whale fishery. Notes the aurora in polar regions (p. 136, 392, 415-418), discusses the connection with weather. States that in Iceland and other countries bordering on the arctic circle the aurora occurs almost every clear night during the winter. Not clear whether this is from his own experience or citation from others.

1820 Crantz, David. The History of Greenland, London. Notes and describes the occurrence of the aurora in the winter (p.46, 306-307, Vol. 1).

1823 Thienemann, L. "Einige Folgerungen aus Beobachtungen uber das Nordlicht, welche in Island, in den jahren 1820 and 1821, anstellte", Ann. Physik, 75 59-67, cf 1827 Thienemann. Conclusions on the form, movement, occurrence, sound of (none), connection with the weather (none), of the aurora drawn from a trip to Iceland in 1820-1821. These are also summarized in Edinburgh Philosophical Journal, 10 366-367 (1824).

1825 Hansteen, Chr. "On the Aurora Borealis and Polar Fogs", Edinburgh Phil. J., 12 83 -93, 235-238. (Taken from the Norwegian in the Christiania J. Nat. Hist.). By comparison of the aurora in Norway, N. America, Greenland, H. concludes that it exists in a ring whose center is the magnetic pole of the earth. The luminous columns are in a direction parallel to the inclination of the magnetic needle. Aurora was frequent from 1720-1790, lately very rare. They have a period of 60-100 years. Discusses the reasons for believing in a magnetic origin of the aurora. The aurora is generally accompanied by strong biting cold.

1827 Thienemann, F.A.L., Reise im Norden Europa's, vorzuglich in Island, in den Jahren 1820 bis 1821, Leipzig, cf 1823 Thienemann. Discussion of the aurora in Iceland (p. 172-180). T.says connected with the clouds, therefore at a height of 2 miles.

1828 Keilhau, M., "Ueber die Nordlichter in Finnmarken" Ann. Phys.14 618-622. Auroral observations during a trip to Finnmark 1828-1829.

1838 Lottin, Victor. Voyage en Islande et au Groenland, Paris. Voyage in 1835-1836. Describes several auroras in Iceland in August, September 1836. Discusses reports of auroral sounds. Compares auroral frequency in Iceland with that at Bossekop in 1838-1839. Describes the characteristics of the auroras. (p.446-459)

1843-1848 Lottin, Victor; Bravais, Auguste; Lilliehöök, C.B; Siljeström, Peter Adam, Voyages en Scandinavie, en Laponie..., Volume: Aurores Boreales, Paris. One of the most often cited books on the aurora during the nineteenth century. Detailed descriptions of auroras during the winter of 1838-1839 in Northern Scandinavia. In Bossekop during 201 days on which aurora could have been seen, 151 were actually observed. Bravais is perhaps best known today for his work in crystallography, memorialized in the Bravais lattice.

1844 Hällstrom, G.G., "De apparitionabus aurorae boreales in septentrionalibus Europae partibus", Acta Soc. Sci. Fennicae, 2 363-376. Catalog of auroras observed at Abo, 1748-1828, and Helsingfors, 1829-1843. Discussion of the annual variation and comparison with Upsala and Petersburg.

1846 Bravais, Auguste, [Report on the results obtained at Bossekop, winter of 1838-1839]. Proces-Verbaux, Soc. Philomatique de Paris, p. 146-150.

1847 Hansteen, C., "Sur les aurores boreales observees a Christiania, de 1837 a 1846", Mem. Acad. Roy. Bruxelles, 20 103-120 of "Observations des Phenomenes Periodiques". Letter from Hansteen to Quetelet with a catalog of observed auroras from 1837-1846. Discusses various aspects of auroral phenomena. Hansteen's results seem to have been communicated through Europe largely through correspondence such as this.

1848 Siljeström, Peter Adam "On those variations of the force and the direction of the terrestrial magnetism which seem to depend on the aurora borealis" Rept. Brit. Assn. Adv. Sci., Swansea, August 1848. One paragraph abstract.

1854 d'Aunet, Leonie, Voyage d'une Femme au Spitzberg, Paris Includes descriptions of auroras observed at Muonioniska and Kengis in Lapland.

1854 Quetelet, A. "Sur les aurores boreales", Bull. Acad. Roy. Bruxelles, 21 282-304. Table of auroras observed at Christiania from Aug. 21, 1846 to April 9, 1853 provided by Hansteen. Table summarizing Christiania auroras from July 1837 to June 1853. There are maxima at the equinoxes, minima at the solstices. Compares with Upsala data 1739-1762 to show that annual variation is the same. Comparison with classical Greek and Roman reports leads Q. to conclude that the magnetic meridian had moved from 24° E. of Greenwich in classical times to about 20° W. of Greenwich in his own times. From a catalog he has prepared he finds the strongest returns of the aurora to come between about 60-90 years.

1859 Fearnley, C. [Method for finding auroral heights]. Forhandlinger i Videnskabs-Selskabet i Christiania, pps. 117149, with figure. Gives a method for determining heights of aurora and applies it to several cases. Obtains heights of the order of 20-40 miles.

1868, 1869, 1874 Angstrom, Anders Jonas. Spectrum of the aurora. Angstrom was the scientist after whom the unit of wavelength was named. He made the first spectral study of the aurora in the fall of 1867, noting the predominant green line, which he verbally communicated to Struve, who confirmed it on his return to St. Petersburg. It was reported to the Upsala Science Society in February 1868, and published in Angstrom' s Recherche sur le spectre solaire, Upsala, 1868. That portion dealing with the auroral spectrum was reprinted and translated in Ann. Physik, 137 161-163 (1869); Ann. de Chim. Phys. 18 481-483 (1869); and in the Phil. Mag. 38 (4th Ser.) 246-247 (1869). Subsequent research appeared posthumously, again reprinted and translated, in Nature 10 210-211 (1874); Jour. de Physique 3 210-214 (1874); and Ann. Physik, Jubilee volume, 424-429 (1874). In the course of these studies he made the first spectral observation of the airglow, noting that the green line was seen in all parts of the sky during a study of the zodiacal light.

1870 Nordenskiöld, A.E., "Meteorologiska Iakttagelser anställda på Beeren-Eiland", Kongl. Svenska VetenskapsAkademiens Handlingar, 8 No. 11, p. 1-20. Meteorological journal for 1865-1866 and 1868 at Bear Island, 72°20' 74°40'N, 18° - 19°17'E. Includes auroral notations.

1871 Forssman, Lars Arvid, "Des relations de l'Aurore Boreale et des perturbations magnetiques avec les phenomenes meteorologiques" Nova Acta Soc. Sci. Upsala, 8 No. 6, (1873). Relates magnetic disturbances and aurora to meteorological parameters.

1875 Wijkander, Aug., "Observations Meteorologiques", Kongl. Svenska Vetenskaps-Akademiens Handlingar, 12 No. 7. Observations at Mossel Bay, Spitzbergen during the winter of 1872-1873. The remarks on pps. 67-92 include auroral notations.

1876 Hildebrandson, H. [Aurora under clouds], Z. Osterr. Ges. Meteolorogie, 11 351. Note stating that auroras were observed under clouds in Hernosand and in Lappland.

1877 Wijkander, Aug., "Observations Magnetiques", Kongl. Svenska Vetenskaps-Akademiens Handlingar, 14 No. 15, p. 1-53, 14 plates. Magnetic observations at Spitzbergen during 1872-1873. Includes notations of auroral activity and comparisons with Swedish results at the same times.

1878, 1884, 1888 Edlund, E., "Recherches sur l'induction unipolaire, l'electricite atmospherique et l'aurore boreale", Kongl.Svenska Vetenskaps-Akademiens Handlingar, 16 No. 1, p. 1-36. Discussion of his theory of unipolar induction and application to atmospheric electricity and the aurora. See also, "Sur l'origine de l'electricite atmospherique, du tonnerre et de l'aurore boreale", Stockholm, 1884, and "Considerations sur certaines theories relative a l'electricite atmospherique", Kongl. Vetenskaps-Akademiens Handlingar, 22 No.11, p. 1-16, 1888.

1882 Tromholt, Sophus, "Om Nordlysets Perioder", Danish Meteorological Institute Yearbook for 1880. Analysis of Greenland auroras. Interprets in terms of North-South oscillatory movements of the aurora diurnally, annually, and with the solar cycle.

1882 Tromholt, Sophus, "Einge Untersuchungen über die vom Monde abhängige Periode des Nordlichtes", Forhandlinger i Videnskabs-Selskabet i Christiania, pps. 1-32, with 3 plates. Discusses auroral frequency as a function of number of days after new moon. The results confirm such earlier studies as those of Cotte (1768-1769) and Dalton (1789). Data used are from Greenland, Iceland and Scandinavia. The results are compared with similarly analyzed data for cloud coverage.

1883 Anon., "Periods of the Aurora", Knowledge, 3 207-208 Short description of Tromholt's work.

1883 Anon., "The Periodicity of Aurorae", The Observatory, 6 116-119. Short description of Tromholt's work.

1883 Terby, F., "Les Periodes de l'Aurore Boreale", Ciel et Terre, 3 553563. Summary of Tromholt's work.

1884 Nordenskiöld, A.E., "Om norrskenen under Vegas ofvervintring vid Berings sund (1878-1879)", in the Vegaexpeditionens veteskapliga iaktagelser, Vol. 1, Stockholm, 1882. French translation in Ann. Chem. Phys. 1-72 (1884). Description and discussion of auroral observations during the wintering of the Vega in Bering Strait, and comparison with observations in Scandinavia and Spitzbergen. Concludes that the aurora (in 1878-1879) forms a ring on the globe, single, double or multiple, whose lower limit is usually at a height of 3/100 of the earth's radius above the surface and whose center is slightly to the north of the magnetic pole. He discusses also such topics as the character of the aurora in different global regions, its spectrum (including lines at 5570 and 6300), and the lack of polarisation of auroral light. A catalog of observed auroras is also given.

1885 Tromholt, Sophus, Under the Rays of the Aurora Borealis, edited by Carl Siewers, Boston: Houghton, Mifflin. A popular work on his stay in northern Scandinavia. Includes a summary of knowledge on the aurora.

1886 Lemström, Selim, Om Polarljuset eller Norrskenet, Stockholm; French translation as L 'Aurore Boreale, Paris. Book.

1896 Paulsen, Adam, "Nordlysets Straalingsteori" in Nyt-Tidsskrift for Fysik og Kemi. Proposed that the aurora resulted from absorption of electrons. Cited in Brekke and Egeland (1979), p. 66-67.

1898 Ekholm, Nils and Arrhenius, Svante, "Ueber den Einfluss des Mondes auf die Polarlichter und Gewitter", Kongl. Svenska Vetenskaps-Akademien Handlingar, 31 No. 2.

1898 Ekholm, Nils and Arrhenius, Svante, "Ueber die nahezu 26tägige Periode der Polarichter und Gewitter", Kongl, Svenska Ventenskaps-Akademien Handlingar, 31 No. 3. In these two papers the authors use data from Sweden, Iceland, Greenland, North America, and Polar stations to see whether there is a lunar periodicity and one similar to that in magnetic data of almost 26 days, the latter presumably of solar origin. They find a variation of about 20% for the lunar and of about 10% for the 26 day periods. The lunar is interpreted in terms of Arrhenius theory of charged particle emission from the sun and their deflection by a charged moon.

1899 Oertel, K., [Review of Ekholm and Arrhenius Papers], Vierteljahrsschrift der Astronomischen Gesellschaft, 34 248-267. A summary and discussion of the two Ekholm and Arrhenius
papers above.

Note: Arrhenius interest in the aurora continued for many years. Other papers of his on the subject were: "Ueber die Ursache der Nordlichter", Ofversigt af Kongl.Vetenskaps-Akademiens Förhandlingar, 1900, No. 5, 545580; "La Cause de l'Aurore Boreale", Rev. Gen. des Sci., 13 65-76 (1902); "On the Electric Charge of the Sun", Terr. Mag. 10 1-8 (1905); "Die Nordlichter in Island und Grönland", Meddelanden f. Kongl. Vetenskaps-Akademiens Nobelinstitut, 1 No. 6, 1-27 (1906); "Nordlicht mit Gewitter am 2 Aug. 1910", Meddelanden f.Kongl. Vetenskaps-Akademiens Nobelinstitut, 2 No. 4, 1-9 (1911).

EFFECTS OF IONOSPHERIC DISTURBANCES ON HIGH LATITUDE RADIO WAVE PROPAGATION

T Røed Larsen

Norwegian Defence Research Establishment
N-2007 KJELLER, Norway

The effects of anomalous high latitude ionization upon radio wave propagation are described for the main types of disturbances: Sudden ionospheric disturbances, relativistic electron events, magnetic storms, auroral disturbances, and polar cap events. Examples of radio wave characteristics for such conditions are given for the frequencies between the extremely low (3-3000 Hz) and high (3-30 MHz) frequency domains.

1. INTRODUCTION

The aim of this review is to survey the various effects that anomalous ionospheric ionization has on radio wave propagation. We will only consider the lower ionospheric region, say below 100 km altitude. The ionized region extends down to 50-60 km, but may be as low as 40 km during the most severe disturbances in the polar areas. It is assumed that only the abnormal ionization effects are of interest in this context. The normal D- and E-region ionization and the propagation modes they sustain are therefore not considered here. Except for brief mention of absorption of HF (3-30 MHz) cosmic radio noise only propagation within the earth-ionosphere waveguide is treated. Furthermore, no references will be given to effects from the ionospheric heating experiments using powerful radio wave transmitters, nor of high-altitude nuclear explosions.

1.1. Variations in the reflecting and absorbing properties of the D-region (60-90 km)

The most important parameter of the ionospheric D-region in changing its reflecting and absorbing properties is the free electron density (N_e) with its time- and spatial variations. The underlying causes for these changes may be very complex, e.g. due

C. S. Deehr and J. A. Holtet (eds.), Exploration of the Polar Upper Atmosphere, 449–462.

to variations in D-region composition or chemistry, or due to changes in the ionizing agents.

The height regions between say 60 and 90 km are normally only lightly ionized: the electron density increases from around 10 to 10^4 electrons/cm^3 over this height interval. Whereas the N_e variations at equator and out to 40-45° latitude are primarily under solar control, at greater latitudes irregular variations become more and more important. From the available data one can conclude that the polar ionosphere exhibits significant time- and spatial variation in the N_e concentrations at all D- and E-region heights. It is probably realistic to also conclude that a comprehensive global and real-time monitoring system of such variations having the necessary capability to "report" the detailed complexities of the ionized atmosphere will not be forthcoming. In the foreseeable future less than optimum solutions will be available, providing real time coverage of limited areas, statistical models, or forecasts to operators of radio wave systems.

The D-region reflects VLF and LF waves from levels around 65-75 km at daytime and 80-90 km during night. For higher frequencies, above say 1 MHz, which are reflected from the E- or F-layers, the D-region will act mainly as an absorbing medium. For the principal polarization at these frequencies (f) the absorption in the case of quasi-longitudinal propagation is proportional to:

$$\frac{N_e \nu}{(f+f_L)^2}$$

where f_L is the longitudinal component of the gyrofrequency, and ν the collision frequency. At lower frequencies ($2\pi(f+f_L) \ll \nu$) absorption is expected to vary like: N_e/ν. At ELF frequencies, especially during nighttime, an appreciable contribution to the wave field also comes from levels above the D-region.

There are several approaches to the problem of deducing the electron density profile from radio propagation data, but this complicated subject will not be discussed here.

2. SUDDEN IONOSPHERIC DISTURBANCES (SIDs)

In this context the interesting effects of solar flares are the solar X-rays (SXR) and the extreme ultraviolet (EUV) emissions which cause disturbances in the lower ionosphere at heights above 60-70 km on the sunlit hemisphere. Emission of energetic particles from flares will be discussed in section 5. The X-ray fluxes in the 1-8 Å band can increase by several orders of magnitude, whereas the EUV variation is less than a few per cent during a flare. The emissions cause enhanced ionization at D-region heights and may change the ion chemistry. The direct ionization rates depend critically upon the X-ray intensities and their spectral distribution.

New satellites with enhanced capabilities to measure the detailed energy spectra of solar X-rays have greatly increased the possibilities of predicting the direct effects upon the atmosphere, and in almost real time (Donnelly 1976, Rothmuller 1978). For accurate prediction of the effects on the radio wave propagation better knowledge of the D-region ion chemistry and the electron loss rates is necessary, as well as more accurate information on the neutral atmosphere (e.g. density, temperature and composition). At low heights, <70 km, Thrane (1978) concludes that only the gross effects of X-ray flares can be predicted due to uncertainties in the ion chemistry.

The effects of solar X-rays and EUV in the atmosphere are generally denoted by the term sudden ionospheric disturbances (SIDs). The monograph by Mitra (1974) on "Ionospheric Effects of Solar Flares" gives a thorough treatment of SIDs, and is highly recommended for an in-depth discussion of these phenomena. The book also provides a valuable bibliography for this purpose.

2.1. Sudden increase/decrease of atmospherics

It has been known for many decades that the field intensity of atmospherics changes during a solar flare. Depending upon the investigating frequency increases or decreases in intensity may be observed. Enhancement of atmospherics (SEA) occurs approximately between 10 and 75 kHz, the latter frequency being dependent upon angle of incidence of the waves. At ELF frequencies, below ~1 kHz enhancement also occurs, whereas between 1 and 10 kHz a signal decrease (SDA) is expected and observed.

2.2. Sudden field and phase anomalies

The detailed effects of SIDs upon VLF and LF phase and amplitude depend upon a number of parameters in addition to the time- and energy characteristics of the SXR emission itself and the D-region chemistry: frequency of VLF waves, path characteristics and propagation distance, solar illumination over the path and season. For specific paths and wave frequencies, however, estimation of the average SXR induced variations should be possible. With real-time modelling of the solar X-ray fluxes even better results should be potentially available. Figure 1 illustrate, however, the complexities of predicting the various SID effects (Rose et al 1971). The phase and amplitude changes at five frequencies in the 21-32 kHz band over two paths of slightly different path lengths along a common great circle trajectory are intricate and exhibit presence of multi-mode propagation effects. Note e.g. the amplitude variation at 31 kHz - the changes have opposite sense due to mode mixing.

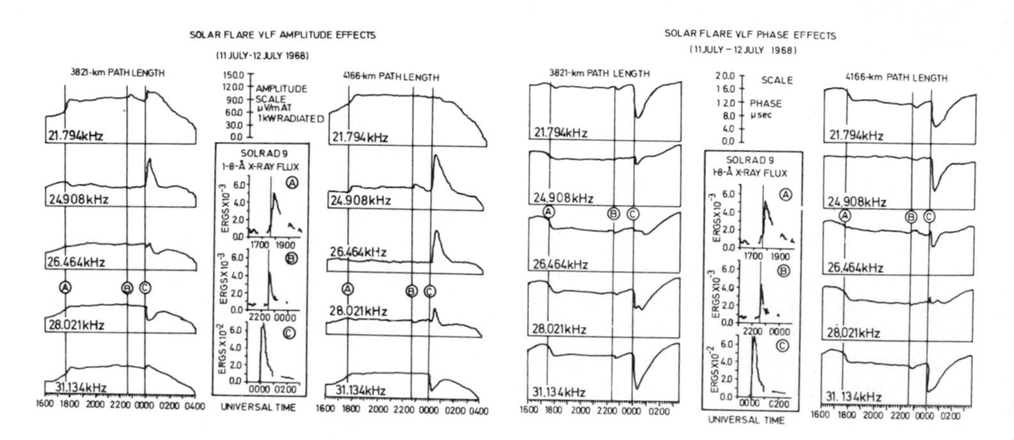

Figure 1. Solar flare effects observed on 11-12 July 1968 on the amplitude and phase at 5 frequencies between 21-32 kHz for two transmission paths (Rose et al 1971)

Detailed information on the SXR emissions are required to model the radio anomalies, as a general result, however, Bain and Hammond (1975) found that fluxes greater than $6 \cdot 10^{-7}$ J $m^{-2}s^{-1}$ in the 0.5-3.0 Å band always produced a SPA on VLF.

2.3. Influences of SFAs and SPAs on operating navigation systems

Several studies have been devoted to effects of solar X-rays on Omega navigation frequencies (10-14 kHz), cf. review by Swanson (1977). In Figure 2 is shown the probability distribution for phase changes at 10.2 kHz on a long distance path for the years 1966-70 (Swanson & Kugel 1973), compared with the typical ("undisturbed") phase deviation distribution, having a standard deviation of 3 centicycles (cec). (The probability values for flare conditions should be reduced by a factor of 4 to correspond to normal, 24 hours daily operation.)

Swanson and Kugel furthermore found from the analysis of 500 events observed over two midlatitude paths (Hawaii and Trinidad to New York) a typical duration of 40 min for the SPA, with a risetime to maximum phase offset of approximately 6 min. Mean maximum phase offset was observed to be 23 cec.

Larsen (1977) analysed Omega phase difference data at 10.2 kHz for two sets of station pairs in 1968: Omega Norway minus Trinidad recorded at Oslo (60°N) and Omega Norway minus Hawaii recorded at Svalbard (78°N). The mean maximum phase offset from expected values was 23 cec (corresponding to 3-4 km error if the disturbances only affected one set of the hyperbolic lines of position (LOPs) - two LOPs are needed for a position fix). The largest position fix error was about 10 km, which appeared on

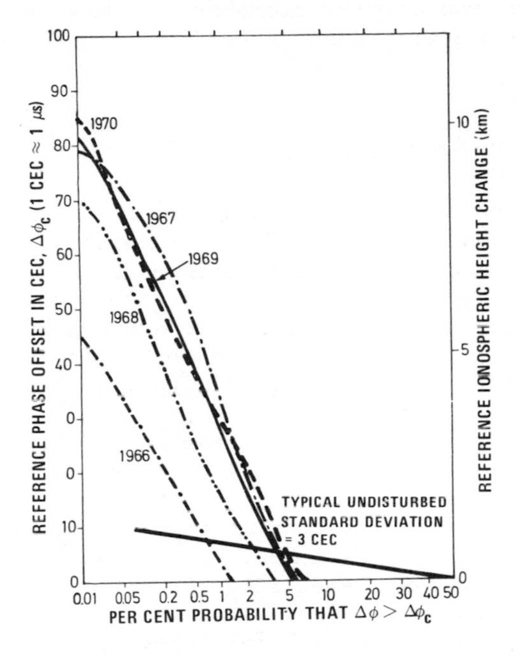

Figure 2. Per cent probability of disturbance level occurrence for observed SPA events on the Hawaii to New York path at 10.2 kHz (Swanson & Kugel 1973)

8 July 1968 1730 UT. Average duration of the SPAs was about 55 mi 55 min. In 1968 near the maximum of the solar cycle the total duration of conditions with SPA effects was less than 1%.

On Loran C frequencies SFA effects have also been detected. However, in areas where the ground wave of the Loran-C pulses are used for navigation, conditions should essentially be unaffected by SIDs unless the increased strength of the Loran-C skywave interfers with the groundwave reception. Possible problems with skywave arrival before the receiver sampling time of the groundwave will be briefly discussed in section 5. Belrose (1968) has described polarization changes in the skywave at Decca frequencies during a flare.

2.4. Short-wave fadeout and sudden cosmic noise absorption

Short-wave fadeout (SWF) are decreases or complete fadeouts of HF signls during the visible flares. Effects are also noted on lower frequencies (LF and MF) and at VHF frequencies enhancements may be observed, due to enhancement of scatter reflections.

Sudden cosmic noise absorption (SCNA) is a decrease in cosmic radio noise received by a riometer which has been a valuable equipment both in case studies and synoptic studies of wide-spread absorption. Using data from several riometer frequencies it is possible to unfold the electron density profile as a function of height.

3. RELATIVISTIC ELECTRON PRECIPITATION EVENTS

Bailey and Pomerantz (1965) concluded from a study of cases with anomalous daytime absorption on HF forward scatter links in Alaska that the absorption was caused by unusual ionization by high energy electrons penetrating below the scattering stratum of the waves. To ionize the atmosphere below the assumed HF scattering heights (~70 km) electron energies of $\gtrsim$400 keV are necessary, hence the term "relativistic electron precipitation" (REP) events are used to identify such disturbances.

The REP events occur mainly between L-values of 4.5 and 6. No events were detected by above authors on HF forward scatter paths having their mid-points beyond L = 10. The events seem to have a large longitudinal extent. Judging from a study of riometer records for a REP event in May 1972, Larsen (1974) concluded that the extent for that case was at least 2000 km.

The REP events, as found by Bailey (1968) occur most frequently in the time period 06 to 18 hours local time. Most events last from 1 to 6 hours, but there are cases where one event seems to extend from one day to the nedt. There is a tendency for a larger number of REP events near the equinoxes approaching 25 cases per month. The local magnetic disturbance during an REP is often small and even though they have no close connection to the SC of storms, they frequently occur during geomagnetically disturbed periods (Bailey 1968).

Abnormal nighttime VLF effects have been detected on a short propagation path in North-Norway in a number of cases when daytime REP activity was detected over Alaska. Belrose (1968) has provided numerous cases of propagation anomalies due to high energy particles.

A number of studies have indicated that REP events are one aspect of magnetospheric substorms even though there are mor substorms than REP events (Rosenberg et al 1972). Other works (Larsen & Thomas 1974, Matthews & Simon 1973, Thorne & Larsen 1976, Imhof et al 1978) provide further information on the disturbances and the energy spectra of the electrons causing the excess ionization.

Apart from Bailey's studies there does not seem to have been systematic investigations of the duration, frequency of occurrence, latitude and longitude coverage of the REP disturbances with the aim of assessing their influence on radiowave propagation. At present, we do not understand the physical mechanisms behind the REPs well enought to predict such events.

4. IONOSPHERIC STORMS AND AURORAL DISTURBANCES

4.1. Propagation at VLF

Several studies have been concerned with radio wave propagation during ionospheric substorms. The effects are usually observed between, say 65° and 70° geomagnetic latitudes, but this range depends upon the substorm activity. At high activity levels an equatorwards movement is expected. Berkey et al (1974) have made a synoptic investigation of the precipitation patterns during 60 substorms during IQSY and IASY, these maps are valuable for analysis of propagation anomalies.

Svennesson (1973) using such maps has studied effects on VLF propagation during substorms. He investigated 16 events during 1969 using 8 receiver sites monitoring 9 transmitters.

His findings are summarized as follows:

- All the observed VLF phase anomalies during the analysed ionospheric substorms could be explained by a depressed ionosphere due to enhanced particle precipitation, no mode conversion effects were observed.
- A good correlation was found between VLF phase anomalies and the ionospheric substorm as observed by riometers in both the day and night ionospheres.
- The amplitudes of the received VLF signals showed in most cases no significant divergence from the normal values during the substorm. The amplitude was low by 0-4 dB, and the signal was noisier than usual during a few of the most severe VLF phase anomalies. This observation of no or very little decrease in the amplitude is consistent with the theoretical model of Wait and Spies (1964) which, for a constant ionospheric conductivity gradient, gives only a very small increase in attenuation of the first order mode due to a decrease in the reflection height.

It is interesting to note that strong effects on VLF signals propagating in the daytime hemisphere were observed during all the analysed events. This was interpreted as being due to energetic electrons (>200 keV).

4.2. Propagation at ELF

Since 1970 a number of ELF propagation studies have been made using the transmissions from US Navy Wisconsin Test Facility (WTF) in the USA. The signal frequencies used lie in the 40-50 and 70-80 Hz ranges. An extensive treatment of this subject is given in the Special Issue on Extremely Low Frequency Communications (IEEE Transactions on Communications, Vol. com-22, No. 4, 1974). The articles in that issue give a comprehensive bibliography for the ELF field.

Anomalies in the received signals have been detected for both mid-latitude and high-latitude paths (Bannister 1975, Davis 1976). Decreases in signal amplitude of more than 6 dB are occasionally observed for several middle latitude paths in the early hours local time (01-03 LT). The causes are believed to be of ionospheric origin, but at present no explanation seem to be completely satisfactory. The effect is probably not one of only signal absorption, but changes in the excitation of the waves into the waveguide must also be invoked (Bannister 1975, Imhof et al 1976, 1977, Booker & Lefeuvre 1977, Booker 1977, Barr 1977).

4.3. Propagation at HF

Auroral phenomena are often accompanied by HF radio "blackout" periods. The cause for such excess ionization is twofold: electrons precipitating into the atmosphere in connection with the aurora, and energetic electrons having higher energies than the

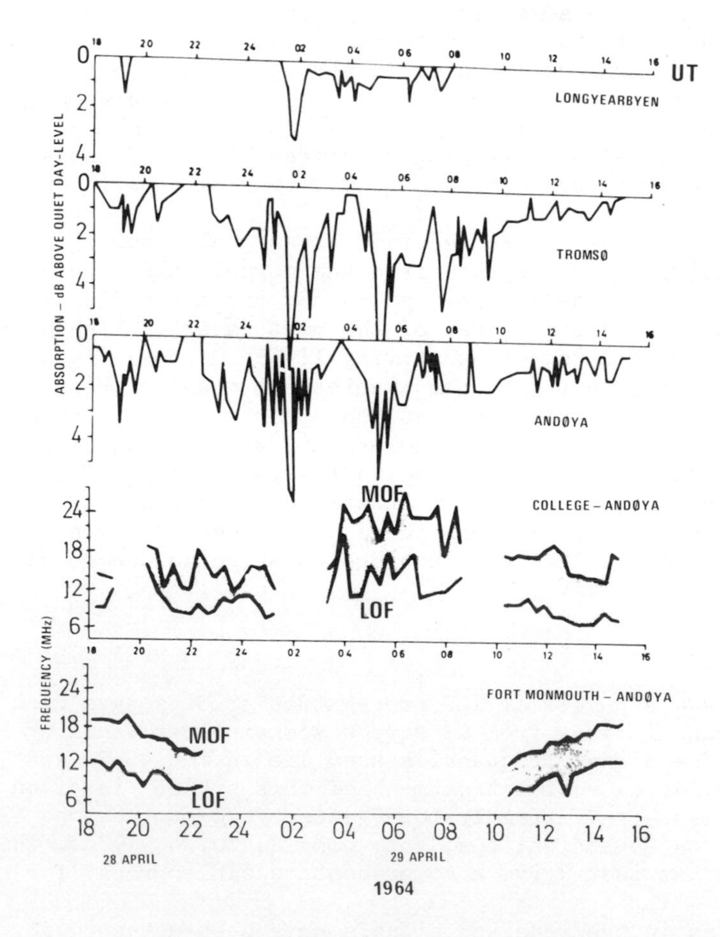

Figure 3. Circuit behaviour during disturbed conditions (Folkestad 1968a)

typical 1-10 keV electrons which cause the visual auroras. Hartz and Brice (1967) showed that the high energy electrons generally precipitated equatorwards of the auroral oval in an almost circular region, around geomagnetic latitudes of 63-67°.

The effects on radiowave propagation are marked. Ordinary HF communication in polar areas is often impossible during "blackout" conditions. The results by Folkestad (1968a) in Figure 3 shows the lowest and highest observed frequencies (LOF and MOF) propagating over two paths between North-Norway and USA. Periods of complete blackouts can be noted during the night hours.

Jones (1978) has reviewed HF communication and the possibilities for quality evaluation of the ionospheric propagation channel. Actual propagation data shown in Figure 4 give sample results of systematic experiments evaluating the quality of HF communication at specific frequencies over two paths in North-Norway (Thrane 1978, private communication). The probability of establishing a connection around local midnight for summer conditions is plotted in % vs the probing frequency. Also plotted is the predicted maximum usable frequency (MUF) for the two paths. The signal transmitted is in digital form and connection, for these purposes, is defined as being established when more than 90% of the bits are properly received. Quite good agreement is obtained between predictions and measurements at the 2-3 highest frequencies. At the lowest frequency (2.5 MHz) reception seems to be limited by low signal to noise ratio.

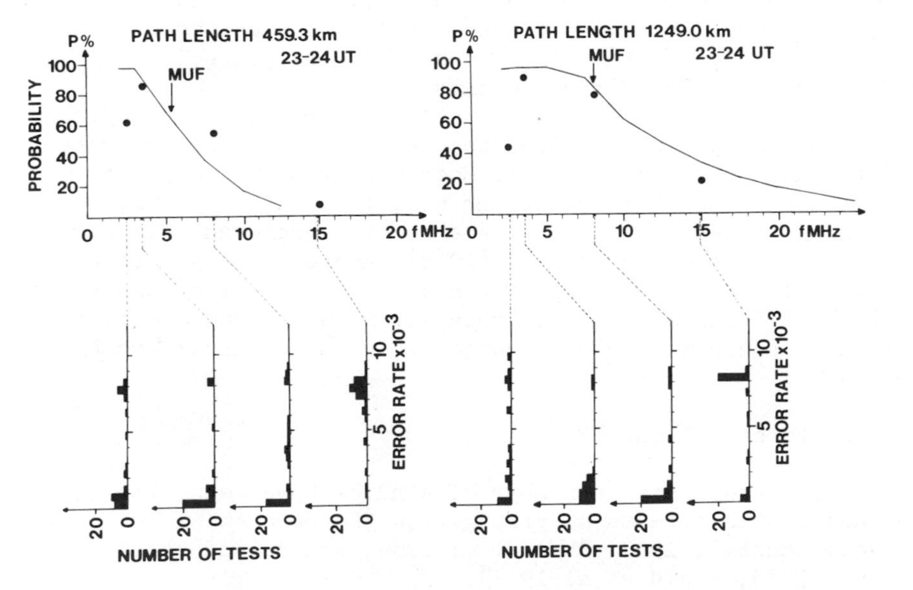

Figure 4. Results from systematic evaluating the quality of digital HF communication for two paths in North-Norway (Thrane 1978, private communication). For explanation, see the text.

5. POLAR CAP EVENTS

Following certain solar flares energetic particles impinge upon the Earth's atmosphere and cause severe and long lasting effects in the polar atmospheres. The energetic protons ionize down at low altitudes (down to ~30 km, Reagan (1975)) over a large spatial area, from the poles to 55-60° geomagnetic latitude. The proton fluxes are quite uniform over these areas, but asymmetries exist, mainly during the initial stages of the events over the central polar areas and at the low latitude cut-off regions. The events are called polar cap events (PCE) or solar proton events (SPE). Historically the acronym PCA (for polar cap absorption) has frequently also been used for the events themselves.

The production of ion pairs during a PCE may increase by several orders of magnitude. For the most intense SPE observed by modern techniques (4 Aug 72) the peak production of ion pairs occurred of 40 km's altitude with a rate of $3 \cdot 10^4$ $cm^{-3}s^{-1}$ (Reagan 1975) almost four orders of magnitude above nominal production at these levels. During such conditions D-region chemistry may actually become less complicated than during undisturbed conditions and the effective electron loss rates decrease.

The frequency of occurrence of SPEs is correlated with the 11 year solar sunspot cycle (Pomerantz & Duggal 1974). More events are observed near the solar maximum periods than during quiet solar conditions, individual SPE events also tend to occur most frequently on days of high solar activity. The most comprehensive list of SPEs available is published in the book edited by Švestka and Simon (1975). A total of 352 cases for the years 1955 to 1969 are documented as "confirmed" events.

The effects of anomalous ionization over the polar regions on radio wave propagation has long been a subject of active research and international collaboration. Most attention has probably been given to the HF and VLF transpolar propagation, but propagation at LF and in recent years ELF has also been explored. For basic reference the review by Belrose (1968), and the books edited by Folkestad (1968b) on "Ionospheric Radio Communication" and Holtet (1974) on "ELF-VLF Radio Wave Propagation" are recommended. Polar problems in connection with HF communication are described by Lied (1967).

5.1. Effects at VLF and LF

At VLF frequencies a number of studies have been carried out to establish the effects on propagation (Potemra et al 1967, 1969, Egeland & Naustvik 1967, Åbom et al 1969, Westerlund et al 1969, Oelberman 1970, Field et al 1972).

In their studies Westerlund et al (1969) concluded that trans-transpolar VLF circuits were sensitive to the effects of proton fluxes (Ep > 25 MeV) as low as ~0.5 protons/cm^2 s ster. Fitchtel and MacDonald (1967) found effects for 0.5 protons/cm^2 s ster above 20 MeV and Oelbermann (1970) obtained VLF effects for fluxes of 1 proton/cm^2 s ster above 20 MeV.

Oelbermann (1970) has reported the results of 5 years of monitoring phase and amplitude for polar and transpolar paths during 30 PCAs from 1965-69.

His main findings are:

- VLF phase (18.6 kHz) is severly advanced and amplitude is seriously attenuated during solar proton events. Phase changes are sometimes measured in cycles, attenuation is often greater than 30 dB.
- VLF field strength is a very sensitive detector of PCEs being sensitive to changes in proton flux on the order of 1 proton/cm^2-s-ster for protons of E > 20 MeV.
- Nighttime transpolar VLF is 10-40 dB stronger than the daytime signal.
- Transpolar VLF signals are stronger in the winter than in the summer by 10-20 dB.
- An apparent increase in solar particle flux at sunsport maximum produces increased absorption of transpolar VLF which results in a reduction of signal strength of from 5 to 10 dB.

Oelbermann (1970) has also detected heavy attenuation on paths which crosses the southern part of the polar cap where the auroral zone dips below the Arctic circle. Svennesson (1973) has made similar observations, cf. section 4.3.

In a study of phase deviations at an Omega frequency of 10.2 kHz Swanson (1974) has derived a probability distribution function of the SPE induced perturbations. Figure 5 shows results of data averaged over a solar cycle and is compared with similar data for SID events (cf. also Figure 2). Phase errors exceeding e.g. 30 centicycles (~30 μs) should on the average occur less than 0.5% of the time. Larsen (1977) found that for solar maximum activity the maximum mean phase deviation was 38 centicycles for the pair Omega Norway minus Omega Hawaii recorded at Svalbard (78°N). During the SPE of 2 November 1969 he deduced position errors of almost 10 km using Omega.

Efforts have been made to reduce the effects of such large positional errors, e.g. Rothmuller (1978) has described an Omega correlation model developed by Naval Ocean Systems Center which translates solar proton flux measured by satellites into phase error corrections. These corrections are in form similar to those now routinely used to account for variations during day, season, etc. The model has as parameters integrated proton flux >10 MeV, time of day and path length through the polar cap (defined as north of 63° geomagnetic latitude) The model has in cases corrected the phase deviations to give nominal 2 km positional accuracy.

In their studies Potemra et al.(1967, 1969) have with good success compared experimental and calculated data on VLF case deviations and HF absorption.

The riometer absorptions during a PCE event can be significant, exceeding the reliable range for the instrument (>20 dB).

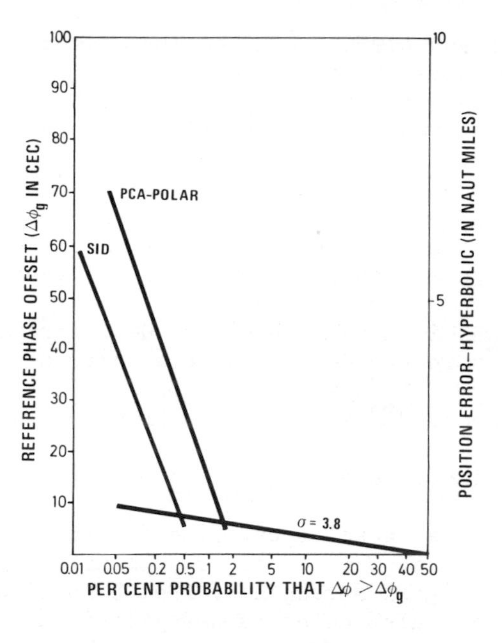

Figure 5. Importance of PCAs and SIDs on Omega propagation (10.2 kHz) averaged over a solar cycle (Swanson 1974)

At 10 MHz absorption can amount to 100-200 dB, resulting in blackout period. Reid (1972) has deduced an empirical relationship between riometer absorption (A in dB) at 30 MHz and the 2π omnidirectional proton flux (J) above 20 MeV:

$$J(>20\ \mathrm{MeV}) = 60\ A^2$$

Belrose (1968) has discussed at great length LF data during disturbed conditions. Larsen and Thrane (1977) has described possible interference from Loran-C skywave upon the groundwave pulse during solar proton events. Due to a lowering of the reflection levels, the skywave may, at some distance from the transmitter, arrive before the navigation information is extracted from the pulse train.

5.2. Effects at ELF

No actual signal measurements of stable ELF transmissions have been made during a solar proton event, but predictions of such effects can be made using available waveguide computer programs and data for previous SPE events.

The result of such a study (Larsen 1979, to be published) is that several dB signal attenuation is expected at 75 Hz, but actual amount depend critically upon the electron and ion density profiles. The details of the positive ion profile at low altitudes is of importance even down to 20-30 km. Attenuation rates are calculated to increase from ~0.9-1.7 dB/Mm to 2-3 dB/Mm for propagation at night over sea water ($\sigma \sim 4$ S/m) and ice ($\sigma \sim 10^{-5}$ S/m), respectively. Simultaneously, the excitation factor also increases by several dB.

6. CONCLUDING REMARKS

The D- and E-region effects on radio wave propagation constitute a many-facetted problem area. Within the limited space available only the major effects could be treated, and the selection of examples discussed are probably biased by the author's interest. A conscientions reader of the references given here and in a more comprehensive review by Larsen (1979) will, however, obtain a fairly complete overview of this interesting area of science.

REFERENCES

Bailey. D.K.: 1968, Rev. Geophys., 6, 289.
Bailey, D.K. and Pomerantz, M.A.; 1965, J. Geophys. Res. 70, 5823.
Bain, W.C. and Hammond, E.: 1975, J. Atmosph. Terr. Phys. 37, 573.
Bannister, P.R.: 1975, J. Atmosph. Terr. Phys. 37. 1203,
Barr, R.: 1977, J. Atmosph. Terr. Phys. 39, 1203.
Belrose, J.S.: 1968, In "Radio Wave Propagation", AGARD-LS-XXIX, NATO-AGARD.
Berkey, F.T., Driatskiy, V.M., Henriksen, K., Hultqvist, B., Jelly, D.H., Shuhuka, T.I., Theander, A., and Ylimiemi, J.: 1974, Planet. Space Sci. 22, 255.
Booker, H.G.: 1977, The relation between ionospheric profiles and ELF propagation in the earth-ionosphere transmission line, Final report 1 Apr. 75 - 30 Nov. 76, Univ. of Ca., San Diego, CA., USA.
Booker, H.G. and Lefeuvre, F.: 1977: J. Atmosph. Terr. Phys. 39, 1277.
Donnelly, R.F.: 1976, p. 178 in Physics of solar planetary environments (Ed. D.J. Williams), American Geophysical Union.
Davies, K.: 1978, In "Operational modelling of the aerospace propagation environment", AGARD Conf. Proc., AG-238, NATO-AGARD.
Davis, J.R.: 1976, J. Atmosph. Terr. Phys. 38, 1309.
Egeland, A. and Naustvik, E.: 1967, Radio Science 2 (new series), 659.
Field, E.C., Greifinger, C., and Schwartz, K.: 1972, J. Geophys. Res. 77, 1264.
Fichtel, C.E. and McDonald, F.B.: 1967, A. Rev. Astron. Astrophys. 5, 351.
Folkestad, K.: 1968a, p. 279 in Ionospheric Radio Communications (Ed. K. Folkestad), Plenum Press, New York.
Folkestad, K.: 1968b, (Ed.) Ionospheric Radio Communication, Plenum Press, New York.
Hartz, T.R. and Brice, N.M.: 1967, Planet. Space Sci. 15, 301.
Holtet, J.: 1974, (Ed.) ELF Radio Wave Propagation, D. Reidel Publ. Co., Dordrecht-Holland.
Imhof, W.L., Larsen, T.R., Reagan, J.B., and Gaines, E.E.: 1976, Analysis of satellite data on precipitating particles in coordination with ELF propagation anomalies. LMSC-D502063. Lockheed Palo Alto Research Laboratory, Palo Alto, Ca., USA.
Imhof, W.L., Larsen, T.R., Reagan, J.B., and Gaines, E.E.: 1977, Analysis of satellite data on precipitating particles in coordination with ELF propagation anomalies. LMSC-D560323. Lockheed Palo Alto Research Laboratory, Palo Alto, Ca., USA.
Imhof, W.L., Reagan, J.B., Gaines, E,E., Larsen, T.R., Davis, J.R., Moler, W.: 1978, Radio Science 13, 717.
Jones, T.B.: 1978, In "Recent Advances in Radio and Optical Propagation for Modern Communications, Navigation and Detection Systems", AGARD-LS-93, NATO-AGARD.
Larsen, T.R.: 1974, p. 171 in ELF-VLF Radio Wave Propagation, (Ed. J. Holtet), D. Reidel Publ. Co., Dordrecht-Holland.
Larsen, T.R.: 1977, In "Propagation Limitations of Navigation and Positioning Systems", AGARD-CP-209, NATO-AGARD.
Larsen, T.R.: 1979, p. 617 in Solar-Terrestrial Predictions Proceedings, Vol. II (Ed. R.F. Donnelly), NOAA, Boulder, USA.
Larsen, T.R. and Thomas, G.R.: 1974, J. Atmosph. Terr. Phys. 36, 1613.
Larsen, T.R. and Thrane, E.V.; 1977, In "Propagation Limitations of Navigation and Positioning Systems", AGARD-CP-209, NATO-AGARD.

Lied, F.: 1967, (Ed.) High frequency radio communication with emphasis on polar problems. AGARDograph 104, Technicision, Maidenhead, England.

Matthews, D.L. and Simon, D.J.: 1973, J. Geophys. Res. 78, 7539.

Mitra, A.P.: 1974, Ionospheric effects of solar flares, D. Reidel Publ. Co., Dordrecht-Holland.

Oelbermann, Jr., E.J.: 1970, J. Franklin Inst. 290, 281.

Pomerantz, M.A. and Duggal, S.P.: 1974, Rev. Geophys. Space Phys. 12, 343.

Potemra, T.A., Zmuda, A.J., Haave, C.R., and Shaw, B.W.: 1967, J. Geophys. Res. 72, 6077.

Potemra, T.A., Zmuda, A.J., Haave, C.R., and Shaw, B.W.: 1969, J. Geophys. Res. 74, 6444.

Reagan, J.B.: 1975, A study of the D-region ionosphere during the intense solar particle events of August 1972, LMSC-D454290, Lockheed Palo Alto Res. Lab., Palo Alto, Ca., USA.

Reid, G.C.: 1972, In Proc. COSPAR Symposium on Solar Particle Event of November 1969,(Ed. J.C. Ulwick), AFCRL-72-0474, 201.

Rose, R.B., Morfitt, D.G., and Bleiweiss, M.P.: 1971, System performance degradation due to varying solar emission activity: SOLRAD application study, Tash II, Techn. rep. TR 1774, Naval Electronics Laboratory Center, San Diego, Ca., USA.

Rosenberg, T.J., Lanzerotti, L.J., Bailey, D.K., and Pierson, J.D.: 1972, J. Atmosph. Terr. Phys. 34. 1977.

Rothmuller, I.J.: 1978, In "Operational Modelling of the Aerospace Propagation Environment", AGARD Conf. Proc. AGARD-AG-238, NATO-AGARD.

Svennesson, J.: 1973, J. Atmosph. Terr. Phys. 35, 761.

Švestka, Z. and Simon, P. (Ed.): 1975, Catalog of solar particle events 1955-1969, D. Reidel Publ. Co., Dordrecht-Holland.

Swanson, E.R.: 1974, Blunders caused by Omega propagation: SPA's amd PCA's. Proc. 2nd Omega Symposium, 5-7 November 1974, Wash, D.C., 202.

Swanson, E.R.: 1977, In "Propagation limitations of navigation and positioning systems", AGARD-CP-209, NATO-AGARD.

Swanson, E.R. and Kugel, C.P.: 1973, Proc. of the fifth annual NASA and Department of Defence, Precise Time and Time Interval (PTTI) Planning Meeting, NASA Publication X-814-74-225, 443.

Thorne, R.M. and Larsen, T.R.: 1976, J. Geophys. Res. 81, 5501.

Thrane, E.V.: 1979, In "Operational Modelling of the Aerospace Propagation Environment", AGARD Conf. Proc., AGARD-AG-238, NATO-AGARD.

Wait. J.R. and Spies, K.P.: 1964, Techn. note No. 300, National Bureau of Standards, Boulder, Co., USA.

Westerlund, S., Reder, F.H., and Åbom, C.: 1969, J. Atmosph. Terr. Phys. 17, 1329.

Åbom, C., Reder, F.H., and Westerlund, S.: 1969, Planet Space Sci. 17, 1329.

THE VARIABILITY AND PREDICTABILITY OF THE MAIN IONOSPHERIC TROUGH

A. S. Rodger and M. Pinnock

British Antarctic Survey, Madingley Road, Cambridge
CB3 OET, UK. (Natural Environment Research Council)

ABSTRACT. Diurnal, seasonal and solar cycle variations in the occurrence and shape of the main ionospheric trough for a single station are briefly described from analyses of night-time ionosonde data from Halley Bay, Antarctica (76°S, 27°W). Five different types of main trough have been identified, each for a particular level of magnetic activity. It is shown that the level of magnetic activity is also important in determining the local time at which the main trough is seen at Halley Bay. Measurements of the movement of the poleward edge of the main trough through the night are presented and this shows some similarities to the known movement of the plasmapause. These results have importance for radio wave propagation in the auroral and subauroral regions.

INTRODUCTION

The main ionospheric trough, often termed the mid-latitude electron density trough, is defined to be the circumpolar region at about L = 4, where the night-time electron concentration at F-region heights is anomalously low. It can be considered as a boundary region between the mid-latitude and polar ionosphere. The region of diminished electron concentration, termed the trough minimum, is bounded by two maxima, one poleward which is relatively constant in electron concentration and the other equatorward where the electron concentration is much more variable. The phenology of the main trough has been extensively studied by satellite techniques (e.g. Muldrew 1965; Rycroft and Thomas 1970; Tulunay and Grebowsky 1978) which in general make measurements well above hmF2, the height of maximum electron

C. S. Deehr and J. A. Holtet (eds.), Exploration of the Polar Upper Atmosphere, 463–469.

concentration of the F2 layer. Most studies have concentrated on periods of high magnetic activity as the main trough is usually more clearly identifiable under these conditions. The relationship of this "top side" trough to other associated phenomena such as the plasmapause, stable auroral red arc (SARC) and the light ion trough have been discussed by Rycroft and Burnell (1970), Wrenn and Raitt (1975), Titheridge (1976) and many others. However, there has been very little study of the main trough below hmF2, despite the significant effects which the variations of electron concentrations below hmF2 associated with the main trough have on h.f. radio wave propagation (Hunsucker 1980).

In this paper, the main features of the daily, seasonal and solar cycle variability in the occurrence and electron concentration gradients (shape) of the main trough at night are briefly described for one location. These have been determined from a detailed study of ground-based vertical incidence ionosonde data from Halley Bay (76°S, 27°W; L = 4.2). (Only night-time data are presented because the main trough cannot normally be detected by ionosondes during periods when there is significant photo-ionisation.) In addition, estimates of the movements of the poleward edge of the main trough are presented. From these observations, it is shown that many features of the main trough are mainly dependent upon the current level of magnetic activity. This finding has significance for prediction of radio wave propagation in the auroral region.

RESULTS AND DISCUSSION

From satellite observations, the main trough is known to move equatorward from high latitudes in the early evening hours. Under most circumstances, Halley Bay, Antarctica, is well placed for studying the main trough and its movements at night-time. Halley Bay ionograms for the months of April in 1958, 1962, 1965, 1972 and the complete year in 1978 have been examined to determine the daily, seasonal and solar cycle variations of occurrence and appearance of the main trough. These years were selected as being representative of different solar activity conditions throughout the solar cycle.

There are three very distinctive characteristics which can be seen on ionogram sequences which allow the main trough to be positively identified as it moves from poleward to equatorward of the observing station. These are (a) a reduction in the maximum plasma frequency of the F2 layer, foF2, and an associated rise in the virtual height of the F layer near the trough minimum; (b) a very steep gradient in electron concentration from the trough minimum towards the pole, often termed the cliff, and (c) considerably more Spread-F on the poleward side of the trough than near the trough minimum or on the equatorward side. These steep gradients in electron concentration can allow an ionosonde, normally providing information at or near vertical incidence only, to record data over considerable horizontal range.

Four separate types of ionogram sequences have been observed in which the main trough was identified moving from poleward to equatorward of Halley Bay at night; these have been labelled types 1 - 4. A further ionogram sequence in which the main trough was observed but remained poleward of Halley Bay and did not move significantly with respect to the station, has been identified; this stationary type is termed type S. In each of these types of patterns, the key features for identification listed above were observed. However, the differences in the ionogram patterns are (a) in the gradient of electron concentration on the equatorial side, (b) in the height of the main trough minimum compared with that of the poleward edge, and (c) in the amount of sporadic E associated with charged particle precipitation observed under the main trough.

F-region iso-ionic contours determined from a type 1 and type 3 main trough ionogram sequence are given in Fig. 1A and B respectively. Fig. 1A (after Bowman 1969) has been constructed from angle of arrival information obtained using ionograms from Ellsworth Station (78°S, 41°W; L = 4.5); the ice shelf on which the station is located being used as a natural interferometer (Bowman 1968). Similar ionogram sequences to those observed at Ellsworth Station are seen at Halley Bay. Fig. 1A shows steep gradients of electron concentration on both equatorward and poleward sides of the trough minimum. The plasma frequency ($f \propto N^{\frac{1}{2}}$) on the equatorward side is very high (~ 9 MHz) and that of the trough minimum lies between 2 and 3 MHz. The contour map in Fig. 1B has been constructed by evaluating the electron concentration profile at 15 minute intervals by the method proposed by Dudeney (1978). In this type, the steep gradient on the equatorward side is absent but the gradients on the poleward side are similar to those in type 1. In Fig. 1B, the height of the trough minimum is similar to that of the poleward edge and the trough minimum is a relatively small feature, sharply in contrast to Fig. 1A. E region features have not been included in the representation for clarity.

The identification of the five types of main trough is based solely on the pattern of E and F region traces seen on the ionograms. However, the analysis has revealed that an ordered set of conditions is associated with each type. These are summarised in Table 1. The type of main trough sequence observed depends upon the local time of occurrence which in turn depends primarily on magnetic activity. When magnetic activity is high (Kp ~ 5), type 1 crossings are seen, occurring in the

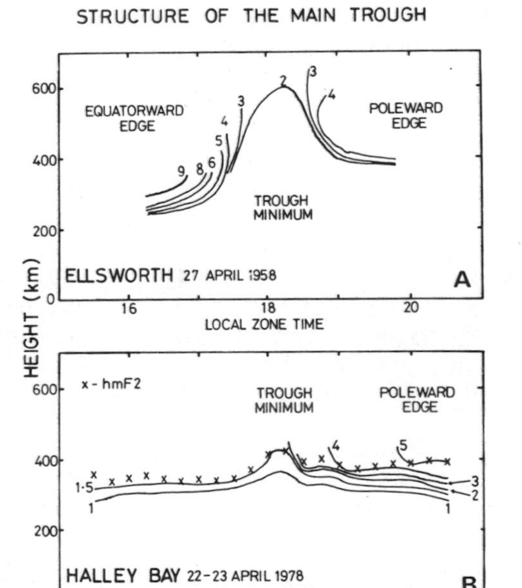

Fig. 1. Iso-ionic contours for type 1 (upper panel) and type 3 (lower panel) main trough. Numbers associated with contours indicate plasma frequency in MHz. Fuller description of Figure in text.

early evening; with lower levels of magnetic activity, the main trough passes over Halley Bay progressively later in the night. The time taken from first observation of the poleward edge of the main trough to the time when the poleward edge is overhead the station increases for each successive type, and for type S the poleward edge is not established overhead. Also sporadic E associated with charged particle precipitation and blackout (Piggott and Rawer 1978; Piggott 1975) is always observed with a type 1 crossing but less frequently as the trough is observed later in the night. The final row of Table 1 gives an indication of ionospheric conditions at Halley Bay after the poleward edge of the trough is overhead. Type 1 and, to a lesser extent, type 2 conditions following a trough crossing are highly variable, however high latitude ridges and trough sequences are observed. The level of magnetic activity required for a particular type of main trough sequence to be observed at Halley Bay is slightly lower at sunspot maximum compared with sunspot minimum.

The main trough was observed at Halley Bay on all nights during the months of April analysed, except under the very quietest magnetic conditions, Kp ≃ 0. The relative frequency of occurrence of each type of main trough crossing with season and

SUMMARY of FEATURES of MAIN TROUGH TYPES

	type 1	type 2	type 3	type 4	type S
TIME OF OCCURRENCE (local zone time)	1600 - 2000	2000 - 2300	2300 - 0100	0100 - 0500	usually after 0100
MAGNETIC ACTIVITY (Kp)	$\geqslant 5$	3 - 5	3 - 5	2	< 2
TIME TO CROSS HALLEY BAY (hours)	2	3	3	4	does not cross
STORM TYPES OF SPORADIC E	always	usually	seldom	none	none
CONDITIONS FOLLOWING TROUGH	blackout high lat. ridges very disturbed	similar to 1 but less severe	very extensive spread-F blackout rare	extensive spread - F	sunrise ends sequence

solar cycle was found to be mainly dependent upon the variations in level of magnetic activity for these epochs.

The virtual height of the poleward edge of the main trough is observed to be remarkably constant above Halley Bay, 360 km $\pm$ 28 km for 26 different occasions, irrespective of the type of main trough seen, season or solar cycle. Thus, assuming this height does not change during the passage of the main trough over Halley Bay, the velocity of the poleward edge can be calculated from the time rate of change of the slant range as the feature moves from oblique to overhead the station. If the assumption of constant virtual height is in error by 100 km, the resultant error in the velocity of the poleward edge will only be about $\pm$ 10%.

The mean velocity of the main trough in the hour immediately preceding the poleward edge establishing itself over Halley Bay ($L = 4.2$) has been calculated for 66 occasions which are roughly evenly distributed in local time throughout the night in the months of April analysed. The velocities, so determined, have been converted into a time rate of change of invariant latitude and hence of L value and are shown in Figure 2. The equatorward movement of the poleward edge of the main trough is rapid in the early evening hours ($\sim$ 2 L shells per hour) which is for magnetically disturbed conditions (see Table 1) and the time taken for the poleward edge to move overhead is about 2 hours. In the early morning, the equatorward movement of the main trough is very slow, hence it takes considerably longer to move overhead of the station at these times which correspond to magnetically quiet conditions. A simple 'tear drop' model of plasmapause

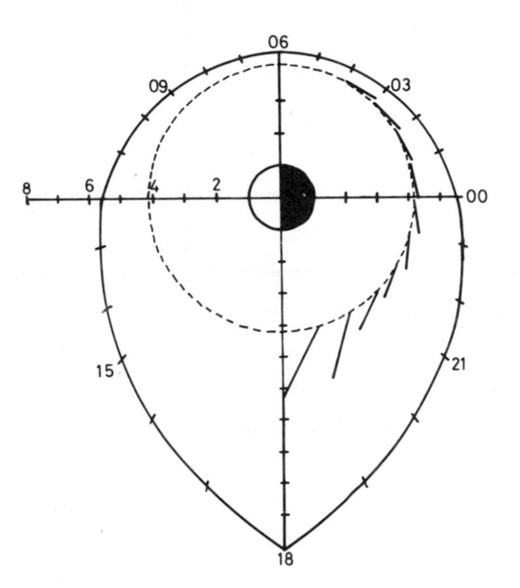

Fig. 2. Bars show the mean velocity of main trough in the local hour immediately preceding the time when the poleward edge of the main trough is established over Halley Bay (L = 4.2), the locus of which is shown by the broken line. Radial axis is L shell. Locus of 'tear-drop' model of the plasmapause for quiet magnetic activity indicated by continuous line.

position (Kavanagh et al. 1968) for very quiet magnetic conditions has been included in Fig. 2, since it is believed that the plasmapause lies close to the geomagnetic flux tube through the main trough (Rycroft and Thomas 1970). The velocities of the plasmapause with changing local time predicted by this model and the movements of the poleward edge of the main trough are very similar at all hours. However, there is considerable disagreement between the positions of the two phenomena especially before midnight. A more compressed 'tear-drop' shaped plasmasphere, dependent upon the level of magnetic activity would be more appropriate, than the quiet time model shown in Fig. 2.

SUMMARY

Detailed analysis of Halley Bay ionograms has shown that there are five different types of main trough sequence, each of which occurs under different conditions of magnetic activity and that magnetic activity determines the local time at which the main trough is seen. These different types can be interpreted as variations of the electron concentrations and the heights of the equatorward edge, the poleward edge and the trough minimum. It is suggested that, in principle, if the level of magnetic activity is known, the time at which the trough occurs, the shape of the trough and the movement of the main trough can be predicted. A practical scheme to do this would significantly increase the reliability of radio wave propagation in the vicinity of the trough. The movements of the main trough are shown to be similar to those of the plasmapause as predicted by the 'tear-drop' model.

ACKNOWLEDGEMENTS

Published by permission of the Director of the British Antarctic Survey, Natural Environment Research Council, Madingley Road, Cambridge CB3 OET, UK.

REFERENCES

Bowman, G.G. 1968, J. Atmos. Terr. Phys., 30, 1115.
Bowman, G.G. 1969, Planet. Space Sci., 17, 777.
Dudeney, J.R. 1978, J. Atmos. Terr. Phys., 40, 195.
Hunsucker, R.D. 1980, this volume.
Kavanagh, L.D., Freeman, J.W. and Chen, A.J. 1968, J. Geophys Res., 73, 5511.
Muldrew, D.B. 1965, J. Geophys. Res., 70, 2635.
Piggott, W.R. 1975, High Latitude Supplement to URSI Handbook of Ionogram Interpretation and Reduction, Rep. UAG-50, World Data Centre A for Solar Terrestrial Physics, Boulder, Colorado.
Piggott, W.R. and Rawer, K. 1978, URSI Handbook of Ionogram Interpretation and Reduction, Rep. UAG-23A, World Data Centre A for Solar Terrestrial Physics, Boulder, Colorado.
Rycroft, M.J. and Burnell, S.J. 1970, J. Geophys. Res., 75, 5600.
Rycroft, M.J. and Thomas, J.O. 1970, Planet. Space Sci., 18, 65.
Titheridge, J.E. 1976, J. Geophys. Res., 81, 3227.
Tulunay, K.Y. and Sayers, J. 1971, J. Atmos. Terr. Phys., 33, 1737.
Tulunay, K.Y. and Grebowsky, J.M. 1978. J. Atmos. Terr. Phys., 40, 845.
Wrenn, G.L. and Raitt, W.J. 1975, Ann. de Geophys., 31, 17.

A REPORT FROM THE INTERNATIONAL SOLAR-TERRESTRIAL PREDICTION WORKSHOP

Boulder 23-27 April 1979

E V Thrane

Norwegian Defence Research Establishment,
N-2007 Kjeller, Norway

ABSTRACT

A brief report is given on the aim and organization of the Solar-Terrestrial Prediction Workshop, held in Boulder, Colorado in April 1979.

INTRODUCTION

The International Solar-Terrestrial Prediction Proceedings and Workshop Program was carried out in the period 1978-80 and represents a major effort in the field. The organizers of the present NATO-ASI therefore thought it would be useful to bring the workshop to the attention of the ASI participants. This paper will therefore briefly review the workshop organization and aims, and point at some of the results obtained. To the author's mind, the workshop proved interesting and successful, both from a practical and a scientific point of view, and as an example of an effective way of organizing international collaboration.

WORKSHOP ORGANIZATION AND AIMS

The Chairman of the Workshop, and the editor of the proceedings was Dr R F Donnelly at the NOAA Space Environment Laboratory, and this laboratory hosted the Workshop, which was held April 23-27 at the College Inn in Boulder, Colorado. A number of US and international organizations were science sponsors of the Workshop, whereas financial support was provided from US sources.

C. S. Deehr and J. A. Holtet (eds.), Exploration of the Polar Upper Atmosphere, 471–480.

The objects of the Workshop were:

1 To determine and document the current state-of-the-art of solar-terrestrial predictions, the application of these predictions, and the future needs for solar-terrestrial predictions.

2 To encourage research, development and evaluation of solar-terrestrial predictions

3 To provide the means for in-depth interaction of prediction users, forecasters and scientists.

These objects were accomplished in the following way; starting one year before the workshop meeting:

1 Review papers were invited from groups that regularly issue predictions, and from groups that use the predictions. These reviews describe the current practice for making predictions.

2 Working groups were established for 14 different topics. These groups worked by correspondence throughout the year before the meeting, and concentrated on deriving recommendations for future developments of solar-terrestrial predictions. A list of the Program Committee members and the working groups and their leaders is given in Table 1.

3 An open call was issued for contributed papers relating to solar-terrestrial predictions. The purpose was to encourage workers in the field to direct their research to pertinent topics in the year before the meeting. Preprints were distributed to and exchanged between the working groups.

4 During the one week meeting arranged in Boulder an intensive study was made of the work accomplished in the preceding year. No formal papers were presented at the meeting, but working sessions were organized. The working groups met separately and jointly to prepare their final reports.

5 The invited reviews, the contributed papers and the working group reports were published in a four volume Proceedings. These volumes contain:

Solar-Terrestrial Predictions Proceedings

Volume I : Prediction Group Reports
Volume II : Working Group Reports and Reviews
Volume III: Solar Activity Predictions
Volume IV : Prediction of Terrestrial Effects of Solar Activity

Table 1

PROGRAM COMMITTEE

F. E. Cook, IPS, Sydney, Australia	A. B. Severny, Crimean Observ., U.S.S.R
A. D. Danilov, Moscow, U.S.S.R.	M. A. Shea, AFGL, Bedford, Mass.
Y. Hakura, Tokyo, Japan	P. Simon, Meudon, France
G. R. Heckman, NOAA SEL, Boulder	W. C. Snoddy, NASA-MSFC, Huntsville
A. P. Mitra, NPL, New Delhi, India	H. Tanaka, Tokyo Observ., Japan
G. A. Paulikas, Aerospace, Los Angeles	J. M. Wilcox, Standford University
N. V. Pushkov, IZMIRAN, U.S.S.R.	D. J. Williams, NOAA, SEL, Boulder

WORKING GROUPS and LEADERS

Long-Term Solar Activity Predictions: Mr. P. S. McIntosh, NOAA ERL SEL, Boulder, Colorado 80303 USA

Short-Term Solar Activity Predictions: Dr. P. Simon, Observatory, Meudon, France

Solar Wind and Magnetosphere Interactions: Dr. C. T. Russell, Institute of Geophysics and Planetary Physics, University of California, Los Angeles, California 90024, USA

Geomagnetic Storms: Dr. S. I. Akasofu, Geophysical Institute, University of Alaska, College, Alaska 99701, USA

Energetic Particle Disturbances: Dr. G. A. Paulikas, The Aerospace Corporation. P. O. Box 92957, Los Angeles, California 90009, USA

Magnetosphere-Ionosphere Interactions: Dr. Richard R. Vondrak, Radio Physics Lab., SRI International, Menlo Park, California 94025, USA

High-Latitude E- and F-Region Ionospheric Predictions: Dr. Robert D. Hunsucker, Geophysical Institute, University of Alaska, College, Alaska 99701, USA

Midlatitude and Equatorial E- and F-Region Ionosphere Predictions: Dr. Charles M. Rush, Institute of Telecommunications Sciences, National Telecommunications and Information Administration, Boulder, Colorado 80303, USA

D-Region Ionospheric Predictions: Dr. Eivind Thrane, Division for Electronics, Norwegian Defence Research Establishment, P. O. Box 25, Kjeller, Norway

Solar-Weather Predictions: Dr. K. H. Schatten, Department of Astronomy, Boston University, 725 Commonwealth Avenue, Boston, Massachusetts 02215, USA

Communications Predictions: Dr. A. P. Mitra, Radio Science Division, National Physical Laboratory, Hillside Road, New Delhi – 110012, India

Subsection on Ionosphere-Reflected Propagation: Dr. B. M. Reddy, Radio Science Division, National Physical Laboratory, Hillside Road, New Delhi 110012, India

Subsection on Trans-Ionosphere Propagation: Dr. John A. Klobuchar, AFGL-PHP, Hanscom Air Force Base, Bedford, Massachusetts 01731, USA

Geomagnetic Applications: Dr. Wallace H. Campbell, USGS Box 25046, Denver Federal Center MS 946, Denver, Colorado 80225, USA

Space-Craft Environment and Manned Space Flight Applications: Dr. A. L. Vampola, The Aerospace Corporation, P. O. Box 92957, Los Angeles, California 90009, USA

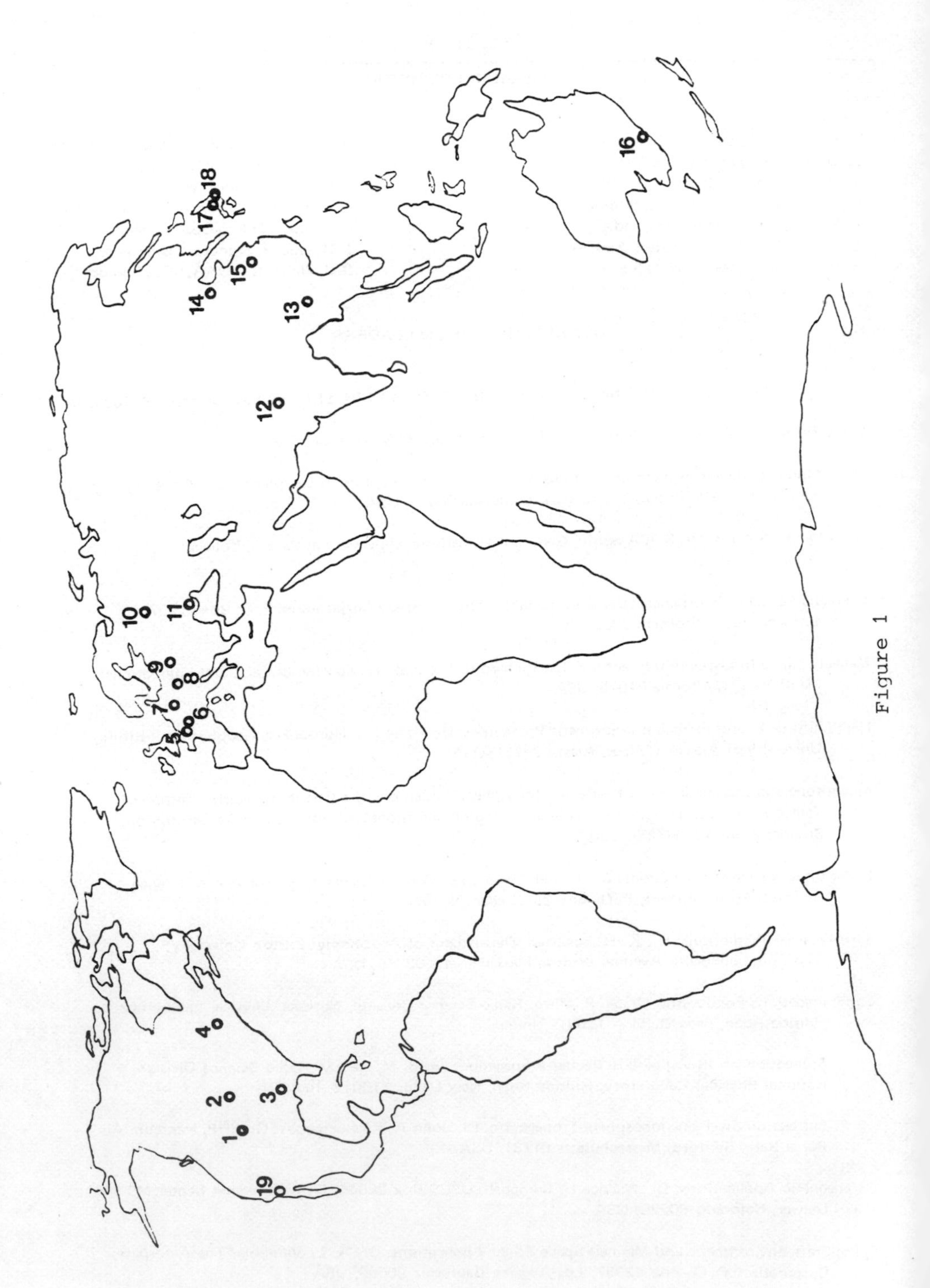

Figure 1

Figure 1. Observatories that reported to the Workshop

1 Space Environment Services Center, NOAA, Boulder, USA
2 Air Force Global Weather Central, Nebraska, USA
3 NASA Marshall Space Flight Center, Alabama, USA
4 Ottawa Magnetic Observatory, Canada
5 The Observatory of Meudon, France
6 Centre National d'Etudes des Telecommunications, Lannion, France
7 The Research Institute of the Deutsche Bundespost, Darmstadt, Germany
8 Geophysical Institute, Czechoslovak Academy of Sciences, Praha
9 Space Research Center, Warsaw, Poland
10 Institute of Applied Geophysics, Moscow, USSR
11 Crimean Astrophysical Observatory, USSR
12 National Physical Laboratory, New Delhi, India
13 Yunnan Observatory, Kunming, China
14 Peking Observatory, Peking, China
15 Purple Mountain Observatory, Nanking, China
16 Ionospheric Prediction Service, Sydney, Australia
17 Radio Research Laboratory, Japan
18 Toyokyawa Observatory, Japan
19 Naval Oceans Systems Center, San Diego, USA

These proceedings are for sale by the Superintendent of Documents, US Government Printing Office, Washington DC 20402. (Order by SD Stock No 003-023-00041-9.)

The workshop proceedings will serve as a valuable reference document with a total of about 2000 pages. It should be noted that the published papers have been subjected to normal refereeing procedures.

PREDICTION GROUPS

Figure 1 shows the geographical location of the prediction groups that responded to the invitation to review their activities for the Workshop. Most of the 19 groups issue predictions of solar phenomena, 10 of them also predict ionospheric and geomagnetic parameters, and a few issue "real time" warnings to the users. It is important to note that Figure 1 does not give a complete picture of the use of solar-terrestrial predictions throughout the world. Many countries, both in South-America, Africa and Europe, use data from the solar observatories to issue their own ionospheric predictions. For example the Norwegian Defence Research Establishment issues monthly frequency predictions to Norwegian civilian and military users of short wave communications, using predicted sunspot numbers and a standard computer code recommended by CCIR (1970).

HOW A PREDICTION IS MADE

It may be of interest here briefly to review the process that leads to a prediction. Heckman (1979) has described the predictions at the Space Environment Services Center in Boulder, and the following summary is based upon his article.

Data acquisition

The first step is, of course, the data acquisition. The most important data is obtained by careful monitoring from the ground of optical emissions from the solar surface. In addition X-ray and EUV intensity measured in satellites and radio emissions measured on the ground are essential for a full description of the sun's electromagnetic radiation. Measurements of energetic particles in the solar wind, and in the magnetosphere are used when satellites are available. The conditions in the ionosphere and magnetosphere are monitored from the ground and from satellites through radio and magnetic measurements.

Cosmic ray data from neutron monitors, and scintillations of radio stars caused by perturbations in the interplanetary plasma are also used, as well as the passage of sector boundaries in the interplanetary magnetic field.

Digesting and summarizing collected information

The second stage in the process of making a prediction is to digest the collected information and summarize it in the form of indices and activity reports. Table 2 (Heckman 1979) lists the indicies and reports compiled by the Boulder Center.

Table 2. Observed indices and activity summaries

- Solar Active Region Summary Report
- Sunspot Number
- Flare (and Other Event) Lists
- Solar Neutral Line Analysis and Synoptic Maps
- Ten-Centimeter Flux
- Solar Proton Events and Proton Flux
- SST Radiation Levels
- Geomagnetic A- and K-indices
- Substorm Log
- Sector Boundaries (at 1 A.U.)

In addition ionospheric conditions are characterized by parameters such as critical frequencies, absorption indicies, noise levels etc.

Predicting future conditions

Based on the compiled information, prediction of future conditions are made, most often in the form of values of the above indicies, but also in the form of qualitative statements describing, for example, ionospheric conditions. The present day predictions are mainly based upon the use of empirical formula, and particularly for short time predictions, upon the judgement and experience of the forecaster. Only in a few areas are physical models sufficiently well developed to be of real value in the predictions. Table 3 lists the predictions issued by the Boulder Center.

In addition, a number of alerts and warnings are issued, predicting solar flares and magnetic disturbances with very short lead times (minutes, hours).

The ionospheric prediction centers use the above information to predict the effects of solar conditions on ionospheric parameters. For example, sunspot number and geomagnetic indicies are used with ionospheric models to predict maximum useable frequencies, propagation modes, absorption, signal levels, circuit reliability and total electron content. The alerts and warnings can indicate the occurrence of Sudden Ionospheric Disturbances, substorm activity and Polar Cap Absorption events.

Table 3. Prediction products (Lead Time given in parentheses)

LONG TERM SOLAR ACTIVITY AND SOLAR RADIATION LEVELS

- Smoothed sunspot number (1 month–10 years)
- Geomagnetic activity and ten-centimeter flux (1 month–10 years)
- General level of solar activity (27 days)

SOLAR ACTIVITY – SHORT TERM

- Solar Flares (1, 2, 3 days)
- Solar proton events (1, 2, 3 days, PFP*)

SOLAR RADIATION LEVELS – SHORT TERM

- Ten-centimeter flux (1, 2, 3 days)

GEOMAGNETIC DISTURBANCE LEVELS

- A, K-indices (1, 2, 3 days)
- Time of sudden commencements (PFP), storm size (PFP)

*PFP: Post Flare Prediction – A prediction of a flare consequence once the flare has occurred.

Distribution to users

Efficient and timely distribution of the predictions and alerts is essential to the user. The distribution systems used by the Boulder Center include telephone, teletype, computer links, WWV shortwave broadcast and mail.

Verification of forecasts

To test the reliability and usefulness of the prediction is an important but difficult task, and great care must be taken to use a proper statistical analysis. For evaluation of flare forecasts, the Boulder Center uses the Brier P-score (Brier 1950). The results indicate that the forecaster's ability to recognize the flare potential of an active solar region is very good, but his success in predicting the time of the flare is less satisfactory.

WHY ARE PREDICTIONS IMPORTANT?

The International Solar-Terrestrial Prediction Workshop has demonstrated that there is world-wide interest in the field, and that considerable effort is directed towards improving present-day forecasting and prediction of solar phenomena and their effects on the terrestrial environment. How are predictions used today, and what are the future prospects? In the following we list the most important uses.

Present uses of long term predictions (time scales of months, years, solar cycles):

- Planning of communication systems using radio waves. Ionospheric predictions are necessary to determine frequency allocations, transmitter powers, antenna types, expected depth of fading, scintillations etc.

- Planning of navigation systems. Important parameters to be estimated are: geographical coverage, ionospheric reflection height variability with resulting expected phase stability and accuracy in positioning.

- Planning of space craft and space missions. The life time of spacecraft against drag is determined by the air density in the upper atmosphere, and the lifetime against electromagnetic and particle radiation is determined by the intensity of solar radiation.

- Planning and design of electrical power stations and power distribution systems, as well as design of pipelines, telephone lines etc. Strong solar activity may induce large electric and magnetic fields that affect such systems, particularly in high latitudes, and can cause power failures, corrosion of pipelines etc.

Present uses of short term forecasters and warnings (time scales of hours, days, weeks)

- Daily and weekly forecasts of ionospheric changes are used to help communicators and navigators to make efficient use of their systems.

- Real - or near real time warnings to communicators and navigators of ionospheric disturbances, black-outs, sudden phase anomalies etc, help the operators to identify the problem and take proper action.

- Forecasts and warnings of geomagnetic disturbances are of importance for the operation of power distribution systems, for geomagnetic surveys, navigation etc.

- Forecasts and warnings of radiation hazards in space are of importance for space missions, particularly for astronauts during extravehicular activity.

- Forecasts of geophysical conditions are needed for scientific experiments, such as launching of instrumented rockets, satellites and balloons.

Possible future developments

In the Workshop Proceedings many detailed recommendations are made on the future development of solar-terrestrial predictions. Based upon the Workshop results the author will try to summarize what, from a personal point of view, should be the general aim in future work.

One immediate goal is to improve the reliability of existing prediction methods by improving the data bases, the distribution system etc.

However, the most important objective is to improve our physical understanding of the sun-earth system to the extent that present day predictions, based on empirical or semi-empirical methods, can be replaced by predictions based upon physical models. The transition will no doubt be gradual and take time and effort, but if successful the consequences will be of great importance. To day the question of possible sun-weather relationships is a highly controversial issue. If we can gain physical insight into the mechanisms that determine long or short term climatic changes, the results will have economic, social and political as well as scientific consequences.

Whatever the outcome of this development, it seems likely that the future will bring improved realtime monitoring of interplanetary space and the earth's environment, and that we can expect greatly improved short time warnings of geophysical events, as well as improved and efficient distribution of the information to the users.

REFERENCES

Brier, G.W. 1950, Mon. Wea. Rev., 78, 1.

CCIR Report 252-2, UIT, Geneve 1970.

Heckman, G., 1979, p. 322, Vol 1 in Solar Terrestrial Prediction Proceedings, US Government Printing Office, Washington DC.

E AND F REGION PREDICTIONS FOR COMMUNICATION PURPOSES AT HIGH LATITUDES

Robert D. Hunsucker

Geophysical Institute
University of Alaska
Fairbanks, Alaska 99701

INTRODUCTION

This paper presents a summary of the report of Working Group C2 [Hunsucker et al. 1980] (High-latitude E and F Region Ionospheric Predictions) at the International Solar-Terrestrial Predictions (ISTP) Workshop held in Boulder, Colorado, April 23-27, 1979. The members of Working Group C2 were: Chairman, R. Hunsucker; Members: R. Allen, P. Argo, R. Babcock, P. Bakshi, D. Lund, S. Matsushita, G. Smith, A. Shirochkov, and G. Wortham. These conference report results will be supplemented by the inclusion of material published since the end of the ISTP Workshop. ISTP WG C2 was concerned about the morphology, phenomenology and parameter prediction of the E and F-layers at latitudes greater than $\Lambda = 60°$. Several survey papers treating this topic have been published in the period starting with the International Geophysical Year (IGY) such as those by Agy [1957], Penndorf and Coroniti [1959], Kirby and Little [1960], Gassmann [1963], Landmark [1964], Hunsucker [1967], Folkestad [1968], Hunsucker and Bates [1969], Bates and Hunsucker [1974], Thrane [1979], and Hunsucker [1980]. Unfortunately <u>most</u> of the E and F layer information described in the preceding <u>reports</u> pertains to the latitude range $\Lambda \simeq 60°$-$75°$. Our knowledge of the main features of ionization below the F2 peak in the region of $\Lambda \simeq 75°$-$90°$ is at best sketchy, making realistic predictions on most HF transpolar paths extremely difficult. The proposed siting of an incoherent scatter radar at $\Lambda \simeq 75°$ [Banks and Evans, 1979] could provide very useful information on F-region morphology up to $\Lambda \simeq 87°$ and E-region morphology up to $\Lambda \simeq 80°$.

C. S. Deehr and J. A. Holtet (eds.), Exploration of the Polar Upper Atmosphere, 481–494.

PHYSICAL PROCESSES IN THE HIGH-LATITUDE E AND F REGIONS

Figure 1 [from Banks and Evans 1979] illustrates schematically the high-latitude processes and regions of interest to ionospheric scientists and communicators.

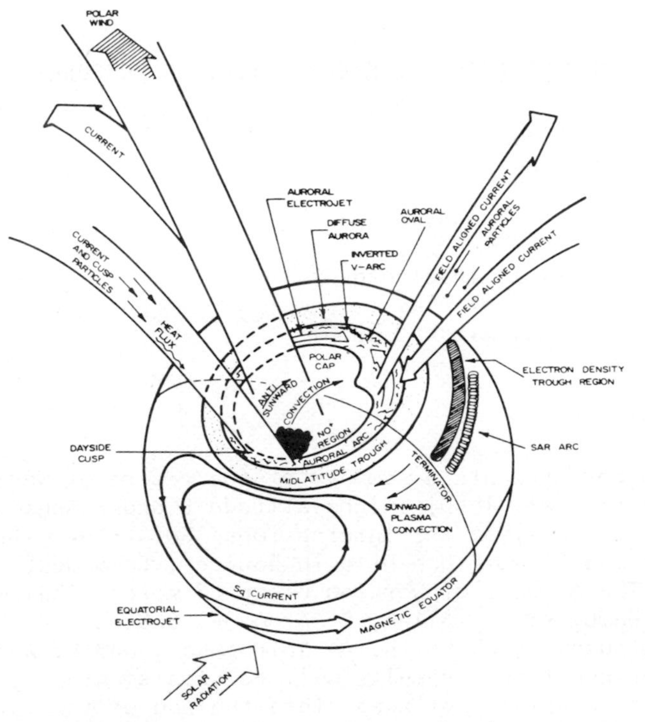

FIGURE 1: A schematic view of high latitude processes of special interest to ionospheric scientists and communicators.

In the mid- and low-latitude E and F regions, the primary source of ionization is solar EUV and UV. The day-to-day variations in the ionosphere essentially reflect changes in the neutral atmosphere density and winds. Electric fields occasionally become an important modulating force during disturbed periods. At high latitudes, the magnetosphere-ionosphere interaction is the dominant dynamic force. There are two primary processes. The precipitation of energetic particles is an important source of ionization, particularly in the E region. The strong electric fields which arise from the interaction of the interplanetary magnetic field (IMF) and the geomagnetic field drive a two-celled ion convection pattern with anti-sunward ion flow across the pole and a return flow along the auroral oval. At E-region heights, the ion convection is restrained by collision with neutrals. The result is a net current flow known as the auroral electrojet which creates Joule heating of the neutral atmosphere. Changes in the neutral density and winds associated with this heat act as feedback mechanisms that change the density and structure of the high-latitude ionosphere.

Each of the main features of the high latitude E and F regions is related to particular aspects of the magnetospheric-ionospheric interactions.

The processes occurring in the auroral oval are the most dynamic and exciting and are clearly the best understood. With the advent of satellite observations, the physical processes that drive the various auroral phenomena have been examined fairly extensively. There is a good understanding of these processes and their relation to the various features of the auroral zone appearing in morphological models. Within the auroral zone, the variations in the F region are probably the least known phenomena, particularly during the disturbed periods. The strong E region created by particle precipitation has limited the ability to observe the F region using ground-based observations. The expanded use of transionospheric propagation in high-latitude regions makes knowledge of F-region variations of increasing importance.

The other main features of the high latitude ionosphere have only recently become the focus of concerted research efforts. The dynamics of the main ionospheric trough and the light ion trough are not well understood. There is no clear consensus on how and why they are formed, and how, or if, they are related and interact. The number of observations of the cusp region is still small enough that there is no clear picture of exactly what the morphology of the energies of the precipitating particles is and how precipitation varies with geomagnetic activity. The dynamics of the cusp as it moves with variations in the IMF are not well understood. The polar cap is probably the least active of the high-latitude regions, but it is also the least observed and the least understood.

Another important factor in the dynamics of the high-latitude ionosphere that is not well understood, and often ignored by ionospheric physicists, is the strong interaction between the ionosphere and the neutral atmosphere. Changes in the neutral density and winds that are driven by ionospheric processes not only are a feedback into the high latitude ionospheric dynamics, but the energetics of auroral processes are an important driving force of the global thermospheric circulation and will indirectly affect the ionosphere at lower latitudes.

Essential Morphology Affecting Communications

From the point of view of the geophysicist, a lot still

remains to be done before we have a clear understanding of the morphological features of the high-latitude E and F regions. The best determined feature of this region is the position of the auroral oval, determined by visual observations, all-sky camera data, and, more recently, satellite photographs. Almost as well defined are three auroral absorption regions, one just equatorward of the diffuse aurora in the midnight sector, another in the pre-noon sector, and a third in the early morning sector during disturbed periods. All have been studied extensively, principally by an international network of riometers, but recently by in-situ particle measurements. Both the auroral oval and the auroral absorption regions are well represented by statistical (computer version) models.

Our knowledge of the main ionospheric trough is in much worse condition. Although it has been studied since the previous solar minimum with ionosonde data, little is known of its variations with solar cycle, its disturbance behavior, the positional variations of the plasmapause, and the variation of ionospheric parameters over this feature. Lack of consistent synoptic data has prevented development of more than first-order statistical models of an F-region trough.

Very much more must be done in studying the localized features of the E and F2 layers, particularly for the polar region of the Southern Hemisphere. The behavior of the F2 region during magnetospheric substorms is so dramatic that we now know it comparatively well on the night side; however, our studies are very incomplete on the day side. For very strong ionospheric disturbances, e.g., PCA or strong aurora, our data from ground-based ionosondes is almost always blacked out; hence, we know even less about the F region during those periods. Recent data from instruments such as incoherent scatter and in-situ probes are beginning to identify and delineate some of the critical parameters of the high-latitude E- and F-region features, but there are not yet sufficient synoptic data to build either descriptive or predictive models.

From the point of view of the system engineer, the morphology should address the effects of the high-latitude ionosphere on his particular system in terms of his system's parameters. He almost always would like these derived from a comprehensive synoptic data base identical to his system. It is very doubtful that this exists, even for standard narrow band HF communication. Even parameters of critical design to certain systems, such as joint availability of various E and F modes, exist only in restricted time-space segments. In terms of those features that are of critical importance to advanced systems, like spread-F fading statistics for the cusp region or the poleward edge of the trough, the scattering cross section of sporadic-E in the midnight sector and slant-E in the polar cap; not very much high quality data exist.

As a result, morphological descriptions of features important to the system designer and user <u>do not exist</u> except in very simplistic qualitative form.

The propagation effects of the main ionospheric disturbances are listed in Table 1 [Thrane 1979]. Figures 2 and 3 illustrate the effects of one auroral oval effect (intense E-region ionization) on radiowave propagation. They display the Chatanika electron density profiles (plot on left side), Chatanika all-sky camera pictures (inset in lower left), and College ionograms at 0706 UT (2106 Alaska Standard Time (AST)) and 0800 UT (2200 AST). Figure 2, showing the simultaneous data for 0706 UT (2106 AST), illustrates the "background level" of E-region ionization of $\approx 0.2 \times 10^6$ electrons/cm^3, the right insert a fE_s of 7.5 MHz and the ASC indicating a discrete auroral arc north of Chatanika. At 0800 UT (2200 AST) in Figure 3, the ASC shows a discrete auroral arc in the beam of the Chatanika radar resulting in an E-region electron density of 1.2×10^6 el/cm^3 and an fE_s of 11.2 MHz. Using the earth-curvature-limited secant factor of 5.0 results in an E-layer maximum-usable-frequency (MUF) of 56 MHz. Maximum observed frequencies (MOF's) of 46 MHz have been recorded on transpolar paths instrumented with forward sounding systems [Hunsucker and Bates 1969].

MODELS FOR HIGH-LATITUDE PROPAGATION PREDICTIONS

Features Which Should Be Modelled

Models are required for long-term prediction of propagation for ionospherically supported systems (VLF, HF, and VHF scatter) and for satellite systems. ISTP Working Group C2 focused on predictions for HF systems, since the D-region (which determines VLF propagation) and F-region irregularities (which cause satellite scintillations) were addressed by other groups. However, the results should be combined.

Among the high-latitude ionospheric features that should be modeled for HF predictions are the following:

a. Location and extent of auroral oval as a function of magnetic activity.

b. E and F-region parameters in auroral oval, mid-latitude trough, daytime cusp, and polar cap.

c. F-region parameters, important for ducting.

d. Profiles of N_e versus height between the E and F2 regions; of importance for prediction of ducted propagation.

TABLE 1: Ionospheric Disturbances

Disturbance	Propagation effects	Time and duration	Possible cause
Sudden Ionospheric Disturbance (SID)	In sunlit hemisphere, strong absorption, anomalous VLF-reflection, F-region effects	All effects start approximately simultaneously. Duration - 1/2 hour	Enhanced solar x-ray and EUV flux from solar flare
Polar Cap Absorption (PCA)	Intense radiowave absorption in magnetic polar regions. Anomalous VLF-reflection	Starts a few hours after flare. Duration one to several days.	Solar protons 1-100 MeV
Magnetic Storm	F-region effects; increase of foF2 during first day, then depressed foF2, with corresponding changes in MUF. E-region effects, storm E_s, D-region effects, enhanced absorption. VLF anomalies.	May last for days with strong daily variations	Interaction of solar low energy plasma with earth's magnetic field, causing energetic electron precipitation.
Auroral Absorption (AA)	Enhanced absorption along auroral oval in areas hundred to thousand kilometers in extent. Sporadic E may give enhanced MUF.	Complicated phenomena lasting from hours to days.	Precipitation of electrons with energies a few tens of keV.
Relativistic Electron Precipitation (REP)	Enhanced absorption and VLF anomalies at sub-auroral latitudes.	Duration 1-2 hours.	Precipitation of electrons with energies of a few hundred keV.
Travelling Ionospheric Disturbances (TID)	Changes of foF2 with corresponding changes of MUF-sometimes periodic.	Typically a few hours.	Atmospheric waves.
Winter anomaly (WA)	Enhanced absorption at midlatitudes.	One to several days.	Probably many causes, such as changes in concentration of minor species, temperature changes, particle precipitation.
Stratospheric Warming	Changes in absorption, VLF anomalies.	Days or weeks, in late winter.	Changes in global circulation pattern.

e. Auroral absorption, PCA behavior (D-region effects).

f. E and F-region irregularities that cause diffuse multipath (spread-F, rapid flutter fading, and off-great-circle propagation).

In the auroral region, models must be detailed enough for raytracing. That is, the shape of the profile and the steep gradients must be represented in three dimensions. Predictions of MUF/LUF using virtual geometry (a priori mode definition and location of reflection points) are meaningless because strong horizontal gradients cause large changes in mode geometry and therefore in D-region crossing points and absorption estimates.

Presently Existing Models

Several models are available for the six features listed above. The ITS-78 (also Miller-Gibbs) model is usually taken as the basic model in the United States and modified either by superposition of more detailed features or by revision of the coefficients themselves. The ITS model is relatively poor in the auroral region. Data accumulated with ground-based and airborne ionosondes by AFGL and RADC (formerly AFCRL) have been used to develop an improved model. This model has recently been further modified by SRI using incoherent-scatter radar (ISR) data from the Chatanika radar near College, Alaska. The SRI model retains the foF2 and hmF2 values in the RADC 1976 polar model and incorporates an auroral E-layer (essentially equivalent to blanketing E_s) and valley from ISR data, as well as an absorption model that varies with magnetic activity [Vondrak et al., 1977].

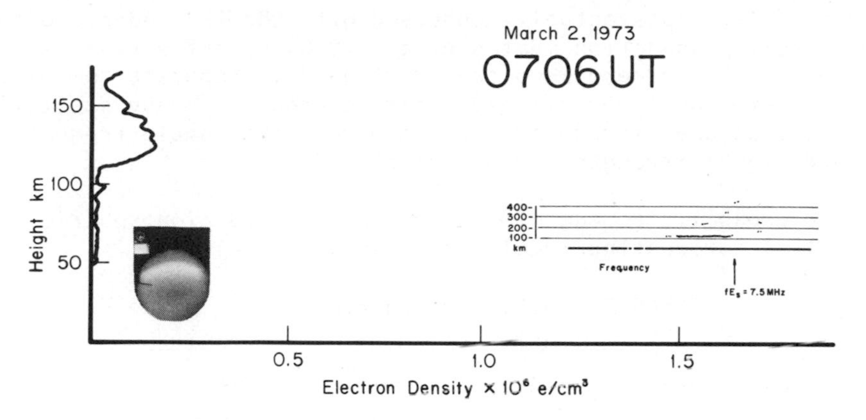

FIGURE 2: Simultaneous Chatanika radar electron density profile, Chatanika all-sky camera picture, and tracing of College ionosonde for 0706 UT 2 March 1973 [Hunsucker 1975].

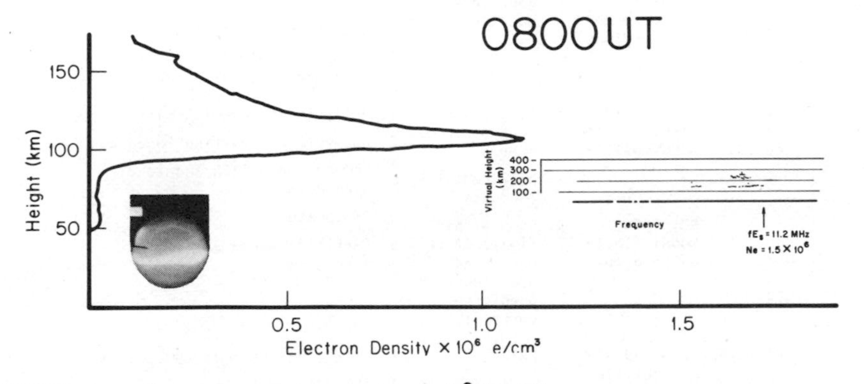

FIGURE 3: Same as above but at 0800 UT.

The USSR has developed statistical models [see, for example, Kovalevskaya and Zhulina, 1979; Troshichev et al. 1979; Avdiushin et al. 1979; and Kerblay and Nosnova 1979]. Differences have been noted between their predictions and those of the United States in the case of auroral absorption and foF2 in the mid-latitude trough during solar maximum. During magnetic disturbances, three distinct regions of auroral absorption have been identified by Soviet scientists, whereas the recent SRI absorption model displays two time/latitude regions of enhanced absorption. Gradients of foF2 in the trough at solar maximum were found to be large by the Soviet group, but relatively small by RADC (the ARCON model by Miller and Gibbs 1975). Further model verification is clearly needed.

The three-dimensional models developed by RADC and SRI are suitable for raytracing. In the original RADC model, the Ne(h) profile between layers must be assumed; the SRI model attempts to match profile shapes actually observed with the ISR radar. Off-great-circle propagation [Bates et al. 1966] by refraction can be predicted by three-dimensional raytracing. Separate computer codes are available for off-great-circle propagation by scattering from field-aligned irregularities; however, the usable frequencies and signal strength are not predicted.

The capabilities and limitations of the U.S. ionospheric propagation forecasting models are listed in Table 2.

TABLE 2: U.S. Ionospheric Models

Feature	Prediction	Adequacy	Techniques for Potential Improvement
1. Auroral oval	1. Feldstein oval for E region 2. (Extent differs for D and F regions)	1. Approximately equivalent to optical predictions 2. Good on 3-hour average	1. DMSP optical for real time (already done) 2. Particle counters (satellite) 3. Chatanika, EISCAT ISR's 4. Backscatter radar (Bates' work) 5. Synoptic analysis of bottom side and topside ionograms
2. E, F2 parameters	ITS coefficients, with RADC modifications SRI foEs model (Chatanika data)	1. foF2 good except during storms 2. h_mF2 contains anomalies 3. Good within limits of data base used if oval location predicted correctly	1. Analysis of more ISR data 2. Analysis of existing ionograms 3. Comparison of ISR and ionogram data 4. Photometers 5. Particle counters
3. Fl layer	1. χ and SSN analytic formula	1. Fair	1. Petrie and Stevens 1965 model 2. ISR data analysis 3. Ionogram analysis (synoptic)
4. Ne(h) profile	1. Analytic, assumed 2. SRI Chatanika model	1. Poor 2. Good within limits of data base	1. ISR data analysis and verification 2. Rockets (3) topside sounding
5. Auroral Absorption	1. Foppiano A.A. model 2. SRI modification of Foppiano model	1. No dependence on magnetic activity 2. Has Kp dependence but has not been verified against independent data	1. Riometer data available from several chains should be used for verification 2. A-1 absorption measurements 3. Oblique path HF signal strength
6. Irregularities (spread-F)	None in HF ray trace codes. Field aligned scatter geometry codes available	General morphology known but not applied in prediction (e.g. Davies, 1972; and Singleton, 1979.	1. Signal strength measurements coordinated with V-I observations 2. Satellite scintillations and spread-F ionogram correlation needed, both simultaneous and statistical.

PREDICTIONS

Present State of Knowledge

The ability to predict the propagation of HF radio waves along auroral and polar paths requires a model of the spatial distribution of electron density in the high-latitude ionosphere; one must then raytrace through this model to calculate the raypath connecting the end points. Because of the very definite spatial structure in the auroral and polar cap regions, the specific propagation predictions may be very sensitive to the positioning of the structures. It is difficult to determine whether washed out "strongly average morphologies" will give more useful results than misplaced features (averaged test results do compare better with averaged morphologies). Many system designers, however, are likely to be concerned with "worst-case" occurrence estimates, which are more accessible through even misplaced structures.

Several different types of systems are dependent upon propagation through the high-latitude ionosphere. Table 3 specifies some of these systems and the type of information needed for the systems to operate optimally.

The present prediction capability is statistical, using averaged synoptic data to define morphologies that ignore small scale (<100 km) structure and transient phenomena (<3 hr.). Therefore critical features such as auroral arcs and substorms are completely neglected. The models do contain auroral E, some auroral E_s, auroral absorption, and the F1 and F2 regions.

HF systems designers and communicators are concerned with the propagation facets mentioned in Table 3. The availability, multimode structure, ducting, absorption, noise, and non-great-circle propagation can be dealt with in some averaged, statistical sense (see, for example, Hatfield, 1979) that is useful to the

TABLE 3: Systems and Prediction Needs

System	Prediction Needs
HF Communication	Availability, dispersion, multi-mode, fading, noise, absorption, scattering cross section, ducting non-great-circle propagation
VHF/UHF Navigation and Communication	Fading (amplitude, phase), $\Delta TEC/\Delta lat$, angular deviation
Remote Sensing (HF)	Refraction, diffraction
VLF Navigation	PCA phase advances; auroral D-region effects

communications system designer. Since the auroral/polar cap models do incorporate varying magnetic activity conditions, the systems designer can test his system under various scenarios.

The accuracy of the predictions available through the use of these polar and auroral models has not been evaluated -- it is expected that on the average they will be significantly worse than the mid-latitude ITS-78 predictions [Barghausen et al. 1969]. This is because of the extreme sensitivity of the high-latitude ionosphere (with its many discrete features) to geomagnetic activity. Therefore, it is not likely that present day prediction capabilities would be very useful to the communication planner (e.g., frequency allocation) on a day-to-day or even month-to-month basis.

Remote sensing systems (HF radars) are also very dependent upon propagation through the polar/auroral ionosphere. The refraction and diffraction of signals by the discrete features will strongly affect the operation of these systems. In this sense, the usefulness of predictions for these systems is on approximately the same level as for HF communications.

Because the very high frequency transionospheric radio signals used in VHF navigation (e.g., GPS) are only slightly refracted (or time delayed) by the polar/auoral E-region, the main effects are F-region refractions and auroral scintillations. The dominant F-region effect is steep spatial gradients; therefore under the highly disturbed oval conditions, knowledge of its location is of primary importance. This is probably best measured in real time (as the southernmost edge of the auroral oval). The scintillation morphology is presently a statistical model (this area is covered in detail by Working Group E2 of the ISTP Workshop).

VLF navigation (e.g., Omega) is very dependent upon the condition of the polar D-region, and is only marginally affected by the E and F-regions. During solar proton events, the polar D-region undergoes increased ionization, which in turn appears as phase advances on transpolar paths. The effect is well understood, and available models give corrections to the phase advances that return the Omega accuracies to their nominal one nautical mile [see, for example, Argo and Rothmuller 1979]. The auroral disturbances happen in the upper D-region (>80 km), and so VLF propagation is only marginally affected.

Some recent advances in the area of HF propagation prediction and forecasting include development of the Ionospheric Communications Analysis and Predictions ("IONCAP") program at ITS/OT, Boulder, and the PROPHET program at NOSC, San Diego. The PROPHET system [Argo and Rothmuller 1979] inputs real time solar/terrestrial geophysical data to make regional propagation forecasts which are tailored to specific system usage. Figure 4 shows input information flow and the output (propagation forecasts). Figures 5 and 6 show two of the possible PROPHET output displays showing predicted time to recovery from a solar flare on the propagation path from a shore station to a ship at sea and a display of LUF, MUF and FOT (optimum traffic frequency) for a given ocean area. Table 4 shows the type of HF propagation predictions obtainable from the ITS/OT IONCAP program.

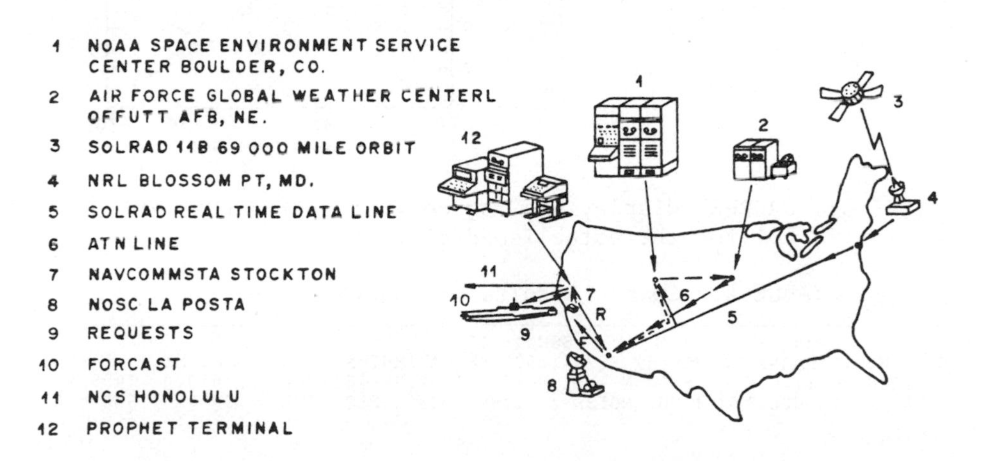

FIGURE 4: Information flow for the PROPHET environmental prediction terminal.

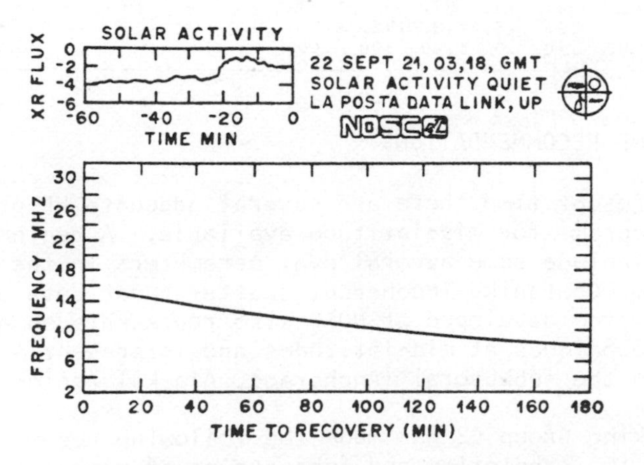

FIGURE 5: PROPHET display predicting time to recovery from a solar flare for the ship DURHAM.

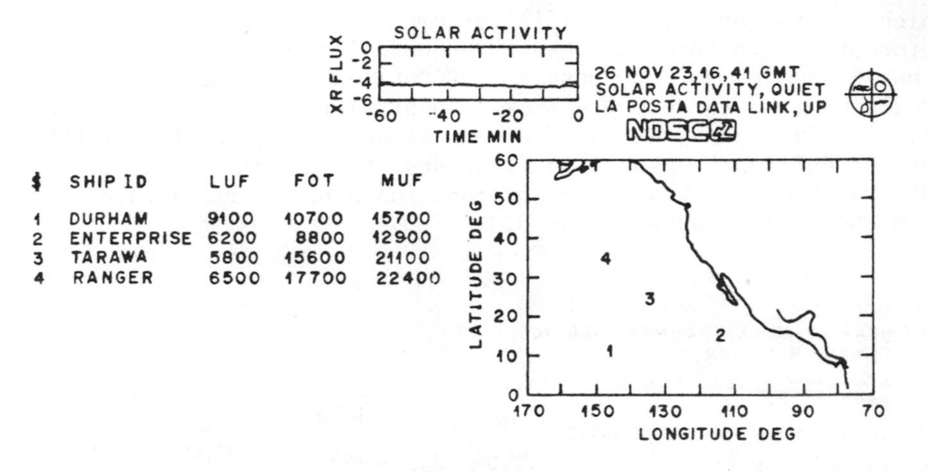

FIGURE 6: PROPHET display showing controller the LUF, MUF and FOT for the established circuits.

TABLE 4: Sample Results of IONCAP Computations

MAR SSN = 10.
MUNICH,W.GERMANY TO MIDLAT RCVR 1500 KM AZIMUTHS N. MI. KM
48.10 N 11.60 E - 48.00 N 31.84 E 82.86 277.98 810.0 1499.9
POWER = 10.000 KW 3 MHZ NOISE = -148.0 D8W PEQ. REL = .90 REQ. SNR = 58.0

UT	MUF												
10.0	14.3	2.5	4.0	6.0	9.0	13.0	19.0	26.0	0.0	0.0	0.0	0.0	FREQ
	1 E	2 E	1ES	2F1	1F1	1F2	1 E	1 E	-	-	-	-	MODE
	6.0	11.5	4.9	32.8	17.2	15.3	6.0	6.0	-	-	-	-	ANGLE
	.50	1.00	1.00	1.00	.92	.67	.00	.00	-	-	-	-	F DAYS
	136.	250.	181.	139.	123.	116.	268.	370.	-	-	-	-	LOSS
	34.	-95.	-21.	24.	44.	57.	-95.	****	-	-	-	-	DBU
	-96	-209	-140	-98	-81	-71	-227	-329	-	-	-	-	S DBU
	69.	-64.	11.	58.	79.	92.	-58.	****	-	-	-	-	SNR
	14.	132.	57.	19.	-6.	-19.	142.	223.	-	-	-	-	RPWRG
	.72	.00	.00	.50	.96	1.00	.00	.00	-	-	-	-	REL

CONCLUSIONS AND RECOMMENDATIONS

At the present time there are several adequate HF propagation prediction programs for mid-latitude available. A beginning has been made to include some auroral oval parameters in the SRI model which uses some Chatanika incoherent scatter radar data as input. The PROPHET system developed at NOSC also represents an advance in forecasting techniques at mid-latitudes and is presently being implemented in the subauroral (Anchorage, Alaska) region.

ISTP Working Group C2 has made the following recommendations for improving the prediction and forecasting of high-latitude E and F-region parameters:

1. Attempt to better understand the basic physical processes governing:

   a. Trough formation and disappearance
   b. Auroral absorption (particle precipitation patterns)
   c. Sporadic-E dynamics
   d. F-region anomalies (spread-F, lacunae, etc.)
   e. Cleft/cusp ionospheric dynamics
   f. Ionosphere/neutral atmosphere interaction

2. Devise a better morphological description of the high-latitude ionosphere (appropriate ionospheric and geophysical parameters):

   a. Main ionospheric trough (F-region); polar orbiting satellites, Chatanika, EISCAT and Millstone incoherent scatter radars, observations of poleward edge

   b. Auroral oval (sporadic and auroral-E); polar orbiting satellites, DMSP, TIROS-N, DE, ionosonde network, incoherent scatter radars, IMS magnetometer chain, HF radars, TRANSIT satellite, GPS satellite, photometers

   c. Cleft/cusp: UHF incoherent scatter radar (E and F-region anomalies), ionosondes, advanced ionospheric sounder, polar orbiting satellites, spectrophotometric observations, HF reception on polar paths

   d. Polar cap (E and F-regions): UHF incoherent scatter radar (E and F-region anomalies), ionosondes, advanced ionospheric sounders, polar orbiting satellites

3. Development and verification of improved long-term and near-real-time models using:

   a. High quality research data now available (Chatanika radar, wideband satellite, Millstone radar)

   b. The large synoptic data base existing (ionosonde network, all-sky camera network, magnetometers, riometers, HF backscatter, solar observations).

REFERENCES

Agy, V. L. and L. R. Teters 1979, Report NTIA-TM-79-22, ITS/NTIA, U.S. Department of Commerce, Boulder, Colorado.

Argo, P. E. and I. J. Rothmuller 1979, Solar-Terrestrial Predictions Proceedings, Vol. 1, 312.

Banks, P. M. and J. V. Evans 1979, An initial feasibility study to establish a very high latitude incoherent scatter radar, Utah State University, Logan, Utah.

Bates, H. F. and P. R. Albee 1966, Scientific Report UAG-R175, Geophysical Institute, University of Alaska, College, Alaska.

Bates, H. F., P. R. Albee and R. D. Hunsucker 1966, J. Geophys. Res., 71, 1413.

Bates, H. F. and R. D. Hunsucker 1974, Radio Science, 9, 455.

Donnelly, R. F. (Ed.) Solar-Terrestrial Predictions Proceedings, Volume 1, 1979: Prediction Group Reports; Volume 2, 1980: Working Group Reports and Associated Topical Reviews, U.S. Department of Commerce, National Oceanic and Atmospheric Administration, Boulder, Colorado.

Folkestad, K. (Ed.) 1968, Ionospheric Radio Communication, Plenum Press, New York.

Heckman, G. 1979, Solar-Terrestrial Predictions Proceedings Volume 1, 322.

Hunsucker, R. D. 1967, QST Magazine, LI, 16.

Hunsucker, R. D. and H. F. Bates 1969, Radio Science, 4, 347.

Hunsucker, R. D. 1975, Radio Science, 10, 277.

Hunsucker, R. D. 1980, Solar-Terrestrial Predictions Proceedings Volume 2, 543.

Hunsucker, R. D., R. Allen, P. Argo, R. Babcock, P. Bakshi, D. Lund, S. Matsushita, G. Smith, A. Shirochkov and G. Wortham 1980, Solar-Terrestrial Predictions Proceedings Volume 2, 513.

Petrie, L. E. and E. E. Stevens 1965, IEEE Trans. Antennas and Propagation, 13, 542.

Tascione, T. F., T. W. Flattery, V. G. Patterson, J. A. Secan, J. W. Taylor Jr. 1979, Solar-Terrestrial Predictions Proceedings Volume 1, 367.

Thompson, R. L. 1979, Solar-Terrestrial Predictions Proceedings Volume 1, 350.

Thrane, E. V. 1979, AGARD Lecutre Series No. 99, Advisory Group for Aerospace Research and Development/NATO, 7 Rue Ancelle 92200 Neuilly Sur Seine, France, 8-1.

Thrane, E. V. et al. 1980, Solar-Terrestrial Predictions Proceedings Volume 2, 573.

Vondrak, R., G. Smith, V. Hatfield, R. Tsunoda, V. Frank and P. Perreault 1977, Chatanika Model of the high-latitude ionosphere for application to HF propagation prediction, RADC-TR-78-7 Rome Air Development Center, Hanscom AFB, MA.

SUBJECT INDEX

(Where a subject is treated over several subsequent pages in the same article only the first page is given.)